D1665933

Antennas and Wireless Power Transfer Methods for Biomedical Applications

Microwave and Wireless Technologies Series

Series Editor: Professor Steven (Shichang) Gao, Chair of RF and Microwave Engineering, and the Director of Postgraduate Research at School of Engineering and Digital Arts, University of Kent, UK.

Microwave and wireless industries have experienced significant development during recent decades. New developments such as 5G mobile communications, broadband satellite communications, high-resolution earth observation, the Internet of Things, the Internet of Space, THz technologies, wearable electronics, 3D printing, autonomous driving, artificial intelligence etc. will enable more innovations in microwave and wireless technologies. The Microwave and Wireless Technologies Book Series aims to publish a number of high-quality books covering topics of areas of antenna theory and technologies, radio propagation, radio frequency, microwave, millimetre-wave and THz devices, circuits and systems, electromagnetic field theory and engineering, electromagnetic compatibility, photonics devices, circuits and systems, microwave photonics, new materials for applications into microwave and photonics, new manufacturing technologies for microwave and photonics applications, wireless systems and networks.

Antennas and Wireless Power Transfer Methods for Biomedical Applications
Yongxin Guo, Yuan Feng and Changrong Liu
February 2024

RF and Microwave Circuit Design: Theory and Applications
Charles E. Free, Colin S. Aitchison
September 2021

Low-cost Smart Antennas
Qi Luo, Steven (Shichang) Gao, Wei Liu
March 2019

Antennas and Wireless Power Transfer Methods for Biomedical Applications

Yongxin Guo
Department of Electrical and Computer Engineering
National University of Singapore
Singapore

Yuan Feng
Department of Electrical and Computer Engineering
National University of Singapore
Singapore

Changrong Liu
School of Electronic and Information Engineering
Soochow University
China

This edition first published 2024

Registered Offices
John Wiley & Sons, Inc., 111 River Street, Hoboken, NJ 07030, USA
John Wiley & Sons Ltd, The Atrium, Southern Gate, Chichester, West Sussex, PO19 8SQ, UK

For details of our global editorial offices, customer services, and more information about Wiley products visit us at www.wiley.com.

Wiley also publishes its books in a variety of electronic formats and by print-on-demand. Some content that appears in standard print versions of this book may not be available in other formats.

Library of Congress Cataloging-in-Publication Data

Names: Guo, Yongxin, author. | Feng, Yuan (Researcher), author. | Liu, Changrong (Professor), author.
Title: Antennas and wireless power transfer methods for biomedical applications / Yongxin Guo, Yuan Feng, Changrong Liu.
Description: Hoboken, NJ : Wiley, 2024. | Series: Microwave and wireless technologies series | Includes index.
Identifiers: LCCN 2023046606 (print) | LCCN 2023046607 (ebook) | ISBN 9781119189916 (cloth) | ISBN 9781119189923 (adobe pdf) | ISBN 9781119189930 (epub)
Subjects: LCSH: Medical electronics. | Antennas (Electronics) | Wireless communication systems.
Classification: LCC R856.A6 G96 2024 (print) | LCC R856.A6 (ebook) | DDC 610.285–dc23/eng/20231214
LC record available at https://lccn.loc.gov/2023046606
LC ebook record available at https://lccn.loc.gov/2023046607

Cover Design: Wiley
Cover Image: © Weiquan Lin/Moment/Getty Images

Set in 9.5/12.5pt STIXTwoText by Straive, Chennai, India

Printed and bound by CPI Group (UK) Ltd, Croydon, CR0 4YY
C9781119189916_060324

Contents

Foreword

The journey of innovation has always been a tale of intersections. When seemingly disparate fields merge, they often pave the way for groundbreaking advancements that possess the potential to redefine human experience. "Antennas and Wireless Power Transfer Methods for Biomedical Applications" is a testament to such an intersection – a melding of the intricate world of antenna technology with the domain of biomedical applications. This exploration is not just an academic endeavor but a beacon that points toward a future where healthcare is seamless, efficient, and unobtrusive.

The significance of antennas and wireless communication in our daily lives is undisputed. However, their application in the biomedical realm elevates their importance to another level altogether. The promise of real-time physiological monitoring, remote treatments, and the potential elimination of cumbersome wires and frequent recharging brings us a step closer to a future where patient care is not just effective but also deeply personalized.

From the opening chapter, the reader is provided a glimpse into the vast expanse of possibilities that emerge when biomedical applications are integrated with modern communication techniques. The subsequent chapters delve deep into the specifics, elucidating the complexities of designing miniaturized and multiband implantable antennas, the art and science behind polarization designs, and differential-fed implantable antennas.

The chapters on wearable antennas and capacitive human body communication further expand the horizon, outlining a world where our very body becomes a medium of communication. Yet, for all these advancements to be feasible, power remains central. The sections on near-field and far-field wireless power transfer demystify the magic behind powering these marvels of technology.

By the time one reaches the concluding chapter, it becomes evident that all these threads of knowledge intertwine to create systems that can revolutionize healthcare, as exemplified by the design intricacies of peripheral nerve implants and neurostimulators.

As you embark on this enlightening journey, it is my hope that you not only grasp the technicalities detailed within these pages but also appreciate the broader vision. This is not just a book but a canvas depicting a future where technology and medicine come together to ensure healthier, longer, and more fulfilling lives for all.

It is with great enthusiasm that I commend this invaluable resource to anyone eager to glimpse into the future of biomedical innovations. Let the discoveries within these pages inspire and propel you into a world of boundless potential.

Preface

It is our privilege to present this comprehensive book, "Antennas and Wireless Power Transfer Methods for Biomedical Applications," a publication that delves into the increasingly essential role that antennas and wireless power transfer technologies play in the realm of biomedical applications. This book provides an in-depth exploration of the latest research and advancements in this intersection of engineering and medical science.

In the contemporary world, the ability to monitor and modulate various physiological and pathological conditions remotely and wirelessly has brought about revolutionary changes in healthcare and treatment modalities. In this context, antennas, with their role in communication, and wireless power transfer methods, enabling remote powering of implantable devices, stand as the twin pillars supporting these advancements.

Chapter 1 introduces the world of biomedical applications and sets the stage for the rest of the book. Here, we traverse the journey of the field, charting out its evolution and contextualizing its relevance in today's healthcare landscape.

In Chapters 2 and 3, we dive deep into the realm of implantable antennas. We investigate their bandwidth enhancement and miniaturization and tackle the issue of polarization design. Chapter 4 brings a specific focus on differential-fed implantable antennas, exploring the particular challenges and opportunities that they present.

We shift our attention to wearable technologies in Chapter 5, examining the role of antennas in on-body and off-body communication. This is followed by Chapter 6, where we venture into the cutting-edge domain of capacitive human body communication, breaking down its mechanisms and outlining its potential applications.

Chapters 7 and 8 tackle the critical aspect of power delivery to these wireless devices. We provide an extensive examination of near-field and far-field wireless power transfer methods and discuss their respective merits and challenges.

Finally, in Chapter 9, we bring all these aspects together through a system design example. This comprehensive application illustrates the integration of these concepts in the design of peripheral nerve implants and neurostimulators.

This book aims to be a beneficial resource for researchers, academicians, professionals, and students engaged in the design and application of antennas and wireless power transfer techniques for biomedical applications. We have endeavored to present the topics in an accessible and reader-friendly manner while maintaining academic rigor and technical depth.

February 2024, Singapore

Yongxin Guo
Yuan Feng
Changrong Liu

Acknowledgment

We would like to express our sincere appreciation for the valuable contributions of our former PhD students and research staff in this field, including Dr. Kush Agarwal, Dr. Zengdi Bao, Dr. Zhu Duan, Dr. Rangarajan Jegadeesan, Dr. Rongxiang Li, Dr. Yan Li, Dr. Han Wang, Dr. Lijie Xu, Dr. Xiaoqi Zhu, and others.

1

Introduction: Toward Biomedical Applications

1.1 Biomedical Devices for Healthcare

The advancement in healthcare and health monitoring technologies has closely paralleled the overarching trajectory of human civilization. In ancient China, for example, practitioners of traditional medicine utilized methodologies such as observation, auditory examination, inquiry, and pulse diagnosis—referred to as "*Wang, Wen, Wen, Qie*"—to determine an individual's health status. These practices, marking the earliest recorded instances of health monitoring, underscored the importance of examining physical manifestations, listening to patients' reported symptoms, inquiring about their medical history, and palpating their pulse in the diagnosis and treatment of various health conditions. Though these methods hinged on subjective assessments, they established an understanding of the crucial linkage between external physical signs and internal health conditions.

With the advent of revolutionary technological and medical breakthroughs, we have embarked on a remarkable journey toward a more precise, quantitative characterization of human health and disease states. This entails harnessing an extensive array of physical, electrical, and chemical indicators in a quest for precise and quantitative comprehension [1] (Figure 1.1). This transition, marking the dawn of modern, data-driven medicine, spurred the development of advanced biomedical devices [2]. These devices integrate sophisticated sensing technologies, data analysis algorithms, and wireless communication capabilities, paving the way for precise and continuous health monitoring [3].

Physical indicators tied to human health include metrics such as heart rate and pulse, which can be gauged through the detection of bodily mechanical movements. Electrical indicators involve signals generated by potential differences within the human body, such as electrocardiograms (ECGs), electroencephalograms (EEGs), and electromyograms (EMGs). These electrical signals reflect the electrical activity of the heart, brain, and muscles, respectively, offering valuable insights into the functionality of these vital organs and our overall physiological state.

Chemical indicators, including metrics such as blood oxygen saturation and glucose levels, provide crucial insights into metabolic activities and bodily functions. These parameters are measured using specialized sensors and analytical techniques, facilitating the early detection and proactive management of a myriad of health conditions, ranging from respiratory disorders and cardiovascular diseases to diabetes.

Antennas and Wireless Power Transfer Methods for Biomedical Applications, First Edition.
Yongxin Guo, Yuan Feng and Changrong Liu.

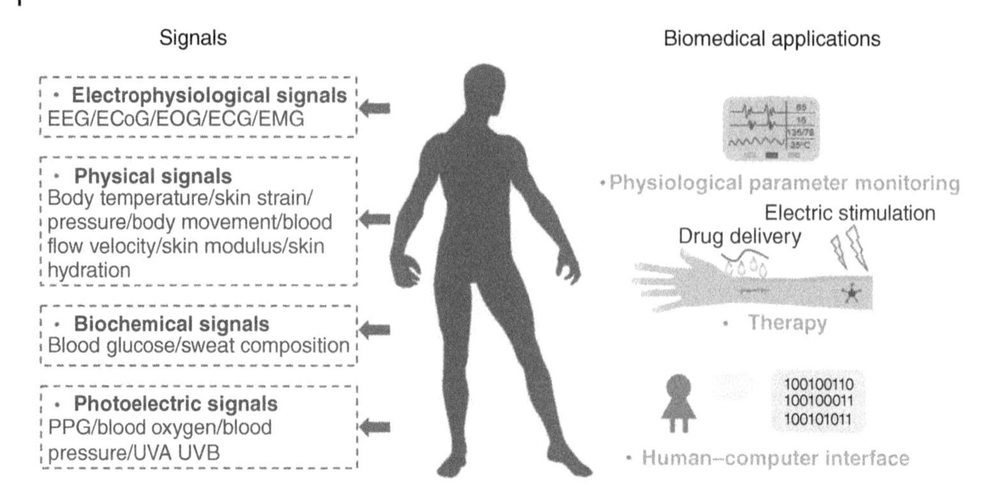

Figure 1.1 Physical, electrical, and chemical indicators for a human body. Source: Chen et al. [1]/Springer Nature/CC BY 4.0.

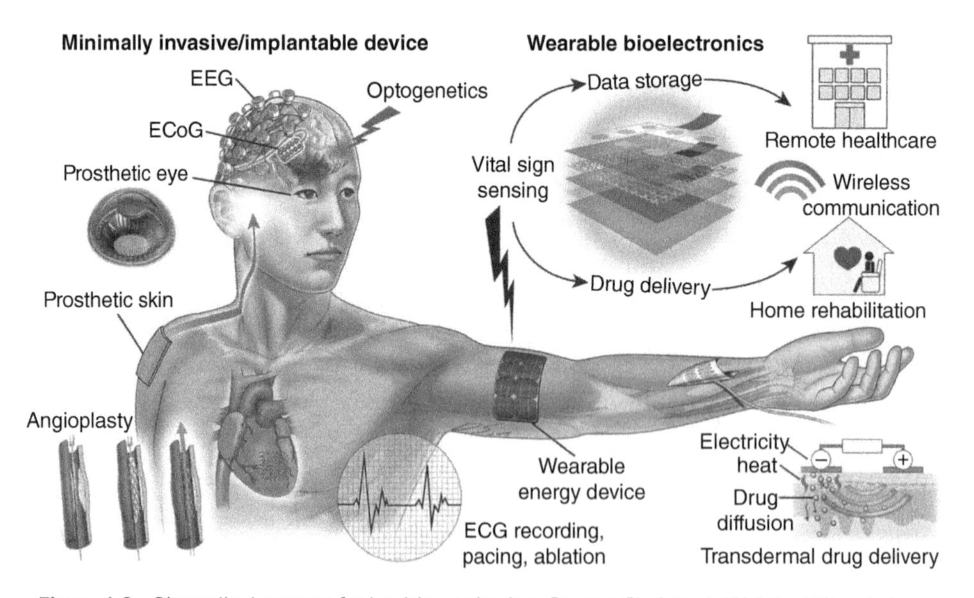

Figure 1.2 Biomedical sensors for health monitoring. Source: Choi et al. [3]/John Wiley & Sons.

The advent of biomedical devices has revolutionized healthcare by integrating these physical, electrical, and chemical indicators into comprehensive health-monitoring systems, as shown in Figure 1.2. Designed to measure, record, and analyze vital signs, these devices equip healthcare professionals with the data necessary to make informed decisions regarding patient diagnosis, treatment, and care. As technology continues to advance, biomedical devices are becoming increasingly miniaturized, accurate, and interconnected. This evolution not only enables individuals to actively monitor their health, but it also fosters the rise of personalized healthcare models, reshaping the healthcare landscape as we know it.

Over the years, the evolution of biomedical devices for healthcare has been marked by substantial advancements, primarily driven by the growing demand for accurate and tailored health-monitoring solutions. Initially, the focus of biomedical devices centered on recording basic vital signs, such as heart rate and blood pressure, using analog tools. However, the emergence of digital technology and the drive toward miniaturization have led to the transformation of these devices into intricate systems capable of monitoring a broad spectrum of physiological parameters [4].

The integration of sensor technologies [4], signal processing algorithms [5], and wireless communication capabilities [6] has spearheaded the development of wearable devices [7], remote health-monitoring systems [8], and implantable medical devices [9]. Wearable devices, such as fitness trackers and smartwatches, have gained significant popularity due to their ability to provide real-time monitoring of vital signs, physical activity, and sleep patterns. These tools empower individuals to keep track of their health and make informed lifestyle decisions.

Remote monitoring systems have brought about a revolution in healthcare, enabling medical professionals to remotely monitor patients' health status and intervene as necessary. These systems typically utilize wearable sensors, home-monitoring devices, and mobile applications, facilitating patients to transmit their health data to healthcare providers for analysis and timely intervention. This approach is especially beneficial for individuals with chronic conditions, the elderly, and those residing in remote locations, as it minimizes the need for frequent hospital visits, thereby enhancing overall healthcare accessibility and outcomes.

Implantable medical devices have also played a pivotal role in the advancement of healthcare. These devices are surgically placed inside a human body to monitor and manage specific medical conditions. Examples of such devices include pacemakers for regulating cardiac rhythm disorders, neurostimulators for controlling chronic pain or movement disorders, and implantable glucose monitors for diabetes management. These devices often incorporate wireless communication capabilities to facilitate data transfer and remote monitoring, enabling healthcare professionals to closely track patients' conditions and adjust treatment protocols accordingly.

The relentless advancements in technology, including miniaturization, improved power efficiency, and enhanced connectivity, have significantly broadened the capabilities of biomedical devices. Further, the integration of artificial intelligence and machine learning algorithms enhances the diagnostic and monitoring abilities of these devices, enabling early detection of irregularities, personalized treatment recommendations, and improved patient outcomes.

In the following sections, we will provide examples of some of the current state-of-the-art wearable and implantable medical devices. These devices showcase the advancements in technology and their potential to revolutionize healthcare.

1.1.1 Wearable Devices

As illustrated in Figure 1.3, wearable devices embody a multitude of forms, merging sophisticated sensing technologies with accessible and user-centric designs [10]. These devices offer an array of capabilities, granting individuals the opportunity to track their health

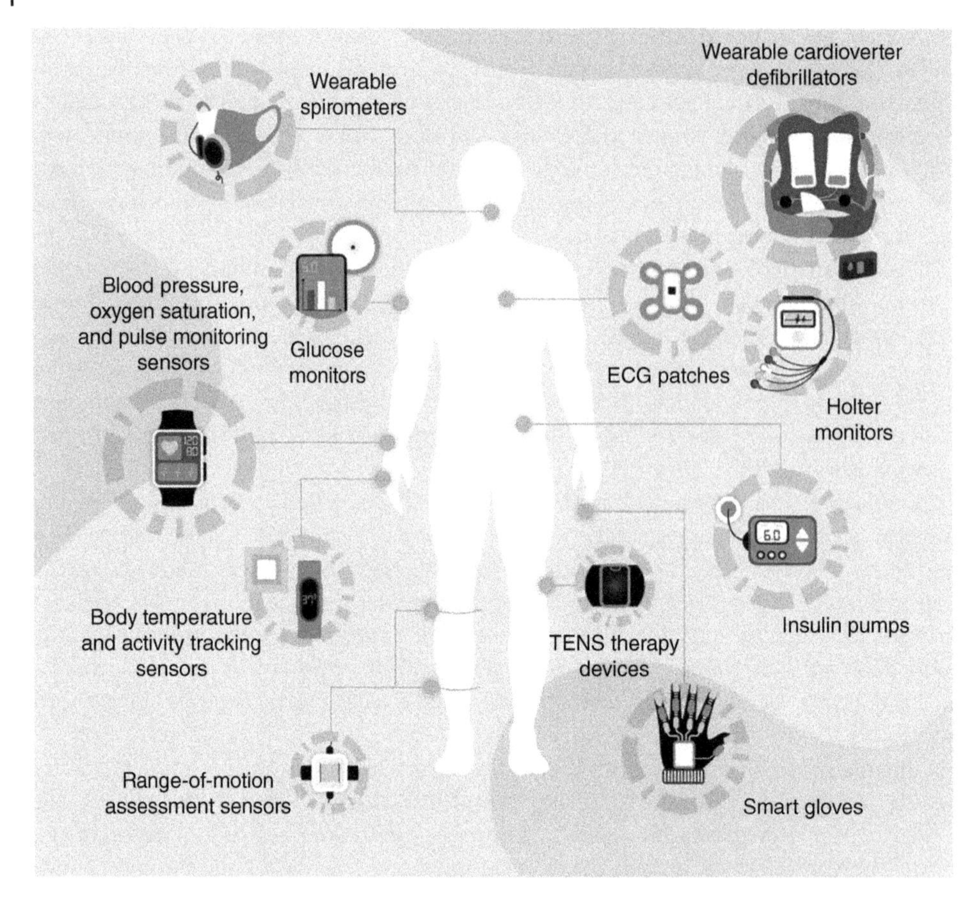

Figure 1.3 Wearable medical devices used in patient care. Source: Ref. [10].

indicators in real-time. Here, we delve into a variety of wearable devices, elucidating their distinct functionalities and application methods.

Wearable spirometers integrated with masks: Specifically designed for individuals managing respiratory conditions such as asthma or chronic obstructive pulmonary disease (COPD), these devices make measuring lung function parameters, including forced vital capacity (FVC) and forced expiratory volume in one second (FEV1), conveniently accessible [11]. The ability to track respiratory health, observe changes in lung function, and adjust medication or treatment plans accordingly equips users with a proactive approach to their health. Additionally, wireless communication technology facilitates data transmission to healthcare providers for remote monitoring and analysis, enabling prompt intervention and personalized care.

Wearable watches and wristbands with integrated blood pressure, oxygen saturation, and pulse monitoring sensors: These devices offer consistent monitoring of vital signs, including blood pressure, oxygen saturation levels, and pulse rate. Throughout the day, users can easily track these parameters, fostering early detection of any potential irregularities. This vital information is particularly useful for individuals managing hypertension, cardiovascular diseases, or respiratory conditions. Furthermore, wireless connectivity supports seamless transmission of vital sign data to healthcare professionals for remote monitoring, providing real-time feedback and proactive condition management [12].

***Body temperature and activity tracking sensors*:** Devices equipped with temperature sensors and accelerometers empower users to monitor body temperature variations and track their physical activity levels [13]. These devices find versatile applications, including fitness tracking, sleep monitoring, and remote patient monitoring. With wireless connectivity, data is seamlessly transmitted to healthcare providers, allowing for remote assessments and personalized care recommendations based on collected data.

***Upper arm wearable continuous glucose monitors (CGMs)*:** For individuals managing diabetes, CGMs are invaluable. These devices consistently monitor glucose levels in interstitial fluid, reducing the need for routine finger pricks [14]. Real-time tracking of glucose levels equips users with better glycemic control and informed decision-making regarding insulin administration, dietary choices, and physical activity. The wireless connectivity of CGMs enables data transmission to smartphones or dedicated receivers, supporting remote monitoring by healthcare providers and timely adjustments to diabetes management plans.

***Thigh and calf wearable devices for range of motion assessment*:** In rehabilitation settings, these devices are useful for assessing joint mobility and tracking range of motion [15]. They typically incorporate sensors to capture muscle activity signals and motion data, evaluating progress for patients with musculoskeletal conditions or those recovering from joint surgeries. These devices' wireless communication capabilities allow collected data to be transmitted to healthcare professionals for remote assessments and guidance, thereby enabling personalized rehabilitation plans and timely adjustments.

***Wearable transcutaneous electrical nerve stimulation (TENS) therapy devices*:** These devices offer non-invasive pain relief by administering electrical stimulation to specific body areas. Frequently utilized for managing chronic pain conditions such as back pain or arthritis, users can adjust the intensity and frequency of the electrical stimulation to their needs. With the integration of wireless communication technology, TENS devices can be remotely controlled and monitored, enhancing pain management convenience and effectiveness [16].

***Insulin pumps*:** These wearable devices provide individuals with type 1 diabetes or insulin-dependent type 2 diabetes with a consistent, precise insulin delivery method throughout the day [17]. Typically worn on the body, these pumps offer continuous insulin infusion, eliminating the need for multiple daily injections. Their wireless connectivity supports remote monitoring of insulin delivery, facilitating personalized insulin dose adjustments, and easing diabetes management.

Smart gloves [18]: Particularly beneficial for rehabilitation purposes, these gloves assist individuals with hand injuries or neurological conditions affecting hand movements. Incorporating sensors that capture hand movements, smart gloves offer real-time feedback and guidance during rehabilitation exercises. The wireless connectivity enables remote monitoring by healthcare professionals and supports personalized rehabilitation plans and progress tracking.

***Holter monitors and ECG patches*:** These wearable devices provide long-term ECG monitoring. While Holter monitors [19] are typically portable devices, ECG patches

[20] are adhesive patches that adhere directly to the skin. Both continuously record the heart's electrical activity over an extended period, facilitating the detection and analysis of abnormal cardiac rhythms or arrhythmias. Their wireless connectivity allows for remote monitoring and immediate intervention in case of critical events.

***Wearable cardioverter defibrillators* (*WCDs*)** [21]: These external devices continuously monitor the heart's electrical activity, prepared to deliver a shock to restore normal heart rhythm in the event of a life-threatening arrhythmia. WCDs are prescribed for individuals at high risk of sudden cardiac arrest, providing temporary protection until definitive treatment can be administered. The wireless connectivity enables remote monitoring by healthcare professionals, ensuring prompt detection of arrhythmias and appropriate intervention.

These wearable devices underscore the transformation of healthcare through technology, empowering individuals to monitor their health metrics, manage chronic diseases, and facilitate remote healthcare collaborations. The incorporation of wireless communication technology enables seamless data transmission, remote observation, and personalized healthcare provision. As the wearable devices field continues to progress, we can look forward to more sophisticated advancements in miniaturization, precision, and connectivity, heralding a new era of personalized and preventive healthcare.

1.1.2 Implantable Devices

As depicted in Figure 1.4, implantable devices are meticulously engineered for surgical implantation within the body, providing enduring therapeutic advantages and substantially enhancing the quality of life for patients. In this segment, we delve into a variety

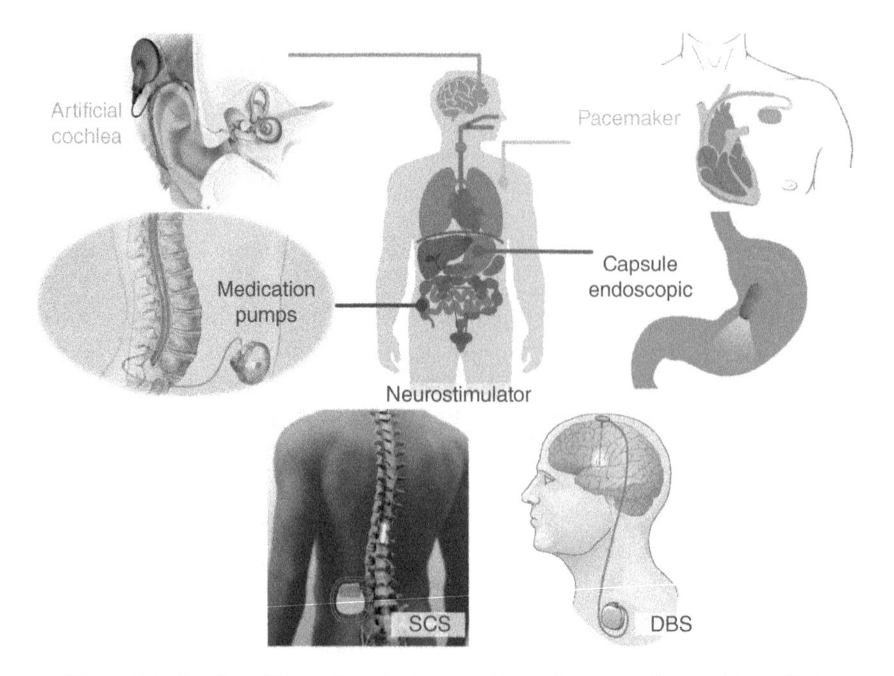

Figure 1.4 Implantable medical devices used in patient care. Source: Mayo Clinic.

of implantable devices, including pacemakers, artificial cochlea, medication pumps, capsule endoscopic systems, and neurostimulators, such as Deep Brain Stimulation (DBS) and Spinal Cord Stimulation (SCS).

***Pacemakers*:** Utilized to manage abnormal heart rhythms or combat bradycardia (a slow heart rate), pacemakers are implantable devices comprising a compact electronic unit and one or more leads that dispatch electrical impulses to the heart muscle, aiding in maintaining a regular heartbeat [22]. They incessantly monitor the heart's electrical activity and deliver stimulation as needed. Tailored to individual requirements, pacemakers can be programmed and adjusted remotely due to advancements in wireless communication technology, thus permitting healthcare providers to ensure optimal device performance.

***Artificial Cochlea*:** More commonly known as cochlear implants [23], these devices are designed to restore auditory sensation in individuals experiencing severe to profound sensorineural hearing loss. The system includes an external speech processor and an internal electrode array that is surgically positioned inside the cochlea. The speech processor captures auditory inputs and converts them into electrical signals. These signals bypass damaged or dysfunctional hair cells in the ear, stimulating the auditory nerve directly, thereby facilitating the perception of sound. Each cochlear implant system is highly personalized, with programming fine-tuned to align with the specific auditory requirements of the individual. Cochlear implants utilize wireless energy and data transmission, enabling efficient communication between the external speech processor and the internal electrode array. The processed audio signals, including speech and environmental sounds, are wirelessly transmitted from the speech processor to the internal electrode array. This electrode array then translates the received signals into electrical impulses, which are delivered to different sections of the auditory nerve, thus simulating the sensation of sound.

***Medication Pumps*:** Also referred to as drug infusion systems, these implantable devices offer controlled and continuous delivery of medication directly into specific body regions or systems [24]. Often used to manage chronic pain, spasticity, or conditions necessitating long-term medication administration, these programmable pumps allow healthcare providers to adjust the dosage and delivery rate based on patient needs. The implantation site varies according to the target area of the medication. Wireless communication technologies enable remote dosage monitoring and adjustment.

***Capsule Endoscopic Systems*:** As a minimally invasive diagnostic tool, capsule endoscopy employs a wirelessly enabled, ingestible capsule embedded with a camera to visualize the interior of the gastrointestinal tract [25]. Upon ingestion, the capsule journeys naturally through the digestive system, capturing images and subsequently transmitting them wirelessly to an external receiver worn by the patient. Given the need for real-time and high-definition image transmission from within the body to the receiver, capsule endoscopic systems pose rigorous demands on wireless communication bandwidth and power consumption. The transmission speed must be sufficiently high to relay quality images without substantial delay or data loss. Additionally, considering that the capsule's voyage through the entire gastrointestinal tract can span several hours, power efficiency becomes critical to ensure the capsule's operation duration and consistent image transmission throughout the procedure. Capsule endoscopic systems provide a non-invasive, patient-centered alternative for diagnosing various gastrointestinal conditions such as Crohn's disease, ulcers, and tumors, particularly in hard-to-reach areas such as the small intestine, traditionally challenging to access with standard endoscopic techniques. By eliminating the necessity for invasive procedures or sedation, capsule endoscopy reduces patient discomfort and minimizes potential complication risks. The wireless communication functionality within these systems facilitates seamless data transmission from inside the body, equipping healthcare professionals with in-depth, accurate visual data necessary for diagnosis and treatment planning.

Neurostimulators: These implantable devices, designed to deliver electrical impulses to particular nerves or brain regions, are employed in the treatment of diverse neurological conditions. Techniques such as DBS [26] can alleviate symptoms associated with movement disorders such as Parkinson's disease or essential tremor. In contrast, SCS [27] is used to manage chronic pain conditions by sending electrical stimulation to the spinal cord. Neurostimulators are highly customizable, offering programmable parameters to meet individual patient needs. Alongside their therapeutic functionality, neurostimulators have evolved to incorporate sophisticated features for monitoring and data transmission. For instance, certain closed-loop neurostimulators [28–30] come with physiological sensing capabilities, including the ability to record EEG signals, making them indispensable tools for both scientific research and clinical applications in the realm of brain–computer interfaces. Consequently, these systems have necessitated the transmission of substantial amounts of physiological data from within the body to an external receiver, similar to the demands of capsule endoscopy systems.

Each of these implantable devices brings distinct advantages to the field of healthcare, representing the continuous advancements in medical technology that provide targeted treatment options, enhanced patient outcomes, and improved quality of life. With the integration of wireless communication technologies, these devices offer expanded functionality and improved usability, enabling remote monitoring, customized adjustments, and personalized care. As the research and development of implantable devices progresses, we anticipate further innovations and enhancements in their capabilities, opening up new possibilities for advanced patient care.

1.2 Wireless Date Telemetry and Powering for Biomedical Devices

In the section above, we delved into the world of wearable and implantable medical devices. Observing from a functional standpoint, we found that wireless communication is integral to these devices, facilitating not only the transmission of data but also enabling the adjustment of device parameters, remote monitoring of device status, and tracking of human physiological indicators. In addition, considering the devices' long-term power needs, wireless powering technologies have emerged as a significant focus in their system integration. However, as these devices continue to decrease in size, both wireless communication and power transfer are met with considerable challenges.

1.2.1 Wireless Data Telemetry for Biomedical Devices

Wireless data telemetry is fundamental in the creation of wireless communication systems for biomedical applications [31]. A wide array of wireless communication techniques have

Figure 1.5 Ultrasonic LSK (load shift keying) modulator for uplink communication in deep implanted medical devices. Source: Mazzilli et al. [32]/from IEEE.

been utilized in these systems, encompassing sound (ultrasound), light (near-infrared), electric fields (near-field capacitive coupling), magnetic fields (near-field inductive coupling), and radio frequency (RF) waves (far-field electromagnetic wave radiation). Each carrier waveform presents distinct advantages and is applied according to specific scenarios, mainly determined by the physical attributes of the carrier.

Ultrasound acts as a conduit for the transmission of mechanical vibrational energy, facilitating the conveyance of both power and information. By employing piezoelectric transducers, energy from ultrasound can be harvested, sparking extensive exploration on its use in wireless power transmission and communication for implantable devices. Notably, in 2014, Mazzilli et al. [32] achieved simultaneous wireless power transmission and uplink wireless communication on a Kinetra neurostimulator manufactured by Medtronic, as shown in Figure 1.5. Implantable wireless communication using ultrasound is especially beneficial for low data transmission rates and deep implantation scenarios. However, attaining high data transmission rates and reliable wireless connections over long distances often proves challenging.

Implantable near-infrared wireless communication technology leverages light waves as carrier waves. Thanks to the broad-spectrum properties of light waves, they hold potential for high data transmission rates in implantable wireless communication [33]. Moreover, as light waves can be physically isolated from low-frequency electromagnetic waves, mitigating mutual interference, wireless data transmission, and power transmission can be independently managed in implantable devices using this technology. Recent advances in near-infrared wireless communication technology for biomedical applications demonstrate the potential for ultra-low power consumption and high-speed transdermal wireless communication, as depicted in Figure 1.6. Yet, the transition of implantable near-infrared wireless communication technology toward clinical applications and widespread adoption depends on the resolution of issues related to short transmission distances and system reliability.

Near-field coupled implantable wireless communication primarily employs near-field magnetic coupled links (inductive links) and capacitive coupled links (capacitive links) for transdermal data transmission, as illustrated in Figure 1.7. Characterized as a short-range

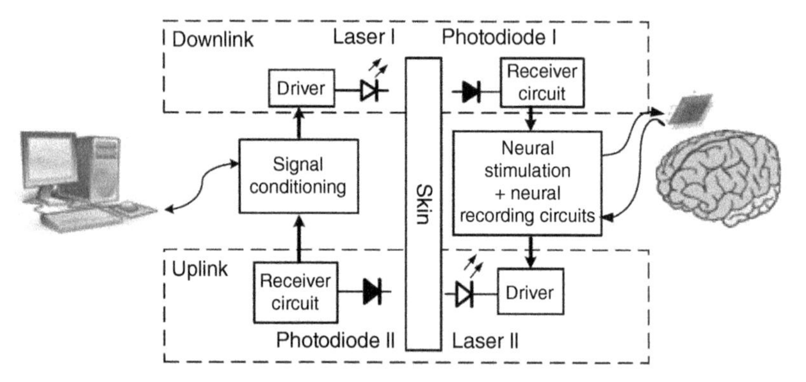

Figure 1.6 Block diagram of bidirectional optical transcutaneous link for brain–computer interface. Source: Liu et al. [33]/John Wiley & Sons.

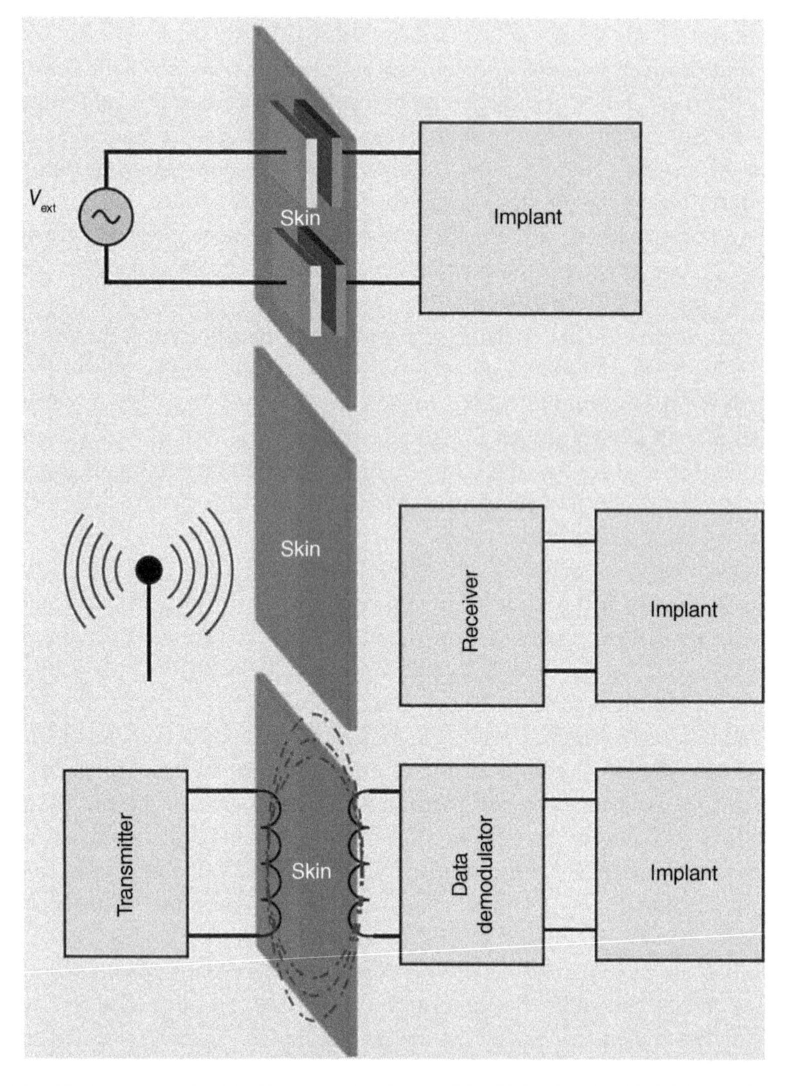

Figure 1.7 Electromagnetic data link: near-field capacitive links, far-field RF links, and near-field inductive links. Source: Bihr et al. [34]/IEEE.

wireless communication, its communication distance is termed as the "touch range" by researchers [35]. Owing to its limited communication distance, it is less suited for scenarios demanding long-distance wireless connections.

Magnetic coupled links, which are conventional implantable wireless communication links, are widely adopted for parameter modifications in various devices such as cardiac pacemakers, neurostimulators, and cochlear implants. These links are often employed for implantable wireless charging, allowing for energy-bearing communication, which makes them a popular choice for most implanted medical devices [36]. The key drawbacks of magnetically coupled links lie in the bandwidth constraints dictated by the low-frequency carrier, which hampers the enhancement of communication speed, and the limited reliable communication range stemming from the nature of magnetic coupling.

Capacitive-coupled links have only recently piqued scholars' interest, with no reported clinical applications to date. In capacitive-coupled links, the electric field between the internal and external electrodes is leveraged for wireless transmission of power and data [37]. Their transmission characteristics are dictated by the impedance features of the electrodes and human tissue [38]. Given that the RF electric field is well confined between the internal and external electrodes—unlike in traditional magnetic coupled links where the electromagnetic flux is dispersed around the implanted coil—capacitive coupled links experience less energy loss. However, the configuration of the internal and external electrodes has a substantial impact on system reliability, posing challenges for the clinical application of capacitive coupling transcutaneous wireless communication technology. As wearable flexible electronic technology advances, the application of this communication technology may progressively come into play.

RF communication technology surpasses distance limitations, facilitating meter-level range wireless communication. The majority of external devices employ RF communication for wireless connections, allowing implantable medical devices and wearable devices equipped with RF communication technology to establish direct connections with other external devices. Hence, in the near term, RF communication technology emerges as the go-to wireless communication solution for implantable medical devices and wearable devices. However, due to interactions with the human body, the application of RF communication technology in biomedicine necessitates particular attention to the key technical area: implantable and wearable antenna technology.

Among the discussed wireless communication methods, RF and near-field coupling (capacitive and inductive) are most frequently employed in biomedical applications. However, antenna design for these communication forms presents several challenges. Miniaturization stands as a primary challenge considering the limited space within implantable devices. Also, due to factors such as tissue absorption, antenna efficiency often proves low. Optimizing the antenna design requires addressing issues such as broadband impedance matching around and post-implantation, polarization matching with external communication devices, and impedance matching with the human body. Integration with RF modules and hardware systems, as well as the choice of antenna type (single-ended or differential), are crucial. Flexibility in design and the use of biocompatible materials for wearable device integration also warrant consideration.

1.2.2 Wireless Power Transmission for Biomedical Devices

Wireless power transmission for biomedical applications aims to address the power supply issue in implantable devices. Traditional implantable medical devices often rely on batteries for power, but the need for frequent battery replacement due to device functionality and limited lifespan can be inconvenient and time-consuming. Therefore, wireless power transmission technology has been introduced to provide continuous power supply to implantable devices.

One prevalent method for wireless power transmission in biomedical devices is inductive coupling [36]. This involves power transfer between two coils, specifically the transmitter coil and the receiver coil. The transmitter coil, typically located outside the body, generates an oscillating magnetic field, which induces a voltage in the receiver coil implanted within the body. This induced voltage is subsequently rectified and used to power the implanted device. Resonant coupling provides an alternative mechanism for wireless power transmission [39]. This method exploits the resonance phenomenon between the transmitter and receiver coils to achieve efficient power transfer. By synchronizing the frequencies of the transmitter and receiver coils, energy transfer can be markedly boosted. This technique enhances power transmission efficiency and bolsters the overall performance of the implanted devices (Figure 1.8).

Nonetheless, several challenges are associated with wireless power transmission for biomedical devices. A significant issue is the limited power transmission distance. As the gap between the transmitter and receiver coils widens, power transfer efficiency drops. Hence, optimizing the design and positioning of the coils is paramount to maximize power transmission efficiency and ensure reliable operation of the implanted devices.

Additionally, factors such as tissue properties and electromagnetic interference can affect the efficiency of wireless power transmission. Human body tissues present varying degrees of conductivity and dielectric properties, which can lead to energy losses and diminish power transmission efficiency. Counteracting these losses and minimizing the effect of electromagnetic interference are crucial design considerations in wireless power transmission systems for biomedical devices.

Moreover, the biocompatibility of materials used in the wireless power transmission system is vital to guarantee the safety and functionality of the implanted devices. The materials

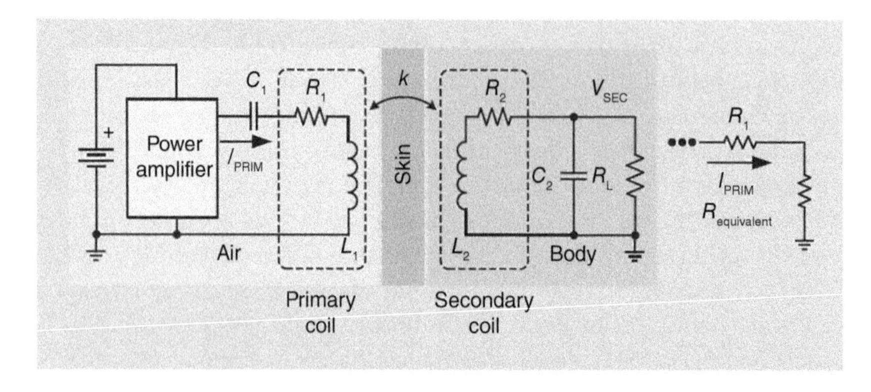

Figure 1.8 Resonant inductive link for implants. Source: Xu et al. [39]/IEEE.

should be biocompatible, nontoxic, and stable within the body to avert any adverse reactions or long-term complications.

Wireless power transmission presents a promising solution for powering biomedical devices. It affords the advantages of extended operation, convenience, and minimized risks tied to invasive surgeries. Overcoming the hurdles of limited power transmission distance, tissue effects, and material biocompatibility is crucial for the successful application of wireless power transmission in biomedical devices. Continued research and advancements in this field are expected to further augment the efficiency and dependability of wireless power transmission systems for biomedical applications.

1.3 Overview of Book

As previously highlighted, biomedical devices have significantly revolutionized our health-monitoring capabilities, disease diagnostic methodologies, and treatment options. In these advancements, wireless technology has played an indispensable role. This book's objective is to offer a comprehensive overview of the application of wireless technology solutions in biomedical devices, laying emphasis on the research contributions of our team in this domain, especially in the areas of antenna design and wireless charging. The book unfolds across the following chapters:

Chapter 2 introduces the design and development of compact, wideband, and multi-frequency band-compatible implantable antennas. It sheds light on the challenges and breakthroughs in antenna design for biomedical applications, tackling topics such as miniaturization, impedance matching, and bandwidth optimization.

Chapter 3 presents designs and techniques for achieving desirable polarization in implantable antennas. It emphasizes the crucial role of polarization matching in ensuring reliable wireless communication in biomedical devices. The chapter discusses various polarization design strategies and their respective influences on system performance.

In Chapter 4, differential-fed implantable antennas are explored. The chapter discusses the benefits and challenges tied to differential-fed antennas and their application in biomedical devices. It encompasses the design principles, impedance matching techniques, and performance evaluation of differential-fed antennas.

Chapter 5 presents wearable antennas designed to facilitate communication between on-body and off-body devices. It explores the design considerations, challenges, and advancements in wearable antenna technologies.

Chapter 6 examines the utilization of capacitive human body communication (HBC) for wireless data transmission. It delves into the principles, modeling techniques, and challenges associated with HBC, covering topics such as channel modeling, modulation strategies, and signal processing techniques for capacitive HBC systems.

Chapter 7 immerses into near-field wireless power transfer techniques for biomedical applications. It discusses the principles, design considerations, and performance evaluation of near-field power transfer systems. The chapter also scrutinizes the challenges associated with power transmission efficiency, coil design, and positioning for reliable and safe wireless charging of implantable devices.

In Chapter 8, the focus is on far-field wireless power transmission for biomedical applications. It explores the principles, system design, and challenges tied to long-distance power transmission, covering topics such as energy-focusing techniques, power optimization, and safety considerations for far-field wireless charging systems.

Chapter 9 illustrates a system design example concentrating on peripheral nerve implants and neurostimulators. It offers a detailed analysis of the design considerations and integration challenges in such systems. The chapter also includes practical examples and case studies to demonstrate the application of wireless technology in peripheral nerve implants and neurostimulators.

Each chapter in the book is fortified with research findings, case studies, and practical insights, providing a thorough understanding of the advancements, challenges, and practical applications of wireless technology in biomedical devices. The book aims to serve as a valuable compendium for researchers, engineers, and healthcare professionals involved in biomedical technology and wireless communication.

References

1 Chen, Y., Zhang, Y., Liang, Z. et al. (2020). Flexible inorganic bioelectronics. *npj Flexible Electron.* 4 (1): 2.

2 Leff, D.R. and Yang, G.-Z. (2015). Big data for precision medicine. *Engineering* 1 (3): 277–279.

3 Choi, S., Lee, H., Ghaffari, R. et al. (2016). Recent advances in flexible and stretchable bio-electronic devices integrated with nanomaterials. *Adv. Mater.* 28 (22): 4203–4218.

4 He, R., Liu, H., Niu, Y. et al. (2022). Flexible miniaturized sensor technologies for long-term physiological monitoring. *npj Flexible Electronics* 6 (1): 20.

5 Khan, S. and Yairi, T. (2018). A review on the application of deep learning in system health management. *Mech. Syst. Sig. Process.* 107: 241–265.

6 Ko, J., Lu, C., Srivastava, M.B. et al. (2010). Wireless sensor networks for healthcare. *Proc. IEEE* 98 (11): 1947–1960.

7 Lu, L., Zhang, J., Xie, Y. et al. (2020). Wearable health devices in health care: narrative systematic review. *JMIR Mhealth Uhealth* 8 (11): e18907.

8 Kalid, N., Zaidan, A.A., Zaidan, B.B. et al. (2018). Based real time remote health monitoring systems: A review on patients prioritization and related "big data" using body sensors information and communication technology. *J. Med. Syst.* 42: 1–30.

9 Joung, Y.-H. (2013). Development of implantable medical devices: from an engineering perspective. *Int. Neurourol. J.* 17 (3): 98.

10 Wearables in Healthcare: The Essence. 2023 Available online: https://www.scnsoft.com/healthcare/medical-devices/wearable [Accessed: 10-06-2023].

11 Tipparaju, V.V., Xian, X., Bridgeman, D. et al. (2020). Reliable breathing tracking with wearable mask device. *IEEE Sens. J.* 20 (10): 5510–5518.

12 Kim, K.J. and Shin, D.-H. (2015). *An Acceptance Model for Smart Watches: Implications for the Adoption of Future Wearable Technology*. Internet Research.

13 Appelboom, G., Yang, A.H., Christophe, B.R. et al. (2014). The promise of wearable activity sensors to define patient recovery. *J. Clin. Neurosci.* 21 (7): 1089–1093.

14 Kang, H.S., Park, H.R., Kim, C.-J., and Singh-Carlson, S. (2022). Experiences of using wearable continuous glucose monitors in adults with diabetes: a qualitative descriptive study. *Sci. Diab. Self-Manage. Care* 48 (5): 362–371.

15 Díaz, S., Stephenson, J.B., and Labrador, M.A. (2019). Use of wearable sensor technology in gait, balance, and range of motion analysis. *Appl. Sci.* 10 (1): 234.

16 Johnson, M.I. (2014). *Transcutaneous Electrical Nerve Stimulation (TENS): Research to Support Clinical Practice*. Oxford University Press.

17 Alsaleh, F.M., Smith, F.J., Keady, S., and Taylor, K.M.G. (2010). Insulin pumps: from inception to the present and toward the future. *J. Clin. Pharmacy Ther.* 35 (2): 127–138.

18 Caeiro-Rodríguez, M., Otero-González, I., Mikic-Fonte, F.A., and Llamas-Nistal, M. (2021). A systematic review of commercial smart gloves: current status and applications. *Sensors* 21 (8): 2667.

19 Galli, A., Ambrosini, F., and Lombardi, F. (2016). Holter monitoring and loop recorders: from research to clinical practice. *Arrhythm. Electrophysiol. Rev.* 5 (2): 136.

20 Lobodzinski, S.S. and Laks, M.M. (2012). New devices for very long-term ECG monitoring. *Cardiol. J.* 19 (2): 210–214.

21 Cheung, C.C., Olgin, J.E., and Lee, B.K. (2021). Wearable cardioverter-defibrillators: a review of evidence and indications. *Trends Cardiovasc. Med.* 31 (3): 196–201.

22 Beck, H., Boden, W.E., Patibandla, S. et al. (2010). 50th Anniversary of the first successful permanent pacemaker implantation in the United States: historical review and future directions. *Am. J. Cardiol.* 106 (6): 810–818.

23 Blume, S. (2019). *The Artificial Ear: Cochlear Implants and the Culture of Deafness*. Rutgers University Press.

24 Mayfield Brain & Spine (2017). *Pain Pump (Intrathecal Drug Pump)*. [EB/OL] https://mayfieldclinic.com/pe-pump.htm.

25 Koulaouzidis, A., Iakovidis, D.K., Karargyris, A., and Rondonotti, E. (2015). Wireless endoscopy in 2020: will it still be a capsule? *World J. Gastroenterol.* 21 (17): 5119.

26 Perlmutter, J.S. and Mink, J.W. (2006). Deep brain stimulation. *Annu. Rev. Neurosci.* 29: 229–257.

27 Verrills, P., Sinclair, C., and Barnard, A. (2016). A review of spinal cord stimulation systems for chronic pain. *J. Pain Res.* 9: 481–492.

28 Qian, X., Chen, Y., Feng, Y. et al. (2017). A platform for long-term monitoring the deep brain rhythms. *Biomed. Phys. Eng. Exp.* 3 (1): 015009.

29 Feng, Y., Li, Y., Li, L. et al. (2019). Tissue-dependent Co-matching method for dual-mode antenna in implantable neurostimulators. *IEEE Trans. Ant. Propag.* 67 (8): 5253–5264.

30 Feng, Y., Li, Y., Li, L. et al. (2020). Design and system verification of reconfigurable matching circuits for implantable antennas in tissues with broad permittivity range. *IEEE Trans. Antennas Propag.* 68 (6): 4955–4960.

31 Nikita, K.S. (2014). *Handbook of Biomedical Telemetry*. Wiley Online Library.

32 Mazzilli, F., Kilinc, E.G., and Dehollain, C. (2014). 3.2 mW ultrasonic LSK modulator for uplink communication in deep implanted medical devices. In: *In 2014 IEEE Biomedical Circuits and Systems Conference (BioCAS) Proceedings*, 636–639. IEEE.

33 Liu, T., Anders, J., and Ortmanns, M. (2015). Bidirectional optical transcutaneous telemetric link for brain machine interface. *Electronics Lett.* 51 (24): 1969–1971.

34 Bihr, U., Liu, T., and Ortmanns, M. (2014). Telemetry for implantable medical devices: Part 3-data telemetry. *IEEE Solid-State Circuits Magazine* 6 (4): 56–62.

35 Hall, P.S. and Hao, Y. (2012). *Antennas and Propagation for Body-Centric Wireless Communications*, 2ee. Artech House, Inc.

36 Zargham, M. and Gulak, P.G. (2012). Maximum achievable efficiency in near-field coupled power-transfer systems. *IEEE Trans. Biomed. Circ. Syst.* 6 (3): 228–245.

37 Sodagar, A.M. and Amiri, P. (2009). Capacitive coupling for power and data telemetry to implantable biomedical microsystems. In: *2009 4th International IEEE/EMBS Conference on Neural Engineering*, 411–414. IEEE.

38 Takhti, M., Asgarian, F., and Sodagar, A.M. (2011). Modeling of a capacitive link for data telemetry to biomedical implants. In: *2011 IEEE Biomedical Circuits and Systems Conference (BioCAS)*, 181–184. IEEE.

39 Xu, H., Handwerker, J., and Ortmanns, M. (2014). Telemetry for implantable medical devices: Part 2-power telemetry. *IEEE Solid-State Circ. Magazine* 6 (3): 60–63.

2

Miniaturized Wideband and Multiband Implantable Antennas

2.1 Introduction

Over the past two decades, the rapid evolution of integrated circuit technology has led to a significant reduction in the size of implantable electronic medicines, scaling down from centimeters to mere millimeters. Concurrently, the functionalities of these implantable electronic medicines have grown more intricate and multifaceted. A quintessential example of this evolution is the pacemaker, as illustrated in Figure 2.1. However, as implantable devices typically operate at relatively low frequencies – often within the medical implant communications service (MICS) band (402–405 MHz) or the medical device radio communications service band (MedRadio, 401–406 MHz) – implantable antennas can take up a significant portion of the biomedical device's space. Consequently, the miniaturization of implantable antenna design becomes an utmost priority and presents itself as one of the most formidable challenges in implantable antenna design.

However, the drive toward miniaturization often leads to a corresponding reduction in the bandwidth of the implantable antenna, which is not conducive to implantable applications. Human tissue, being a nonhomogeneous and lossy dielectric, presents a range of relative permittivity values, varying dramatically from 5 to 80 [1]. Additionally, the relative permittivity within the same tissue can fluctuate with age, vary across different areas of the human body, and differ from one patient to another [2]. Due to these shifts in relative permittivity, miniaturized narrowband implantable antennas run the risk of detuning [3], potentially compromising robust communication for implants and impacting the lifespan of implantable electronic medicine. In clinical applications, pacemaker manufacturers have reported considerable variations in antenna impedances when pacemakers are implanted in different pocket locations, complicating the selection of suitable tuning circuit parameters [4]. To address these challenges, it is critical to design a wideband antenna, which can mitigate the detuning issue associated with implantable antennas and guarantee robust communication. Furthermore, for applications demanding high-rate implantable data transfer, such as implanted brain–computer interfaces (BCI), a wideband implantable antenna is indeed preferred.

Presently, certain radio frequency (RF) modules employed in implantable medical devices leverage multiband designs to facilitate multitasking capabilities. For instance, the ZL70103, a medical implantable RF module, includes a 2.45 GHz wake-up receiver and a

Antennas and Wireless Power Transfer Methods for Biomedical Applications, First Edition.
Yongxin Guo, Yuan Feng and Changrong Liu.

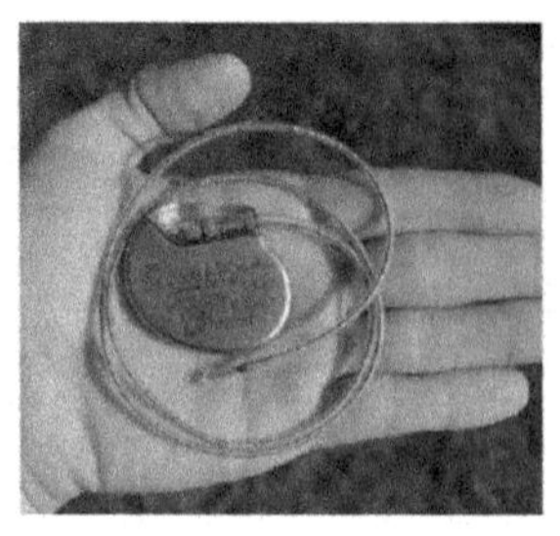

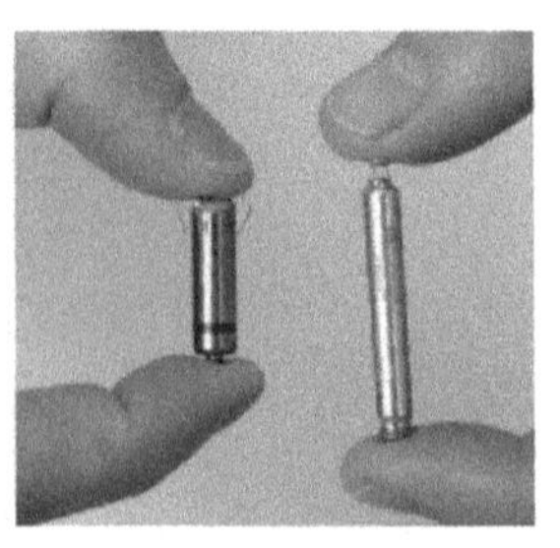

Figure 2.1 Depiction of the progression of pacemakers from the 1950s (left) to today (right). Longevity ranged from a few months on early devices to over 10 years for current devices. ECG monitoring, MRI compatible, leadless technologies equipped in current devices. Source: Professor Marko Turina/Wikimedia Commons/CC BY 3.0; Steven Fruitsmaak/Wikimedia Commons/CC BY 3.0.

400 MHz telemetry transceiver. As such, the multiband design of the implantable antenna holds significant importance.

In this chapter, we will examine the design methods of miniaturized implantable antennas and the strategies to enhance the bandwidth of implantable antennas, including aspects of wide bandwidth and multiband design.

2.2 Miniaturization Methods for Implantable Antenna Design

This section will commence with a review of various methods employed in the miniaturization of implantable antenna design. Subsequently, we will provide examples of implantable designs that correspond to each of these methods.

2.2.1 Use of High-permittivity Dielectric Substrate/Superstrate

Utilizing a high-permittivity dielectric substrate or superstrate is the most straightforward approach to decrease the dimensions of implantable antennas, as such a substrate or superstrate can shift the resonant frequency toward the lower spectrum. At the same time, the employment of a superstrate is necessary to isolate the antenna from the lossy surrounding tissues. In reference [5], it has been experimentally verified that a dielectric superstrate applied on a patch antenna significantly lowers the resonant frequency. For an implantable antenna, a biocompatible superstrate can also protect the tissues surrounding the antenna [6]. This superstrate layer functions as a buffer between the radiating patch and human tissues, mitigating RF power at the sites of lossy human tissues. Furthermore, a superstrate can generally enhance both the bandwidth and the gain efficiency of the antenna [7].

Table 2.1 showcases various dielectric materials that have been used in implantable antenna design. As illustrated in Table 2.1, Rogers RO3210/RO3010/6002 has been extensively leveraged as substrates or superstrates for implantable antenna design, characterized

Table 2.1 Materials for implantable antenna design.

Materials	Relative dielectric constant	References
Macor	$\varepsilon_r = 6.1$	[8]
Rogers 6002	$\varepsilon_r = 10.2$	[9]
Rogers 3210	$\varepsilon_r = 10.2$	[7, 10–15]
$MgTa_{1.5}Nb_{0.5}O_6$	$\varepsilon_r = 28$	[16]
ARLON 1000	$\varepsilon_r = 10.2$	[17]
Rogers TMM10	$\varepsilon_r = 9.2$	[18]
Rogers 3010	$\varepsilon_r = 10.2$	[19–31]

by their relative dielectric constant of 10.2. To push the boundaries of miniaturization further in implantable antennas, substrates with significantly higher relative permittivity have been put into use.

In study [16], $MgTa_{1.5}Nb_{0.5}O_6$ (ε_r = 28, denoted by ANMg) is employed for significant size reduction. The antenna's design, displayed in Figure 2.2, is divided into three sections: the central radiating monopole part, the inductive dual inverted L-type loading part, and the coupling dual back-to-back short L-type part. The antenna's total length ($w/2 + d + d_1 + d_2 + d_3$) is designed as a $3\lambda_g/4$ resonator of 402 MHz. Compared to similar implantable antennas with superstrates, this non-superstrate antenna boasts advantages in miniaturization and low-profile design, wide measured impedance bandwidth, and gain stability within 0.3–0.5 GHz. Over a period of 60 days (Figure 2.3), the antenna demonstrates stable characteristics when directly loaded on high-dielectric-constant ceramic substrate tissue.

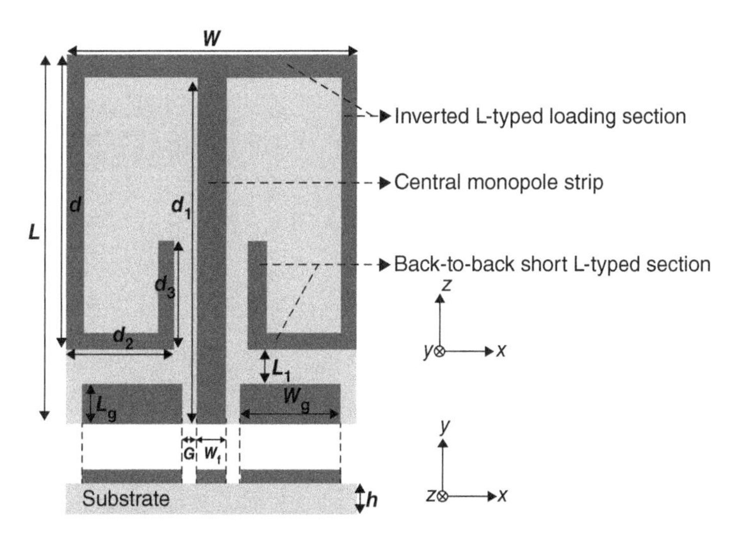

Figure 2.2 Geometry of the antenna ANMg. Source: Reproduced with permission of ©2010 IEEE.

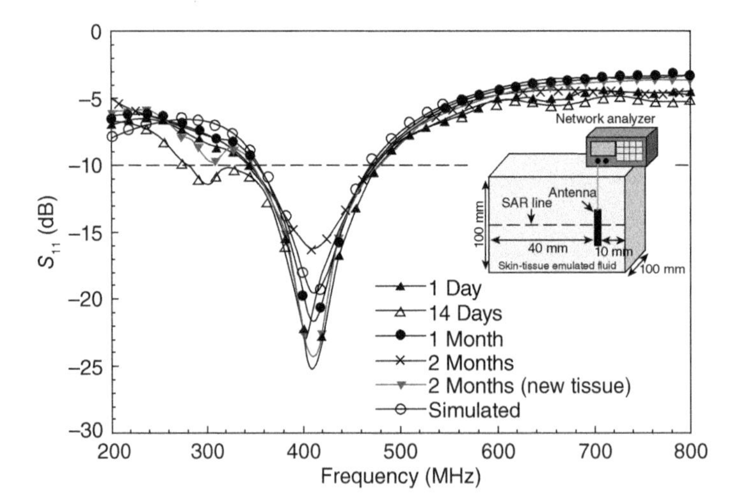

Figure 2.3 Simulated and measured S_{11} of the ANMg antenna immersed into the emulated human skin tissue for a period of 60 days. Source: Reproduced with permission of ©2010 IEEE.

2.2.2 Use of Planar Inverted-F Antenna Structure

The space-saving properties of the Planar Inverted-F Antenna (PIFA) have spurred widespread discussions regarding their use in implantable devices to achieve miniaturization. Figure 2.4 displays two kinds of implantable antennas designed for such devices [6], both operating within the 402–405 MHz band. High-permittivity material ($\varepsilon_r = 10.2$) was used as the substrate and superstrate layers to reduce the antennas' dimensions. The size of the spiral PIFA structure ($32 \times 24 \times 8\ \text{mm}^3$) is substantially smaller than the spiral antenna ($40 \times 32 \times 8\ \text{mm}^3$). The reason lies in the fact that the resonant length of a microstrip patch antenna is half-wavelength, while for PIFA it is a quarter-wavelength. Thus, compared

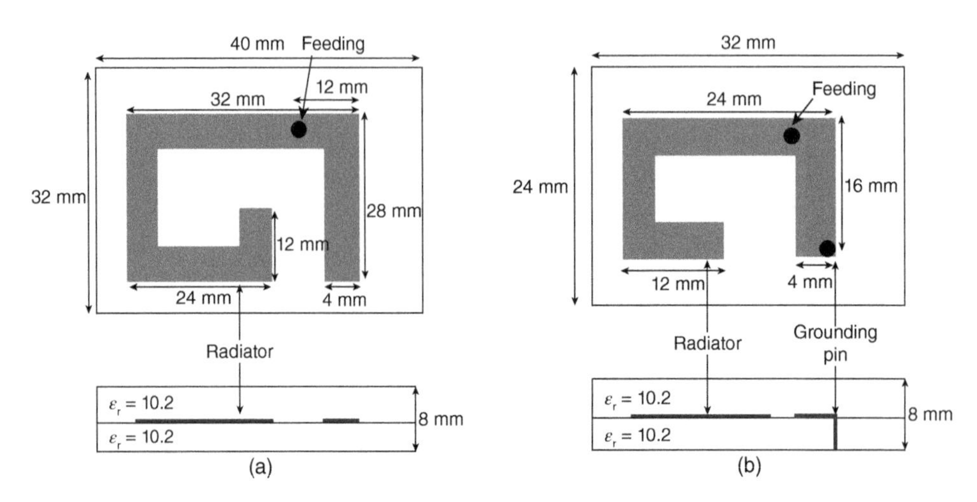

Figure 2.4 Two kinds of antennas designed for the implantable device [6]: (a) spiral microstrip antenna, (b) spiral PIFA structure. Source: Reproduced with permission of ©2004 IEEE.

Table 2.2 Performance comparisons of different antennas.

Reference	Antenna	Frequency	Dimensions(mm^3)	Miniaturization
[6]	PIFA	402 MHz/2.4 GHz	$32 \times 24 \times 4$	PIFA
[7]	PIFA	402 MHz/2.4 GHz	$22.5 \times 22.5 \times 2.5$	Meandered PIFA
[8]	PIFA	402 MHz	$26.6 \times 19.6 \times 6$	Spiral
[11]	PIFA	402 MHz	$8 \times 8 \times 1.9$	Stacked PIFA
[14]	PIFA	402 MHz/ 433 MHz/ 868 MHz/ 915 MHz	$\pi \times 6^2 \times 1.8$	Meandered PIFA
[16]	Monopole	402 MHz	$18 \times 16 \times 1$	Ceramic substrate
[17]	Patch	402 MHz/2.4 GHz	$15 \times 15 \times 3.81$	Split ring resonator
[18]	PIFA	402 MHz/2.4 GHz	$8 \times 13.1 \times 5.2$	Multilayer PIFA
[19]	PIFA	402 MHz/2.4 GHz	$16.5 \times 16.5 \times 2.54$	Spiral PIFA
[20]	PIFA	402 MHz/2.4 GHz	$19 \times 19.4 \times 1.27$	PIFA, open-end slots
[21]	Dipole	402 MHz/2.4 GHz	$10 \times 10 \times 0.675$	Inductive loading
[22]	Dipole	402 MHz	$16.5 \times 15.7 \times 1.27$	Inductive loading
[23]	Dipole	402 MHz/542 MHz	$27 \times 14 \times 1.27$	Spiral arm
[24]	PIFA	402 MHz/2.4 GHz	$13.5 \times 15.8 \times 0.635$	PIFA
[25]	Patch/slot	402 MHz	$10 \times 16 \times 1.27$	Meandered slot
[26]	Slot	402 MHz	$10 \times 11 \times 1.27$	Meandered slot
[27]	PIFA	402 MHz	$12.5 \times 12.5 \times 1.27$	Meandered PIFA
[32]	PIFA	402 MHz	$22.5 \times 22.5 \times 1.27$	Double L-strips PIFA
[10]	PIFA	402 MHz	$\pi \times 7^2 \times 1.9$	Stacked PIFA
[33]	Cavity slot	2.4 GHz	$1.6 \times 2.8 \times 4$	H-shaped slot

to a microstrip antenna, a PIFA is a more effective choice for size reduction. PIFAs have, as a result, been extensively studied by many research groups. Table 2.2 compares the performance of various implantable antennas, demonstrating the prevalence of the PIFA structure in the design of implantable antennas.

Concerning PIFAs, two distinct types were designed and studied to determine the superior shape for implantable antenna design [8]. Figure 2.5 showcases the two distinct PIFA structures intended for implantable antenna design. To facilitate a comparison of their resonant frequencies, the spiral and serpentine antennas were made identical in all aspects, with the width = 2.8 mm, total length = 98 mm, and parameters A, B, C, and D measuring 7, 8.4, 7.7, and 4.9 mm, respectively. The key distinction between the two antennas lies in their resonant frequencies. In the same physical length and under the same simulation environment, the spiral PIFA registers a lower resonant frequency (402 MHz) compared to the serpentine PIFA (475 MHz), as depicted in Figure 2.6. Research findings from [8] indicate that the spiral antenna exhibits stronger coupling at the antenna's center, whereas the serpentine antenna shows coupling to adjacent arms. As such, the serpentine antenna is electrically shorter than the spiral PIFA.

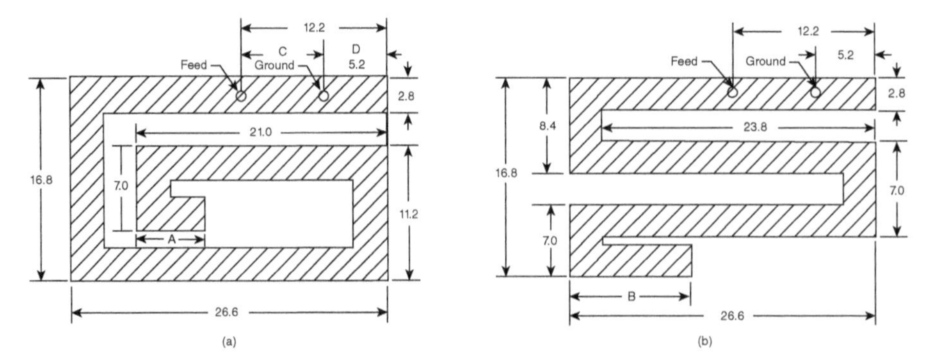

Figure 2.5 Two kinds of PIFAs designed for the implantable device [8]; (a) spiral PIFA, (b) serpentine PIFA structure. Source: Reproduced with permission of ©2004 IEEE.

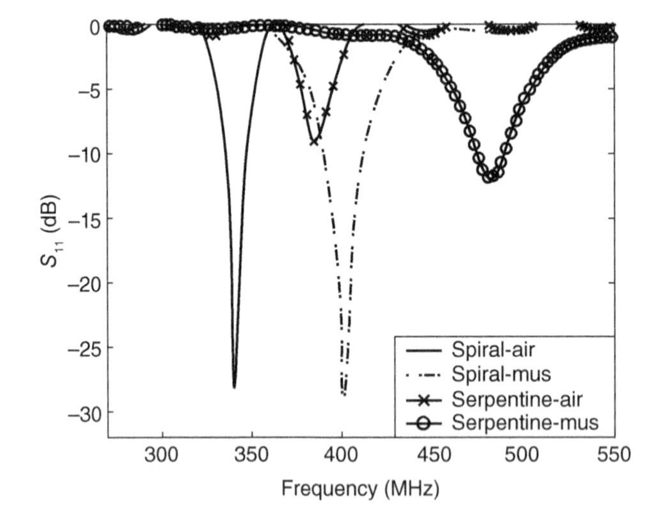

Figure 2.6 Comparison between $|S_{11}|$ of spiral and serpentine antennas in 2/3 muscle and air [8]. Source: Reproduced with permission of ©2004 IEEE.

2.2.3 Lengthening the Current Path of the Radiator

Employing meandered or spiraled line/slot is another potent approach to achieve size reduction. By extending the current path of the radiator, the antenna's resonant frequency can be shifted to a lower frequency, thus resulting in size reduction. To better comprehend the operating principle, an example from our previous work [27] is provided. The design process began with a square ring antenna, which resonates at a lower frequency compared to a square patch antenna of the same size. However, the square ring's resonant frequency is significantly higher than the MICS band. Given that patch meandering can considerably reduce the antenna size, several narrow slits were introduced into the non-radiating edges of the square ring. The resonance frequency can be fine-tuned by adjusting the size and the number of turns in the meanders. Figure 2.7 illustrates the reflection coefficients

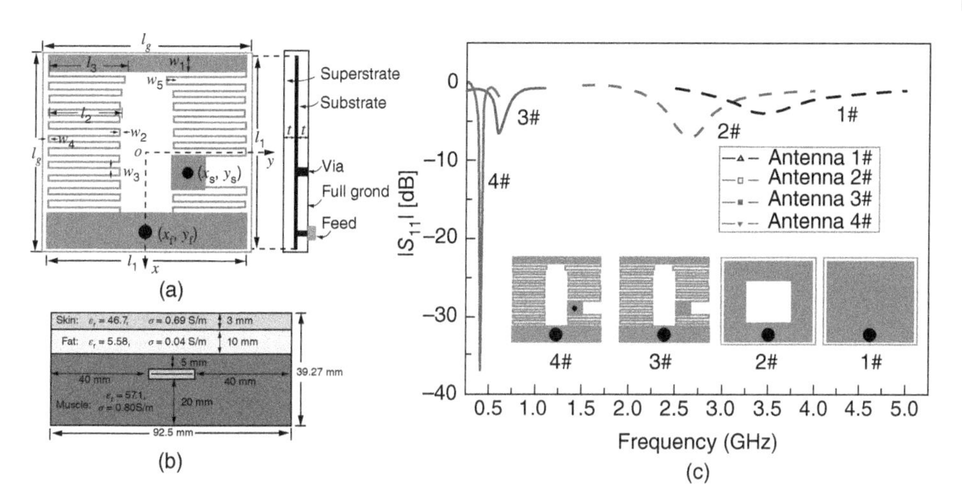

Figure 2.7 (a) Top view and side view of the proposed implantable antenna, (b) three-layer cubic phantom model for the design. Dimensions (in millimeters): $l_g = 12.5$, $l_1 = 12$, $l_2 = 4.3$, $l_3 = 4.7$, $w_1 = 1$, $w_2 = w_3 = w_4 = 0.2$, $w_5 = 0.4$, $t = 0.635$, $(xs,ys) = (1,2.5)$, and $(xf,yf) = (5,0)$, (c) comparison of reflection coefficients for the four antennas embedded in the three-layer phantom model. 1#: square patch; 2#: square ring; 3#: meandered square ring; 4#: meandered square ring with shorting pin [27]. Source: Reproduced with permission of ©2015 IEEE.

($|S_{11}|$) responses versus frequency for the four aforementioned antennas embedded in a phantom. In other words, ANTENNA 1#: square patch; ANTENNA 2#: square patch with a central square slot (abbreviated to square ring); ANTENNA 3#: meandered square ring; ANTENNA 4#: meandered square ring with a shorting pin. From Figure 2.7, it is evident that the resonant frequency can be drastically lowered from 3.43 GHz to 402 MHz through the meandering and shorting design. Furthermore, the impedance matching of the proposed antenna (4#) is much superior to that of the square patch antenna.

Additionally, in [14], size reduction was achieved by introducing open slots on the radiator to elongate the current path, and the radiator was stacked for further size reduction. In [19], a miniaturized dual-band implantable PIFA was designed using two spiral arms. In [25], size reduction was effected by a meandered slot and open slots on the ground to extend the current path.

From Table 2. 2, it is seen that this method has been widely used by many research groups to design miniaturized implantable antennas. To further reduce the dimensions of implantable antenna, stacking the radiator is a good solution to lengthen the current path, as reported in [10, 11].

Figure 2.8 lists the parametric model of a miniature PIFA [14]. The model consists of a 6 mm radius ground plane and two 5 mm radius vertically stacked, meandered patches. Also, the antenna can be easily optimized at desired resonant frequency by changing slot lengths (A_l, B_l, C_l, D_l, E_l, A_u, B_u, C_u, D_u, E_u, F_u, X_s, and Y_s), as can be seen in Figure 2.9. Longer meanders assist in lengthening the effective current path, thus achieving lower resonance frequencies.

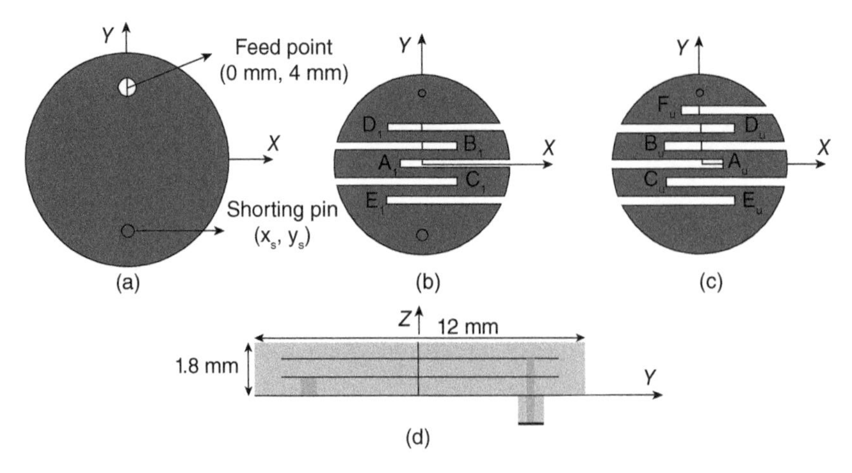

Figure 2.8 Parametric PIFA model; (a) ground plane, (b) lower patch, (c) upper patch, and (d) side view [14]. Source: Reproduced with permission of ©2012 IEEE.

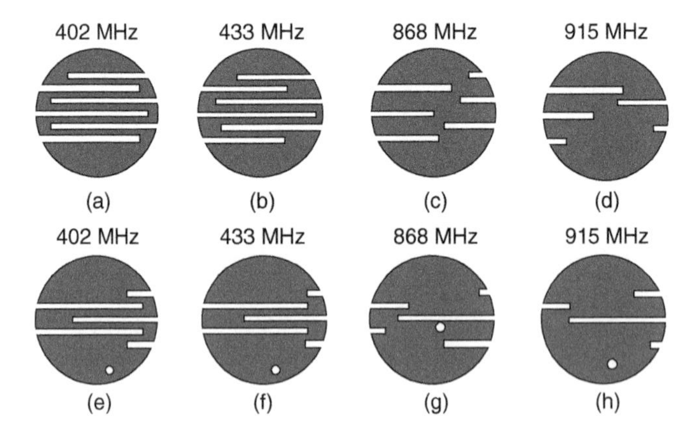

Figure 2.9 Geometries of the: (a)–(d) upper and (e)–(h) lower patches at different resonant frequencies. Source: Reproduced with permission of ©2012 IEEE.

2.2.4 Loading Technique for Impedance Matching

The loading method can be leveraged to enhance impedance matching within the desired frequency band. Typically, inductive or capacitive loading can be employed to balance the imaginary part of the impedance, thereby achieving optimal impedance matching at the desired frequency. In [21], an inductive loop is placed on one of the dipole arms for impedance matching, as depicted in Figure 2.10. The inductive loop is positioned on the upper side of the substrate and connected to the left arm via "via1." Figure 2.11 exhibits a Smith chart that compares impedance matching with and without the inductive loading. As seen from the chart, with inductive loading, the impedance at 402 MHz and 2.45 GHz can be well matched at 50 Ω. It is noteworthy that the inductively loaded loop may induce capacitive coupling, but this does not significantly impact the antenna's performance.

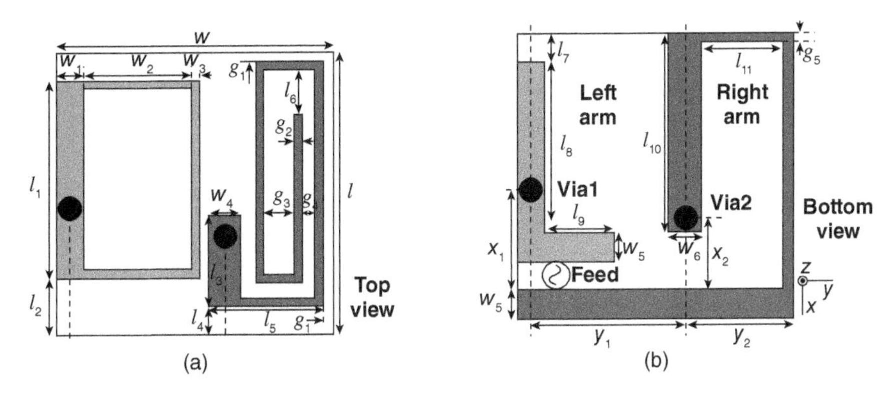

Figure 2.10 Geometry of the proposed dipole antenna: (a) top view, (b) bottom view. Source: Xu et al. [21]/IEEE.

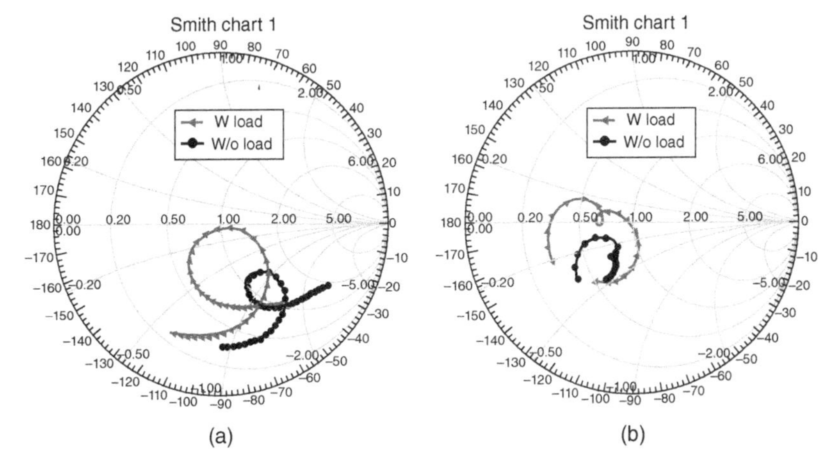

Figure 2.11 Impedance matching of the dipole antenna with and without inductive loading: (a) at MICS band, (b) at ISM band. Source: Xu et al. [21]/IEEE.

In [29], capacitive loading resulted in an antenna size reduction of about 72%, replacing the standard implantable circularly polarized (CP) microstrip patch design at a fixed operating frequency. Besides inductive or capacitive loading, split-ring resonator (SRR) loading is another type of loading for impedance matching and size reduction. In [17], an SRR and a spiral are both short-circuited to the ground plane, enabling size reduction.

Figure 2.12 illustrates the structure of the multilayer antenna discussed in [17]. This antenna comprises three layers. The feeding layer includes a ground plane and a microstrip line for electromagnetic coupling. The radiation layer comprises two short-circuited coupled SRRs. The shorting pins are directly connected to the ground plane to introduce effective shunt inductance. The SRRs are highly compact and suitable for designing a dual-band antenna. A superstrate layer is placed above the SRR radiating layer to prevent a short circuit between the radiating layer and human tissues. From the electric field distribution studies in [17], it can be inferred that the spiral primarily defines the excitation of the lower operating frequency (MICS), while the inner ring mainly defines the higher operating frequency

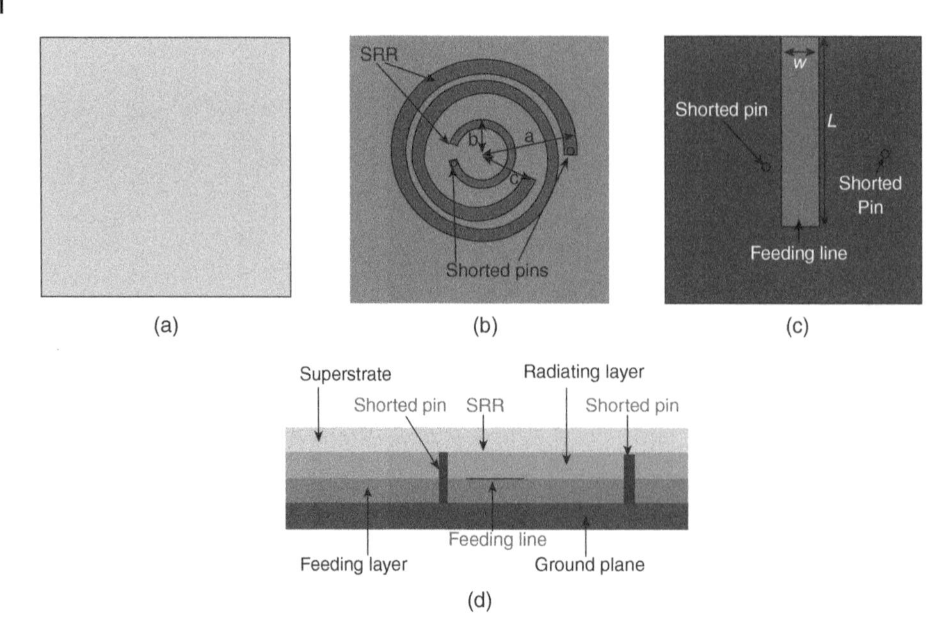

Figure 2.12 Description of the multilayer antenna [17]: (a) superstrate layer, (b) SRR radiating layer, (c) feeding layer, (d) side view. Source: Reproduced with permission of ©2010 IET.

(Industrial, Scientific, and Medical [ISM]). Note that a coupling effect exists from the ring to the spiral, determining the operating frequency of both bands.

2.2.5 Choosing Higher Operating Frequency

Various frequency bands can be employed for medical implants, as outlined in Table 2.3. These bands include the Medical Device Radio Communication Service (MedRadio, 401–406 MHz) and the ISM band (433–434.8 MHz, 902–928 MHz, 2.4–2.48 GHz, and 5.725–5.875 GHz). The formerly known Medical Implant Communication Service (MICS) band (402–405 MHz) is the most frequently employed for medical implant communications. Other frequency bands have also been proposed for biotelemetry in implantable

Table 2.3 Typical operating bands for biomedical applications.

Frequency band	Frequency range
Medical Implant Communication Services (MICS)	402–405 MHz
Medical Device Radio Communication Service (MedRadio)	401–406 MHz
Wireless Medical Telemetry Services (WMTS)	1395–1400 MHz
Industrial Scientific Medical (ISM)	433–434.8 MHz 902–928 MHz 2.4–2.48 GHz 5.725–5.875 GHz
Impulse Radio Ultrawideband (IR-UWB)	3.1–10.6 GHz

devices. In [34], an impulse radio ultra-wideband (IR-UWB) pulse operating at a center frequency of 4 GHz and a bandwidth of 1 GHz was selected as the excitation for the implantable antenna. In [35, 36], the capsule antenna and implantable antenna were designed for the Wireless Medical Telemetry Service (WMTS) band (1395–1400 MHz).

A higher operating frequency results in a shorter wavelength, thereby enabling the design of antennas with a smaller volume. Moreover, higher operating frequencies with potential wide bandwidth are more suitable for high data-rate communication. However, higher operating frequencies can lead to increased biological tissue loss in the human body and greater path loss in free space. In reality, various factors such as device dimensions, operating frequency, and transmission distance must be taken into account within the overall framework of device requirements. In the context of implantable BCI, miniaturization and data bandwidth are primary considerations. Typically, the BCI device is implanted in the head, a location where the human body experiences less loss. Therefore, high-frequency IR-UWB communication is the ideal choice for BCIs. The design of the IR-UWB antenna and circuit system has garnered significant attention [37–39], as illustrated in Figure 2.13.

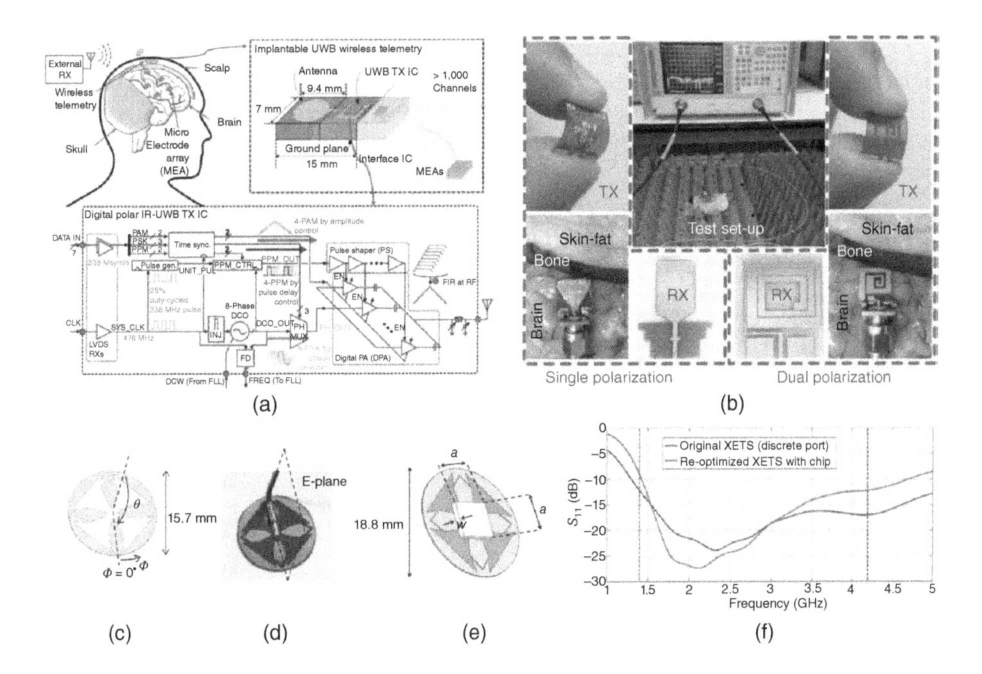

Figure 2.13 (a) Conceptual diagram of intra-cortical neural sensing with 1.05 cm^2 implantable UWB wireless telemetry (top) and block diagram of proposed UWB transmitter (TX). Source: Song et al. [37]/IEEE, (b) the flexible, polarization-diverse UWB antennas for BCI systems. Source: Bahrami et al. [38]/from IEEE. (c)–(f) high data rate impulse radio communication antenna, it is a uniplanar antenna composed of two crossed exponential tapered slots (XETS), intersected by a square-looking slot. Source: Felicio et al. [39]/with permission of IEEE, (c) simulation model, (d) antenna prototype, (e) re-designed XETS antenna with the chip attached, (f) simulated input reflection coefficient of the redesigned XETS antenna, including the biocompatible material and the electronic chip model, compared to the original bare XETS antenna.

2.3 Wideband Miniaturized Implantable Antenna

As previously noted, broadband design is advantageous for implantable antennas, addressing the detuning issue and enhancing the data rate of the in-body to off-body communication channel. This section will collate and discuss methods to broaden the impedance bandwidth of implantable antennas. Additionally, the design of detuning-immune narrowband implantable antennas will also be discussed.

2.3.1 Introducing Adjacent Resonant Frequency Points

It is recognized that by introducing a resonant frequency close to the desired frequency point, the impedance bandwidth of antennas can be significantly broadened. Inspired by this concept, different types of antennas boasting wide bandwidth have been explored and designed for biomedical applications. These include linear wire antennas (such as the meandered dipole [22] and folded dipole antenna [40]), PIFA [41], 3D folded antennas [42], slot antennas [43], loop antennas [44, 45], and microstrip patch antennas [46].

2.3.1.1 Linear Wire Antenna

For example, in [22], by introducing a strip connected to a simple dipole, a new resonance is introduced in addition to its fundamental resonance. This approach enhances the bandwidth, and, after optimization, the simulated bandwidth of the dipole antenna increases from 25.7% to 37.8% compared to the antenna without the strip.

As can be seen from Figure 2.14, a simple meandered dipole is fed at three different positions, labeled P_1, P_2, and P_3, with respect to the 50-Ω reference. The antenna is simulated in a one-layer-skin model with dimensions of 180 mm × 180 mm × 180 mm and it is located at the center of its x–y plane with a depth of 3 mm. The dielectric properties of the skin tissue at 402 MHz are found to be $\varepsilon_r = 46.74$, $\sigma = 0.69$ S/m. The reflection coefficients of the dipole fed at different positions are compared in Figure 2.15. When the feeding position changes from P_1 to P_3, the matching becomes better gradually. Thus, an off-center-fed dipole is adopted for the design.

To broaden the bandwidth of the dipole antenna, a meandered strip is attached to the dipole, as depicted in Figure 2.16. The strip consists of two segments with different widths. The width of the narrower part is set to be 0.3 mm, while the width of the wider part,

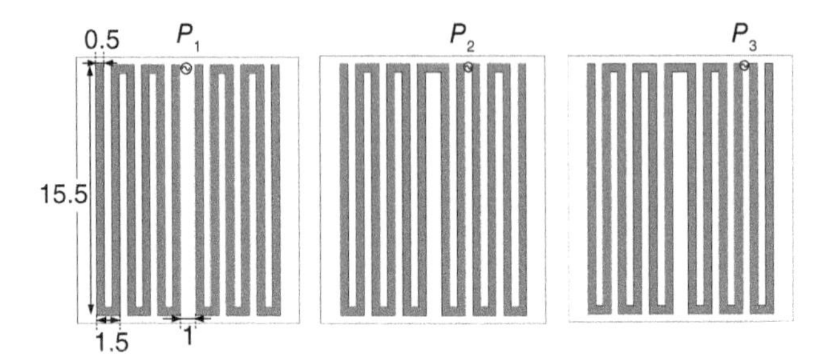

Figure 2.14 Configuration of an implantable dipole fed at different positions (unit: mm). Source: Xu et al. [22]/IEEE.

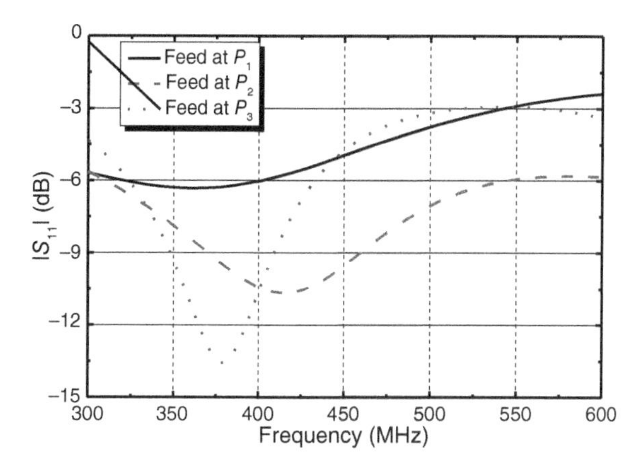

Figure 2.15 Comparison of the $|S_{11}|$ for the dipole fed at different positions. Source: Xu et al. [22]/IEEE.

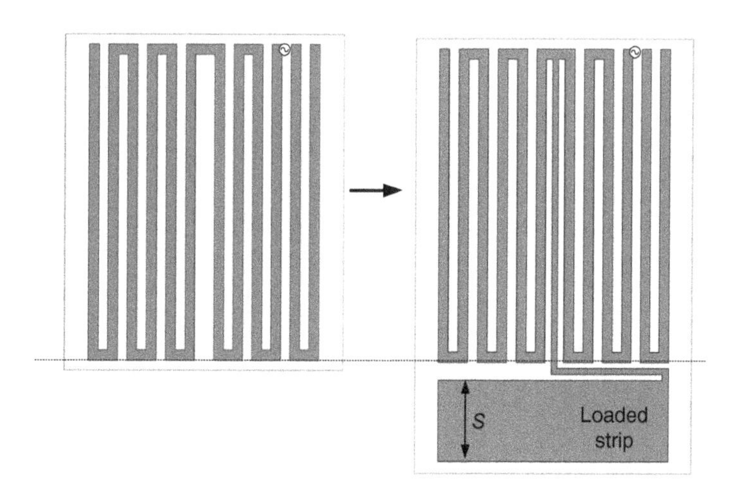

Figure 2.16 An off-center-fed dipole loaded with a strip. Source: Xu et al. [22]/IEEE.

labeled S, is adjusted to alter the electrical length of the entire strip. As shown in Figure 2.17, the introduction of the meandered strip induces another resonance, thus enhancing the bandwidth.

Based on the above analysis, a wideband planar dipole is designed as shown in Figure 2.18. It consists of a meandered dipole and a two-width strip connected to its center. One end of the dipole arm is modified into a spiral shape for improved impedance matching. It should be noted that the port is situated on the same plane as the radiation part and the superstrate is slightly shorter than the substrate for soldering purposes, which, according to the simulation, will not impact the result. A compact volume of 329 mm^3 (16.5 mm × 15.7 mm × 1.27 mm) is achieved. After final optimization, the detailed parameters of the planar dipole antenna are $l = 16.5$ mm, $w = 15.7$ mm, $w_1 = 1.5$ mm, $w_2 = 2.2$ mm.

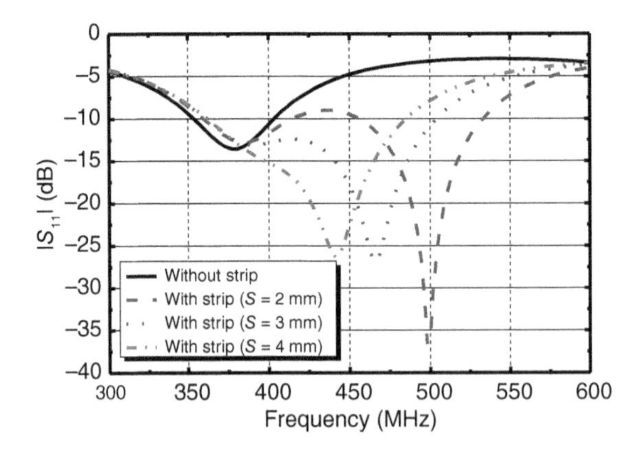

Figure 2.17 Comparison of the $|S_{11}|$ with and without the strip. Source: Xu et al. [22]/IEEE.

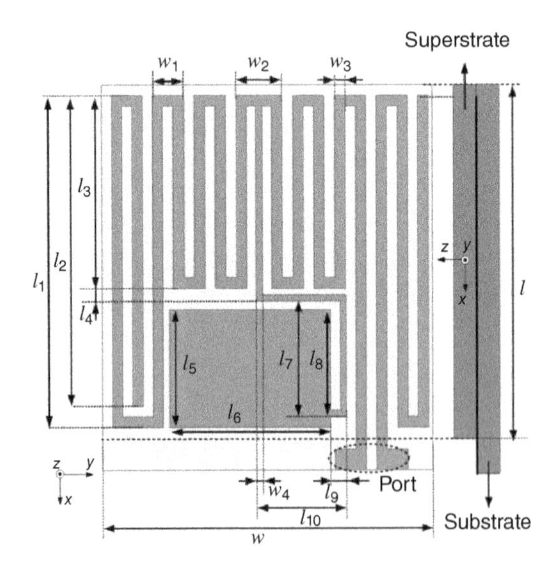

Figure 2.18 Configuration of the dipole antenna in planar form. Source: Xu et al. [22]/IEEE.

$w_3 = 0.5$ mm, $w_4 = 0.3$ mm, $l_1 = 15.5$ mm, $l_2 = 14.5$ mm, $l_3 = 9$ mm, $l_4 = 0.6$ mm, $l_5 = 5.5$ mm, $l_6 = 7.7$ mm, $l_7 = 5.4$ mm, $l_8 = 5$ mm, $l_9 = 0.75$ mm, $l_{10} = 4.3$ mm.

The $|S_{11}|$ of the planar dipole compared with that of the dipole without strip is displayed in Figure 2.19. For a dipole without the strip, the simulation result shows that a bandwidth of 25.7% can be achieved with $|S_{11}|$ less than −10 dB, covering from 338 to 438 MHz. While for the dipole antenna with bandwidth enhancement, a wide bandwidth of 37.8% can be achieved, covering from 348 to 510 MHz. Note that bandwidth can be tuned by adjusting l_5 of this configuration to meet different practical needs. Moreover, the radiation efficiency of the wideband planar antenna is also presented in Figure 2.19, which is below 2×10^{-4} over the frequency band, and the peak gain at its boresight is −35 dBi at 402 MHz.

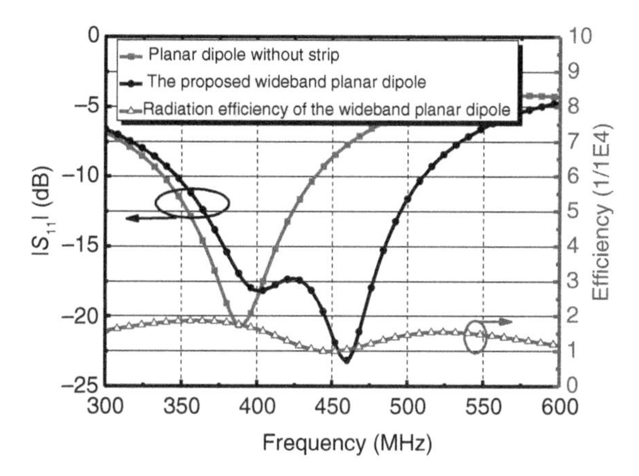

Figure 2.19 Simulated $|S_{11}|$ of the planar dipole together with its radiation efficiency as compared with the $|S_{11}|$ of a dipole without the strip. Source: Xu et al. [22]/IEEE.

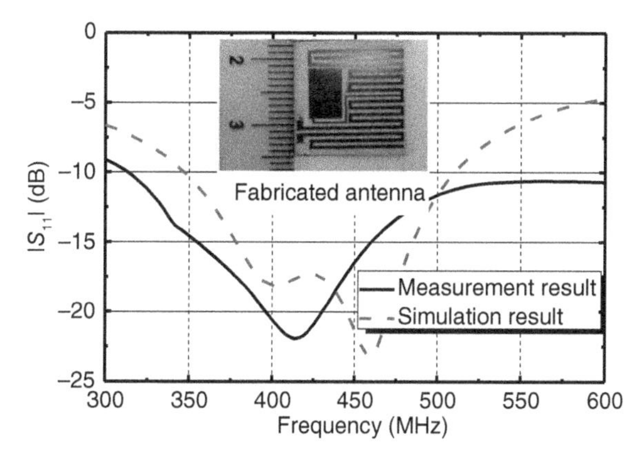

Figure 2.20 Measured result of the planar dipole antenna compared with the simulated result. Source: Xu et al. [22]/IEEE.

To validate the wideband performance of the proposed dipole, it is fabricated and measured in skin-mimicking gel. Since the dielectric properties of skin at 402 MHz are found to be $\epsilon_r = 46.74$ and $\sigma = 0.69$ S/m, the recipe of gel is 56.18% sugar, 2.33% salt, and 41.49% deionized water. The measured results together with the simulated ones are compared in Figure 2.20. As seen from the graph, wide bandwidth performance can be achieved during measurement.

In [42], a 3D wire antenna with broadband performance was designed specifically for dental implants. The proposed design began with a reference PIFA featuring a Hilbert-shaped fractal geometry. This design extended the current path to achieve miniaturization, as shown in Figure 2.21(a). Taking into account the wear and tear caused by chewing or biting actions, the reference PIFA was folded into a 3D shape, and the fractal radiators were tucked into the two sides of the molar. However, the bandwidth of the reference PIFA was found to be unsatisfactory, as displayed in Figure 2.21(c). In order to

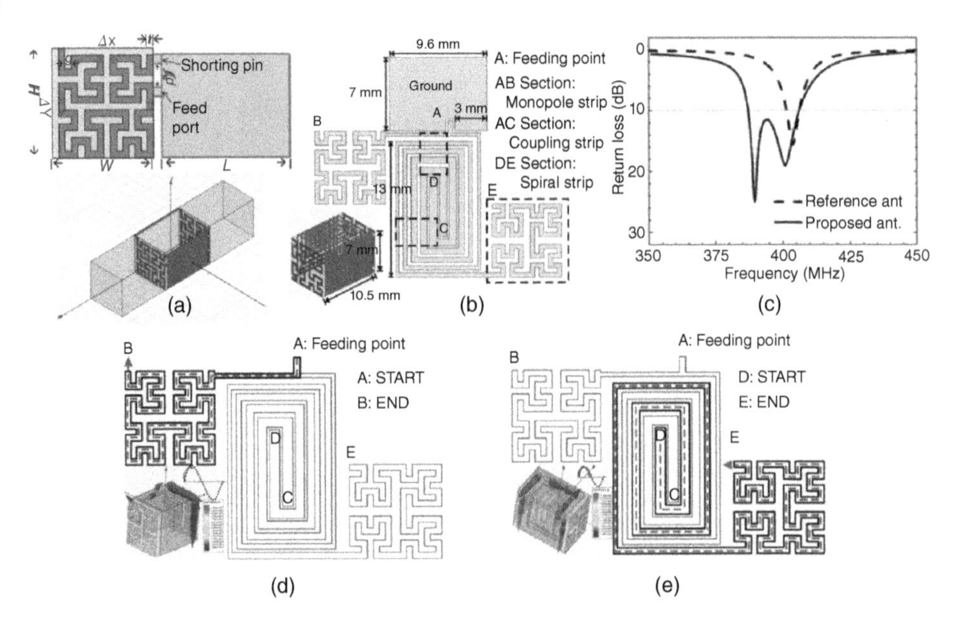

Figure 2.21 A folded 3D broadband antenna for dental implants. Source: Yang et al. [42]/IEEE: (a) the architecture of a reference antenna for dental implants, the simulation was performed using both the anterior and posterior teeth, (b) the 2D layout and 3D structure and dimensions of the broadband antenna, (c) return loss of the proposed broadband antenna and the reference antenna, (d) simulated current distribution of the antenna at 400.8 MHz, (e) simulated current distribution of the antenna at 389.3 MHz.

increase the bandwidth, a second resonant mode in close proximity to the first mode was introduced with a compact form factor. The metal microstrip structure of the reference antenna was adjusted by incorporating an Archimedean spiral between the two Hilbert fractal parts to combine the different modes and enhance inductive coupling. The revised structure and dimensions of the proposed antenna are displayed in Figure 2.21(b). As a result, the proposed broadband antenna, with a bandwidth of 18.1 MHz, demonstrates a 346% improvement compared to the reference antenna, which only has a bandwidth of 5 MHz. Figure 2.21(d) and (e) illustrate the current paths and distributions of these dual bands. The first current flows from the end of the Hilbert component (Point B) to the antenna-feeding section (Point A), with maximum amplitude at the feeding point. This represents the quarter-wavelength mode of 400.8 MHz. The second current flows from the center of the spiral section (Point D) toward the end of the other Hilbert portion (Point E), with the maximum amplitude at the midpoint of this path and minimum amplitude at the start and endpoints. This represents the half-wavelength mode of 389.3 MHz.

2.3.1.2 Slot Antenna

Not only wire antennas, but also any type of antennas capable of exciting two adjacent modes can be considered for broadband implantable antenna designs. However, the limitations of various application scenarios should be accounted for in each different design. In [43], a slot antenna with wideband characteristics was proposed for implantable applications. This dual-ring topology excites two closely spaced resonant frequencies to

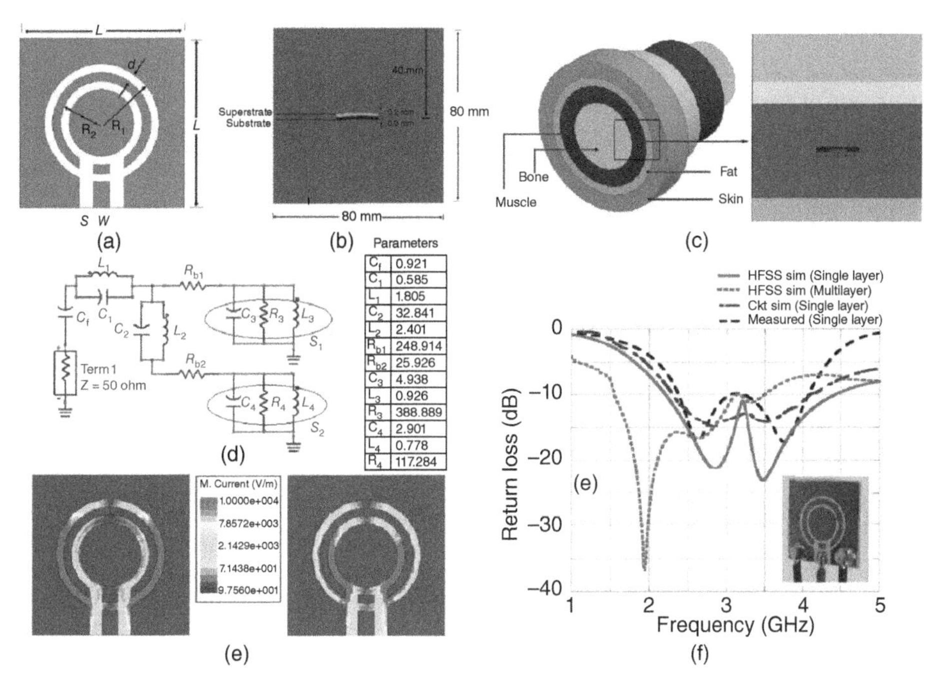

Figure 2.22 The broadband implantable slot antenna. Source: Das and Mitra [43]/IEEE: (a) top view of the antenna, (b) antenna placed at the center of a cubic single-layer human muscle model, (c) antenna placed in the four-layer cylindrical human arm model, (d) equivalent circuit model for the implantable slot antenna, (e) magnetic current distribution for the dual-ring slot antenna, left: at the first resonating frequency, right: at the second resonating frequency, (f) return loss characteristics of the antenna and the prototype antenna (inset).

achieve a larger bandwidth, as shown in Figure 2.22(a). An equivalent circuit model of the implantable slot antenna can be used to analyze the wideband characteristics, as depicted in Figure 2.23(d). Coupling elements are inserted in the circuit diagram to model the interaction between individual slot sections and the extension of the CPW feed with the CPW feed line. Capacitance C_f is used for the feed section, while the LC-coupling circuit with L_1 and C_1 models the interaction between the feed section and the CPW line. This feed line directly excites the outer ring and couples to the inner ring with another LC-coupling circuit, composed of inductor L_2 and capacitor C_2. Now the sections marked S_1 and S_2 correspond to two distinct resonating frequencies, generated by the dual-ring configuration of the antenna. It is observed that the first resonance is contributed by the outer ring slot, while the second resonance mainly arises from the inner ring slot. Here, resistances Rb1 and Rb2 represent losses due to the biomedical environment [43]. Figure 2.22(e) displays the magnetic current distribution for the dual-ring slot antenna, which provides solid support for the wideband characteristic benefiting from two overlapping resonating frequency modes. This ring-slot antenna was designed for hand implants. When comparing the antenna simulation results in different models, as shown in Figure 2.22(b) and (c), it becomes clear that the complex dielectric characteristics of the tissues greatly influence the impedance bandwidth of the implantable antenna, as depicted in Figure 2.22(f).

2.3.1.3 Loop Antenna

In [44], a loop antenna was specifically designed for a capsule endoscope. The loop structure is carefully engineered to conform to the inner wall of the capsule, as depicted in Figure 2.23(a). This antenna operates at a frequency of 433 MHz and exhibits a full-wave loop configuration. It is worth noting that the antenna comprises two sections with oppositely phased currents and two points where the current is zero, as illustrated in Figure 2.23(b). In order to explore the antenna's performance, parametric studies were conducted, focusing on the influence of slots. The inclusion of slots resulted in an additional resonance at approximately 900 MHz, thereby expanding the overall bandwidth of the antenna.

2.3.1.4 Microstrip Patch Antenna

In [46], a broadband microstrip patch antenna was specifically developed for pacemakers. The proposed antenna design is a differential-fed microstrip patch antenna, as depicted in Figure 2.24(a). The pacemaker, enclosed within a metal case, utilizes a low-profile microstrip patch antenna mounted on the surface of the case, as shown in Figure 2.24(b). The choice of this antenna configuration is well-suited for pacemaker applications.

In this design, the differential signals effectively suppress even-order modes, such as TM_{20} and TM_{22}, which results in a null in the broadside direction. This suppression is advantageous as it helps maintain the radiation peak of the proposed microstrip patch antenna at the boresight (Z-axis) for efficient communication with an in-vitro monitor. Moreover, the TM_{10} and TM_{30} modes were precisely tuned and resonated within the operational range to enhance the overall bandwidth of the antenna, as demonstrated in Figure 2.24(c)–(e).

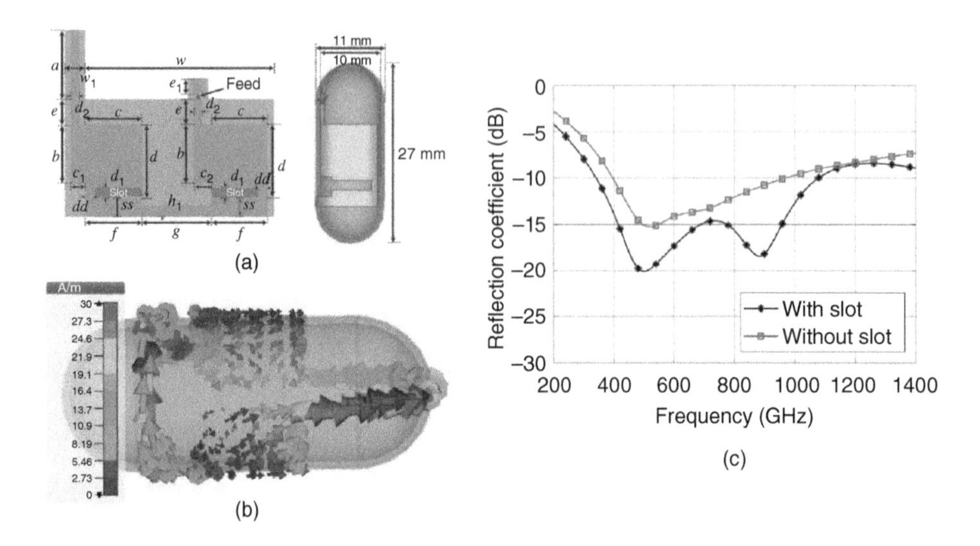

Figure 2.23 The broadband loop antenna for capsule endoscope. Source: Miah et al. [44]/IEEE: (a) the layout of the antenna and its placing around the capsule, (b) the current distribution of the capsule antenna, a full-wavelength mode at 433 MHz, (c) the return loss characteristics of the antenna, the added slot creates another resonance that appears at around 900 MHz.

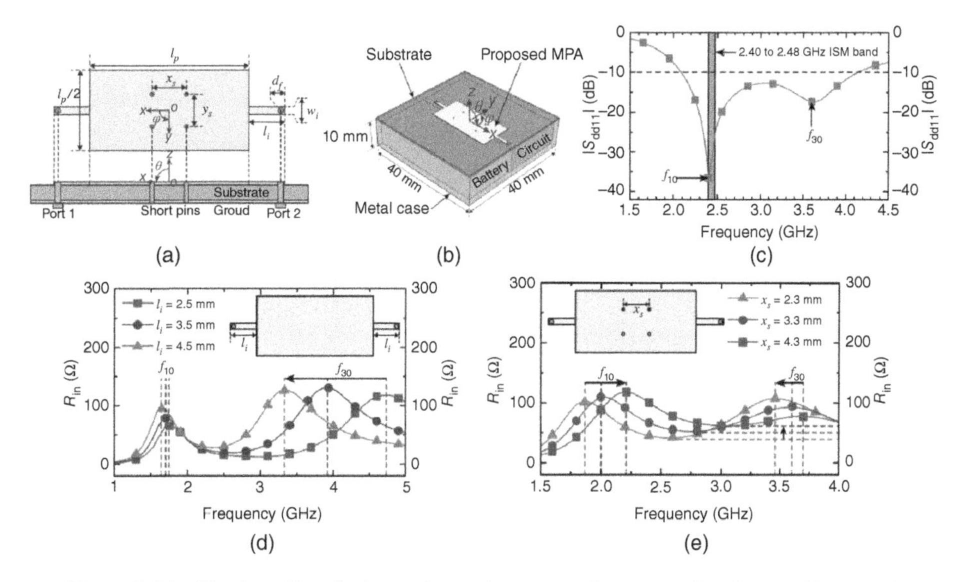

Figure 2.24 The broadband microstrip patch antenna for pacemaker. Source: Yang et al. [46]/IEEE/CC BY 4.0: (a) geometrical schematic of the proposed implantable microstrip patch antenna, (b) the microstrip patch antenna in the complete pacemaker model, (c) the return loss characteristics of the antenna, (d) and (e) parameters studies of the microstrip patch antenna, simulated input resistance R_{in} of the differential-driven microstrip patch antenna.

2.3.2 Multiple Resonance and Wideband Impedance Matching

The wideband miniaturized implantable antenna in the aforementioned design is achieved through the implementation of a double resonant structure. This structural configuration, characterized by multiple resonance characteristics, offers significant advantages in the design of broadband miniaturized implantable antennas.

In [47], the conformal capsule antenna exhibits a wide impedance bandwidth ranging from 1.64 to 5.95 GHz and demonstrates stable impedance-matching characteristics even when subjected to changes in tissue conditions. This conformal capsule antenna adopts a wrapped slot-loop structure fed by a coaxial cable, with the introduction of a slot in the inner patch to achieve an additional resonance. As depicted in Figure 2.25(a) (top right inset), the center core of the coaxial cable is connected to the inner patch of the antenna via a via, while the metallic shield of the coaxial cable is connected to the outer patch of the antenna through another via.

The capsule shell comprises a cylindrical section at the center, with hemispherical caps at the top and bottom. As shown in the bottom right inset (cross-section of the capsule), the center cylinder and bottom cap are joined together, and the top cap can be connected to the bottom part. The dimensions of the capsule shell, including its inside diameter (10 mm) and inside height (25 mm), are consistent with those of a typical capsule used in endoscope applications. To fabricate the capsule, photosensitive resin (PR) is employed, utilizing 3D printing technology. The thickness of the capsule shell is set at 0.5 mm.

The antenna itself is first printed on a flexible 10 mil-thick single-sided Rogers 5880 substrate and is subsequently wrapped and inserted into the capsule. This fabrication

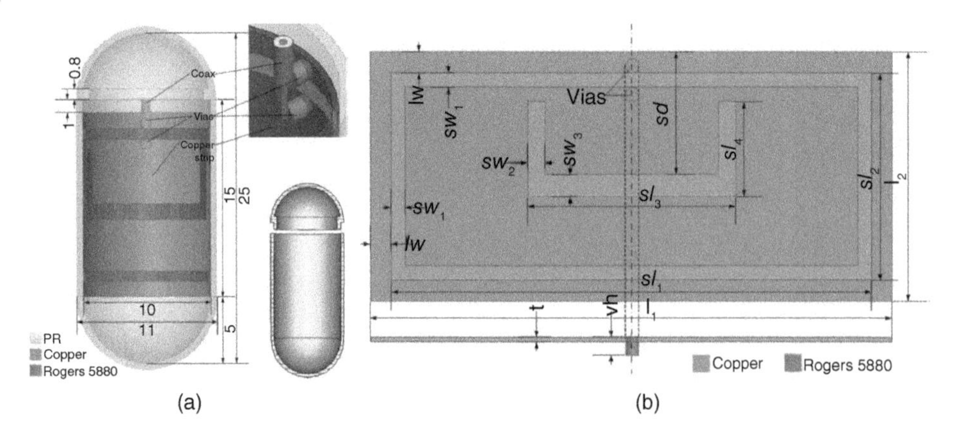

Figure 2.25 (a) Geometry of the proposed conformal capsule antenna. The top right inset shows the feeding structure: the inner patch is connected to the center core of a coax through a via and the outer patch is connected to the metallic shield of the coax through another via. The bottom right inset shows the cross-section of the capsule shell, (b) geometry of the proposed capsule antenna before it is wrapped (front view and top view). The optimized values for the parameters are $l_1 = 30$, $l_2 = 14$, $lw = 1.2$, $sl_1 = 27.6$, $sl_2 = 11.6$, $sl_3 = 12$, $sl_4 = 4.1$, $sw_1 = 0.8$, $sw_2 = 1$, $sw_3 = 1.3$, $sd = 8.1$ (Units: mm).

approach offers cost-effectiveness and does not necessitate highly precise manufacturing techniques. The geometry of the proposed capsule antenna prior to wrapping is illustrated in Figure 2.25(b).

To investigate the broadband characteristics and operating mechanism of the antenna depicted in Figure 2.25, we examined two sets of conformal inner-wall capsule antennas as examples, as illustrated in Figure 2.26. Each set consists of a complementary structure: (i) a loop antenna and a slot-loop antenna, and (ii) a loop + U patch antenna and a slot-loop + U-slot antenna. Figure 2.27 demonstrates that the loop antenna has a bandwidth of only 15.6%, whereas the slot-loop antenna, which complements the loop antenna, exhibits a much wider bandwidth of 34.4%. Although the loop and slot-loop antennas are not strictly complementary and are not operating in a homogeneous medium, this phenomenon can

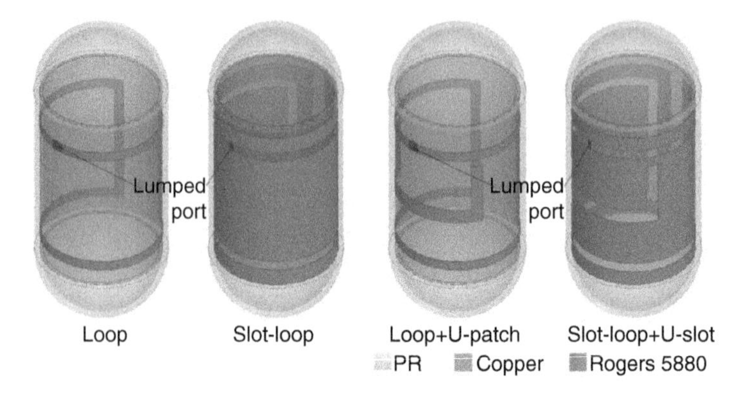

Figure 2.26 Two sets of conformal inner-wall capsule antennas.

be qualitatively explained using the theory of complementary antennas. According to the theory of complementary antennas, we can derive the following relationship:

$$Z_l \cdot Z_s = \frac{\eta^2}{4} \tag{2.1}$$

where Z_l and Z_s are the input impedances of an antenna and its complement, respectively, and η is the effective wave impedance.

The input impedances of the loop antenna and slot-loop antenna, obtained through HFSS simulations, are illustrated in Figure 2.27(b). The impedance of the loop antenna exhibits significant changes with frequency, whereas the impedance of the slot-loop antenna shows relatively smooth variations. Additionally, the slot-loop antenna demonstrates a second resonance around 4 GHz. In Figure 2.27(b), the calculated Z_s is obtained using Eq. (2.1), where Z_l represents the impedance of the loop antenna, and the wave impedance η is assumed to be $143 + j17.4\,\Omega$. The profile of Z_s closely resembles the impedance profiles of the slot-loop antenna. Moreover, Figure 2.27(a) showcases an additional resonance introduced by incorporating an extra U-slot at the center of the slot-loop antenna. Conversely, adding a parasitic U-patch has almost no impact on the S_{11} of the loop antenna. While the parasitic patch could be modified to tune the antenna, it is observed that it is easier to tune the slot antennas to introduce additional resonances. Based on these findings, it can be concluded that slot antennas offer wider bandwidths and can be readily tuned to accommodate additional resonances in other applications of conformal capsule antennas. For the additional resonance around 2.7 GHz observed in the proposed slot-loop + U-slot antenna, further parametric studies can provide a more comprehensive explanation. Figure 2.28 clearly demonstrates that the resonance around 2.7 GHz emerges as the length of the vertical section of the U-slot (sl_4) is increased, as depicted in Figure 2.25(b). Hence, it can be inferred that the U-slot at the center of the antenna introduces the additional resonance around 2.7 GHz.

The proposed slot-loop +U-slot antenna has a very wide impedance bandwidth. Therefore, it can be immune to the changes in dielectric properties of surrounding tissues. As shown in Figure 2.29, its impedance-matching characteristic is very stable and relatively insensitive to the variations of the surrounding tissues. We placed the antenna into different tissues, such as Muscle (ε_r = 52.729, σ = 1.7388, @ 2.4 GHz), Stomach (ε_r = 62.158,

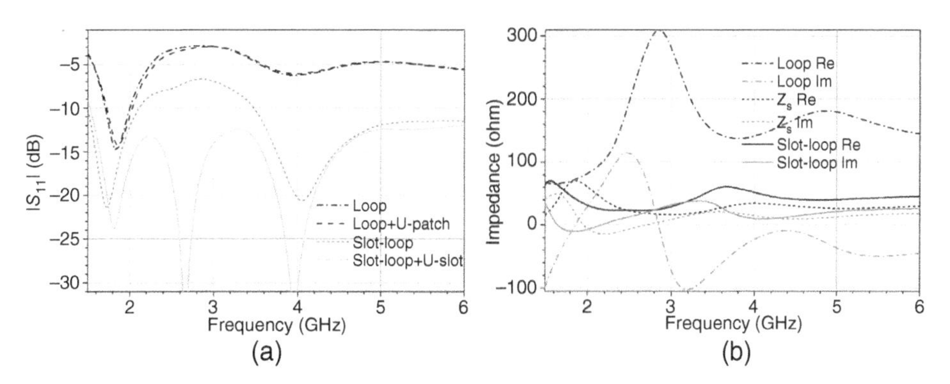

Figure 2.27 (a) $|S_{11}|$ of the antennas in Figure 2.26, (b) Input impedance of the loop and slot-loop antennas in Figure 2.26, and Z_s calculated using (2-1).

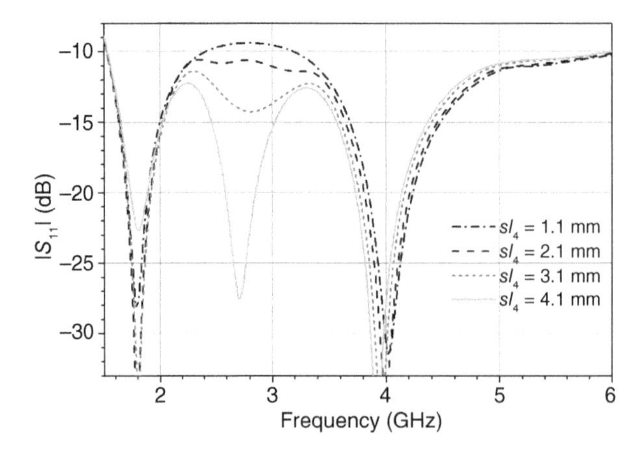

Figure 2.28 Effects of length of the vertical section of the U-slot sl_4 on $|S_{11}|$ of the proposed capsule antenna.

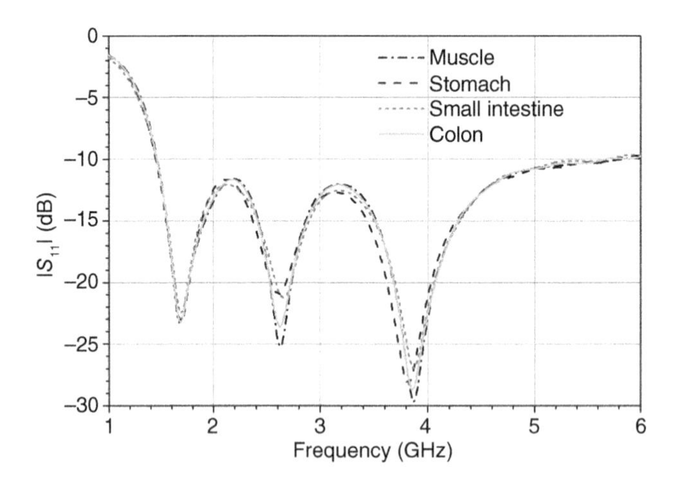

Figure 2.29 $|S_{11}|$ of the proposed capsule antenna in dispersive tissues simulated using CST.

$\sigma = 2.2105$, @ 2.4 GHz), Small intestine ($\varepsilon_r = 54.425$, $\sigma = 3.1731$, @ 2.4 GHz), and Colon ($\varepsilon_r = 53.879$, $\sigma = 2.0383$, @ 2.4 GHz), and got a stable impedance bandwidth.

The detuning of the implantable antenna is not solely influenced by changes in the surrounding tissues. Practical considerations such as the presence of a battery and internal circuits within the capsule, as well as variations in the thickness of the capsule shell due to manufacturing reasons, can also significantly impact the performance of the conformal capsule antenna. In order to assess the influence of these factors on the proposed antenna, we conducted a study, and the results demonstrate that the antenna's environmental sensitivity is greatly reduced, thanks to its broadband characteristics.

In Figure 2.30, we employed a model where the battery was represented as a perfect electric conductor (PEC) cylinder coaxial with the capsule shell. Our findings reveal that among the location, height, and radius of the battery, the radius has the most significant effect on the antenna's performance. Assuming a battery height of 5 mm and located at the bottom

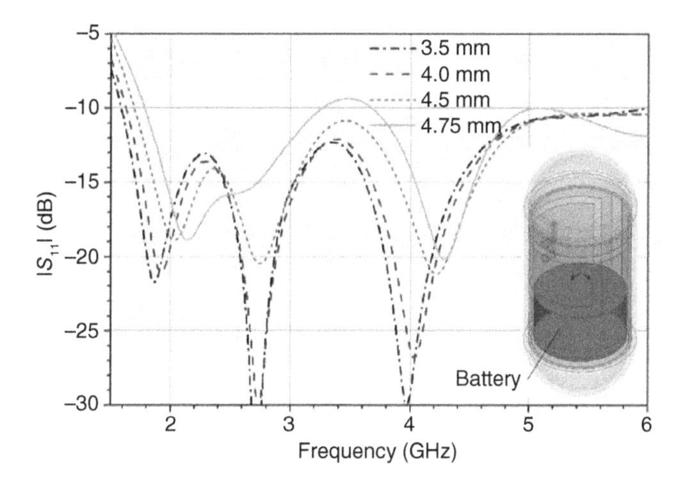

Figure 2.30 Effects of radius of the battery located inside of the capsule on $|S_{11}|$ of the proposed capsule antenna. Inset shows the location of the battery inside the capsule.

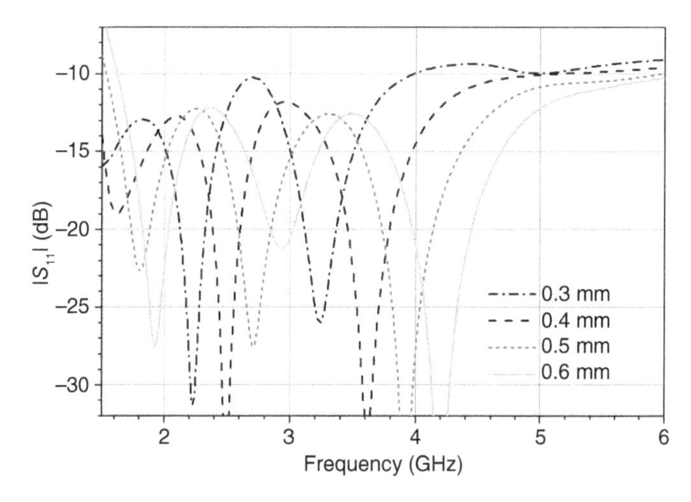

Figure 2.31 Effects of thickness of the capsule shell on $|S_{11}|$ of the proposed antenna.

of the capsule, Figure 2.30 illustrates that when the battery comes into contact with the antenna substrate (solid line, curve of 4.75 mm), the S_{11} of the antenna slightly exceeds −10 dB around 3.5 GHz. Conversely, if the battery is slightly moved away from the antenna substrate, for example, with a radius 0.25 mm smaller (dotted line, curve of 4.5 mm), the impact on the antenna bandwidth is minimal. Furthermore, the simulations indicate that placing the battery farther from the feed point has an even lesser effect on S_{11}. Given that a battery size of 5 mm in height and 4.5 mm in radius is typically adequate for a typical capsule system, we can conclude that the proposed antenna will maintain stable performance when a battery and its associated circuits are inserted into the capsule.

Figure 2.31 demonstrates that the three resonances shift toward lower frequencies in a monotonic manner as the shell thickness is reduced. Consequently, employing a thinner

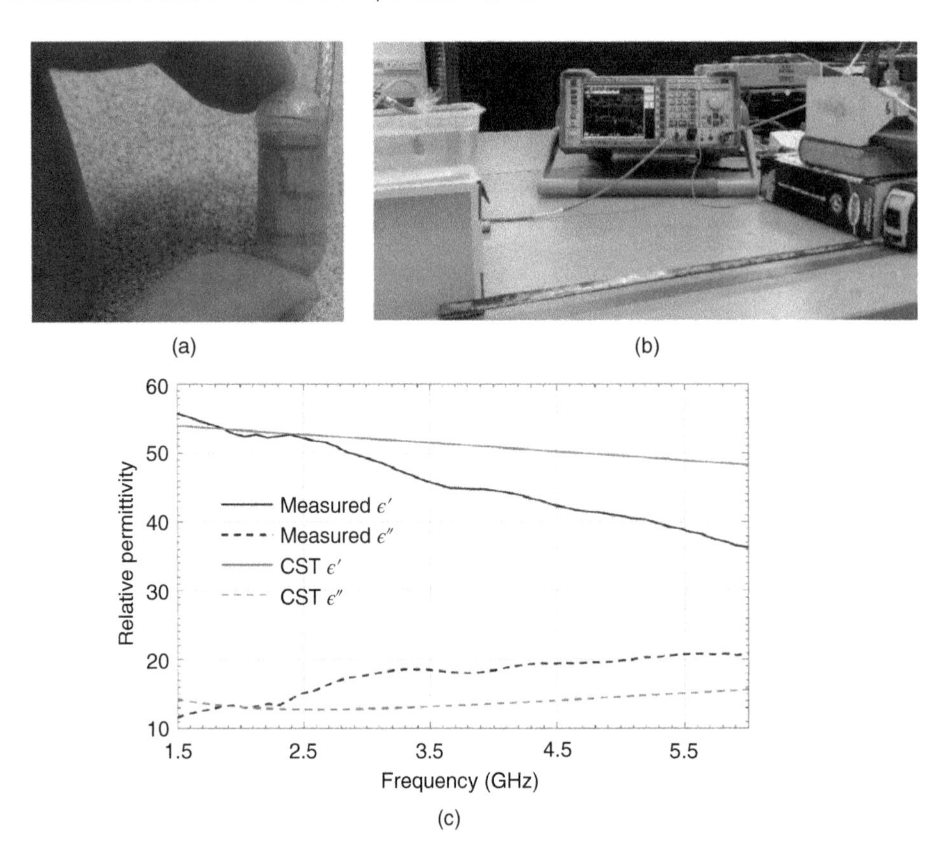

Figure 2.32 (a) Fabricated capsule antenna fed by a coaxial cable, (b) setup of communication link measurement, (c) measured relative permittivity of used muscle phantom and relative permittivity of muscle in CST.

capsule shell assists in achieving antenna miniaturization. Moreover, the simulations suggest that the bandwidth be further enhanced by using a thinner capsule while appropriately tuning the antenna.

The fabricated prototype of the proposed antenna is showcased in Figure 2.32(a), while Figure 2.32(b) illustrates the measurement setup for the communication link. The proposed antenna, positioned within the muscle phantom, is connected to port-1 of a Rohde & Schwarz ZVL vector network analyzer (VNA). To prevent the solution from leaking into the capsule, the capsule is wrapped in a plastic cover with negligible thickness. Given the wideband nature of the proposed antenna, a wideband horn antenna is chosen for excitation and connected to port-2 of the VNA. The distance between the proposed antenna and the horn aperture is 50 cm, which corresponds to four free-space wavelengths at 2.4 GHz. The dielectric properties of the acquired phantom are measured using an HP 85070A dielectric probe kit, as depicted in Figure 2.32(c). The capsule antenna is situated at the center of the muscle phantom, which measures 11.5 cm × 11.5 cm × 5 cm.

Figure 2.33 presents the measured $|S_{11}|$ of the proposed antenna, which remains better than −10 dB across the frequency range spanning from 1.64 to 5.95 GHz (113.6%). These

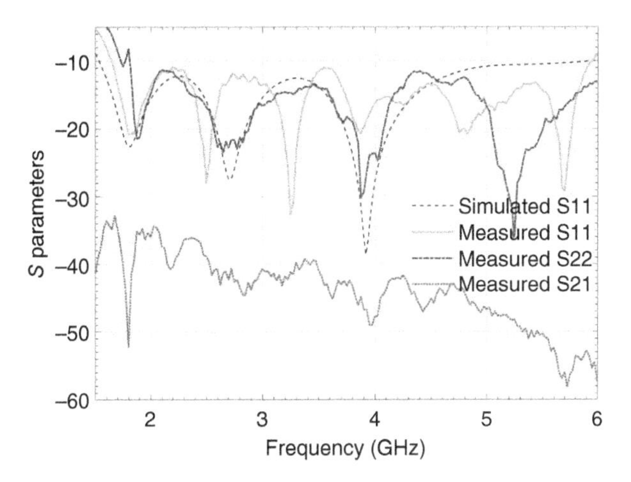

Figure 2.33 Simulated $|S_{11}|$ of the proposed capsule antenna and measured S parameters of the capsule antenna and a horn antenna.

measurements align well with the simulation results, despite differences in the plot profiles. The observed shift in the two upper resonances and the appearance of additional resonances within the range can be attributed to two factors. First, fabrication errors contribute to the presence of air gaps or bubbles between the capsule shell and the antenna surface, as well as between the capsule shell and the plastic wrap. These factors were not accounted for in the simulations. As mentioned earlier, the antenna bandwidth is sensitive to the choice of shell thickness. Additionally, in the measurement setup, the coaxial cable is soldered to the vias, slightly deviating from the setup used in the simulations. It is worth noting that the feed configuration also significantly impacts the impedance match. Second, as demonstrated in Figure 2.32(c), the phantom used exhibits dielectric properties that vary significantly with frequency. In contrast, the simulations assume either smooth or constant variations in these properties. Despite the differences between the phantom's properties and those assumed in the simulation, the proposed capsule antenna adequately covers the desired frequency range. Therefore, from an impedance matching standpoint, the antenna proves relatively insensitive to surrounding environments, thanks to its wideband impedance matching design.

The measured $|S_{21}|$ at 2.45 GHz is approximately −36 dB, and as a general trend, it decreases with increasing frequency. Meanwhile, the gain of the horn antenna at 2.45 GHz is 7.9 dBi and increases with frequency. The decline of $|S_{21}|$ with frequency primarily stems from the increase in free-space loss and the loss induced by the muscle phantom, as its conductivity rises with frequency, as depicted in Figure 2.32(c).

For patch antennas, introducing slots in the radiator and ground plane is an effective method to achieve multiple resonant frequency points. This approach is also well-suited for designing wideband implantable antennas. In [48], an ultra-miniaturized implantable patch antenna with ultra-wideband (UWB) operation is designed for leadless pacemakers using slotting techniques. Figure 2.34 illustrates the structure and layout of this UWB antenna, which has overall dimensions of 6 mm × 7 mm × 0.254 mm. The ground plane features a meander-shaped design with several slots, serving the purposes of compactness,

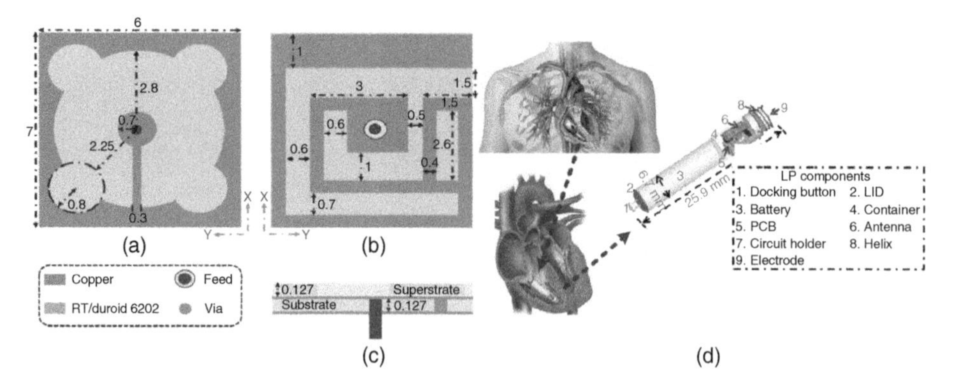

Figure 2.34 Structure and layout of the UWB antenna (units: mm): (a) front view, (b) rear view, (c) side view, (d) detailed architecture of the leadless pacemaker and its components. Source: Faisal et al. [48]/IEEE.

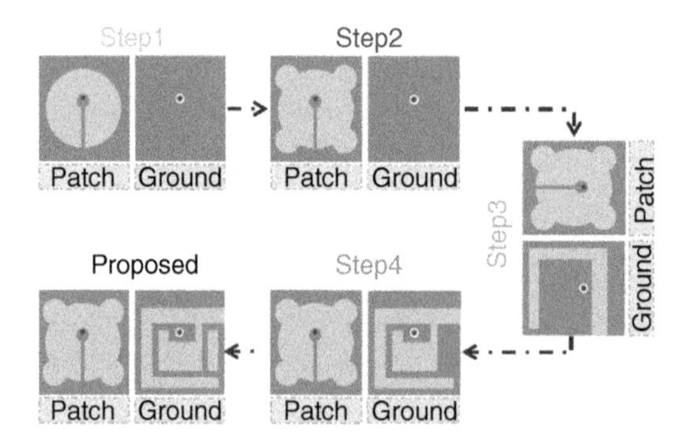

Figure 2.35 Stepwise designing modifications. Source: Faisal et al. [48]/IEEE.

tuning, and bandwidth enhancement. The main radiator is connected to the feed point through a rectangular metallic strip. By carefully selecting the dimensions of the ground and radiator slots, multiple resonances are generated, and these resonances are combined to achieve an ultra-wide bandwidth.

As depicted in Figure 2.35, the design of the proposed UWB antenna involved five steps. In the initial prototype (Step 1), the antenna consisted of a rectangular patch with a circular cut in the radiator. Figure 2.36 shows the $|S_{11}|$ response, indicating a single resonance at 4 GHz with a −10 dB bandwidth of 200 MHz due to the circular slot in the radiator.

In Step 2, the antenna was modified by introducing a via and four semicircular slots in the patch. This modification resulted in the generation of two strong resonances at approximately 1.6 and 4.6 GHz, enabling dual-band operation.

Moving to Step 3, the resonances from Step 2 were shifted to lower frequencies by adding an open-ended slot in the ground plane below the substrate. This effect can be attributed to the increased capacitance caused by the etching slots in the ground plane. Additionally, a third resonance mode at 2.15 GHz was generated in this step.

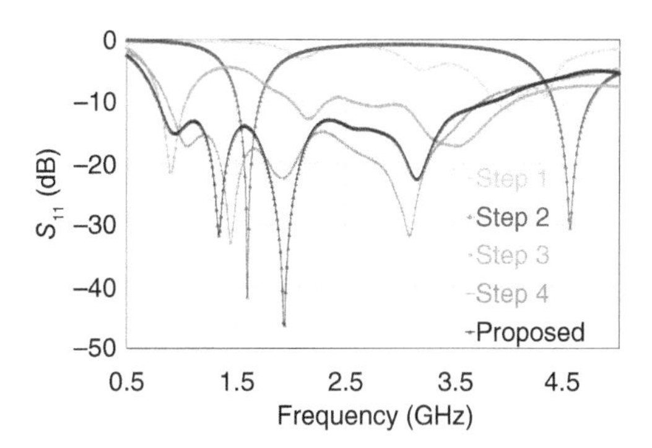

Figure 2.36 $|S_{11}|$ comparison for various designing steps. Source: Faisal et al. [48]/IEEE.

Step 4 focused on improving impedance matching by introducing a second slot in the ground plane. This allowed the resonances from Step 3 to be combined, resulting in an ultra-wide bandwidth of 2940 MHz (880–3820 MHz).

In the final step (Step 5), adjustments were made to shift the overall bandwidth toward the left side to cover the 868–868.6 MHz ISM band. This was achieved by adding a rectangular slot in the ground and moving the via toward the top left corner of the antenna. The final design achieved a bandwidth of 3040 MHz (0.79–3.83 GHz).

Furthermore, in [20], the inclusion of open-end slots in the ground plane resulted in a wide bandwidth that encompassed the MICS band at 402 MHz, as illustrated in Figure 2.37. The open-end slots, arranged in a configuration consisting of a main slot along the y-axis and two parasitic slots of lengths L_1 and L_2 along the x-axis, played a crucial role in tuning the resonant frequencies to achieve an optimal bandwidth at the lower band.

Figure 2.38 depicts the simulation setup for the dual-band antenna. The antenna was simulated within a one-layer skin model with a thickness of 60 mm. It was positioned 100 mm away from the skin's edges on both the left and right sides, and 3 mm from the top surface of the skin. The dielectric properties at the frequencies of 402 MHz and 2.45 GHz were determined to be $\varepsilon_r = 46.74$, $\sigma = 0.69$ S/m and $\varepsilon_r = 38.06$, $\sigma = 1.44$ S/m, respectively.

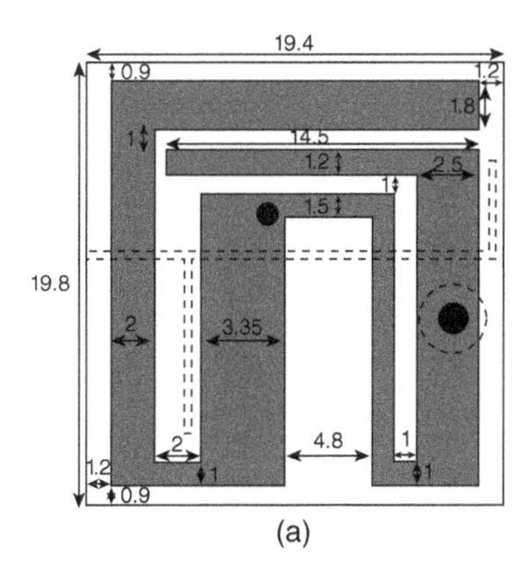

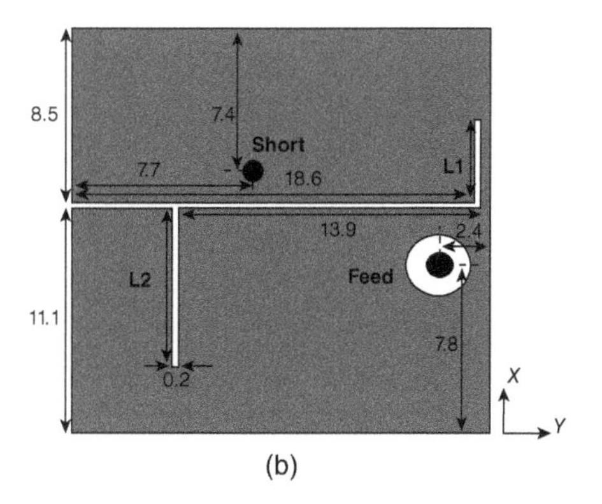

Figure 2.37 Dual-band antenna with open-end slots: (a) top view, (b) bottom view.

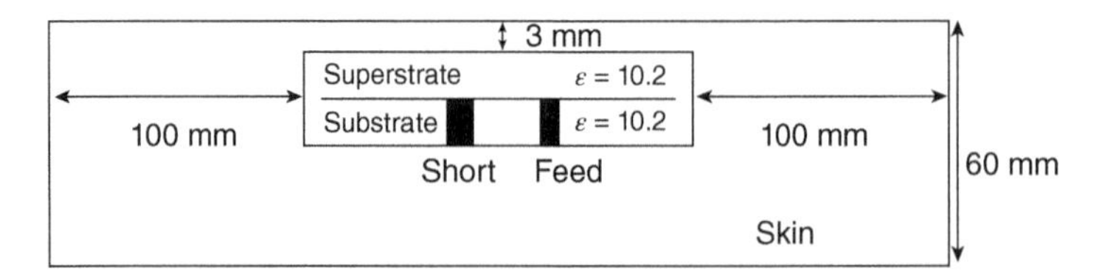

Figure 2.38 Simulation environment of the dual-band antenna with open-end slots.

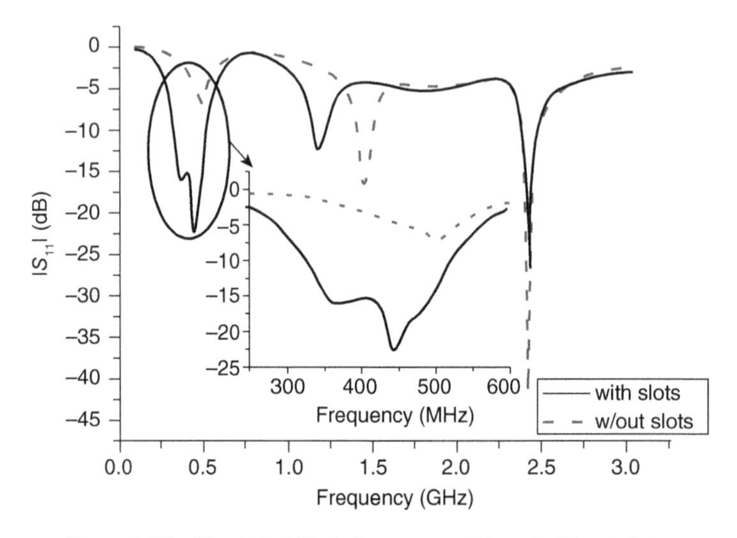

Figure 2.39 Simulated $|S_{11}|$ of antennas with and without slots.

The antenna consists of substrate and superstrate layers made of Rogers 3010 ($\varepsilon_r = 10.2$, $\tan\delta = 0.0035$), with both layers having a thickness of 0.635 mm.

Figure 2.39 illustrates the simulated $|S_{11}|$ of the antenna in both cases: with and without the slot cuts in the ground plane. Without the slots, the antenna resonates at frequencies of 502 MHz and 2.42 GHz. Although the desired ISM band at 2.45 GHz is well matched, the matching at the lower band needs improvement. However, with the inclusion of the open-end slots in the ground plane, the bandwidth at the lower band is significantly broadened, while the bandwidth at the upper band remains largely unchanged. In a realistic implantation scenario, it is advisable to avoid directly placing the medical device below the slotted ground in order to minimize the impact of the slots on the antenna's performance.

For the lower band, three closely spaced resonant frequencies, designated as f_1, f_2, and f_3, can be observed with the slots cut in the ground plane. Compared with Figure 2.40(b), it is seen from Figure 2.40(a) that f_1 is generated and shifts down distinctly with the increasing L_1. And from both Figure 2.40 (a) and (b), f_3 shifts down with the increasing L_1 and L_2. When L_1 and L_2 are chosen to be 3.2 mm and 5.3 mm, f_1, f_2, and f_3 are designed at 380, 444, and 480 MHz, respectively. It should be mentioned that coating biocompatible materials is typically employed for real implantable applications. The resonant frequencies will shift slightly with biocompatible materials added. By adjusting the lengths of the slots, frequency shifting can be compensated and a wide bandwidth at the lower band can still be achieved.

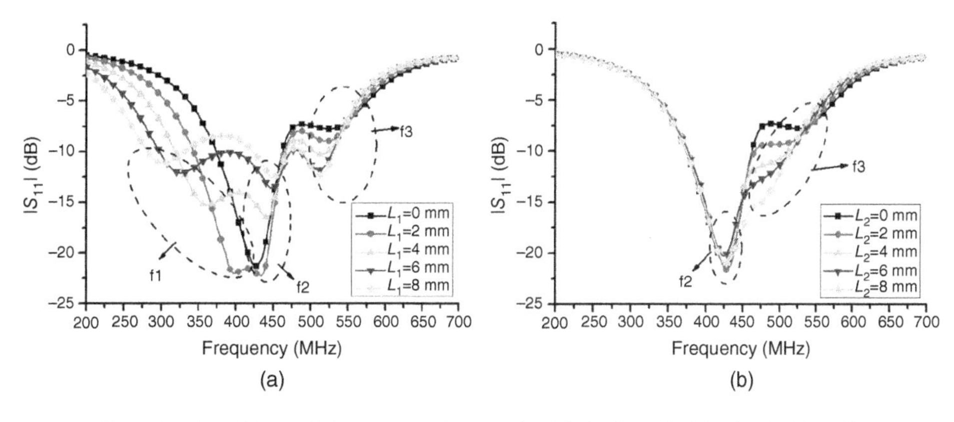

Figure 2.40 Effects of the open-end slots (a) with L_1 ($L_2 = 0$), (b) with L_2 ($L_1 = 0$).

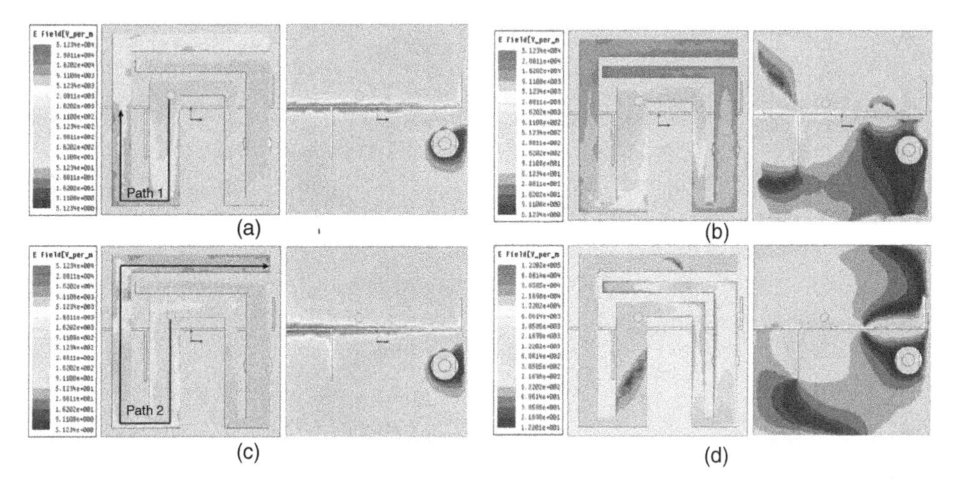

Figure 2.41 Electric-field intensity distributions of the proposed antenna at (a) 380 MHz, (b) 444 MHz, (c) 480 MHz, and (d) 2.42 GHz.

Figure 2.41 displays the electric-field intensity distributions of the antenna and the ground at different frequencies. At 444 MHz, shown in Figure 2.41(b), the antenna functions as a $\lambda/2$ resonator, with the slots playing a role in increasing the length of the current path. The slots are highly excited at 380 and 480 MHz compared to 444 MHz. To simplify the analysis, we divide the antenna into the left strip and the right strip from the short end. The right strip of the antenna serves as a coupling element to excite the slots and does not participate in the resonance at f_1 and f_3. Another coupling occurs between the slotted ground plane and the left strip of the antenna element, facilitated by parasitic capacitance. This coupling capacitance is added to the slot resonator to lower the resonant frequency of the slots. As shown in Figure 2.41(a) and (c), energy couples from the slots to the left strip along path 1, forming a $\lambda/4$ resonator at 380 MHz, and along path 2, forming a $\lambda/2$ resonator at 480 MHz. For the ISM band at 2.42 GHz, as depicted in Figure 2.41(d), it is achieved by exciting the harmonic mode in the right strip of the antenna.

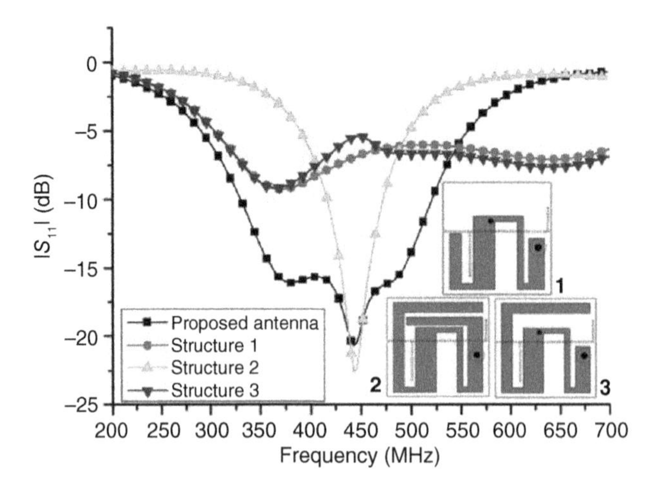

Figure 2.42 Simulated $|S_{11}|$ of three modified structures compared to the proposed antenna at the lower band.

To further verify the operating principle at the lower band, three modified structures are proposed and compared to the proposed antenna as shown in Figure 2.42. Since the right strip of the antenna element does not take part in resonance at f_1 and f_3, it is simplified in structures 1 and 3 as shown in the figure. For the left strip in structure 1, the part from the open-end to the slot is cut off to distinguish f_1 from f_3. The curves of $|S_{11}|$ show that structure 1 resonates at f_1 and structure 3 resonates at both f_1 and f_3. Note that the $|S_{11}|$ value of -6 dB is used as the resonance criteria here, which is similar to that for the mobile phone antennas. For structure 2, the short is removed since the current is weakest at the short point ($\lambda/2$ resonator) and others remain the same. The simulated $|S_{11}|$ shows that structure 2 resonates at f_2.

Similarly, there are numerous examples of wideband implantable antennas achieved through the combination of multiple resonators [49–52]. These coupled multi-resonant modes help improve the impedance matching of different frequency bands, although their qualitative explanation is primarily based on the loading of parasitic parameters. In Section 2.2.3, we introduce the loading technique as a means to achieve miniaturization of implantable antennas. Additionally, capacitance loading [53], inductive loading [54], and dielectric loading [55] are other methods used to realize wideband implantable antennas. The loading technique enhances the wideband impedance matching characteristics of implantable antennas. For example, in [54], a broadband 3D conformal antenna with a closed loop, which can be viewed as an additional inductance aiding impedance matching, was designed for central venous catheter (CVC) applications. Figure 2.43 illustrates the proposed antenna, which is an extended monopole with shorting and inductive loading loops. The design process shown in Figure 2.44 and the Smith chart depicted in Figure 2.45 provide insights into the realization of broadband impedance matching for the proposed antenna.

First, the design process began with a single monopole, as depicted in Figure 2.44(a). The impedance at 403 MHz, shown in Figure 2.45(a), indicates that the monopole length is too short, resulting in resonant frequencies above the MedRadio band. As a result, the matching

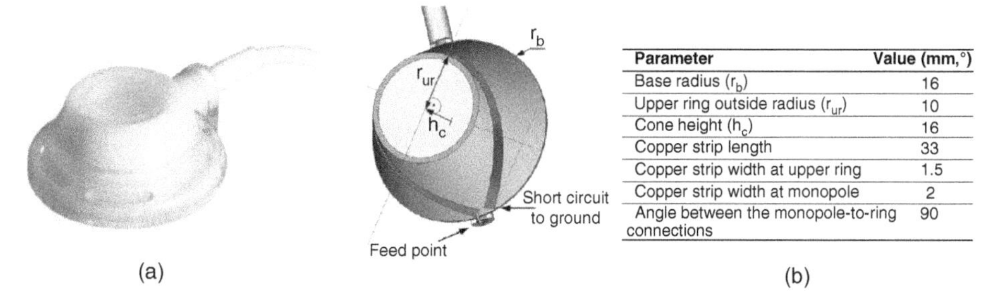

Parameter	Value (mm,°)
Base radius (r_b)	16
Upper ring outside radius (r_{ur})	10
Cone height (h_c)	16
Copper strip length	33
Copper strip width at upper ring	1.5
Copper strip width at monopole	2
Angle between the monopole-to-ring connections	90

Figure 2.43 (a) Implanted MRI-compatible CVC, (b) top view of the final antenna simulation model without the protection layer. Source: Schmidt et al. [54]/IEEE.

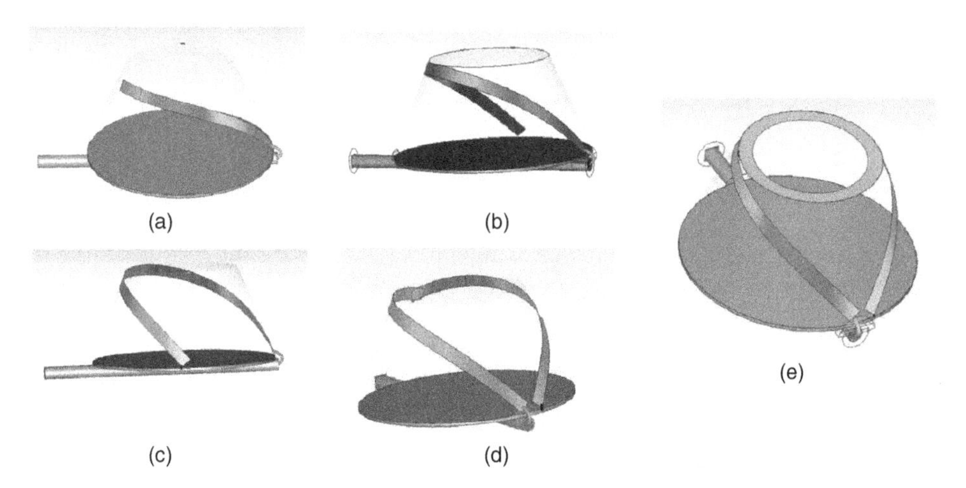

Figure 2.44 Topologies used during the optimization process: (a) short monopole, (b) long monopole, (c) long monopole with short circuit to ground, (d) extended monopole with short circuit, and (e) Extended monopole with short circuit and loop. Source: Schmidt et al. [54]/IEEE.

conditions are suboptimal. Second, by extending the monopole, as shown in Figure 2.44(b), the impedance at 403 MHz moves in a clockwise direction toward the inductive region of the Smith chart. Third, to take advantage of the cone surface and implement a miniaturization technique, a short circuit is added, as illustrated in Figure 2.44(c). This addition causes the impedance at 403 MHz to be mirrored to the capacitive region of the Smith chart with respect to the center, as shown in Figure 2.45(a). Subsequently, further extension of the monopole, depicted in Figure 2.44(d), does not lead to a significant improvement in the impedance matching at 403 MHz. The impedance still resides in the capacitive region of the Smith chart, as shown in Figure 2.45(a). Therefore, the addition of a closed loop was considered as an inductive loading loop to match the impedance at 403 MHz to the desired 50 Ω, as shown in Figure 2.44(e) and Figure 2.45(a).

The Smith chart in Figure 2.45(b) illustrates that the geometric changes made to the antenna have a minimal impact on the impedance matching for the 2–3 GHz range. This can be attributed to the resistive properties of the lossy medium surrounding the antenna.

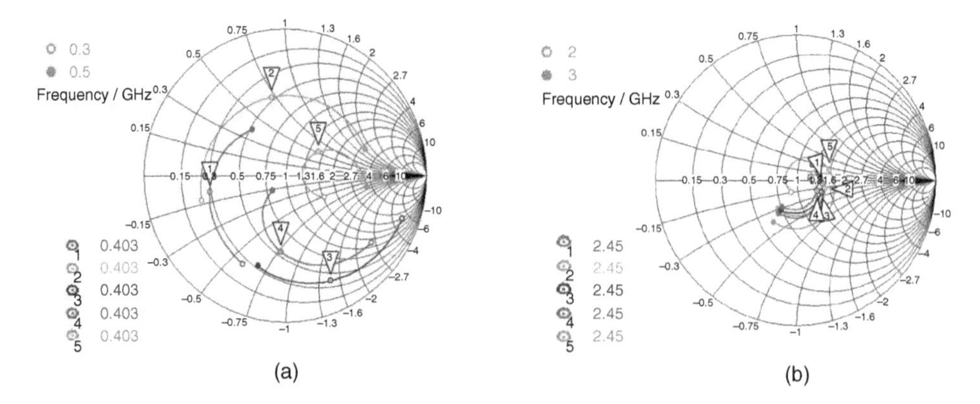

Figure 2.45 Smith chart showing the complex impedance matching between (a) 0.3–0.5 GHz and (b) 2.0–3.0 GHz referring to the design steps toward the final antenna model 1) Short monopole, 2) long monopole, 3) long monopole with short circuit to ground, 4) extended monopole with short circuit, and 5) extended monopole with short circuits and loop. Source: Schmidt et al. [54]/IEEE.

After conducting additional minor optimizations, the final topology was selected, where the monopole extends over both sides of the truncated cone shell, as depicted in Figure 2.43(b).

To ensure insulation between the metal antenna components and the surrounding tissues, the proposed CVC antenna should be coated. A thin coating is preferred to achieve miniaturization and a wide bandwidth, as it brings the high permittivity media closer to the antenna. However, the radiation efficiency decreases due to the proximity of the lossy media, requiring a trade-off. In this case, the proposed CVC antenna was coated with a 0.2 mm biocompatible synthetic film made of silicone rubber, the same material as the CVC. Figure 2.46 illustrates the simulation results of the proposed CVC antenna with a 0.2 mm biocompatible synthetic film within a 6 cm^3 cube phantom. As a result, the proposed CVC antenna achieved a wide bandwidth ranging from 0.27 GHz to up to 4 GHz.

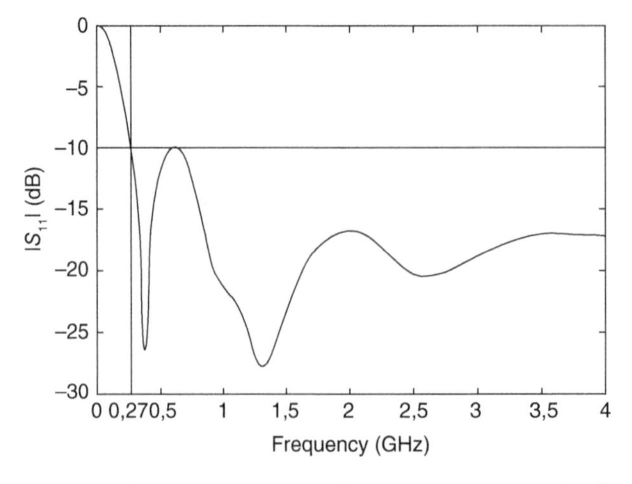

Figure 2.46 Simulated $|S_{11}|$ versus frequency of the CVC antenna in a 6 cm^3 skin phantom.

2.3.3 Advanced Technology for Detuning Problem

Future implantable devices face the challenge of operating in unpredictable and dynamic in-body environments. Wideband implantable antenna design aims to address the issue of antenna detuning caused by changes in the dielectric properties of human tissue. In addition to broadband solutions, there are novel approaches to tackle the detuning problem in implants, such as human tissue-insensitive implantable antennas [40, 56, 57] and tunable matching circuit technology [58–60].

In [56], stable impedance matching of the antenna in different tissues is achieved in part by utilizing a 0.5 mm thick alumina superstrate with high permittivity. This superstrate provides sufficient decoupling from the surrounding tissues in most cases.

In [57], the proposed capsule antenna maintains good impedance matching in both air and body tissues. However, achieving these characteristics comes at a cost. The design sacrifices radiation efficiency due to increased current symmetry and the high wave-impedance contrast between the capsule environment and tissues.

In [40], the compact printed meandered folded implantable dipole antenna offers two operational modes. It functions as a broadband antenna with adjacent resonant frequency points to address the detuning problem.

These innovative approaches demonstrate efforts to overcome the challenges of antenna detuning in implantable devices. Further research and development in this area are essential for the successful deployment of implantable devices in dynamic in-body environments.

In [58], a method for automatic matching network design and synthesis was proposed for implantable devices. The architecture of the proposed single-step automatic matching network is illustrated in Figure 2.47, which includes impedance measurement and a tunable matching network. This system has been applied in pacemakers [59].

In the proposed architecture, a capacitor C is used as a generic detector inserted between the power module and the tunable matching network. The sensed signals v_1 and v_2 are attenuated for linearity, down-converted to a lower intermediate frequency, and analyzed by

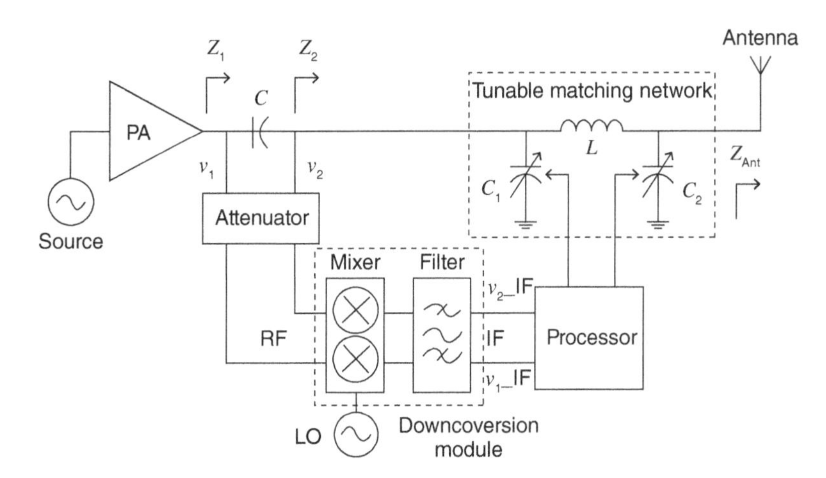

Figure 2.47 Architecture of the automatic matching system. Source: Po et al. [58]/with permission of IEEE.

a processor. The advantage of this architecture, as shown in Figure 2.47, is that the different modules required for antenna impedance tuning, such as the down-conversion module and the baseband processor, are already included in common radio transceiver architectures. Consequently, only minimal additional hardware is needed. Furthermore, the power consumption of radio communication modules is primarily determined by the power amplifier during transmission and the low-noise amplifier during reception. The additional power consumption from the computing unit and extra hardware is relatively small compared to the power consumption of the power amplifier. As integrated circuit technology continues to advance, tunable circuits like these can be easily integrated into implantable systems. The design of the system can help alleviate the complexities associated with implantable antenna detuning. In Chapter 9, a detailed example design based on [60] will be presented.

2.4 Multiband Miniaturized Implantable Antennas

In the preceding sections, we discussed the methods for miniaturization and wideband design of implantable antennas. However, in addition to these aspects, multiband design is also crucial for implantable antennas to enable multiple functionalities such as wake-up receivers, telemetry, and wireless power transmission. In this section, we will present several examples of multiband antenna designs to illustrate the methods employed for achieving multiband functionality in implantable devices.

2.4.1 Compact PIFA With Multi-current Patch

The basic principle of multiband antenna design is to create multiple current paths. A compact PIFA with multi-current paths, in [19], is a prime example. This dual-band implantable antenna shown in Figure 2.48 is fabricated on a two-layer Rogers 3010 substrate ($\varepsilon_r = 10.2$, $\tan\delta = 0.0023$), each layer thickness of 50 mil. The simulation model is based on one-layer skin model.

To explain the antenna operation mechanism, we consider two special cases for the proposed antenna. As shown in Figure 2.49, case 1 has the same two spiral resonators as the original one in Figure 2.48, except the short-pin, while case 2 only has one spiral resonator at the left. For both cases 1 and 2, the same ground plane is employed as the original one in Figure 2.48. Case 1 is a dual-band microstrip antenna. As seen from the simulated $|S_{11}|$ in Figure 2.50, it has two resonant frequencies: 400 MHz and 2.4 GHz. Case 2 is a dual-band PIFA. Simulated $|S_{11}|$ shows that this case also has two resonant frequencies: 430 MHz and 2.4 GHz. The impedance bandwidth at lower resonant frequency of both antennas is about 25 MHz. The size for case 2 is obviously smaller than that for case 1, because of its PIFA antenna structure. Simulated impedance bandwidth of the proposed antenna shown in Figure 2.50 covers from 393 to 447 MHz at MICS band and from 2.34 to 2.48 GHz at ISM band for $|S_{11}|$ less than −10 dB.

Current distributions of both cases are shown in Figure 2.51. From Figure 2.51, we can conclude that the path from the feeding point to the end of the left metal line mainly contributes to the ISM band for both cases. For case 1, the entire path from the left end to the right end contributes to the MICS band. For case 2, the entire path from the short-pin to

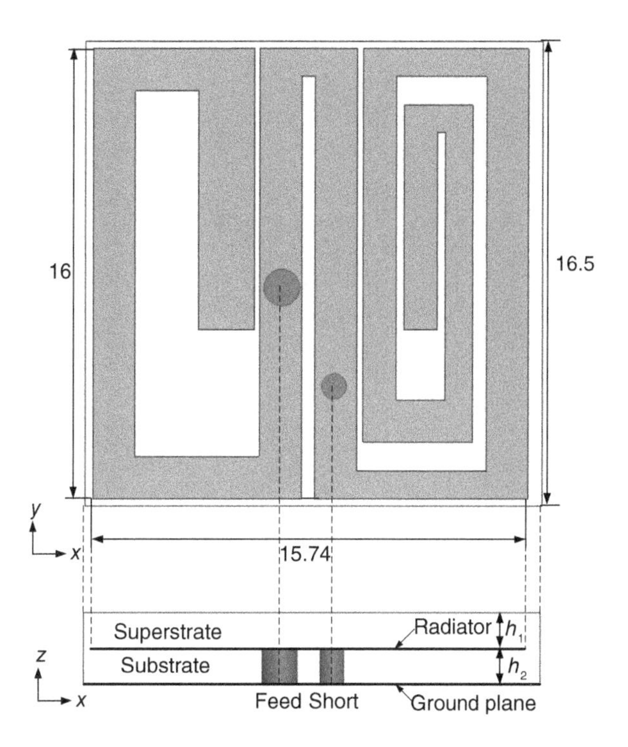

Figure 2.48 Geometry of dual-band implantable antenna. Source: Adapted from Liu et al. [19].

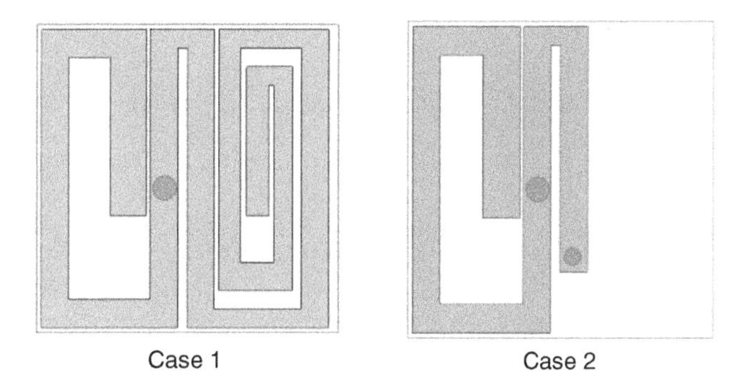

Figure 2.49 Two types of dual-band antennas.

the left end contributes to the MICS band. It is noted that case 1 is operating at a half-wave resonance, while case 2 is operating at a quarter-wave resonance.

Simulated peak gain for antenna 1 at 402 MHz is about −30 dBi, while the peak gain is about −31.5 dBi for case 2 at 430 MHz. Case 1 radiates slightly better than case 2 at lower resonant frequency because case 1 has more radiation area. At ISM band, both radiators have higher gain.

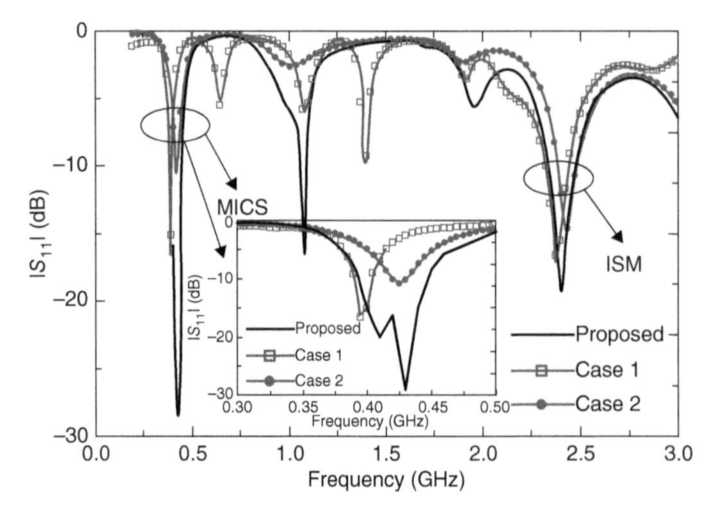

Figure 2.50 Comparison $|S_{11}|$ of different antenna cases.

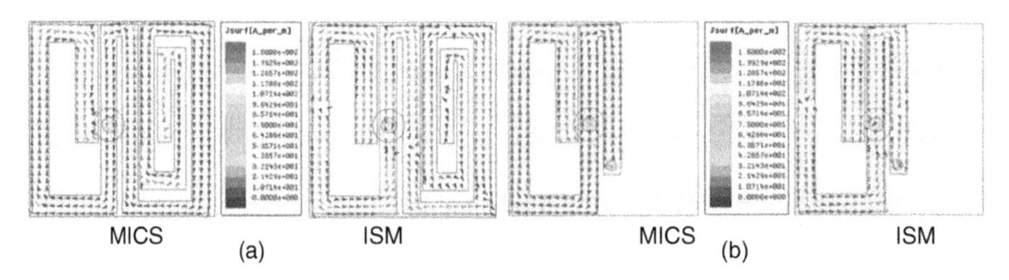

Figure 2.51 Current distributions of both antennas: (a) Case 1: patch, (b) Case 2: PIFA.

In order to improve the bandwidth at the MICS band, another spiral resonator was added in case 2 to form the proposed antenna. The added spiral path was coupled by the left spiral as in case 2 to produce another quarter-wave resonance at 430 MHz, and also the half-wave resonance is formed at 400 MHz, which is the same as case 1. Impedance bandwidth was improved by introducing two neighboring resonant frequencies at MICS band.

Figure 2.52 shows the electric current distributions of the proposed antenna at different resonant frequencies. At 400 MHz, the electric current flows along the same direction from

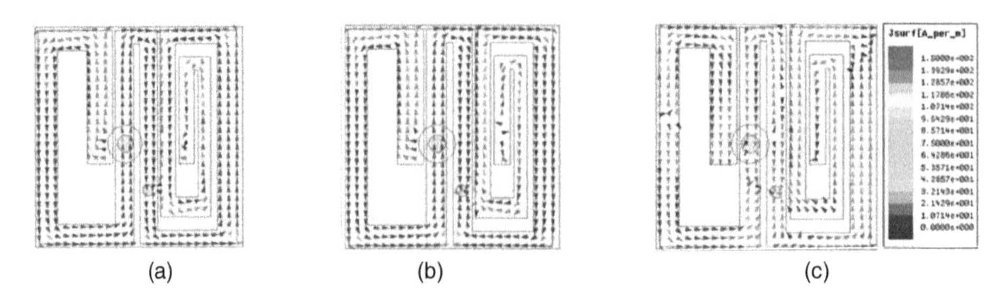

Figure 2.52 Electric current distributions at (a) 400 MHz, (b) 430 MHz, and (c) 2.45 GHz.

the left end to the right end of the entire metal trace, which indicates that the antenna is a half-wave resonator at this frequency. At 430 MHz, the current directions are opposite on either side of the shorting pin, which indicates either part is a quarter-wave resonator at this frequency. The current distribution of the proposed antenna at 2.4 GHz is similar to that in cases 1 and 2 shown in Figure 2.51.

The fabricated antenna with its superstrate is seen in Figure 2.53(a). For the measurement of the multiband antenna, we need to consider the dielectric properties of different frequencies and make different phantoms. Broadband tissue phantoms are hard to get. The antenna was measured by use of the gel mimicking techniques. The recipes for both MICS and ISM bands are given in Figure 2.53(b).

Figure 2.54 (a) shows the comparison of the simulated and measured $|S_{11}|$ for the antenna embedded in the MICS-band material. As seen from the results, the measured impedance bandwidth below −10 dB is in the frequency range of 430–490 MHz. The frequency shifting could be also caused by the air gap between substrate and superstrate. The measured relative bandwidth at MICS band is about 13%. A good agreement is obtained between simulation and measurement for the relative bandwidth. Figure 2.54(b) shows the comparison of the simulated and measured $|S_{11}|$ for the antenna embedded in the ISM band material. The measured impedance bandwidth is in the frequency range of 2.47–2.58 GHz, or 4.36%. A similar bandwidth is obtained from the simulation.

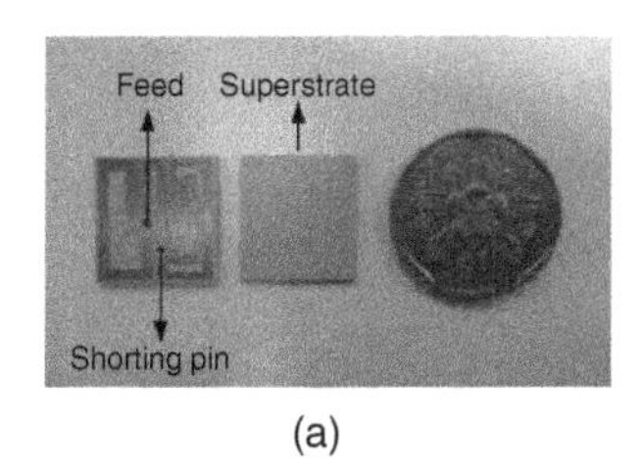

(a)

Materials	MICS band	Materials	ISM band
Sugar	56.18%	DGBE	5.1%
Salt	2.33%	Triton X-100	36.7%
Deionized	41.49%	Deinized	58.2%

(b)

Figure 2.53 (a) Photograph of fabricated implantable antenna, (b) recipes for skin-mimicking materials for MICS and ISM bands.

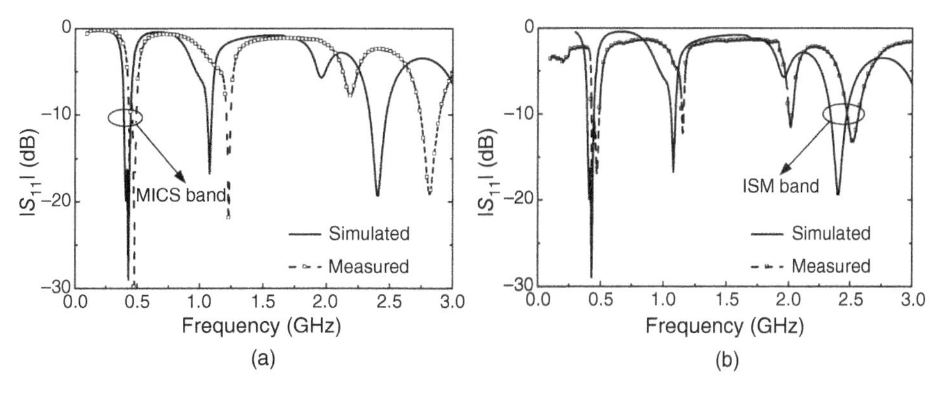

Figure 2.54 (a) Simulated and measured $|S_{11}|$ using skin-mimicking gel for MICS Band, (b) simulated and measured $|S_{11}|$ using skin-mimicking gel for ISM band.

Utilizing the two spiral antennas coupled with each other, a miniature dual-band implantable antenna has been designed and fabricated to achieve a wide bandwidth and good radiation performance. The simulated impedance bandwidths of the proposed implantable antenna are 12.58% at MICS band, and 5.8% at ISM band. The measured impedance bandwidths are 13% at MICS band, and 4.36% at ISM band.

2.4.2 Open-end Slots on Ground

In Section 2.3.2, we discussed how slots in the ground plane play a crucial role in creating multiple resonant frequency points, tuning resonant frequencies, and enhancing impedance matching. This concept extends beyond broadband implantable antennas and can also be applied to the design of multiband implantable antennas.

In [61], a compact spiral-shaped implantable antenna with multiband characteristics was designed for leadless pacemakers. Figure 2.55 illustrates the detailed geometric model of the proposed spiral-shaped implantable antenna, as well as the architecture and dimensions of the leadless pacemaker. The radiating patch of the antenna features a spiral structure with symmetrical upper and lower arms, as depicted in Figure 2.55(a). The ground plane of the antenna incorporates an open-end slot, which contributes to significant miniaturization, bandwidth enhancement, and frequency tuning, as shown in Figure 2.55(b). The substrate and superstrate used in the design are Rogers RT/duroid 6010, with dielectric constants (ε_r) of 10.2 and a tangent delta (tanδ) of 0.0035. The antenna is excited using a 50 Ω coaxial feed with a diameter of 0.4 mm, as depicted in Figure 2.55(c) and (d). The introduction of a ground slot and appropriate feed position enables the antenna to operate in multiple frequency bands with enhanced bandwidth.

The impact of the open-end slot can be observed by comparing the $|S_{11}|$ characteristics at different stages of the proposed antenna design, as shown in Figure 2.56. Initially, the antenna with a spiral patch featuring two symmetrical arms and a full ground plane resonated at 624 MHz with a narrow bandwidth of 2361 MHz, along with a weak resonance mode at 1620 MHz. To achieve tuning at the desired MICS frequency band without increasing the antenna size, a horizontal slot was introduced in the ground plane, shifting the lower frequency band to 570 MHz while experiencing a slight shift of the

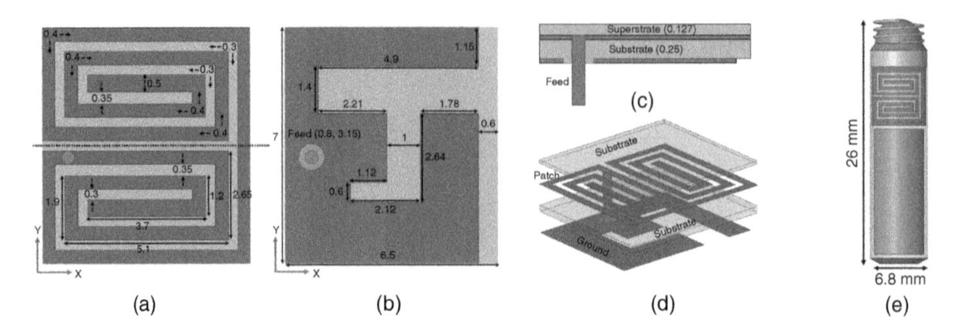

Figure 2.55 (a)–(d) Footprint of the proposed multiband implantable antenna (units: mm): (a) radiating patch, (b) ground plane, (c) side view, (d) exploded view, and (e) leadless pacemaker. Source: Shah et al. [61]/with permission of IEEE.

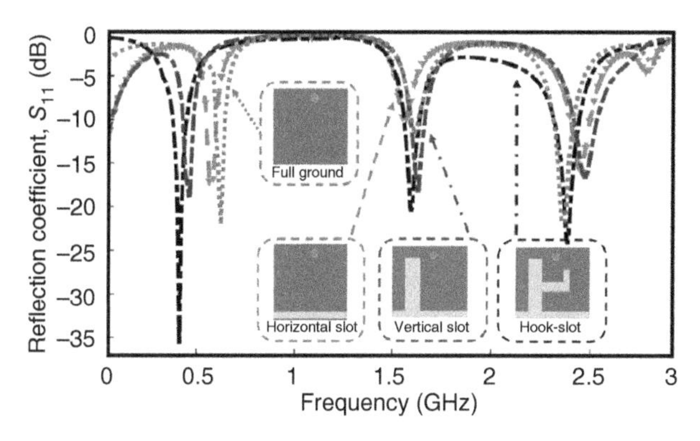

Figure 2.56 Effects of the ground slots on the performance of the antenna. Source: Shah et al. [61]/with permission of IEEE.

higher ISM band toward the right side of the frequency spectrum. By further incorporating a vertical slot, as depicted in Figure 2.56, the current path on the ground was elongated, resulting in a shift of the lower frequency band to 450 MHz, covering the 433 MHz ISM band with a low −10 dB bandwidth. Additionally, the resonance depth at 1620 MHz was increased ($|S_{11}| < -15$ dB). Finally, with the introduction of a hook-shaped slot in the ground, the current path on the ground became significantly longer and more balanced compared to the previous step, leading to a considerable improvement in impedance matching at all frequency bands, successfully achieving the desired MICS, midfield, and ISM bands.

Similarly, the same method of incorporating open-end slots in the ground plane has been utilized in the design of multiband implantable antennas in [62–64]. The use of open-end slots is a common approach in patch implantable antenna design to achieve multiband operation, broadband performance, and miniaturization. Its significance should not be overlooked.

2.4.3 Single-layer Design

The multiband implantable antennas discussed earlier all employ a two-layer structure. In general, a multilayer structure is advantageous for achieving multiple current paths and impedance matching, facilitating multiband design and miniaturization [15]. However, the multilayer structure may not be suitable for all applications, particularly for conformal capsule antennas. Designing a single-layer multiband antenna, especially with miniaturization considerations, can be challenging. In this section, we introduce a single-layer tri-band inverted-F antenna (IFA) designed specifically for conformal capsule applications [65].

The single-layer tri-band IFA is illustrated in Figure 2.57. The configuration of the capsule remains the same as described in Section 2.3.2, as shown in Figure 2.57(a). The proposed capsule antenna is initially printed on a flat 10 mil-thick single-sided Rogers 5880 substrate. Subsequently, the antenna is wrapped and inserted into the capsule body, adhering to the inner wall of the capsule shell. The capsule cap is then connected. The geometry of the unwrapped antenna is depicted in Figure 2.57(b). Two vias, shown in both Figure 2.57(a)

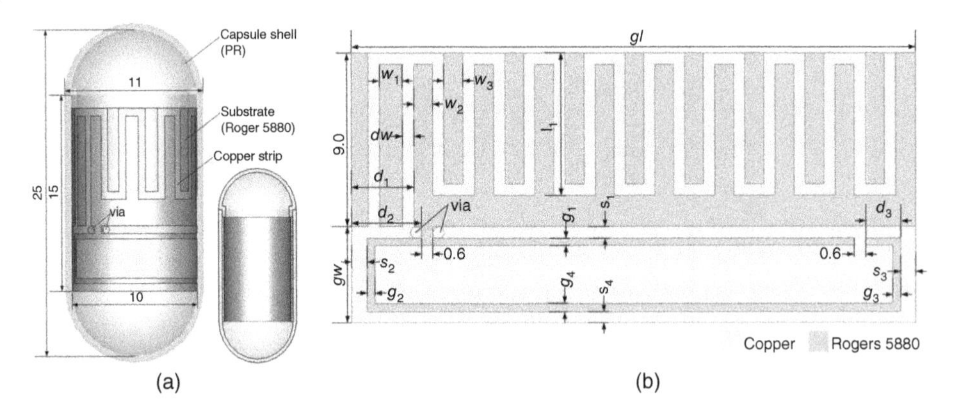

Figure 2.57 Geometry of the proposed tri-band conformal capsule antenna: (a) front view of the antenna conformed to capsule shell, the inset shows the cross-section of the capsule antenna, (b) geometry of the antenna before it is wrapped. Optimized values of these parameters: $gl = 29.8$, $gw = 5$, $dw = 0.6$, $l_1 = 7.4$, $w_1 = 1.2$, $w_2 = 1$, $w_3 = 1$, $d_1 = 3.3$, $d_2 = 3.7$, $d_3 = 0.9$, $s_1 = 0.6$, $s_2 = 0.8$, $s_3 = 0.8$, $s_4 = 0.8$, $g_1 = 0.4$, $g_2 = 0.4$, $g_3 = 0.4$, $g_4 = 0.4$. (Units: mm).

and (b), are utilized for feeding the antenna, connected to the center core and metallic shield of the coaxial cable, respectively. The fabrication method employed in this study is straightforward and effective. However, it should be noted that if a more advanced 3D printer is used, the antenna could be directly printed on the inner wall of the capsule, eliminating the need for the Rogers 5880 substrate and vias. Additionally, it is important to mention that the inside of the capsule shell is filled with air. In simulations, the capsule is positioned at the center of a cubic muscle phantom. As the three operating bands are significantly separated, the dielectric properties of the muscle phantom are set to be frequency-dependent in these simulations.

IFAs are commonly used in various applications, and antenna engineers often modify the geometry of the "F" structure to achieve additional resonances. However, in biomedical applications, it is desirable to have antennas that are simple and compact. Starting with a typical IFA, as shown in case 1 of Figure 2.58(a), by introducing meandering in the "F" structure, the resonance of the IFA can be shifted to lower frequencies. However, as the "F" structure is elongated further to shift the resonance to around 400 MHz, the impedance matching of the IFA deteriorates, as depicted in Figure 2.58(b). This degradation can be attributed to the fact that the ground plane of the IFA becomes too small compared to the operating wavelength as the frequency is lowered.

Figure 2.58(c) illustrates the current distribution on the antenna (case 1) at its first resonant frequency, which is 361 MHz. It can be observed that the currents on the meandered "F" structure flow in the same direction. In contrast, the current path on the ground plane is very short, resulting in a highly unbalanced quarter-wavelength monopole. Therefore, in order to achieve a dual-/tri-band conformal capsule antenna with an IFA and introduce additional resonances at 918 MHz and/or 2.45 GHz, it is necessary to shift the resonance of the quarter-wavelength monopole mode (the lowest resonance) to the MedRadio band while maintaining good impedance matching to achieve miniaturization of the antenna.

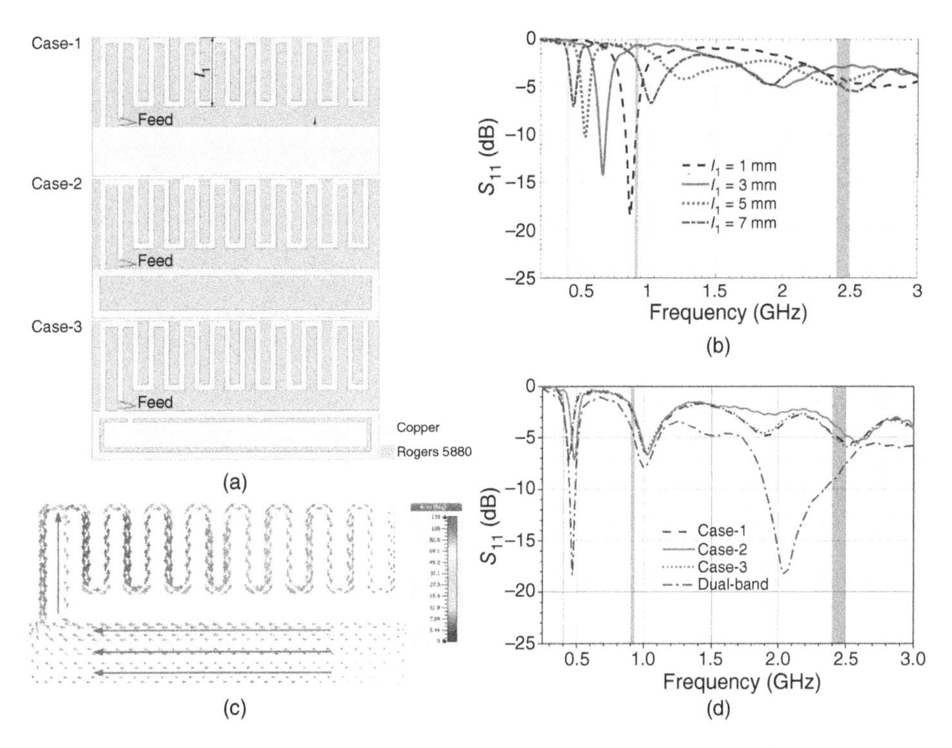

Figure 2.58 (a) Geometries of three IFAs before they are wrapped, which have been tested ($l_1 = 7$ mm), (b) simulated $|S_{11}|$ of the IFA shown as case-1 in (a) as a function of l_1, (c) time snapshot of current distribution on the IFA shown as case-1 in (a) at its lowest resonant frequency 361 MHz, (d) simulated $|S_{11}|$ of the IFAs shown in (a) and the dual-band IFA shown in Figure 5, when $l_1 = 7$ mm.

Case 2 and case 3, as shown in Figure 2.58(a), represent two modifications of the ground plane of the IFA that have been tested. However, as shown in Figure 2.58(d), these modifications do not introduce any additional resonances, and the impedance matching at the lowest resonant frequency does not improve either.

By modifying the ground of the IFA, we have found that the geometry depicted in Figure 2.59(a) can effectively generate additional resonances at 2.05 GHz, as shown in Figure 2.58(d). This modification also significantly improves the impedance matching at the lowest resonance. In comparison to the conventional IFA shown in case 1 of Figure 2.58(a), the ground of the dual-band IFA is changed to a hollow loop, and its feed point is relocated. The current distributions on the antenna at the two resonant frequencies, 411 MHz and 2.015 GHz, before it is wrapped, are illustrated in Figure 2.59(b) and (c), respectively.

From Figure 2.59(b), it can be observed that the currents on the “F” structure remain relatively unchanged, while the current path on the ground becomes much longer compared to Figure 2.58(c). As a result, the antenna becomes more balanced than the antenna shown in case 1 of Figure 2.58(a), resulting in a considerable improvement in impedance matching at the low resonant frequency. In Figure 2.59(c), it can be noted that the currents on the “F” structure change direction multiple times, while the currents on the loop-shaped

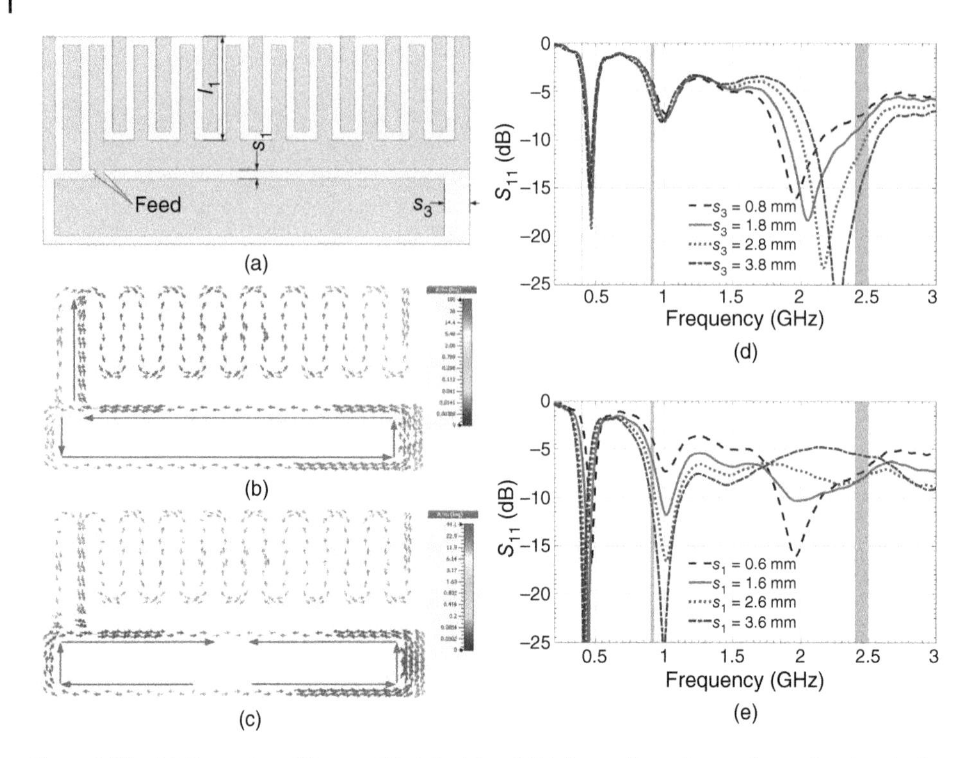

Figure 2.59 (a) Geometry of the modified dual-band IFA for conformal-capsule applications before it is wrapped, (b)–(c) time snapshots of the current distributions on the dual-band IFA shown in (a) at its lowest resonant frequency 411 MHz (b) and its third resonant frequency 2.015 GHz (c), (d)–(e) effects of s_3 and s_1 on $|S_{11}|$ of the dual-band IFA shown in (a).

ground flow in two opposite directions. Therefore, the resonance at 2.015 GHz is caused by the mode of a full-wavelength loop antenna.

Figure 2.59(d) presents the simulation results of $|S_{11}|$ for the dual-band IFA as a function of s_3. By adjusting the perimeter of the loop-shaped ground, the resonant frequency near 2 GHz of the dual-band IFA can be easily tuned. Furthermore, it can be observed from Figure 2.58(d) that both the IFAs shown in Figure 2.58(a) and the dual-band IFA exhibit another resonance around 1 GHz, which is associated with the second-order resonance of the one occurring at 450 MHz. However, the impedance matching at this higher resonance is not satisfactory.

Figure 2.59(e) demonstrates that the impedance matching near this resonance can be improved for the dual-band IFA by adjusting s_1, and the resonance can be tuned to the desired frequency of 915 MHz by properly adjusting the other geometrical parameters, although the resonance near 2 GHz disappears. Thus, the dual-band IFA shown in Figure 2.59(a) can effectively cover frequencies of 403 MHz/915 MHz or 403 MHz/2.45 GHz.

To achieve the tri-band characteristic, the dual-band IFA is further modified by adding a patch inside and connecting it to the loop-shaped ground, as shown in Figure 2.57. The final version of the IFA successfully covers the three desired bands. The current

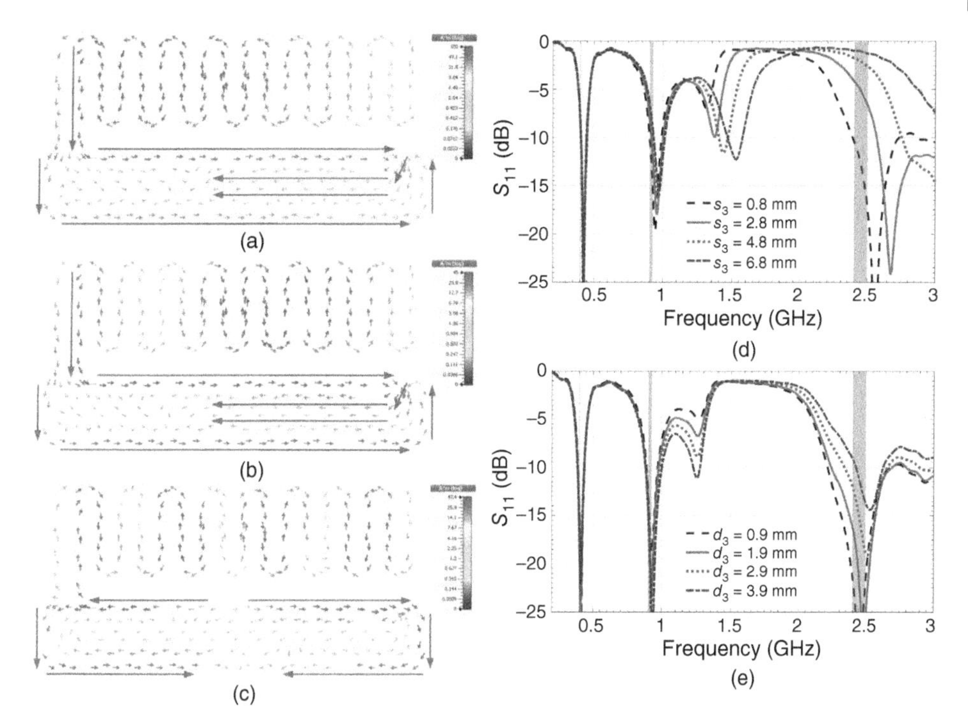

Figure 2.60 (a)–(c) Time snapshots of current distributions on the proposed tri-band IFA at (a) 352 MHz, (b) 830 MHz, and (c) 2.29 GHz before it is wrapped, (d)–(e) effects of s_3 and d_3 on $|S_{11}|$ of the proposed tri-band conformal capsule antenna.

distributions on the tri-band IFA at its three resonant frequencies are shown in Figure 2.60 before it is wrapped.

Figure 2.60(a) displays the current distribution of the resonance at 352 MHz. It can be observed that the currents on the meandered "F" structure flow in the same direction, indicating the mode of a quarter-wavelength monopole. The currents on the ground follow a different path compared to the dual-band IFA shown in Figure 2.59(b). However, this current path is longer than that shown in Figure 2.58(c), enabling a good impedance match near the lowest resonant frequency for this design.

Figure 2.60(b) shows the current distribution of the resonance at 830 MHz. The currents on the "F" structure reverse directions once, indicating the mode of a half-wavelength monopole. The currents on the ground flow in a similar manner to the lowest resonance shown in Figure 2.60(a). The current distribution corresponding to the resonance occurring at 2.29 GHz, shown in Figure 2.60(c), is similar to that shown in Figure 2.60(b) and is associated with the mode of a full-wavelength loop antenna. By exciting these three modes, the proposed IFA successfully achieves the desired tri-band characteristic.

To gain a better understanding of the characteristics of the proposed tri-band conformal capsule antenna, parametric studies were conducted. Figure 2.60(d) presents the simulated $|S_{11}|$ of the proposed tri-band conformal capsule antenna as a function of the parameter s_3. Similar to the case of the dual-band IFA shown in Figure 2.59(d), the resonant frequency near 2.5 GHz of the tri-band antenna shifts upward as s_3 is increased. This is expected since the structures corresponding to these resonant frequencies in the dual-band and tri-band IFAs are the same. Furthermore, a new resonance appears near 1.5 GHz as s_3 increases. The effects of the connecting position between the loop-shaped ground and the patch inside the ground, denoted as d_3, are presented in Figure 2.60(e). It can be observed that this parameter primarily affects the resonances near 915 MHz and 2.5 GHz, and an additional resonance near 1.25 GHz appears as d_3 increases.

In summary, by properly adjusting all the mentioned parameters, good impedance matching characteristics can be achieved simultaneously at the three desired frequency ranges.

A prototype of the tri-band conformal capsule antenna is fabricated, as shown in Figure 2.61(a). In practical applications, the capsule antenna is fed by the transceiver inside the capsule, and the capsule shell is hermetic. In the experiment, a hole is drilled in the capsule cap to measure the impedance match of the antenna, allowing the coaxial cable to enter the capsule and feed the antenna.

The performance of the capsule antenna is measured in three frequency ranges: 0.25–0.65 GHz, 0.65–1.6 GHz, and 1.6–3.5 GHz. The measurements are conducted by placing the antenna at the center of three fluids that mimic the muscle at the respective frequencies. The fluids, approximately 11.5 cm × 11.5 cm × 5 cm in size, are prepared using deionized water, diethylene glycol butyl ether (DGBE), and Triton X-100. The dielectric properties of the fluids are measured using an HP 85070A dielectric probe kit.

The capsule antenna is connected to port-1 of a Rohde & Schwarz ZVL VNA, and the measured $|S_{11}|$ results are shown in Figure 2.61(b). It can be observed that the measured $|S_{11}|$ characteristics agree well with the simulated ones. Although the three measured resonant frequencies shift slightly to the left compared to the simulated ones, the three desired frequency bands are reasonably covered. Furthermore, it is noted that all three measured bands are wider than the corresponding simulated ones. Apart from fabrication errors, these discrepancies may have been caused by a small solution leak into the capsule during the experiment, which is difficult to avoid due to fabrication limitations.

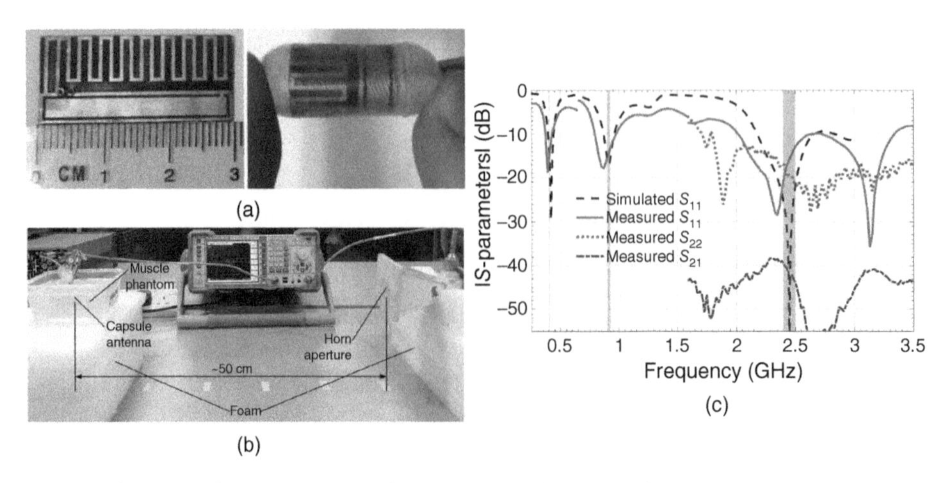

Figure 2.61 (a) Fabricated prototype of the proposed tri-band conformal capsule antenna, (b) the measurement setup, and (c) simulated and measured S-parameters of the proposed tri-band conformal capsule antenna and a horn.

In the experiment, a horn is connected to port-2 of the VNA, with the horn aperture set to be 50 cm (four wavelengths in air at 2.45 GHz) away from the capsule antenna. The operating frequency range of the horn is 1.5–18 GHz, and its gain increases with frequency, reaching approximately 7.9 dBi at 2.45 GHz. The $|S_{21}|$ measurement is affected by various factors, including free-space loss, gains, and impedance mismatch losses of the horn, as well as the capsule antenna. From Figure 2.61(c), it is observed that the $|S_{21}|$ is approximately −42 dB at 2.45 GHz and reaches its peak at around −38 dB around 2.35 GHz, where the capsule antenna is well matched.

2.5 Conclusions

The miniaturization of implantable antennas is crucial for achieving minimally invasive implantation of implants. This not only simplifies surgical procedures and reduces postoperative infections but also enables the widespread use of miniature implantable sensors for long-term and real-time monitoring of human physiological signals and disease treatment. In this chapter, we introduce methods for designing miniaturized implantable antennas and provide a summary of bandwidth enhancement and multiband antenna design. The design of miniaturized implantable antennas should be tailored to the specific application scenario and take into account the structural design of the implants. The biocompatibility of antenna materials should also be carefully considered. The permittivity of human tissue is frequency-dependent, and for broadband and multiband antenna designs, the dispersive properties of human tissue should be taken into account. Accurate human tissue permittivity curves with frequency variations should be used in simulation studies. In antenna measurements, it is challenging to find phantoms that precisely match the dielectric properties of human tissue over a wide frequency range. Segmental measurements are often necessary. Additionally, due to the lossy nature of human tissue, broadband antennas may exhibit altered characteristics when measured in the presence of tissue. It is important to evaluate the gain bandwidth of the antenna, not just the impedance bandwidth. Furthermore, addressing the detuning issue of antennas in human tissue requires optimizing the antenna design. In the future, online tuning techniques will likely be employed in implantable systems to address this challenge at the system level. In conclusion, achieving miniaturization, addressing dispersive properties of human tissue, considering gain bandwidth, and optimizing antenna designs are essential considerations for the successful development of implantable antennas.

References

1. Rahmat-Samii, Y. and Kim, J. (2005). Implanted antennas in medical wireless communications. *Synthesis Lectures on Antennas* 1 (1): 1–82.
2. Hasgall, P., Di Gennaro, F., Baumgartner, C. et al. (2018). IT'IS Database for thermal and electromagnetic parameters of biological tissues, Version 4.0. IT'IS.
3. Vidal, N., Curto, S., Villegas, J.L. et al. (2012). Detuning study of implantable antennas inside the human body. *Prog. Electromagn. Res.* 124: 265–284.

4 Li, P., Zhang, L., Liu, F., Amely-Velez, J. (2015). Optimized matching of an implantable medical device antenna in different tissue medium using load pull measurements. In: Microwave Measurement Conference, 2015 86th ARFTG: 1–4.

5 Bhattacharyya, A. and Tralman, T. (1988). Effects of dielectric superstrate on patch antennas. *Electron. Lett.* 24: 356–358.

6 Kim, J. and Rahmat-Samii, Y. (2004). Implanted antennas inside a human body: simulations, designs, and characterizations. *IEEE Trans. Microwave Theory Tech.* 52 (8): 1934–1943.

7 Karacolak, T., Hood, A.Z., and Topsakal, E. (2008). Design of a dual-band implantable antenna and development of skin mimicking gels for continuous glucose monitoring. *IEEE Trans. Microwave Theory Tech.* 56: 1001–1008.

8 Soontornpipit, P., Furse, C.M., and Chung, Y.C. (2004). Design of implantable microstrip antennas for communication with medical implants. *IEEE Trans. Microwave Theory Tech.* 52 (8): 1944–1951.

9 Soontornpipit, P., Furse, C.M., and Chung, Y.C. (2005). Miniaturized biocompatible microstrip antenna using genetic algorithm. *IEEE Trans. Antennas Propag.* 53 (6): 1939–1945.

10 Lee, C.-M., Yo, T.-C., Luo, C.-H. et al. (2007). Compact broadband stacked implantable antenna for biotelemetry with medical devices. *Electron. Lett.* 43 (12): 1–14.

11 Liu, W.-C., Chen, S.-H., and Wu, C.-M. (2009). Bandwidth enhancement and size reduction of an implantable PIFA antenna for biotelemetry devices. *Microwave Opt. Technol. Lett.* 51: 755–757.

12 Liu, W.-C., Yeh, F.-M., and Ghavami, M. (2008). Miniaturized implantable broadband antenna for biotelemetry communication. *Microwave Opt. Technol. Lett.* 50: 2407–2409.

13 Liu, W.-C., Chen, S.-H., and Wu, C.-M. (2008). Implantable broadband circular stacked PIFA antenna for biotelemetry communication. *J. Electromagn. Waves Appl.* 22: 1791–1800.

14 Kiourti, A. and Nikita, K.S. (2012). Miniature scalp-implantable antennas for telemetry in the MICS and ISM bands: design, safety considerations and link budget analysis. *IEEE Trans. Antennas Propag.* 60 (8): 3568–3575.

15 Huang, F.-J., Lee, C.-M., Chang, C.-L. et al. (2011). Rectenna application of miniaturized implantable antenna design for triple-band biotelemetry communication. *IEEE Trans. Antennas Propag.* 59 (7): 2646–2653.

16 Chien, T.F., Cheng, C.-M., Yang, H.-C. et al. (2010). Development of nonsuperstrate implantable low-profile CPW-fed ceramic antennas. *IEEE Antennas Wirel. Propag. Lett.* 9: 599–602.

17 Sánchez-Fernández, C.J., Quevedo-Teruel, O., Requena-Carrión, J. et al. (2010). Dual-band microstrip patch antenna based on short-circuited ring and spiral resonators for implantable medical devices. *IET Microwaves Antennas Propag.* 4 (8): 1048–1055.

18 Merli, F., Bolomey, L., Zürcher, J.-F. et al. (2011). Design, realization and measurements of a miniature antenna for implantable wireless communication systems. *IEEE Trans. Antennas Propag.* 59 (10): 3544–3555.

19 Liu, C., Guo, Y.X., and Xiao, S. (2012). Compact dual-band antenna for implantable devices. *IEEE Antennas Wirel. Propag. Lett.* 11: 1508–1511.

20 Xu, L.J., Guo, Y.X., and Wu, W. (2012). Dual-band implantable antenna with open-end slots on ground. *IEEE Antennas Wirel. Propag. Lett.* 11: 1564–1567.

21 Xu, L.J., Guo, Y.X., and Wu, W. (2014). Miniaturized dual-band antenna for implantable wireless communications. *IEEE Antennas Wirel. Propag. Lett.* 13: 1160–1163.

22 Xu, L.J., Guo, Y.X., and Wu, W. (2015). Bandwidth enhancement of an implantable antenna. *IEEE Antennas Wirel. Propag. Lett.* 14: 1510–1513.

23 Duan, Z., Guo, Y.X., Xue, R.F. et al. (2012). Differentially-fed dual-band implantable antenna for biomedical applications. *IEEE Trans. Antennas Propag.* 60 (12): 5587–5595.

24 Duan, Z., Guo, Y.X., Xue, R.F. et al. (2014). Design and in vitro test of a differentially fed dual-band implantable antenna operating at MICS and ISM bands. *IEEE Trans. Antennas Propag.* 62 (5): 2430–2439.

25 Liu, C., Guo, Y.X., and Xiao, S. (2012). A hybrid patch/slot implantable antenna for biotelemetry devices. *IEEE Antennas Wirel. Propag. Lett.* 11: 1646–1649.

26 Xu, L.J., Guo, Y.X., and Wu, W. (2013). Miniaturised slot antenna for biomedical applications. *Electron. Lett.* 49 (17): 1060–1061.

27 Li, H., Guo, Y.X., Liu, C. et al. (2015). A miniature-implantable antenna for MedRadio-band biomedical telemetry. *IEEE Antennas Wirel. Propag. Lett.* 14: 1176–1179.

28 Liu, C., Guo, Y.X., and Xiao, S. (2014). Design and safety considerations of an implantable rectenna for far-field wireless power transfer. *IEEE Trans. Antennas Propag.* 62 (11): 5798–5806.

29 Liu, C., Guo, Y.X., and Xiao, S. (2014). Capacitively loaded circularly polarized implantable patch antenna for ISM-band biomedical applications. *IEEE Trans. Antennas Propag.* 62 (5): 2407–2417.

30 Xu, L.J., Guo, Y.X., and Wu, W. (2015). Miniaturized circularly polarized loop antenna for biomedical applications. *IEEE Trans. Antennas Propag.* 63 (3): 922–930.

31 Liu, C., Guo, Y.X., and Xiao, S. (2014). Circularly polarized helical antenna for ISM-band ingestible capsule endoscope systems. *IEEE Trans. Antennas Propag.* 62 (12): 6027–6039.

32 Lee, C.-M., Yo, T.-C., Huang, F.-J., and Luo, C.-H. (2008). Dual-resonant π-shape with double L-strips PIFA for implantable biotelemetry. *Electron. Lett.* 44 (14): 837–838.

33 Xia, W., Saito, K., Takahashi, M., and Ito, K. (2009). Performances of an implanted cavity slot antenna embedded in the human arm. *IEEE Trans. Antennas Propag.* 57 (4): 894–899.

34 Thotahewa, K.M.S., Redouté, J.M., and Yuce, M.R. (2013). SAR, SA, and temperature variation in the human head caused by IR-UWB implants operating at 4 GHz. *IEEE Trans. Microwave Theory Tech.* 61 (5): 2161–2169.

35 Izdebski, P.M., Rajagopalan, H., and Rahmat-Samii, Y. (2009). Conformal ingestible capsule antenna: a novel chandelier meandered design. *IEEE Trans. Antennas Propag.* 57 (10): 900–909.

36 Rajagopalan, H. and Rahmat-Samii, Y. (2012). Wireless medical telemetry characterization for ingestible capsule antenna designs. *IEEE Antennas Wirel. Propag. Lett.* 11: 1679–1682.

37 Song, M., Huang, Y., Shen, Y. et al. (2022). A 1.66 GB/s and 5.8 pJ/b transcutaneous IR-UWB telemetry system with hybrid impulse modulation for intracortical brain-computer interfaces. In: *2022 IEEE International Solid-State Circuits Conference (ISSCC)*, 394–396.

38 Bahrami, H., Mirbozorgi, S.A., Ameli, R. et al. (2016). Flexible, polarization-diverse UWB antennas for implantable neural recording systems. *IEEE Trans. Biomed. Circuits Syst.* 10 (1): 38–48.

39 Felicio, J.M., Fernandes, C.A., and Costa, J.R. (2016). Wideband implantable antenna for body-area high data rate impulse radio communication. *IEEE Trans. Antennas Propag.* 64 (5): 1932–1940.

40 Magill, M.K., Conway, G.A., and Scanlon, W.G. (2017). Tissue-independent implantable antenna for in-body communications at 2.36–2.5 GHz. *IEEE Trans. Antennas Propag.* 65 (9): 4406–4417.

41 Kiourti, A., Costa, J.R., Fernandes, C.A., and Nikita, K.S. (2014). A broadband implantable and a dual-band on-body repeater antenna: design and transmission performance. *IEEE Trans. Antennas Propag.* 62 (6): 2899–2908.

42 Yang, C.L., Tsai, C.L., and Chen, S.H. (2013). Implantable high-gain dental antennas for minimally invasive biomedical devices. *IEEE Trans. Antennas Propag.* 61 (5): 2380–2387.

43 Das, S. and Mitra, D. (2018). A compact wideband flexible implantable slot antenna design with enhanced gain. *IEEE Trans. Antennas Propag.* 66 (8): 4309–4314.

44 Miah, M.S., Khan, A.N., Icheln, C. et al. (2019). Antenna system design for improved wireless capsule endoscope links at 433 MHz. *IEEE Trans. Antennas Propag.* 67 (4): 2687–2699.

45 Li, H., Wang, B., Guo, L., and Xiong, J. (2019). Efficient and wideband implantable antenna based on magnetic structures. *IEEE Trans. Antennas Propag.* 67 (12): 7242–7251.

46 Yang, Z., Zhu, L., and Xiao, S. (2020). An implantable wideband microstrip patch antenna based on high-loss property of human tissue. *IEEE Access* 8: 93048–93057.

47 Bao, Z., Guo, Y.-X., and Mittra, R. (2017). An ultrawideband conformal capsule antenna with stable impedance matching. *IEEE Trans. Antennas Propag.* 65 (10): 5086–5094.

48 Faisal, F., Zada, M., Yoo, H. et al. (2022). An ultra-miniaturized antenna with ultra-wide bandwidth for future cardiac leadless pacemaker. *IEEE Trans. Antennas Propag.*

49 Alrawashdeh, R.S., Yi, H., Kod, M., and Sajak, A.A.B. (2015). A broadband flexible implantable loop antenna with complementary split ring resonators. *IEEE Antennas Wirel. Propag. Lett.* 14: 1506–1509.

50 Jiang, Z., Wang, Z., Leach, M. et al. (2019). Wideband loop antenna with split-ring resonators for wireless medical telemetry. *IEEE Antennas Wirel. Propag. Lett.* 18 (7): 1415–1419.

51 Zhang, H., Li, L., Liu, C. et al. (2017). Miniaturized implantable antenna integrated with split resonate rings for wireless power transfer and data telemetry. *Microwave Opt. Technol. Lett.* 59 (3): 710–714.

52 Wang, M., Liu, H., Zhang, P. et al. (2021). Broadband implantable antenna for wireless power transfer in cardiac pacemaker applications. *IEEE J. Electromagn. RF Microwaves Med. Biol.* 5 (1): 2–8.

53 Cui, W., Liu, R., Wang, L. et al. (2019). Design of wideband implantable antenna for wireless capsule endoscope system. *IEEE Antennas Wirel. Propag. Lett.* 18 (12): 2706–2710.

54 Schmidt, C., Casado, F., Arriola, A. et al. (2014). Broadband UHF implanted 3-D conformal antenna design and characterization for in-off body wireless links. *IEEE Trans. Antennas Propag.* 62 (3): 1433–1444.

55 Dissanayake, T., Esselle, K.P., and Yuce, M.R. (2009). Dielectric loaded impedance matching for wideband implanted antennas. *IEEE Trans. Microwave Theory Tech.* 57 (10): 2480–2487.

56 Nikolayev, D., Zhadobov, M., Le Coq, L. et al. (2017). Robust ultraminiature capsule antenna for ingestible and implantable applications. *IEEE Trans. Antennas Propag.* 65 (11): 6107–6119.

57 Nikolayev, D., Zhadobov, M., and Sauleau, R. (2019). Immune-to-detuning wireless in-body platform for versatile biotelemetry applications. *IEEE Trans. Biomed. Circuits Syst.* 13 (2): 403–412.

58 Po, F.C.W., de Foucauld, E., Morche, D. et al. (2011). A novel method for synthesizing an automatic matching network and its control unit. *IEEE Trans. Circuits Syst. I Regul. Pap.* 58 (9): 2225–2236.

59 Po, F.C.W., de Foucauld, E., Vincent, P. et al. (2009). Afast and accurate automatic matching network designed for ultra low power medical applications. In: *2009 IEEE International Symposium on Circuits and Systems (ISCAS)*, vol. 2009, 673–676. IEEE.

60 Feng, Y., Li, Y., Li, L. et al. (2020). Design and system verification of reconfigurable matching circuits for implantable antennas in tissues with broad permittivity range. *IEEE Trans. Antennas Propag.* 68 (6): 4955–4960.

61 Shah, I.A., Zada, M., and Yoo, H. (2019). Design and analysis of a compact-sized multi-band spiral-shaped implantable antenna for scalp implantable and leadless pacemaker systems. *IEEE Trans. Antennas Propag.* 67 (6): 4230–4234.

62 Gani, I. and Yoo, H. (2016). Multi-band antenna system for skin implant. *IEEE Microwave Wireless Compon. Lett.* 26 (4): 294–296.

63 Zada, M. and Yoo, H. (2018). A miniaturized triple-band implantable antenna system for bio-telemetry applications. *IEEE Trans. Antennas Propag.* 66 (12): 7378–7382.

64 Zada, M. and Yoo, H. (2019). Miniaturized dual band antennas for intra-oral tongue drive system in the ISM bands 433 and 915 MHz: design, safety, and link budget considerations. *IEEE Trans. Antennas Propag.*

65 Bao, Z.D., Guo, Y.X., and Mittra, R. (2017). Single-layer dual-/tri-band inverted-f antennas for conformal capsule type of applications. *IEEE Trans. Antennas Propag.* 65 (12): 7257–7265.

3

Polarization Design for Implantable Antennas

3.1 Introduction

Real-time and long-term monitoring of human physiological signals is crucial for disease diagnosis, with implantable devices serving as ideal tools for such monitoring. However, the stability of the connection between the implantable device and the external terminal greatly influences the quality of the data collected. The connection between the implantable and external antennas is weak due to the loss of human tissue. Further complicating matters, the unpredictability of the implantable device's orientation and the environmental complexity can cause polarization mismatching, which degrades the quality of the wireless link [1]. While some implantable antenna designs do not consider polarization mismatching, it is critical to assess the polarization loss for a wireless communication link. If both the implantable and external antennas use linear polarization, the communication link may be impossible if the two antennas exhibit orthogonal polarizations. Moreover, changes in the positions of the implantable antennas in mobile patients could cause significant polarization mismatching or even render the link inoperative.

Communications via far-field radio frequency (RF) linked telemetry can be hindered by multipath distortion, which creates nulls in the transmission pattern of either the external or implantable device. In an RF-linked environment, reflections of a transmitted wave caused by walls or other objects result in a standing wave pattern [2]. Areas where the standing wave pattern produces a signal amplitude below the noise floor are referred to as nulls or null areas. For wireless-access applications, circularly polarized (CP) radiation is desirable as it can reduce multipath interference and improve bit-error rates compared to linear polarization.

CP-implantable antennas offer several key advantages [3]:

(1) A CP antenna is very effective in combating multipath interference or fading
(2) A CP antenna can reduce the "Faraday rotation" effect due to the ionosphere
(3) The use of CP antennas requires no strict orientation between the transmitting and receiving antennas

Given these benefits, CP radiation is highly desirable for biomedical applications. However, the design of a CP implantable antenna faces various challenges, as it must consider numerous factors, including compact size, polarization purity, radiation efficiency, and safety issues.

Antennas and Wireless Power Transfer Methods for Biomedical Applications, First Edition.
Yongxin Guo, Yuan Feng and Changrong Liu.

In this chapter, we will discuss the polarization characteristics of implantable antennas, provide design schemes for miniaturized CP antennas, and explore the performance of CP antennas in live animals.

3.2 Compact Microstrip Patch Antenna for CP-implantable Antenna Design

Microstrip patch antennas are widely used in the design of compact-implantable antennas due to their low profile, ease of fabrication, cost-effectiveness, and ability to conform to curved surfaces, as discussed in Chapter 2. CP patch antennas can be realized using either a single-feed or multi-feed technique.

In the context of miniaturizing implantable antennas, the single-feed CP antenna is generally preferred over the multi-feed CP antenna. This is because the multi-feed CP antenna requires a phase quadrature feed network, which tends to be complex and requires additional space within the implant. On the other hand, the single-feed technique involves perturbing the shape of the patch to excite two orthogonal modes with a phase difference of $\pm 90°$.

In this section, we will discuss two types of compact microstrip patch antennas that are commonly used for miniaturized-implantable antenna designs.

3.2.1 Capacitively-loaded CP-implantable Patch Antenna

To design a small-size patch antenna, many techniques have been developed for a single-fed CP patch antenna that operates in free space [4]. These include employing shorting pins/walls [4], cutting slots in the radiator patch [5–13], or the ground plane [13–15], embedding tails along the edge [16], loading metamaterial structures [17], or transforming the patch antenna into a wire mesh and then squeezing it in according fashion [18], or using inductive/capacitive loading on patch [19, 20].

In order to design a CP microstrip patch antenna in a human body environment, characteristics of an implantable patch antenna should be studied [21]. In this section, capacitive loading technique for a small CP-implantable patch antenna will be discussed [22].

3.2.1.1 An Implantable Microstrip Patch Antenna with a Center Square Slot

To study the impedance characteristics of an implantable patch antenna with a center square slot, a simple implantable antenna is proposed, as shown in Figure 3.1 [23]. The antenna design is based on a two-layer Rogers 3010 substrate with a dielectric constant (ε_r) of 10.2 and a thickness of 25 mil for each layer. The loss tangent ($\tan \delta$) is 0.003. The feed position is located at the center of the antenna along the y-axis, denoted as point "la." The patch size is 9.5 mm × 9.5 mm, and the ground plane size is 10 mm × 10 mm. The feeding position is fixed at coordinates (3.7 mm, 0 mm).

To simulate the antenna in a human body environment, a cubic skin phantom model, as shown in Figure 3.2, is used. The antenna structure is placed inside the skin phantom, with a distance of 4 cm from the side of the structure to the edge of the skin, 4 mm from the top of the superstrate to the edge of the skin, and 20 mm from the bottom of the substrate

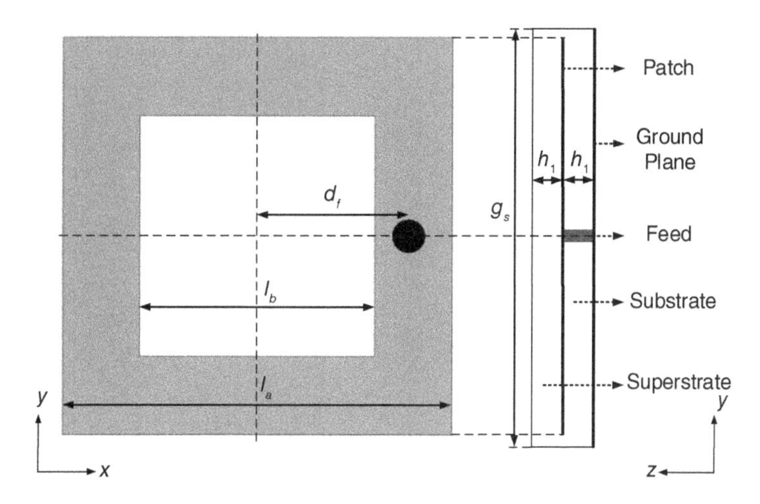

Figure 3.1 Geometry of an implantable patch antenna with a center square-slot.

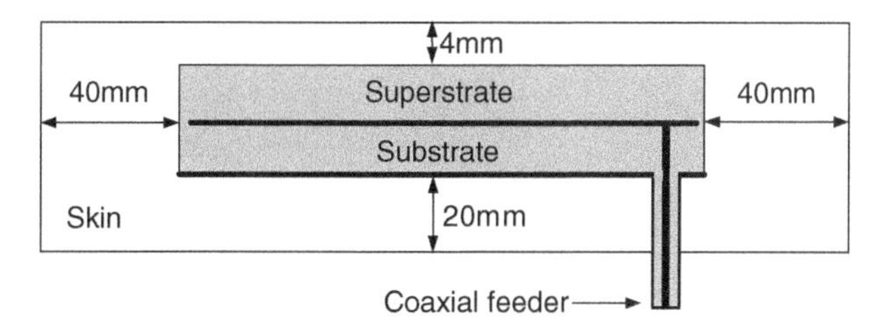

Figure 3.2 Cubic skin phantom geometry for the design of an implantable antenna.

to the edge of the skin. The skin's electrical properties at 2.4 GHz, including the relative permittivity ($\varepsilon_r = 38$) and conductivity ($\sigma = 1.44$ S/m), are considered in the simulation [24]. The size of the skin phantom cube is $90 \times 90 \times 25.27$ mm^3.

These design parameters and simulation models provide the basis for analyzing the impedance characteristics of the implantable patch antenna and its performance within a human body environment.

Figure 3.3 shows the simulated reflection coefficient values and input impedance characteristics with different lengths (l_b) of the center square-slot. It can be seen that the resonant frequency of the microstrip patch antenna is about 4.23 GHz without a center square slot ($l_b = 0$ mm). The resonant frequency will shift from 4.23 GHz to 3.07 GHz when l_b is increased from 0 to 6 mm. The impedance matching of the antenna with a center square-slot is better than that of a regular microstrip patch antenna. From Figure 3.3(b), the input impedance including real and imaginary parts of impedance will increase quickly with increased l_b.

In typical compact microstrip patch antennas with center square-slots on the radiators operating in a free space environment, the presence of a center square-slot results in a longer excited patch surface current path compared to a regular-size square microstrip antenna. With an increasing side length of the center square slot, the minimum input impedance

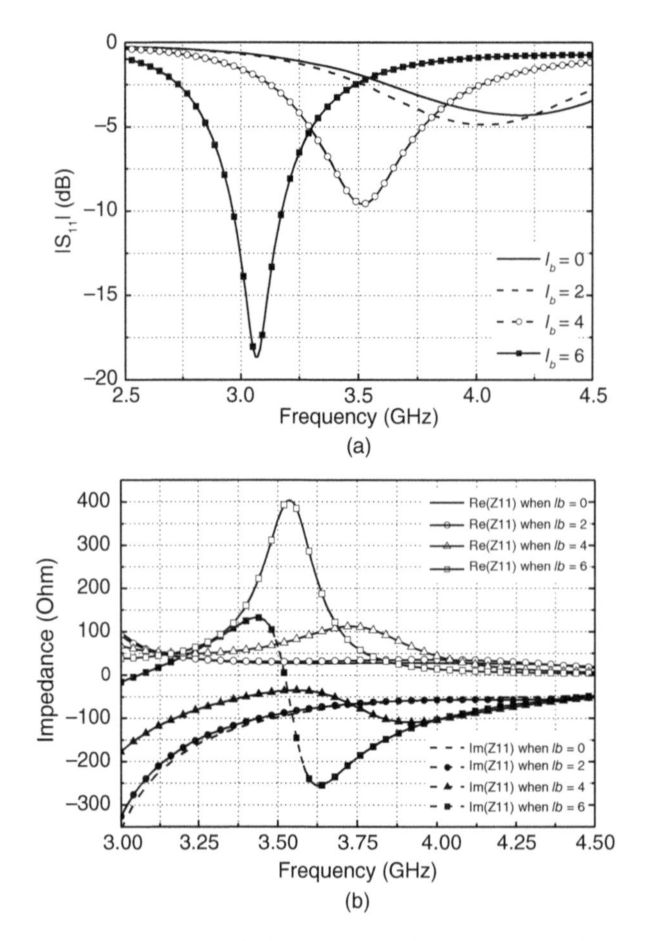

Figure 3.3 Simulated reflection coefficients and input impedance of the implantable patch antenna with different sizes of center square-slot.

inside the patch increases rapidly. To achieve good impedance matching, an impedance transformer is commonly employed [10–12]. The impedance transformer helps to match the antenna impedance with the characteristic impedance of the feeding network or transmission line, ensuring efficient power transfer and minimizing signal reflections.

In this study, the lengthening of the center square-slot leads to a downward shift in the resonant frequency and an increase in the input impedance. Due to the large initial imaginary part of the input impedance, good impedance matching can be achieved without the need for an impedance transformer. The inclusion of a center square-slot allows for effective size reduction and improved impedance matching in the design of the implantable microstrip patch antenna. It should be noted that the initial configuration of the antenna is with linear polarization, and achieving CP operation can be easily accomplished by introducing a pair of perturbation elements at the inner edge of the patch antenna with a center square-slot. This method is commonly used in the design of single-fed CP microstrip patch antennas.

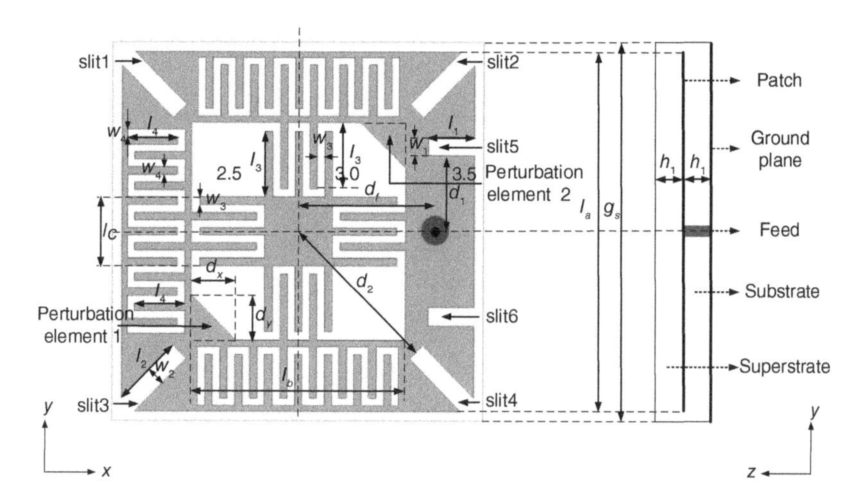

Figure 3.4 Geometry of the CP-implantable antenna with capacitive loading.

3.2.1.2 Compact-implantable CP Patch Antenna with Capacitive Loading

Based on the study of the initial topology that the patch antenna would have good size-reduction and good impedance matching with suitable size of a center square-slot, a compact implantable CP patch antenna with capacitive loading is designed, as shown in Figure 3.4.

With the capacitive loading [19] on the three sides of the patch, the patch size can be reduced with a fixed operation frequency. The capacitive couplings at the center of patch are to enhance the capacitive effect. Moreover, slits 1–6 are employed to lengthen the current path, thus lowering the resonant frequency. For simplicity, slits 1–4 have identical dimensions, and slits 5 and 6 have identical dimensions. CP operation was achieved (shown in Figure 3.5) by protruding a pair of perturbation elements 1 and 2 with the same dimensions of $d_x \times d_y$ (shown in Figure 3.4) to split the fundamental resonant TM_{11} mode into two near-degenerated orthogonal modes with equal amplitudes and 90° phase difference. The determined geometry parameters are $g_s = 10$ mm, $l_a = 9.5$ mm, $l_b = 5.75$ mm, $l_c = 1.8$ mm, $l_1 = 1.25$ mm, $l_2 = 1.97$ mm, $l_3 = 1.75$ mm, $l_4 = 1.35$ mm, $h_1 = 0.635$ mm, $d_1 = 2$ mm, $d_2 = 4.5$ mm, $d_x = d_y = 1.2$ mm, $d_f = 3.7$ mm, $w_1 = w_2 = 0.5$ mm, $w_3 = w_4 = 0.2$ mm. Cubic skin phantom given in Figure 3.2 is used to optimize the proposed antenna.

As shown in Figure 3.5, the simulated impedance bandwidth of the CP patch antenna covers from 2.36 to 2.55 GHz (~7.74%) for reflection coefficient less than −10 dB and axial ratio (AR) bandwidth covers from 2.44 to 2.48 GHz (~1.63%) for AR <3 dB. The simulated results mean that the lowering in the center frequency can correspond to an antenna size reduction of about 72% by using the proposed design in place of the regular implantable CP microstrip patch design at a fixed operating frequency. The simulated peak gain is ~−22 dBi at 2.45 GHz. The main polarization of this proposed antenna is right-handed circularly polarization (RHCP) with cross-polarization discrimination (XPD) ~22 dB at main radiation direction.

Figure 3.6 shows the fabricated CP-implantable antenna with its superstrate. The antenna was measured by using a homogeneous mixture solution proposed in [23]. The recipe

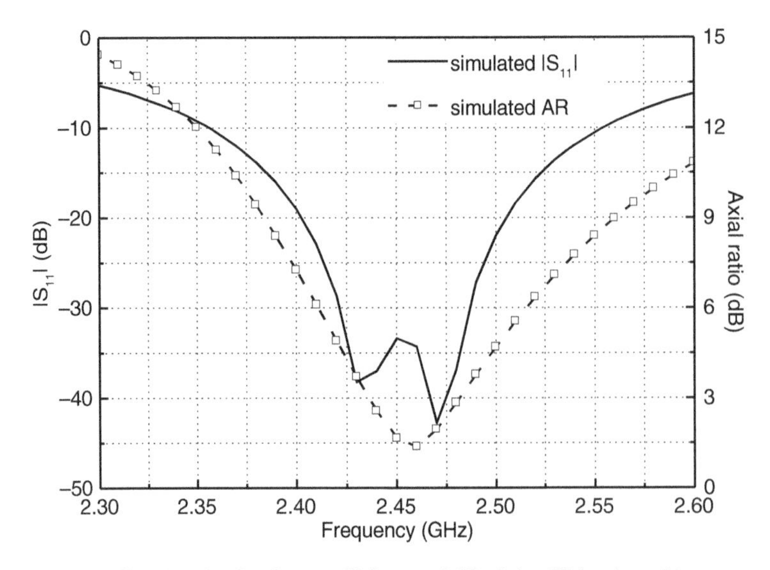

Figure 3.5 Simulated reflection coefficient and AR of the CP-implantable antenna.

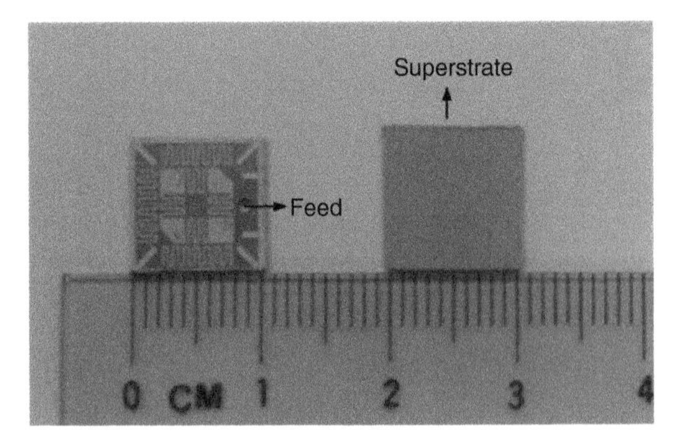

Figure 3.6 Photograph of fabricated CP-implantable patch antenna.

is 58.2% deionized water, 36.7% Triton X-100 (*polyethylene glycol mono phenyl ether*), and 5.1% *diethylene glycol butyl ether* (DGBE). In the measurement, a similar volume of skin material was chosen to demonstrate the design concept. *S*-parameters were carried out by Rohde & Schwarz ZVA50 vector network analyzer (VNA). Figure 3.7 illustrates the communication link of transmitter (Tx) and receiver (Rx), where the implantable CP antenna is the Tx antenna and a dipole antenna is designed as the external Rx antenna to demonstrate the polarization property of the proposed antenna.

Figure 3.8 shows the measured *S*-parameters of the implantable CP antenna and the external dipole antenna. It can be seen from Figure 3.8 that the measured $|S_{11}|$ of the proposed antenna is less than −10 dB in the frequency range from 2.41 to 2.67 GHz (10.2%). The resonant frequency shift could be due to the air gap between the substrate and superstrate. Besides this, difference between simulation and measurement could

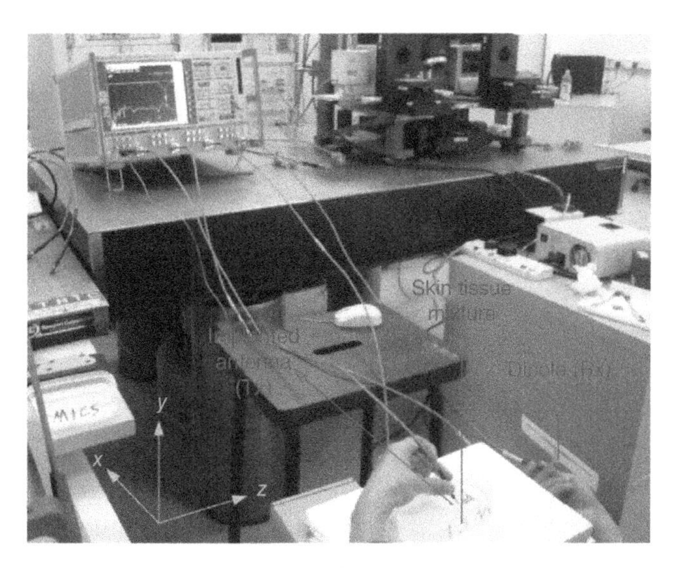

Figure 3.7 Illustration of the measurement setup.

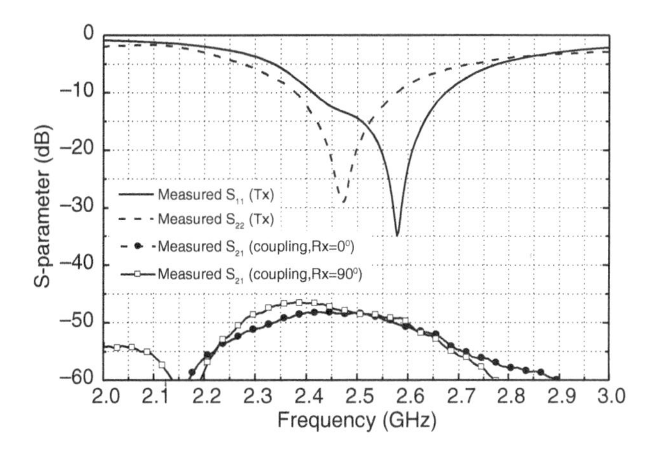

Figure 3.8 Measured S-parameters of the implantable CP patch antenna and external antenna.

be due to the fabrication tolerance. The dipole antenna has an impedance bandwidth of ~8.06% (2.38–2.58 GHz). The communication link of the proposed antenna and the dipole antenna was measured when the dipole was placed at phi = 0° and 90°, respectively. Note that the distance between the Tx and the Rx is ~200 mm. The CP purity of the proposed antenna can be calculated by comparing the communication link levels for two orthogonal polarizations. Good polarization purity can be achieved at around 2.5 GHz ($|S_{21}|$ [phi = 0°] ≅ $|S_{21}|$ [phi = 90°]).

3.2.1.3 Communication Link Study of the CP-implantable Patch Antenna

The impedance matching and polarization performance of the compact-implantable CP microstrip patch antenna was demonstrated by the measurement results shown in

Figure 3.8. Besides, radiation performance should be discussed for communication link study. For accurate simulation, the implantable CP patch antenna was evaluated within the Gustav voxel human body. Numerical analyses are performed using the CST Microwave Suite [24]. Assuming far-field communication, the link power budget can be described in terms of [25]:

$$\text{Link margin (dB)} = \text{Link } C/N_0 - \text{Required } C/N_0$$
$$= P_t + G_t - L_f + G_r - N_0 - E_b/N_0 - 10\log_{10}B_r + G_c - G_d \quad (3.1)$$

where P_t is the transmit power, G_t is the transmit antenna gain, L_f is the path loss (free-space), G_r is the receive antenna gain, and N_0 is the noise power density.

According to the free-space reduction in signal strength with distance between the transmitter and receiver, the path loss in free-space can be calculated as follows:

$$L_f\ (\text{dB}) = 20\ \log(4\pi d/\lambda) \quad (3.2)$$

where d is the distance between Tx and Rx. Considering the impedance mismatch loss,

$$L_{imp}\ (\text{dB}) = -10\ \log(1 - |\Gamma|^2) \quad (3.3)$$

where Γ is the appropriate reflection coefficient.

As for the compact-implantable CP patch antenna, note that the antenna is operated at 2.46 GHz with good circularly polarization, while $|S_{11}|$ at 2.46 GHz is ~−8.3 dB, when the antenna was embedded in the CST Gustav voxel human chest. The impedance mismatch loss is ~0.695 B at 2.46 GHz.

For the Rx antenna, one exterior linear polarized (LP) dipole antenna with realized gain of 2.15 dBi and another exterior CP antenna with realized gain of 2.15 dBi are considered. Note that the Tx-implantable antenna is with circularly polarization, the polarization mismatching losses are ~0 dB and ~3 dB when the Rx dipole antenna is with circularly polarization and linear polarization, respectively. The impedance mismatch loss of the receiver and Rx antenna is neglected in this condition. The link margin can be calculated using the parameters listed in Table 3.1 [25].

Figure 3.9 shows the calculated link margin with different distance between the Tx and Rx. In order to make wireless communication possible, the link margin should be better

Table 3.1 Parameters of the link budget.

Transmission		**Receiver**		**Signal quality**	
Frequency (GHz)	**2.46**	**Rx antenna gain G_r (dBi)**	**2.15**	**Bit rate B_r (kb/s)**	**7.0**
Tx power (dBm)	−40	Polarization	CP/LP	Bit error rate	1.0×10^{-5}
Tx antenna gain G_t (dBi)	−20.4	Temperature T_0 (k)	293	E_b/N_0 (ideal PSK) (dB)	9.6
EIRP (dBm)	−60.4	Boltzmann constant k	-1.38×10^{-23}	Coding gain G_c (dB)	0
		Noise power density N_0 (dB/Hz)	−199.95	Fixing deterioration G_d (dB)	2.5

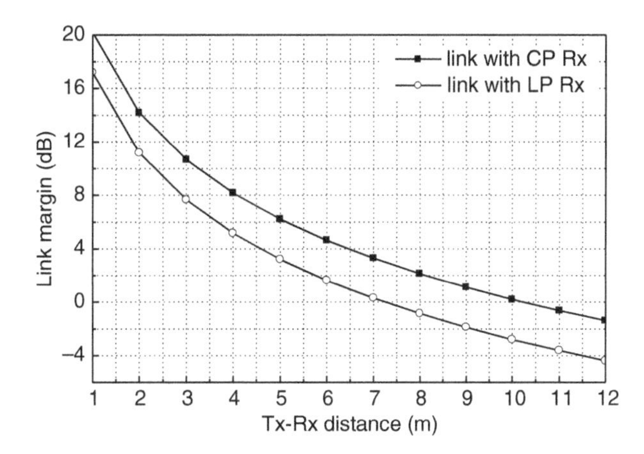

Figure 3.9 Calculated link margin of the CP-implantable patch antenna.

than 0 dB. According to the calculated results, it is possible to use wireless communication when the distance of Rx and Tx is less than 10 m with the Rx antenna in CP operation and less than ~7 m when with Rx antenna in LP operation. The short-range communication ability was demonstrated by the calculated results.

3.2.1.4 Sensitivity Evaluation of the Implantable CP Patch Antenna

To evaluate the sensitivity of the implantable CP patch antenna, whether it could be used in other specific biomedical applications or would the measurement environment affects the performance of the patch antenna. The CP patch antenna was studied in different environments. It can be summarized as follows:

A. Effect of Different Phantoms

In order to evaluate whether the CP patch antenna can be used in other implant positions or other biomedical applications, the antenna is embedded in the below-mentioned four different body phantoms:

(1) **Phantom 1**: skin phantom ($90 \times 90 \times 25.27\ mm^3$, shown in Figure 3.2)
(2) **Phantom 2** [26]: three-layered phantom ($180 \times 60 \times 60\ mm^3$)
(3) **Phantom 3** [27]: cylinder muscle phantom ($4/3 \times \pi \times 403\ mm^3$)
(4) **Phantom 4** [28]: scalp phantom ($80 \times 50 \times 5\ mm^3$)

Figure 3.10 shows the detailed body phantoms (2, 3, and 4). Figure 3.11 reports the simulated results with different phantoms. The resonant frequencies of 2.45 GHz, 2.60 GHz, 2.40 GHz, and 2.40 GHz were found for the skin, three-layered, muscle, and scalp phantoms, respectively. The resonant frequency shift is mainly due to the change of relative dielectric constant, especially in three-layered phantom as the relative dielectric constant of fat is very small compared with other tissues. Similar resonant frequencies are obtained between muscle and scalp phantoms. Moreover, the proposed antenna embedded in muscle and scalp phantoms can still maintain good impedance matching and circularly polarization, as shown in Figure 3.11(b). The results show that the proposed antenna can be optimized for other specific biomedical applications.

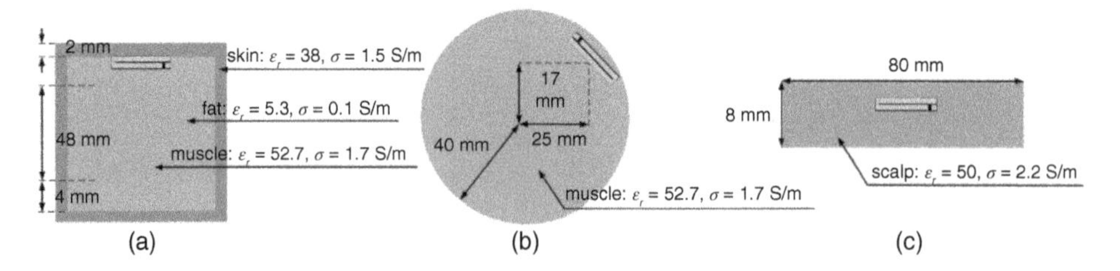

Figure 3.10 Cross views of different body phantoms: (a) the three-layered phantom, (b) the cylinder muscle phantom, and (c) the scalp phantom.

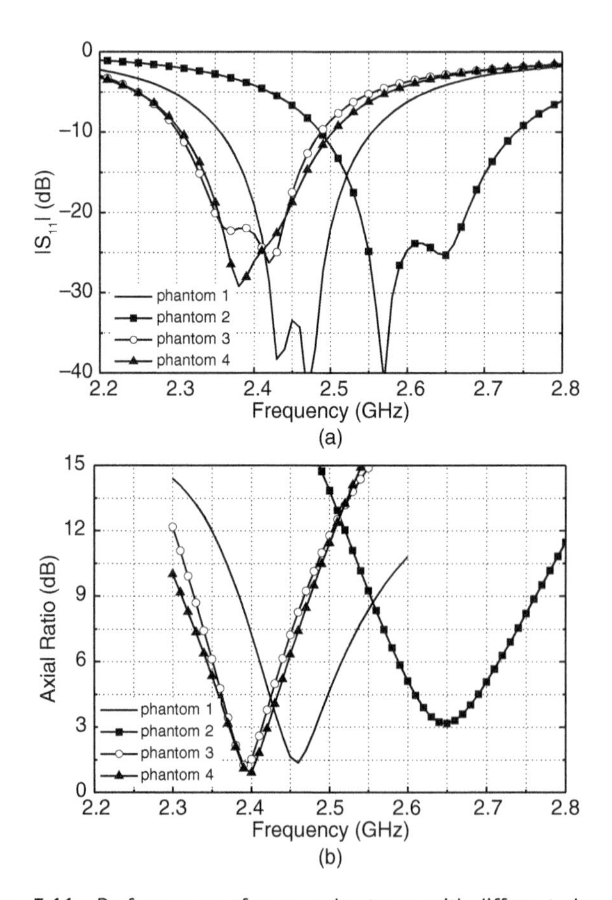

Figure 3.11 Performances of proposed antenna with different phantoms: (a) reflection coefficients, (b) axial ratio.

B. Effect of Coaxial Cable

The measurement setup shown in Figure 3.7 did not consider the effect of coaxial cable. It should be mentioned that there could be small coupling between the coaxial cable and the implantable antenna when testing. Referring to [27], four different simulation models are listed in Figure 3.12 to evaluate the feeding coaxial cable effect.

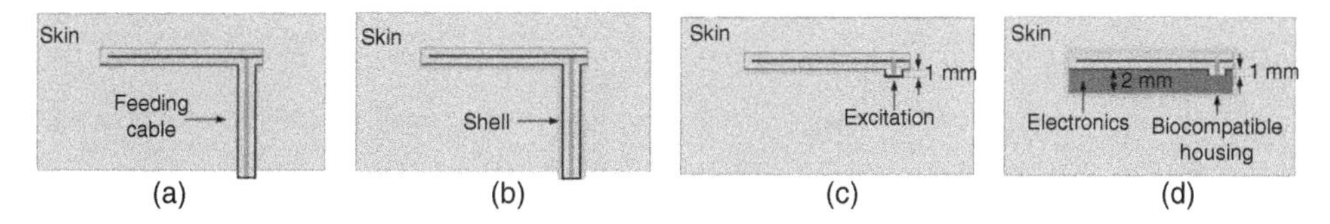

Figure 3.12 Description of different models: (a) case 1, (b) case 2, (c) case 3, and (d) case 4.

(1) **Case 1**: The coaxial cable is in direct contact with the skin, as shown in Figure 3.12(a)
(2) **Case 2**: A biocompatible material (Silastic MDX4-4210 Biomedical-Grade Base Elastomer ($\varepsilon_r = 3.3$, $\sigma = 0.01$ S/m, thickness = 0.05 mm) [29]) shell surrounds the cable and ground plane, as shown in Figure 3.12(b)
(3) **Case 3**: The length of coaxial cable is reduced to 1 mm, as shown in Figure 3.12(c)
(4) **Case 4**: The whole implant device consists of implantable antenna, biocompatible housing, and other electronics, as shown in Figure 3.12(d); the other electronics are set as perfect electric conductors (PEC) in HFSS

Simulated reflection coefficients for the four different cases are shown in Figure 3.13. The results indicate that a small coupling exists between the currents flowing on the external metal of the cable and the body phantom [27]. The impedance matching of the proposed antenna in all the simulation models is good. AR performance in case 4 means that the packaging of the whole implant system would have a little effect on the polarization property. After the discussions about the cable effect, the test in Figure 3.7 has demonstrated the antenna's performance.

C. Biocompatible Insulation

In this design, Rogers RO3010 is used as substrate and superstrate. However, implantable antennas need to be biocompatible with the surrounding tissue, as Rogers RO3010 is not a biocompatible material. There are two typical approaches to address the biocompatibility issue for practical applications. One is to design antennas directly on biocompatible materials such as Macor, Teflon, and Ceramic Alumina [30]; the other one is to encase the implantable antenna with a thin layer of low-loss biocompatible coating [29]. This discussion is based on the simulation model shown in Figure 3.2.

For the first approach (shown in Figure 3.14(a)), Rogers RO3010 ($\varepsilon_r = 10.2$) is replaced by biocompatible alumina (Al_2O_3) ceramic ($\varepsilon_r = 9.8$). Figure 3.15(a) shows that the resonant frequency is shifted from 2.45 GHz to 2.5 GHz. This is reasonable, as the relative dielectric constant of alumina ceramic is a little smaller than that of RO3010. For further realistic applications, biocompatible metals such as silver palladium (Ag/Pd) can be used instead of copper to achieve antenna structure on the biocompatible materials. In further study, the implantable CP patch antenna with biocompatible materials can be optimized to the desired resonant frequency.

The other one is to encase the proposed implantable antenna with a thin layer of low-loss biocompatible coating as shown in Figure 3.14(b). A biocompatible material Silastic MDX4-4210 Biomedical-Grade Base Elastomer ($\varepsilon_r = 3.3$, $\sigma = 0.01$ S/m) [29] is used here to evaluate the coating effect.

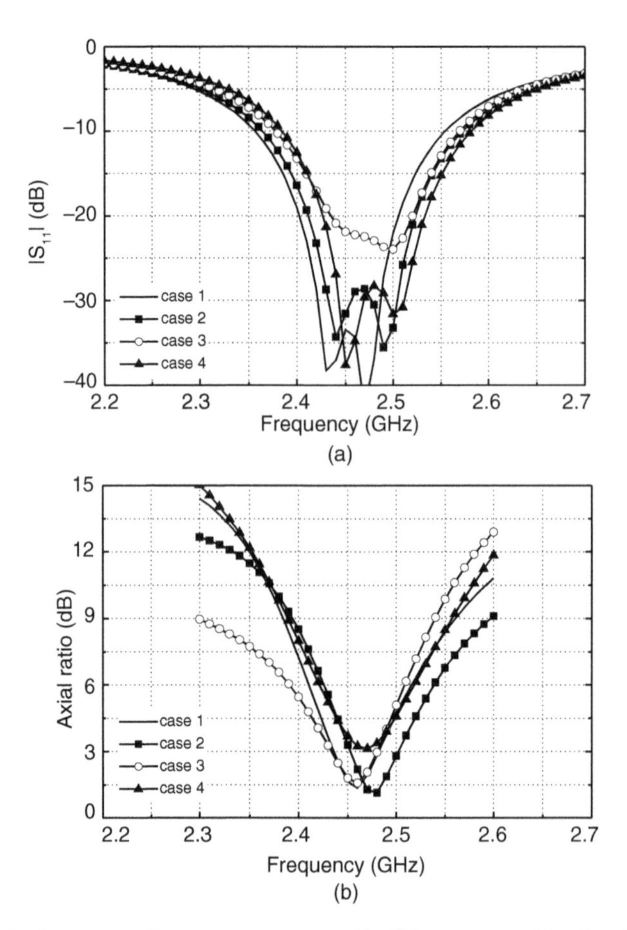

Figure 3.13 Performances of proposed antenna with different cases: (a) reflection coefficients comparison, (b) AR comparison.

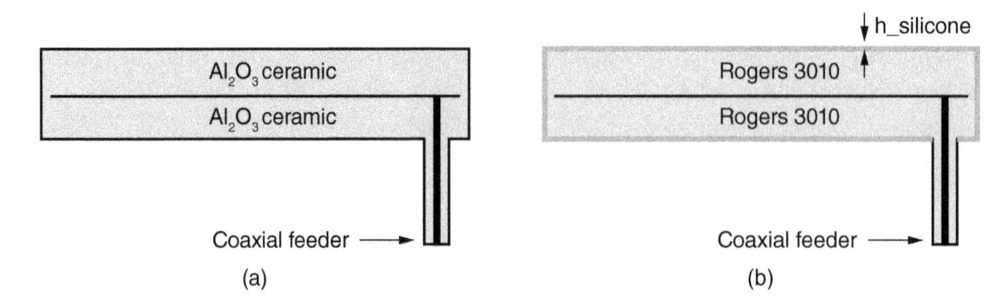

Figure 3.14 Two approaches to address the biocompatibility issue for practical applications: (a) antenna with biocompatible material, (b) antenna with encased biocompatible material.

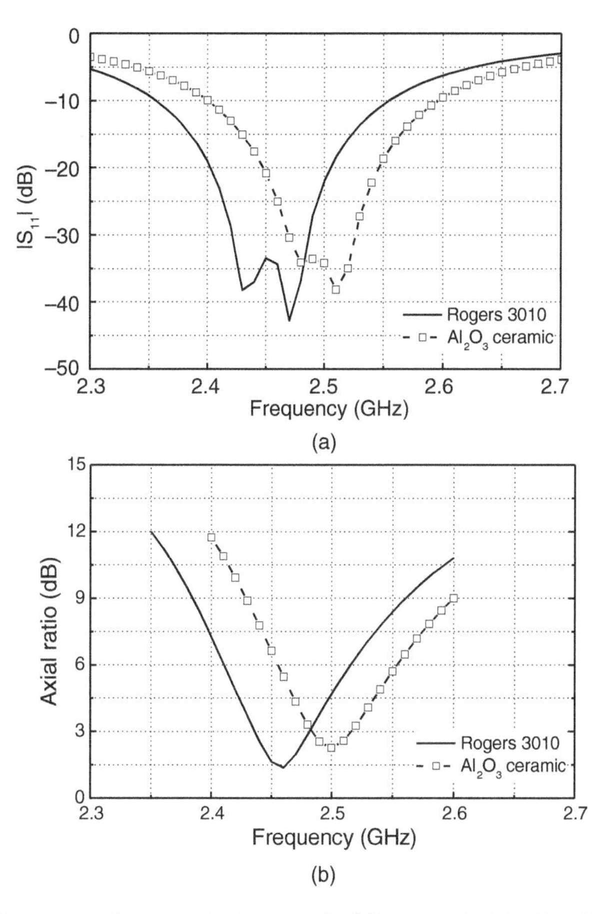

Figure 3.15 Performances of proposed antenna with different materials: (a) reflection coefficients comparison, (b) AR comparison.

It is important to note that the thickness of the encased biocompatible material can have an impact on the performance of the antenna, and it should be considered in practical antenna designs [30, 31]. Figure 3.15 illustrates that as the thickness of the encased biocompatible material decreases, the frequency shifts become less pronounced. Even with a thickness of 0.05 mm or 0.01 mm, good polarization purity can still be achieved, albeit with a slight shift in the resonant frequency.

3.2.2 Miniaturized Circularly Polarized-implantable Annular-ring Antenna

In contrast to the square-shaped antennas discussed in Section 3.2.1, circular antennas are more suitable for implantable or ingestible medical devices with a circular shape, such as wireless capsule endoscope systems. Additionally, circular ring patches occupy a smaller area compared to square or circular patches operating at the same frequency. In this section, a circular-shaped CP annular-ring antenna operating in the ISM band (2.4–2.48 GHz) is proposed for implantable medical devices. Circularly polarization is

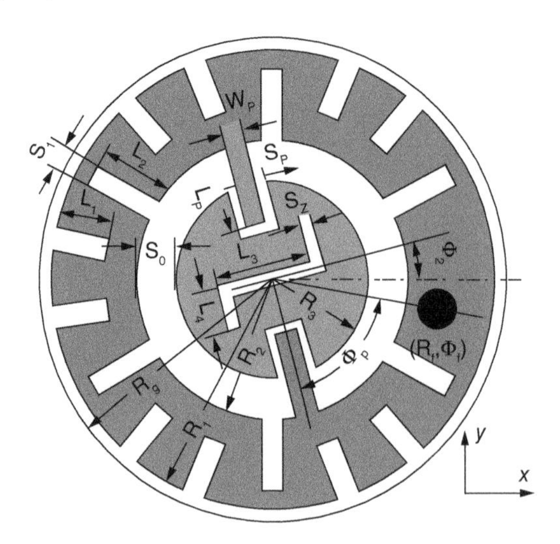

Figure 3.16 Geometry of the proposed implantable antenna.

achieved by incorporating a pair of open stubs on the inner boundary of the annular ring. Moreover, miniaturization of the antenna is accomplished by embedding two types of rectangular slots onto the annular ring.

Figure 3.16 shows the top view of the proposed antenna. The shaded radiating element consists of an annular ring of outer/inner radius R1/R2 and a central circular patch of radius R3. In order to minimize the antenna's size, two types of rectangular slots with size L1 × S1 and L2 × S1 are cut in the annular ring. To realize circularly polarization, a pair of open stubs protruding in the inner boundary of the annular ring are used to excite two near-degenerate orthogonal modes with equal amplitude and 90° phase difference. A Z-shaped slot on central circular patch is designed to further optimize the impedance matching and AR. The optimized dimensions are: Rg = 5.5 mm, R1 = 5.2 mm, R2 = 3.2 mm, R3 = 2.2 mm, Rf = 4.0 mm, φf = 10°, φp = 65°, φz = 15°,L1 = 1.27 mm, L2 = 1.62 mm, L3 = 2.23 mm, L4 = 1.04 mm, Wp = 0.5 mm, Lp = 1.9 mm, Sp = 0.2 mm, Sz = 0.3 mm, S0 = 1.0 mm, and S1 = 0.5 mm.

The proposed antenna is composed of two layers of Rogers 3010 substrates, with each layer having a thickness of 25 mil. A superstrate is used for dielectric loading and corrosion protection of the radiating element. The antenna is fed by a 50-Ω coaxial cable. In the simulation, the antenna is positioned at the center of a human skin model with the same configuration as the antenna proposed in Section 3.2.1, as shown in Figure 3.2.

The initial dimensions of the proposed antenna, designed to operate in the dominant TM11 mode, are determined based on [32]. To further reduce the antenna's size, two types of rectangular slots are introduced on the annular ring. Figure 3.17 displays the simulated $|S_{11}|$ values of three different antenna cases, along with the proposed antenna, all having the same outer radius of the metal radiating patch and substrate size. It can be observed that the resonant frequencies shift downward from 4.47 GHz, 3.34 GHz, and 3.34 GHz to 2.42 GHz, indicating that the size of the proposed antenna is reduced by 84.7% compared to the original circular patch by incorporating the inner and outer rectangular slots. When the sizes

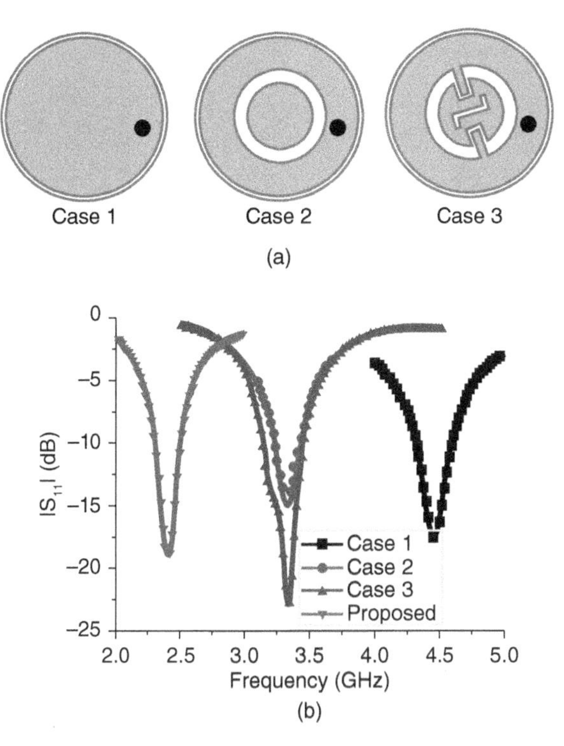

Figure 3.17 $|S_{11}|$ of the proposed antenna considering three different cases: (a) three different cases, (b) $|S_{11}|$ of the antenna.

of the inner or outer rectangular slots increase, the resonant frequency of the proposed antenna shifts further downward. Additionally, the inner rectangular slots have a greater influence on the resonant frequency of the antenna compared to the outer slots. In other words, the rectangular slots significantly impact the antenna size at a given resonant frequency.

To illustrate the CP operation mechanism of the proposed antenna, Figure 3.18 depicts the surface current distributions at 2.45 GHz for four phases of 0°, 90°, 180°, and 270°. The predominant surface current in the azimuth plane rotates clockwise, which is accountable for left-hand circularly polarization (LHCP). According to the analysis, the sizes (Lp, Wp) and positions (φp) of the open stubs in the inner boundary are dominant influential factors on the antenna performance. Figure 3.19(a) shows AR curves of the proposed antenna against the stub length Lp. When Lp = 1.8 mm, the antenna has the lowest AR, whereas the central frequency of 3-dB ARBW shifts up. When Lp = 2.0 mm, the lowest AR becomes larger. Therefore, an optimal AR at 2.45 GHz can be obtained for Lp = 1.9 mm. Figure 3.19(b) shows AR curves of the proposed antenna against the open stub position φp. We can conclude from Figure 3.19(b) that the larger the angle φp, the smaller the AR at 2.45 GHz, and the wider the 3-dB ARBW. Hence, $\varphi p = 65°$ is chosen for a good CP performance.

It is worth noting that the inclusion of a Z-shaped slot can further optimize the performance of the antenna. Figure 3.20 illustrates the AR curves of the proposed antenna with

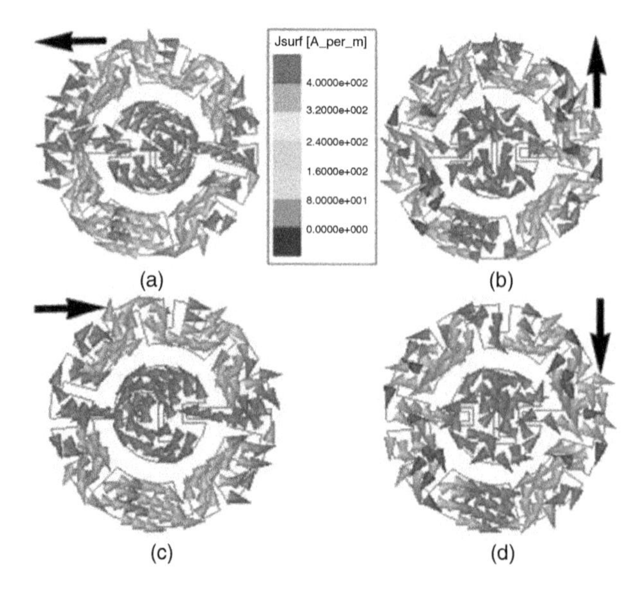

Figure 3.18 Simulated surface current distributions of the proposed antenna at 2.45 GHz: (a) $t = 0°$, (b) $t = 90°$, (c) $t = 180°$, and (d) $t = 270°$.

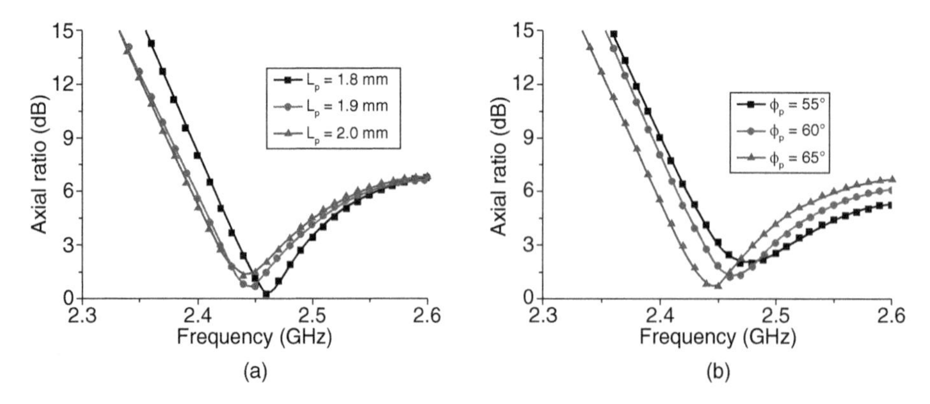

Figure 3.19 AR of the proposed antenna with different parameters: (a) open stub lengths, (b) open stub positions.

different slot lengths (L3) of 2.13 mm, 2.23 mm, and 2.33 mm. From the graph, it can be observed that the optimal AR curve is achieved when L3 is set to 2.23 mm at a frequency of 2.45 GHz. When L3 is decreased to 2.13 mm, the lowest AR value increases. On the other hand, when L3 is increased to 2.33 mm, the 3-dB axial-ratio bandwidth (ARBW) becomes narrower. Additionally, the slot length L3 can also be used to improve the impedance matching of the antenna. The optimal impedance matching is achieved when L3 is set to 2.23 mm. The detailed simulated results are not provided here for the sake of brevity.

As a result of these optimizations, the proposed antenna achieves optimized CP performance. It exhibits a 3-dB ARBW of 2.49%, ranging from 2.419 to 2.48 GHz, with the lowest AR value being 0.7 dB at 2.45 GHz.

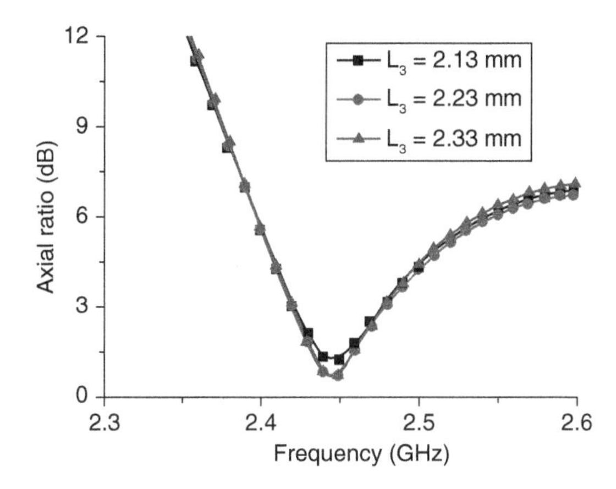

Figure 3.20 AR of the proposed antenna with different *Z*-shaped slot lengths.

3.3 Wide AR Bandwidth-implantable Antenna

For a single-layer microstrip patch antenna, the CP characteristic is achieved by two orthogonal modes with a ±90° phase difference. It usually results in a narrower bandwidth. For implantable medical devices, it is known that a broadband implantable system has better tolerance to different human tissue environments and can provide high data rate exchange with external apparatus, which is essential for applications such as cochlear implants, neural signal recording, and other high-resolution imaging purposes. In this section, two implantable patch antennas with wide AR bandwidth will be discussed.

3.3.1 Miniaturized CP-implantable Loop Antenna

3.3.1.1 Configuration of the CP-implantable Loop Antenna

The CP antenna proposed in [33] was designed to cover the frequency range from 902 to 928 MHz for biomedical applications. The antenna is intended to be implanted into a human arm and placed close to the skin surface. Since skin and muscle have a higher dielectric constant compared to fat, they play a significant role in reducing the size of the implant antenna. When comparing skin with muscle, it is evident that an implantable antenna placed in the skin would be closer to the external receiver, thereby minimizing power loss caused by the tissue. Hence, a skin model is preferred, and its dimensions are selected to simulate a 2/3 equivalent human arm. In Figure 3.21, a simplified one-layer skin model measuring 60 mm × 180 mm × 60 mm is created to mimic the human environment during simulation. The dielectric properties of the skin tissue are set to be frequency-dependent according to [21]. At 915 MHz, the dielectric properties of the skin are characterized by a relative permittivity of 41.35 and a conductivity of 0.87 S/m. The implantable antenna is embedded in the skin tissue at a depth of h, with its upper surface facing the skin surface. The antenna design utilizes Rogers 3010 substrate with a thickness of 0.635 mm.

Figure 3.22 illustrates the configuration of the implantable CP loop antenna, which consists of a square loop connected with four LC loadings. The center of the configuration

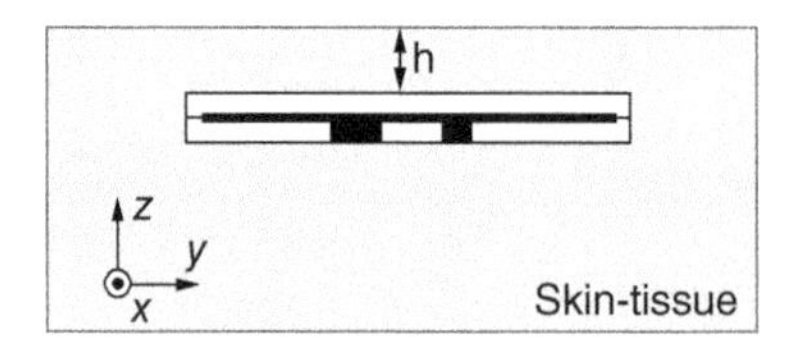

Figure 3.21 Simulation environment of the antenna embedded in the skin-tissue with depth of h.

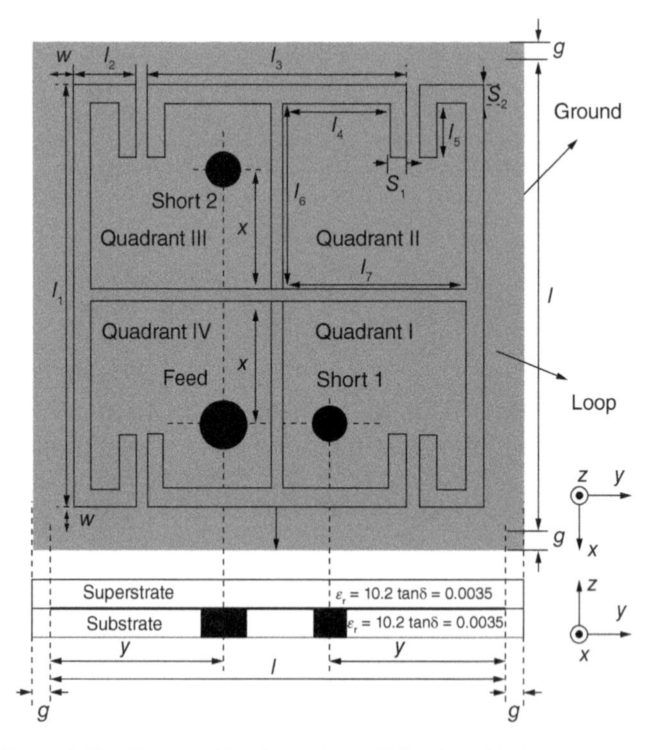

Figure 3.22 Top and side views of the CP-implantable loop antenna.

is set as the origin, and the coordinate system is also depicted in Figure 3.22. As shown in the figure, four small patches are connected to the loop using four high-impedance lines located in different quadrants. Additionally, two shorting pins with a diameter of 0.9 mm are positioned in quadrants I and III. It is important to note that all the loadings exhibit symmetry around the origin. The proposed antenna is fed in quadrant IV. Through optimization, when the depth h is set to 3 mm, a compact size of 13 mm × 13 mm with a ground plane measuring 14 mm × 14 mm is achieved. The detailed parameters are optimized as follows: $x = 3.35$ mm, $y = 5$ mm, $g = 0.5$ mm, $l = 13$ mm, $l_1 = 11.6$ mm, $l_2 = 1.8$ mm, $l_3 = 7.4$ mm, $l_4 = 3.05$ mm, $l_5 = 1.5$ mm, $l_6 = l_7 = 5.15$ mm, $w = 0.7$ mm, $s_1 = s_2 = 0.5$ mm.

The antenna is designed with RHCP, and LHCP can be achieved by mirroring the positions of the feed and shorts along the x-axis. Figure 3.23 illustrates the simulated $|S_{11}|$ and AR for both the RHCP and LHCP configurations, with respect to the 50-Ω reference. In the

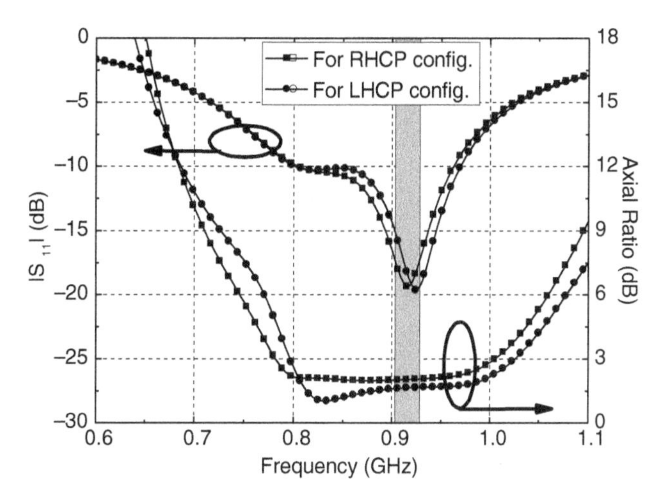

Figure 3.23 Simulated and AR (when $h = 3$ mm).

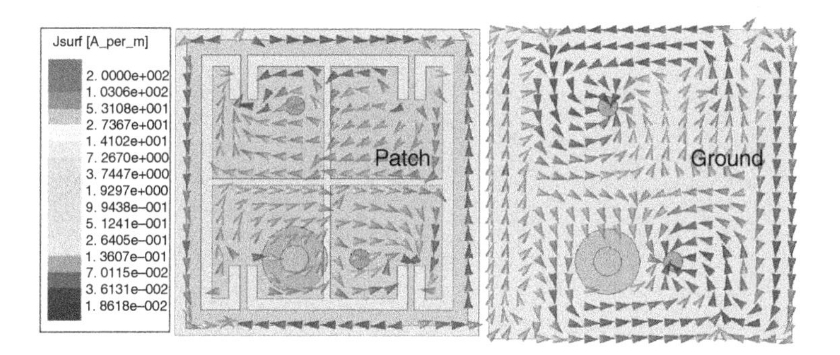

Figure 3.24 The current distributions of the loop antenna at 828 MHz.

optimized RHCP configuration, a wide bandwidth of 18.2% is achieved, with $|S_{11}|$ below −10 dB and AR below 3 dB. The bandwidth covers the frequency range from 802 MHz to 963 MHz, with a center frequency of 882.5 MHz.

As observed from the $|S_{11}|$ curve of the proposed antenna, there is an additional resonance at 828 MHz, in addition to the main resonance at 915 MHz. Figure 3.24 illustrates the current distributions on the antenna and ground plane at 828 MHz. It is evident from Figure 3.24 that the presence of the two shorts primarily contributes to the resonance at 828 MHz. Strong currents are observed on the ground plane at this resonance, indicating that changes in the ground size would affect the impedance matching of the antenna.

The simulated radiation patterns at 915 MHz are depicted in Figure 3.20 for both configurations. The maximum realized gain values at the boresight direction are both −32 dBi. It should be noted that these realized gain values are simulated for the antenna positioned within a 60 mm × 180 mm × 60 mm one-layer-skin model, at a depth of 3 mm. Both the gain values and radiation patterns may vary depending on the size of the phantom (Figure 3.25).

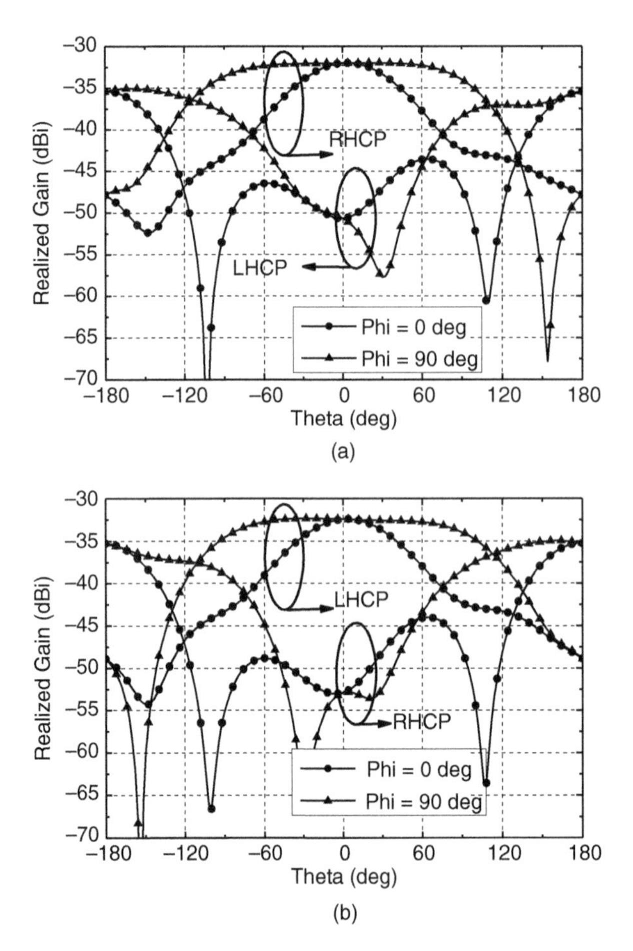

Figure 3.25 Simulated radiation patterns at 915 MHz (when $h = 3$ mm): (a) RHCP configuration, (b) LHCP configuration.

3.3.1.2 Principle of the CP-implantable Loop Antenna

To validate the operational principle of the proposed antenna, let us consider the RHCP configuration as an example and analyze its miniaturization and CP properties.

As shown in Figure 3.22, the miniaturized loop antenna can be recognized as a loop antenna connected with four high-impedance lines at different quadrants. Figure 3.26(a) shows a host transmission line loaded with a high-impedance line and a square patch. The corresponding equivalent circuit is shown in Figure 3.26(b), where the T-network consists of two inductors, L_1 and C_1, which stand for the inductance and capacitance of the host transmission line, L_2, which is the inductance of high impedance line and C_2, which is the total capacitance induced by the loading. It is obvious that the *LC* loading would bring in slow wave effect and cause miniaturization of the structure.

Figure 3.27 shows three antenna configurations that are established with a fixed size of 13 mm × 13 mm, namely *case* 1 to *case* 3 to understand the miniaturization mechanism of the proposed antenna. Three cases can be described below:

(1) **Case 1**: a simple square loop antenna loaded with two shorts at its corner symmetrically ($w = 1.8$ mm), as shown in Figure 3.21(a)
(2) **Case 2**: the width of the loop stays unchanged as case 1, and four-square patches are added and connected to the loop with four high impedance lines at different quadrants ($w = 1.8$ mm), as shown in Figure 3.21(b)
(3) **Case 3**: the width of the loop w decreases while the length of the patches a keeps increasing ($x = y = 3$ mm), as shown in Figure 3.21(c)

To facilitate a comparison with the proposed antenna, three additional cases are simulated in the same environment, and all three cases are excited by a coaxial feed at the corner of quadrant IV. Figure 3.27(a) shows the results for the loop antenna without shorts, where a one-wavelength resonance with linear polarization is observed at 1.93 GHz. Upon loading the two shorts, another resonance with linear polarization appears close to the original one.

Moving on to case 2, as shown in Figure 3.27(b), we can view it as a host transmission line loaded with a high impedance line and a square patch, as depicted in Figure 3.26(a). The corresponding equivalent circuit is illustrated in Figure 3.26(b), where the T-network

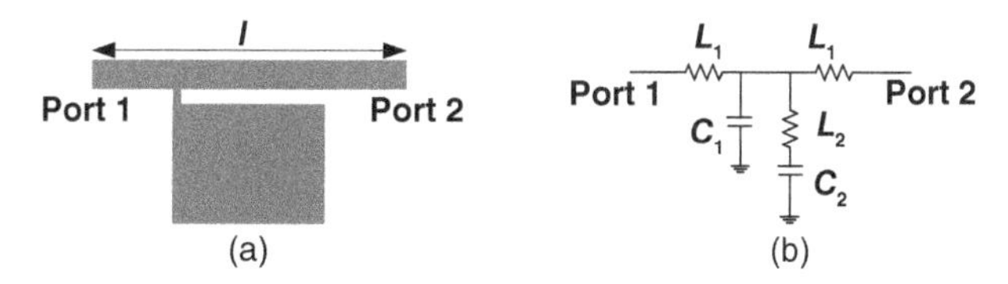

Figure 3.26 Transmission line with LC loading: (a) structure, (b) equivalent circuit.

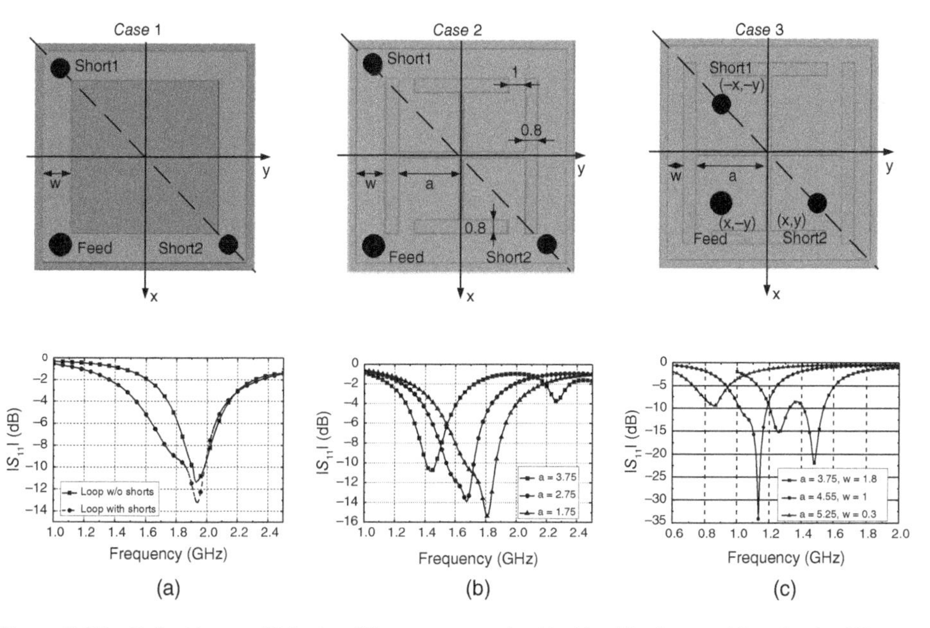

Figure 3.27 Reflection coefficients of three cases embedded in skin tissue with a depth of 3 mm.

represents the inductance and capacitance of the host transmission line. It is evident that the LC loading introduces a slow wave effect and contributes to the miniaturization of the structure. The four LC loadings have identical dimensions, and the length of the patch is denoted as a. It can be observed that as a increases, the resonant frequencies shift to a lower band. Comparing case 2 to case 1, when the length of the square patch a is increased to 3.75 mm, the resonant frequency shifts from 1.93 GHz to 1.42 GHz, resulting in a 26% size reduction.

Moving further to case 3 in Figure 3.27(c), the width of the loop w decreases while the length of the patches a continues to increase. By tuning a and w, the desired frequency band can be covered. Compared to case 1, the center frequency can be adjusted from 1.93 GHz to 0.86 GHz (when $a = 5.25$ mm and $w = 0.3$ mm), achieving a 55.4% size reduction.

As seen in Figure 3.23, the AR performance of the proposed antenna remains consistent at around 2 dB within the desired frequency band. Therefore, we will focus on analyzing the antenna's performance at 915 MHz. Circularly polarization can be achieved when the magnitudes of the two components are identical and the time-phase difference between them is odd multiples of $\pi/2$. At the frequency of 915 MHz, with an input power of 1 W, the simulated magnitudes of Ex and Ey are 139.6 mV/m and 137.1 mV/m, respectively. The simulated phases are 125.4 degrees and 22.1 degrees, respectively, when phi = 0 degrees and theta = 0 degrees. Using these values, we can calculate the AR as follows:

$$\text{AR} = \frac{\text{major axis}}{\text{minor axis}} = \frac{\text{OA}}{\text{OB}} \tag{3.4}$$

where,

$$\text{OA} = \left[\frac{1}{2}\left\{E_x^{\,2} + E_y^{\,2} + [E_x^{\,4} + E_y^{\,4} + 2E_x^{\,2}E_y^{\,2}\cos(2\Delta\phi)]^{1/2}\right\}\right]^{1/2} \tag{3.5}$$

$$\text{OB} = \left[\frac{1}{2}\left\{E_x^{\,2} + E_y^{\,2} - [E_x^{\,4} + E_y^{\,4} + 2E_x^{\,2}E_y^{\,2}\cos(2\Delta\phi)]^{1/2}\right\}\right]^{1/2} \tag{3.6}$$

Thus, the AR is calculated to be 1.26, which is approximately 2.04 dB, indicating the presence of circularly polarization.

3.3.1.3 Antenna Measurement and Discussions

The fabricated antenna is shown in Figure 3.28(a), with a compact size of $13 \times 13 \times 1.27\ \text{mm}^3$. The measurement setup is depicted in Figure 3.28(b). To validate the antenna performance, measurements are conducted under two scenarios: one in a plastic container filled with pork, and the other in a plastic container filled with skin-mimicking gel [34]. It should be noted that, according to the simulation, the size of the tissue does not significantly affect the reflection coefficient of the implantable antenna. Therefore, the dimensions of the pork and skin-mimicking gel used during the measurements are nearly the same, measuring 72 mm × 135 mm × 30 mm.

Figure 3.29 presents a comparison between the measured and simulated $|S_{11}|$ values. It is evident from the figure that two resonances are clearly observed in the measurement results. The measured antenna bandwidths in both pork and skin-mimicking gel are 29.4% (709–954 MHz) and 27.8% (737–975 MHz), respectively, covering the desired frequency range. The discrepancy between the measured and simulated results can be attributed to the following reasons:

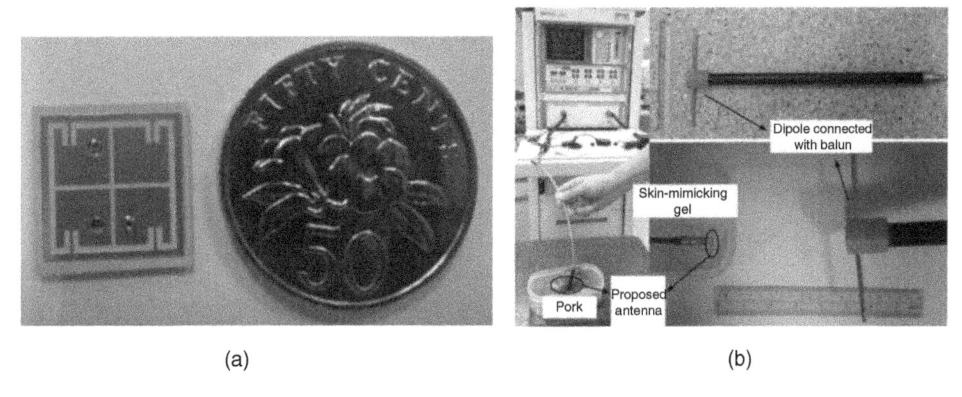

Figure 3.28 (a) The photograph of fabricated implantable CP loop antenna, (b) measurement setup.

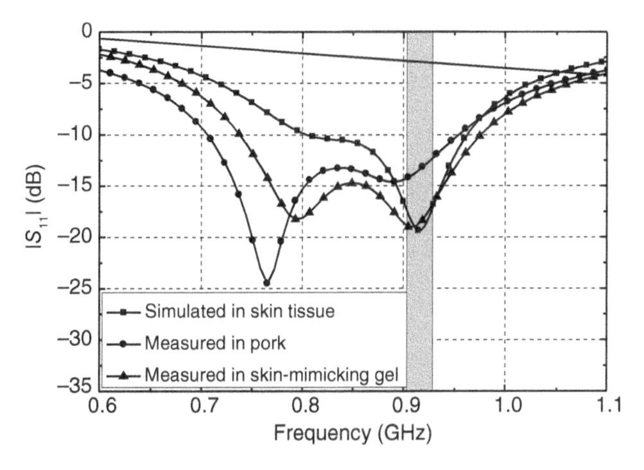

Figure 3.29 Comparison of the simulated and measured $|S_{11}|$.

(1) the fabrication discrepancy.
(2) tissue sinking into the gap between substrate and superstrate during measurement.
(3) the differences in dielectric properties between simulation and measurement.
(4) the cable connected to the antenna during measurement.

To demonstrate the CP property of the proposed antenna, a linearly polarized dipole connected with a balun is placed outside the tissue gel to serve as a receiver. The dipole is placed 150 mm away from the implanted antenna as shown in Figure 3.28(b). The measurement of $|S_{21}|$ was carried out in four directions, including two pairs of orthogonal directions, 0 deg (Tx and Rx antennas are at the same polarization) and 90 deg (Tx and Rx antennas are at the orthogonal polarization), +45 deg and −45 deg, respectively, and the measurement results are given in Figure 3.30. From the comparison of $|S_{21}|$ in the desired band, it can be seen that the received power has a maximum difference only up to 5 dB, which suggests CP property of the proposed antenna.

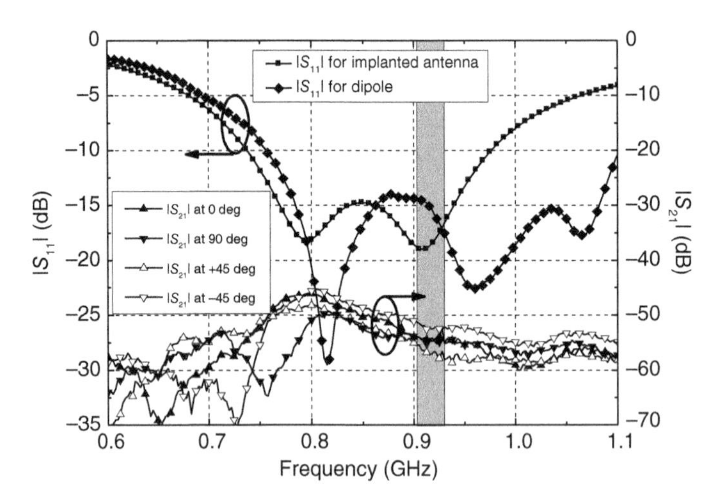

Figure 3.30 Measured results of communication link at four orientations.

3.3.1.4 Communication Link of the Implantable CP Loop Antennas

In the case of up-link communication (from the implant to the base station), an antenna in free space with a center frequency of 915 MHz lying at the x–y plane is viewed as the receiver antenna, and the implantable antenna is considered to be at the transmitter side. The distance between the transmitter and the receiver is labeled d. The communication link margin (LM) can be calculated as follows:

$$\mathrm{LM} = P_\mathrm{r} - \mathrm{PL} + G_\mathrm{r} - \mathrm{RNF} - \mathrm{SNR} \tag{3.7}$$

where P_r is the radiated power of the transmitter antenna, PL is the path loss and RNF is the receiver noise floor, G_r is the gain of the receiver antenna and SNR is the signal-to-noise ratio of the receiver.

The values of RNF and SNR are from [35] and the various parameters of the link budget are listed in Table 3.2. Note that the $|S_{11}|$ at 915 MHz is around −19 dB, thus the impedance mismatch loss can be neglected. Besides, since the maximum output power for the transmitter chip is −19 dBm [36], it is considered to be the input power of the transmitter antenna P_t, and the loss for propagating through the tissue is included in the Tx antenna gain G_t of −32 dBi as in the EM simulation. For the receiver antenna, supposing it is well matched

Table 3.2 Parameters of the link budget.

Operating frequency	915 MHz
Tx power P_t	−19 dBm
Tx antenna gain G_t	−32 dBi
Rx antenna polarization	CP
Rx antenna gain G_r	2 dBi
Receiver noise floor *RNF*	−101 dBm
SNR (BER = 1E−5)	14 dB

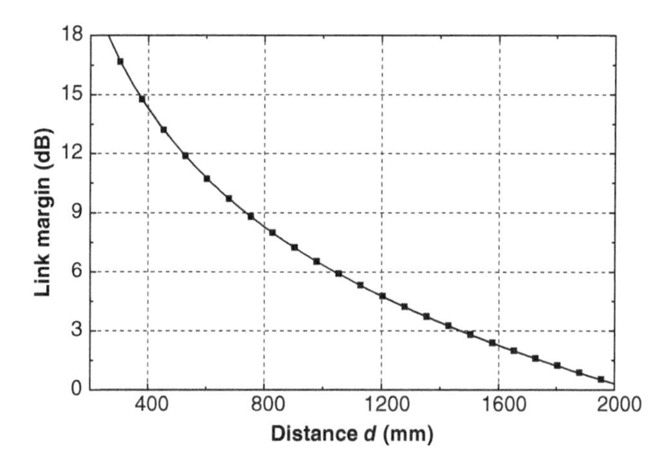

Figure 3.31 Variation of link margin with respect to the distance *d* between two CP antennas.

to 50 Ω with CP property, the impedance mismatch and polarization mismatch losses can be approximated as 0 dB. It should be noted that when the receiver antenna has linear polarization property, the polarization mismatch loss is about 3 dB.

According to (3–7), the LM varying with distance *d* can be calculated as shown in Figure 3.31. The LM should be better than 0 dB in order to realize wireless communication. From Figure 3.31, it can be seen that when the distance between two antennas is within 2 m, up-link communication can be established in this case.

3.3.2 Ground Radiation CP-implantable Antenna

Ground radiation antennas, which utilize the ground plane as the primary radiator, are well-suited for miniaturized implants due to their low profile and single metal layer construction. In the design of ground radiation antennas, reactive components are often incorporated for impedance matching and frequency tuning purposes. Taking inspiration from these concepts, we propose a CP ground radiation antenna that exhibits both a low profile and a wide AR bandwidth for implant applications. The antenna is designed to operate within the ISM band (2.4–2.48 GHz) and achieves circularly polarization through an asymmetric structure relative to its coplanar waveguide (CPW) feed. The tuning of the antenna is achieved by incorporating two capacitances, and their effects and tuning mechanisms are studied by varying their capacitance values [37].

In this design, simulations are conducted using a single-layer human tissue model that mimics the electrical properties of skin at 2.4 GHz [21]. The model has a relative dielectric constant of 37.88 and a conductivity of 1.44 S/m. The dimensions of the tissue box used in the simulations are $120 \times 120 \times 75\ \text{mm}^3$. The proposed implantable antenna is positioned at the center of the tissue box, 4 mm below the top surface and 71 mm away from the bottom surface.

For the design of the CP ground radiation antenna, a 0.508-mm-thick Taconic substrate is utilized, which has a permittivity of 2.95 and a dielectric loss tangent of 0.0028. The antenna configuration is square-shaped with dimensions of $10.4 \times 10.4 \times 0.508\ \text{mm}^3$.

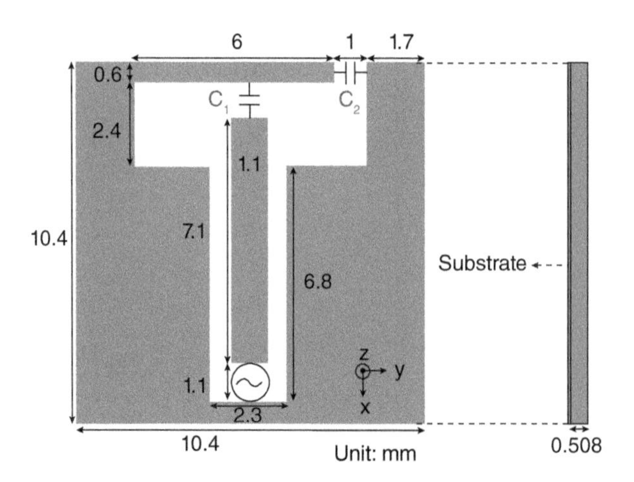

Figure 3.32 Configuration of the proposed CP-implantable antenna (top and side views).

A small clearance of 2.4 mm × 7 mm (approximately $0.024\lambda \times 0.057\lambda$ at 2.45 GHz) is incorporated within the ground plane to provide flexibility for the realization of a multifunctional platform for high-integration systems. The CPW feeding structure has a gap width of 0.6 mm, which is adjusted for impedance-matching purposes. Two capacitors, C1 and C2, are positioned in different regions of the antenna to achieve the desired resonant frequency and CP characteristics (Figure 3.32).

The proposed CP antenna is derived from a conventional linearly polarized ground radiation antenna, as illustrated in Figure 3.33(a). By introducing a symmetrical structure with respect to the CPW feed and adding a capacitor C1 between the CPW line and the upper strip, two in-phase current paths are formed. When analyzing the current distributions in this configuration, it can be observed that the horizontal currents along the y-direction on both sides of the antenna cancel each other out. However, the parallel vertical currents along the x-direction generate radiations with a polarization direction aligned with the x-axis.

By introducing an additional capacitor C2 at the right side of the upper strip, the antenna structure becomes asymmetrical, causing the currents on the two paths to be out of phase. Upon analyzing the current distributions of the modified structure at different points in time within one period, it is observed that similar current distributions, as shown in Figure 3.33(b), can still be established, resembling the original configuration in Figure 3.33(a). Additionally, at another point in time within the same period, current distributions as shown in Figure 3.33(c) are observed, where the currents on both paths are oriented along the y-axis, resulting in linear polarization along the y-axis. These current distributions demonstrate that the modified structure is a viable option for CP antenna designs.

It is worth noting that the proposed antenna can be accurately represented by an equivalent circuit, as depicted in Figure 3.34(a). The reflection coefficients obtained from HFSS simulations and the equivalent circuit models are compared in Figure 3.34(a) and demonstrate a close resemblance. This structure exhibits characteristics similar to a shunt RLC resonator centered around its resonant frequency. To validate this behavior, the input impedance of the structure is simulated and compared with that of a shunt

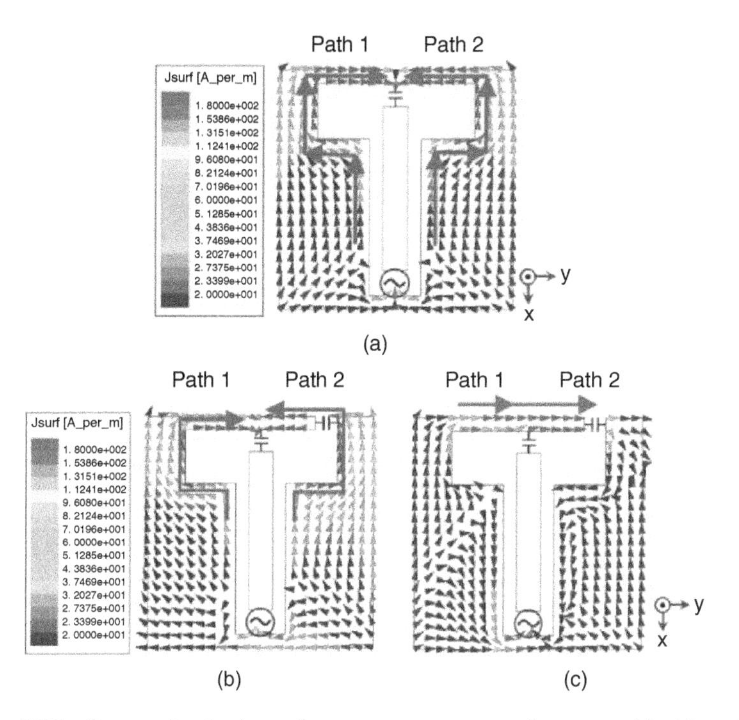

Figure 3.33 Current distributions of antennas at resonance frequency: (a) without C2, (b) and (c) with C2.

RLC resonator, as shown in Figure 3.34(b). The values of the elements in the shunt RLC resonator are determined using the following extraction method:

$$C = \frac{1}{4\pi}\frac{d(Im(Y_{\text{in}}))}{df}\bigg|_{f=f_0} \tag{3.8}$$

$$L = \frac{1}{(2\pi f_0)^2 C} \tag{3.9}$$

$$R = Z_{\text{in}}(f_0) \tag{3.10}$$

It can be observed from Figure 3.34(a) that within the required bandwidth (2.4–2.48 GHz), the structure in Figure 3.32 indeed acts like a shunt resonator.

To determine the optimal combination of C1 and C2 for achieving the best CP performance, a thorough investigation is conducted. Table 3.3 presents five combinations of C1 and C2 based on available capacitor values from Murata Electronics. All of these combinations enable the proposed antenna to resonate around 2.45 GHz. The table also includes the corresponding values of AR for each combination. It can be observed that the combination of C1 = 2.2 pF and C2 = 1.6 pF yields the lowest AR at the resonance and is selected for fabrication and measurement. From the simulated curves in Figure 3.35, it is evident that the antenna achieves an impedance bandwidth of 445 MHz ($|S_{11}|<-10$ dB), ranging from 2.230 to 2.675 GHz. Additionally, an AR bandwidth (<3 dB) of 251 MHz (approximately 10.2%) is obtained at the main radiation direction (theta = 0°).

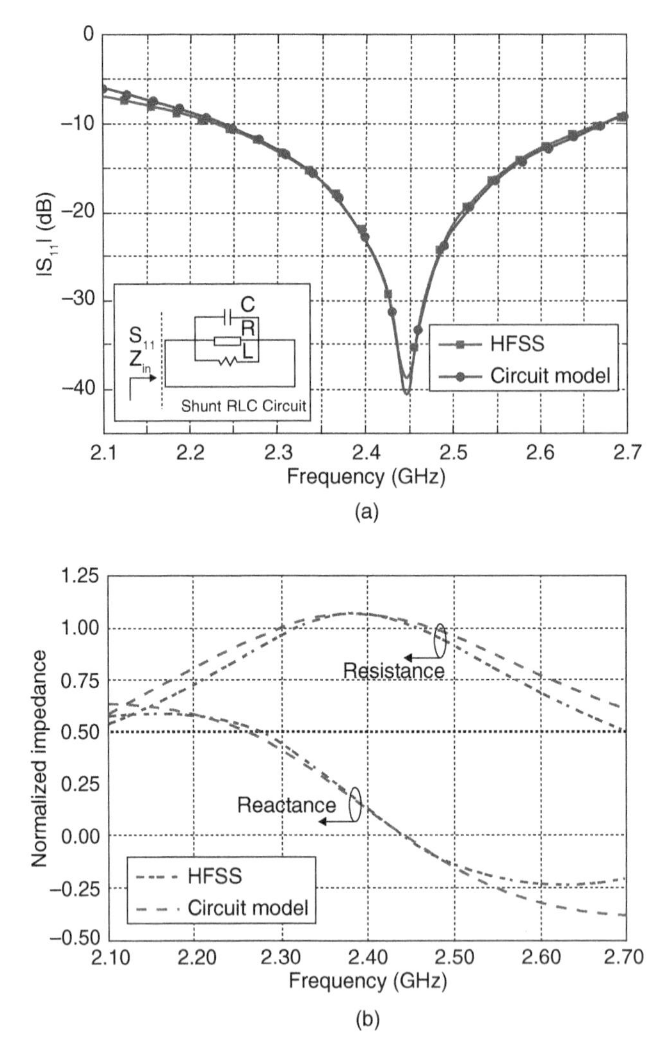

Figure 3.34 (a) Simulated $|S_{11}|$ results of the proposed CP ground radiation antenna (C1 = 2.2 pF and C2 = 1.6 pF), (b) simulated input impedance of the proposed CP ground radiation antenna (C1 = 2.2 pF and C2 = 1.6 pF).

Clearly, from the current distribution shown in Figure 3.33, we can conclude that C1 primarily contributes to frequency tuning, while the AR characteristic is mostly influenced by C2. In order to achieve optimal performance in different human tissue environments, it is necessary to conduct a parametric study of the values of C1 and C2.

The fabricated prototype of the CP ground radiation antenna is depicted in Figure 3.36(a). To validate its performance, in vitro measurements were conducted using a solid phantom. The antenna was placed inside a rectangular polystyrene container measuring $120 \times 120 \times 75\,\text{mm}^3$, filled with a skin-mimicking phantom consisting of 53% sugar and 47% deionized water by weight, as described in [34]. To solidify the phantom, 1 g of dry agarose was added per 100 mL solution with a heating procedure. It is important to note

Table 3.3 Resonant frequency and AR performance of C1 and C2 combinations.

C1 and C2 (pF)	1.8 and 2	2 and 1.8	2.2 and 1.6	2.4 and 1.5	2.7 and 1.2
Resonant frequency (GHz)	2.46	2.45	2.45	2.43	2.46
Axial ratio (dB)	3.25	1.88	0.74	1.02	3.85

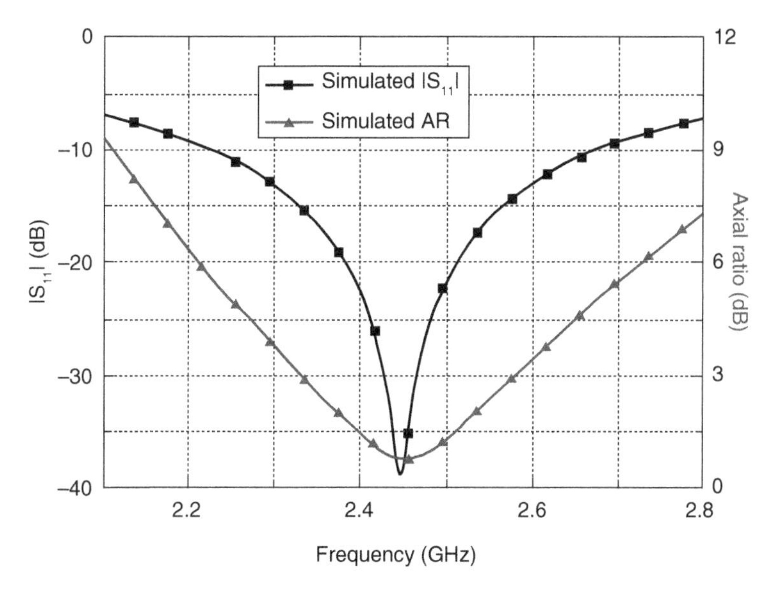

Figure 3.35 Simulated reflection coefficients and AR (at main radiation direction) of the proposed ground radiation antenna in a single-layer tissue model.

that this skin-mimicking phantom has a higher conductivity than real human skin [34]. As shown in Figure 3.36(b), the proposed antenna was positioned 4 mm below the top surface of the phantom to maintain consistency with the simulation setup.

The reflection coefficients of the proposed antenna were measured using a SubMiniature version A connector with a 125-mm-long cable. To minimize the potential effects of the cable on small antennas, sellotape was used to surround the cable. This implementation of sellotape mimics the function of a vacuum shell, as recommended in [38], to prevent direct contact between the outer conductor of the cable and the lossy body phantom. For verification, the case of the proposed antenna excited by a long cable was also simulated. In the simulation setup, a vacuum shell with a thickness of 0.1 mm was modeled around the cable. As shown in Figure 3.37, there is no significant difference between the cases with and without the cable. Figure 3.37 also illustrates the measured and simulated $|S_{11}|$. The measured center frequency of 2.440 GHz closely matches the simulated value of 2.447 GHz. The achieved bandwidth for $|S_{11}|<-10$ dB ranges from 2.150 to 2.771 GHz, which is broader than the simulated bandwidth. It should be noted that the impedance bandwidth is widened in the measurement due to the higher conductivity of the actual skin-mimicking material, which decreases the quality factor of the proposed antenna compared to the value used

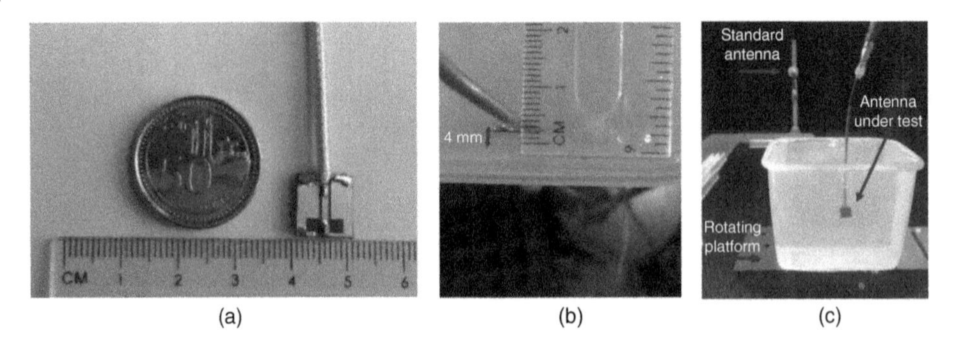

Figure 3.36 (a) Fabricated CP ground radiation antenna, (b) location of the proposed CP antenna inside the solid phantom, and (c) far-field measurement setup.

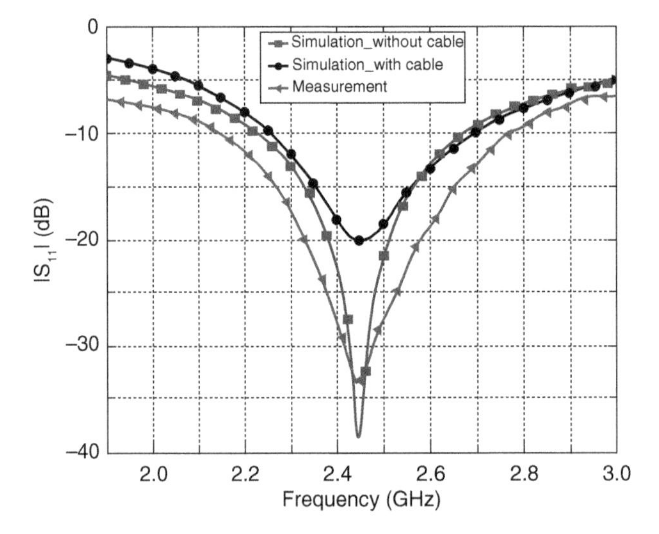

Figure 3.37 Simulation and measurement result comparison of reflection coefficients.

in the simulation. The discrepancy between the measured and simulated $|S_{11}|$ results can also be attributed to fabrication tolerances, soldering procedures, and other factors.

The far-field measurement was conducted inside an anechoic chamber using an orbit measurement system, as depicted in Figure 3.36(c). The absolute power gain of the antenna was measured using the conventional gain comparison method with a standard LHCP spiral antenna. A maximum measured gain value of −21.1 dBic was obtained at 2.45 GHz, which is similar to the simulated value of −19.8 dBic.

Both the simulated and measured radiation patterns, shown in Figure 3.38, demonstrate that the proposed antenna exhibits acceptable backward radiation (within $-180° < \theta < -90°$ and $90° < \theta < 180°$), which is at least 10 dB lower than the radiation toward its maximum radiation direction. Additionally, a good front-to-back ratio of approximately 30 dB is achieved. Therefore, the proposed CP antenna ensures limited backward radiation when used in biomedical applications.

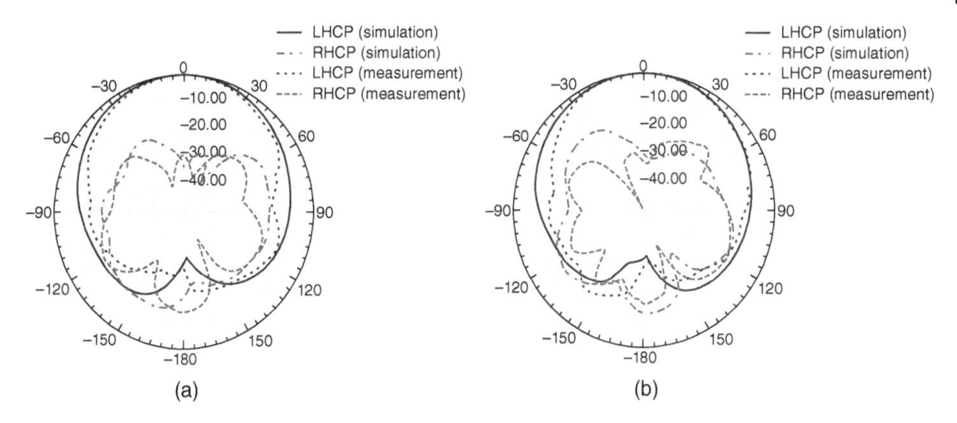

Figure 3.38 Simulated and measured radiation patterns at 2.45 GHz: (a) Phi = 0°, (b) Phi = 90°.

3.4 Application Base Design of CP-implantable Antenna – Capsule Endoscopy

In the previous section, we discussed various types of planar antennas with CP radiation for implantable applications. However, when it comes to practical applications, implants or implantable medical devices come in various shapes due to application limitations. Designing CP-implantable antennas that can seamlessly integrate with these devices is essential.

With the rapid advancement of wireless technology, there is a growing number of clinically employed implantable or ingestible devices, including pacemakers, implantable defibrillators, capsule endoscopes, and implanted therapeutic devices. Among these devices, a capsule endoscopy plays a significant role in wireless communication links from inside the body to external devices.

A capsule endoscopy is a device designed to enable doctors to visualize the most inaccessible parts of the gastrointestinal (GI) tract. The capsule itself is typically sized at 26 mm × 11 mm and consists of an optical dome, lens, light emitting diodes, complementary metal-oxide-semiconductor (CMOS) imager, transmitter, and antenna, as shown in Figure 3.39 [39]. Images captured by the capsule camera are recorded and transmitted using the capsule antenna. The patient wears a data recorder to capture the transmitted images. After the procedure, the images are downloaded to a computer, where they are reviewed and interpreted by a gastroenterologist with specialized training.

Considering the varied positions and postures of capsule antennas within the human body, the utilization of CP antennas is an excellent solution to ensure a reliable wireless communication link.

3.4.1 Axial-mode Multilayer Helical Antenna

The helical antenna, known for its ability to generate CP waves, presents itself as a promising candidate for CP capsule antenna design. In a study conducted at the wireless

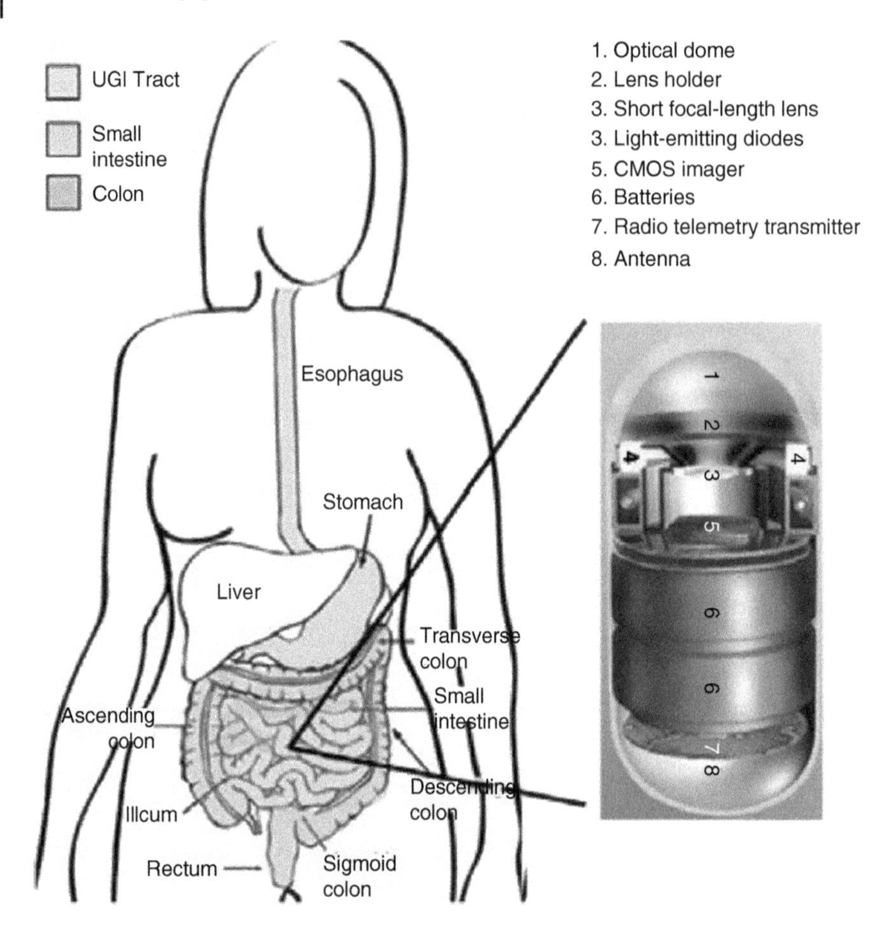

Figure 3.39 Complete digestive track that the biotelemetric capsule would be able to map and the typical components found within an imaging capsule system. Source: Izdebski et al. [39]/IEEE.

medical telemetry service (WMTS) band [40], an inverted conical helical antenna was proposed. The detailed structure of this antenna can be observed in Figure 3.40.

Typically, a capsule system is characterized by a compact size of $26 \times 11\,\mathrm{mm}^2$. The inverted conical helical antenna employed in this study features a wire radius of 0.2 mm and a pitch of 1 mm. The radius of the first turn is set at 0.75 mm, incrementing by 1 mm with each subsequent turn. The ground plane radius is slightly smaller than the radius of the capsule, measuring 4.5 mm. Additionally, an additional metallic plate with a radius of 4 mm is positioned above the ground plane for tuning purposes. The gap between the two plates and the radius of the upper metallic plate are utilized to control the frequency and achieve impedance matching. The overall length of the helical antenna arm is 33.11 mm. It is worth noting that the helical antenna operates in the normal mode with linear polarization, but to achieve circularly polarization, it should be operated in the axial mode.

However, the installation of the proposed helical antenna as detailed in [40] may encounter difficulties in achieving stable performance. To solve this issue, a subsequent work [41] introduced a multilayer helical antenna design. This updated design not only

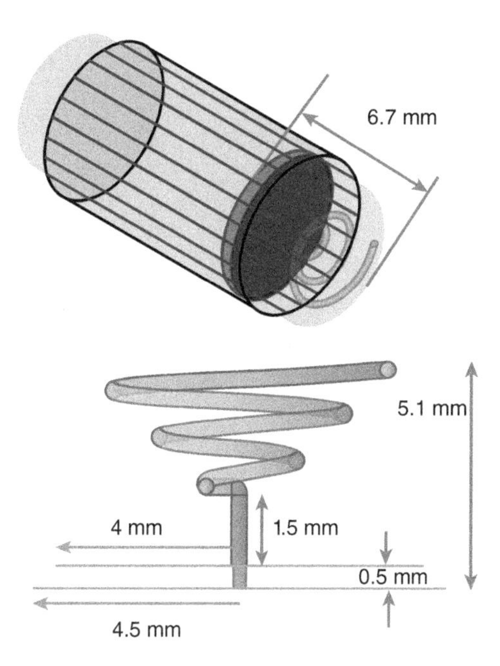

Figure 3.40 Normal-mode inverted conical helical antenna. Source: Rajagopalan et al. [40]/IEEE.

ensures easier installation but also helps in achieving broadband circularly polarization, making it more viable for practical applications.

3.4.1.1 Antenna Structure

To develop a compact and straightforward CP antenna for ingestible wireless capsule endoscope systems, a multilayer helical antenna is presented, as illustrated in Figure 3.41. The proposed antenna comprises three distinct open loops situated on different layers of Rogers RO3010, which has a dielectric constant (ε_r) of 10.2, each layer thickness of 50 mil, and a loss tangent ($\tan\delta$) of 0.0022. The helical antenna is fed via a coaxial cable, a method that simplifies the integration of a transceiver under the ground plane and minimally impacts the antenna's radiation performance. Figure 3.41(a) shows the top and side views of the proposed antenna. The substantial thickness of the substrate increases the angle of each loop, subsequently reducing the metal layers needed to maintain its circularly polarization, predominantly in the main lobe. The specific configuration of the three metal layers is shown in Figure 3.41(b). M1 has a sweep angle of α, and M2 and M3 have a sweep angle of β. The inner and outer radii of the open loops are r1 and r2, respectively. Each loop features two circular catch pads with a radius r3 = 0.5 mm, utilized for connecting the metal vias between loops. The substrate's radius is chosen as 5.5 mm, aligning with the typical dimensions of commercial capsules, approximately 11 mm by 26 mm.

To facilitate initial optimization and demonstrate the design concept, we utilize a one-layer muscle phantom model as shown in Figure 3.42. Furthermore, this one-layer muscle phantom will be employed for measurements, and the acquired results will be compared with the simulated ones to evaluate the design concept. The muscle model measures $100 \times 100 \times 100\,\text{mm}^3$. The electrical properties of the muscle (ε_r = 52.79,

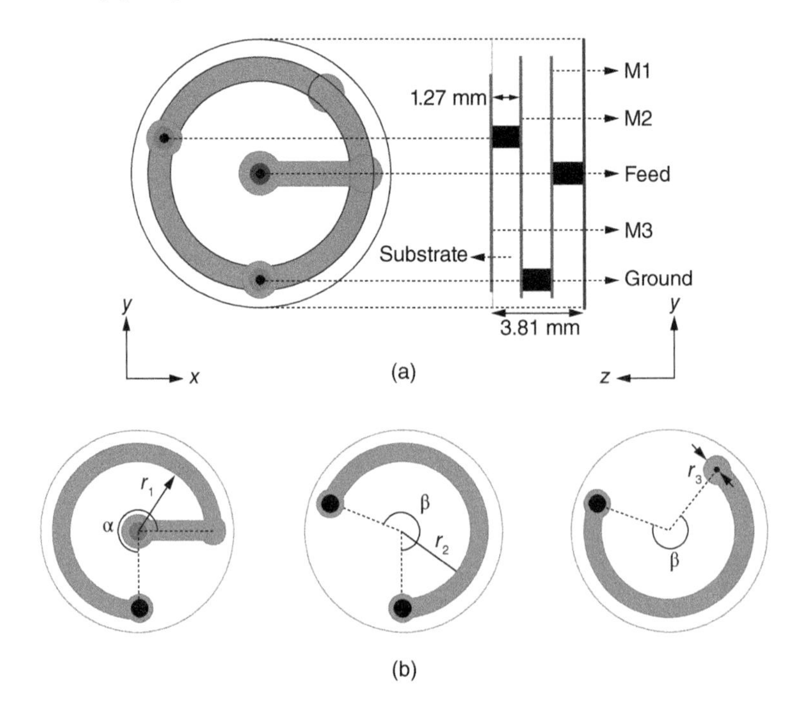

Figure 3.41 Configuration of the helical antenna: (a) top view and side of the proposed antenna, (b) top view of three metal layers.

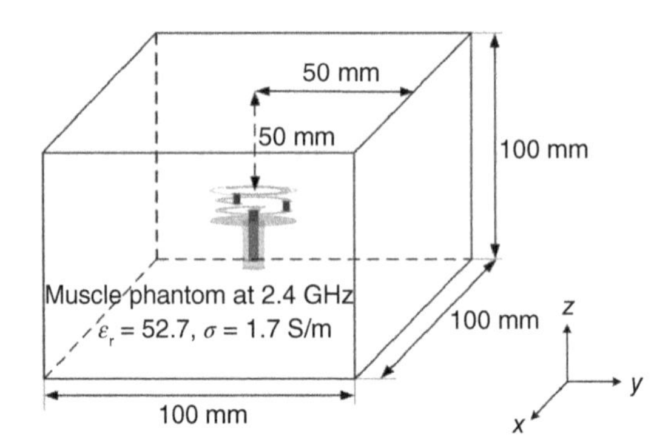

Figure 3.42 Numerical calculation model using muscle phantom.

$\sigma = 1.705$ S/m [21]) at 2.4 GHz are used in the one-layer muscle simulation model. The helical antenna is positioned at the phantom's center, with the distance between the antenna's top and the phantom's top being 50 mm.

Figure 3.43 shows the simulated reflection coefficient and AR at the main radiation direction (theta = 0°) of the proposed antenna. The simulated impedance bandwidth can cover around 1.85–2.8 GHz ($|S_{11}| < -10$ dB). And the AR bandwidth is from 2.0 GHz to 2.8 GHz (AR <3 dB). The determined geometry parameters are: $r_1 = 3.8$ mm, $r_2 = 4.3$ mm, $r_3 = 0.5$ mm, $\alpha = 270°$, $\beta = 250°$. The radii of the connecting via and inner conductor of

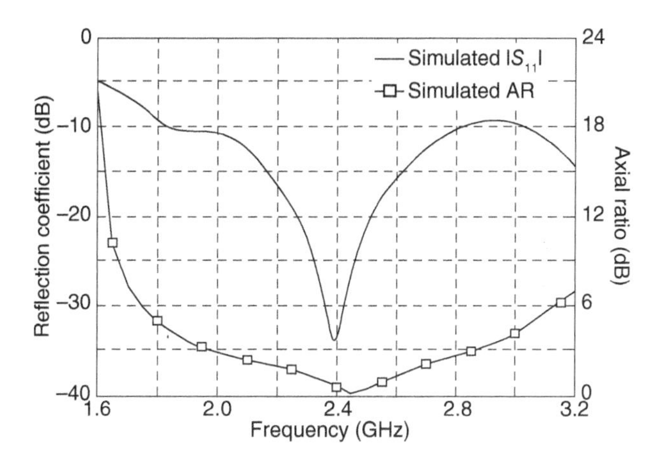

Figure 3.43 Simulated reflection coefficient and AR (at main radiation direction: theta = 0°) of the proposed capsule antenna in the one-layer muscle phantom.

the coaxial cable are set as 0.45 mm and 0.65 mm, respectively. The peak realized gain is ~−32 dBic at 2.4 GHz. The primary polarization is RHCP with XPD ~27 dB at main radiation direction.

3.4.1.2 Conformal Capsule Antenna Design Including Biocompatibility Shell Consideration

The initial multilayer helical antenna was designed with a cylindrical structure, making it highly suitable for a conformal capsule antenna. However, the biocompatibility of the antenna is a crucial consideration since it needs to be swallowed and pass through the human body. To address this, various biocompatible materials can be used for coating the capsule antenna:

(1) PEEK (Polyetheretherketones, $\varepsilon_r = 3.2$, $\tan\delta = 0.01$) [27]
(2) Silastic MDX4-4210 Biomedical-Grade Base Elastomer ($\varepsilon_r = 3.3$, $\tan\delta = 0.01$) [29]
(3) Polyethylene ($\varepsilon_r = 2.25$, $\tan\delta = 0.001$) [39]

PEEK has been selected as the capsule shell and biocompatible coating material. According to studies on biocompatible coating methods [30, 31], the thickness of the biocompatible coating material plays a significant role in determining the resonant frequency of the implantable antenna. Additionally, the effective dielectric constant can also influence the resonant frequency. In this design, the thickness of the PEEK coating is set to be 0.05 mm.

In this context, two conformal capsule antenna design cases are presented in Figure 3.44. Case 1 involves the initial multilayer helical antenna with a biocompatible material coating, resulting in a conformal structure with a height of 3.86 mm. On the other hand, Case 2 includes the initial helical antenna with an added split sphere superstrate and biocompatible coating, which offers better conformality for a wireless capsule system with a height of 6.77 mm, significantly higher than that of Case 1.

Figure 3.45 illustrates the comparison results between the two cases for the conformal capsule antenna design. In Case 1, the resonant frequency for achieving an AR of less than three shifts up substantially, while the resonant frequency for the reflection coefficient

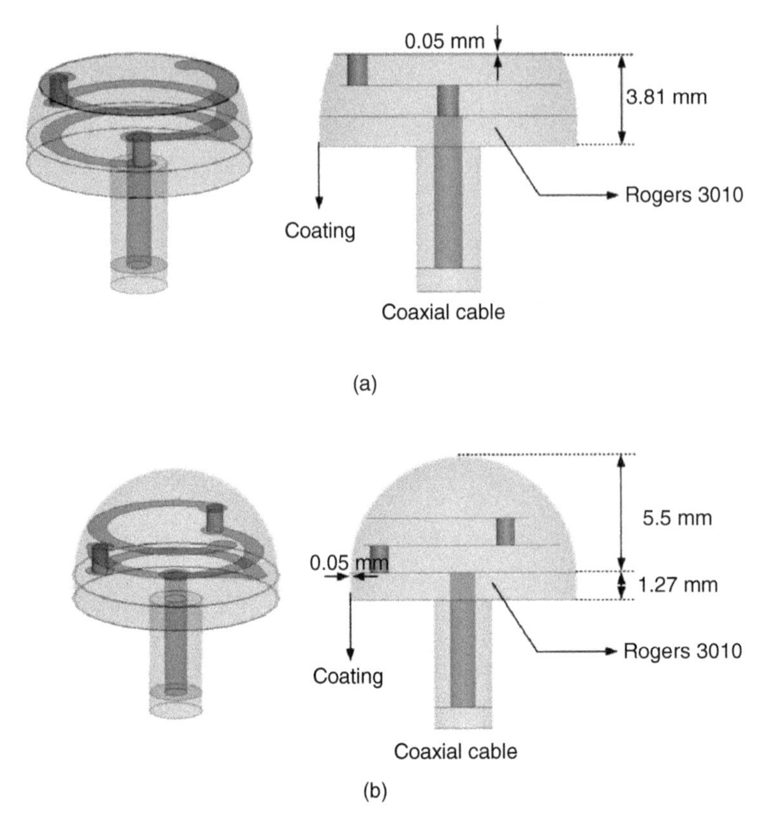

Figure 3.44 Two cases for the conformal capsule antenna design: (a) case 1, (b) case 2.

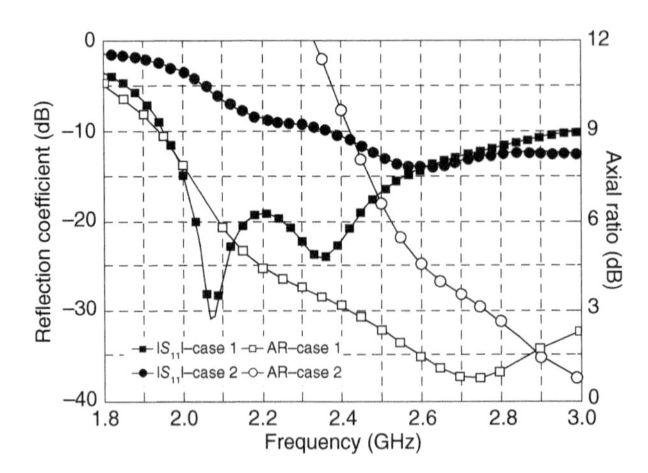

Figure 3.45 Simulated results of two cases for the conformal capsule antennas embedded in the one-layer muscle tissue in Figure 3.42.

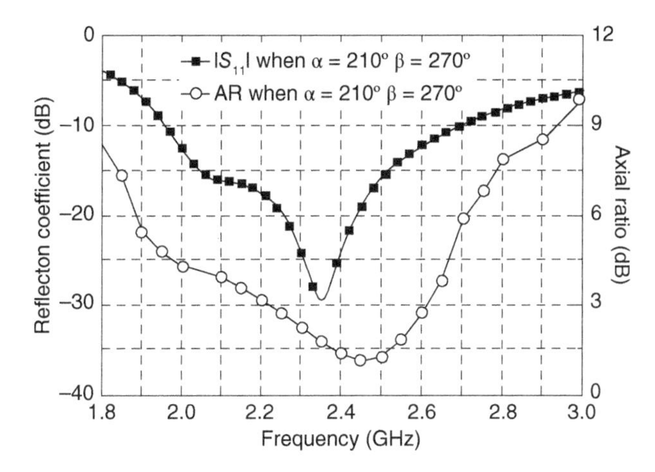

Figure 3.46 Optimized results of case 1 in the muscle tissue, when $\alpha = 210^{\circ}$, $\beta = 270^{\circ}$.

below −10 dB remains almost unchanged. For Case 2, the corresponding frequencies for the reflection coefficient and AR shift to 2.6 GHz and 3.0 GHz, respectively. This outcome is reasonable as the dielectric constant of the biocompatible coating ($\varepsilon_r = 3.2$) is much smaller than that of the antenna substrate ($\varepsilon_r = 10.2$). Therefore, the thickness of the biocompatible coating should be taken into consideration for optimization purposes.

The significant frequency shift observed in case 2 may pose challenges in optimizing the antenna to achieve the desired resonant frequency, as it could lead to the extension of the helical antenna's radius beyond the radius of the capsule system. Considering the antenna's profile and the complexity of optimization, case 1 is selected for the wireless capsule endoscopy (WCE) application. It is worth noting that the edges of case 1 can be redesigned to be less sharp, as reported in [27]. Through parametric studies of the multilayer helical antenna, the resonant frequency of the reflection coefficient and AR can be optimized to approximately 2.45 GHz.

Figure 3.46 depicts the optimized reflection coefficient and AR results when $\alpha = 210°$, $\beta = 270°$. Compared with Figure 3.45, the resonant frequency of the conformal helical antenna is optimized to the desired frequency. The antenna can cover the entire 2.4 GHz ISM-Band with good impedance matching and CP purity.

3.4.1.3 Wireless Capsule Endoscope System in a Human Body

After addressing the biocompatibility issue of the proposed antenna, the next step involves integrating the antenna with other electrical components to create a functional wireless capsule endoscope system. It is important to consider the effects of these additional components on the overall performance of the system. In the previous design, a coaxial cable was used, which comes into direct contact with the muscle, resulting in a small coupling between the currents flowing on the external metal of the cable and the muscle phantom.

Figure 3.47 illustrates the wireless capsule endoscope system, taking into account the presence of electrical components and the coaxial cable. A very short coaxial cable is utilized as a feeding cable, with the feeding port located at the top of the transceiver. For simplicity,

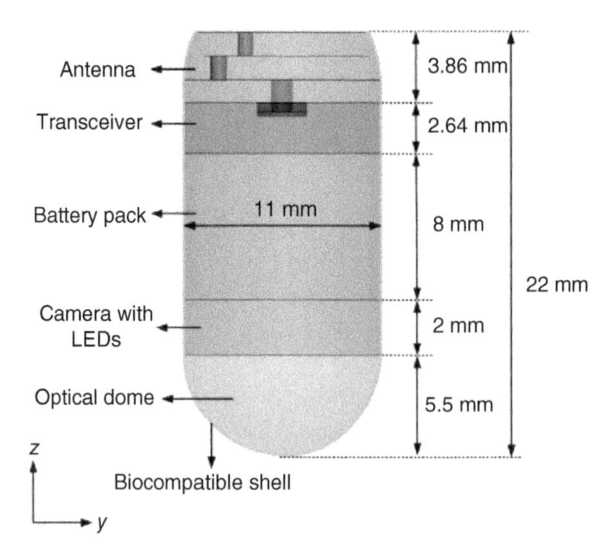

Figure 3.47 The conceptual application of the helical antenna in a wireless capsule endoscope system.

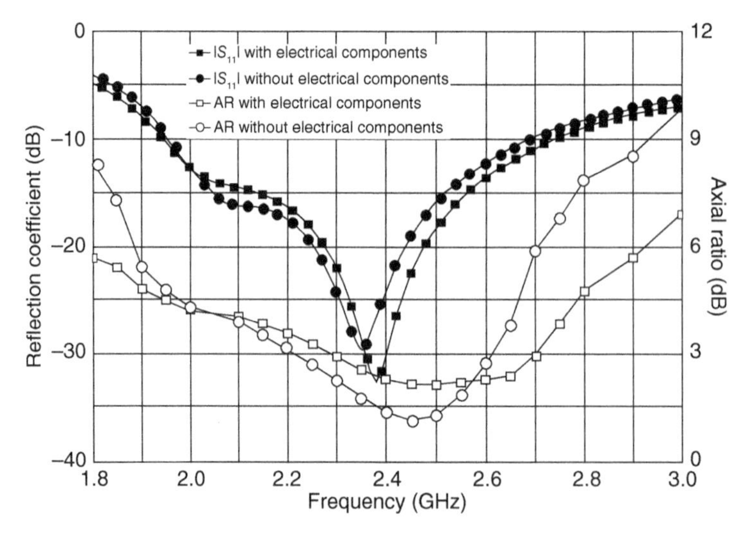

Figure 3.48 Performance comparisons of the proposed antenna with/without electrical components.

we assume the transceiver, battery pack, and other electrical components are made of a PEC. The optical dome is filled with vacuum to optimize the system's performance.

Figure 3.48 shows the performance comparisons of the proposed antenna with/without electrical components. The resonant frequency would have a very small shift and the capsule antenna in the wireless capsule endoscope system can maintain its operating frequency. The common bandwidths can cover 2.3–2.7 GHz for $|S_{11}| < -10$ dB and AR < 3 dB. The results demonstrate that the proposed antenna is very suitable for potential

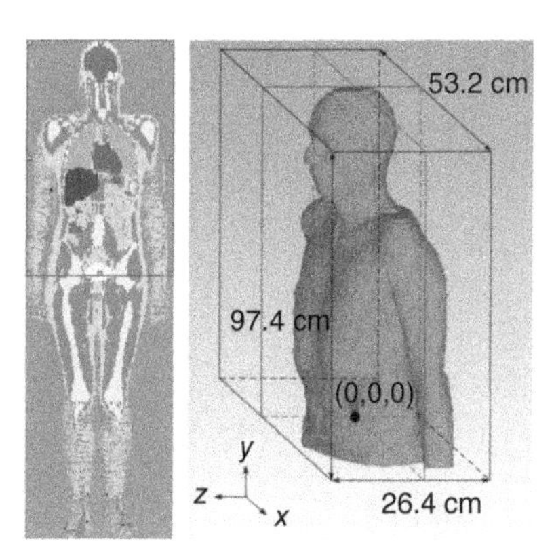

Figure 3.49 Three-dimensional Voxel Gustav human body used for the capsule antenna design.

high-data-rate communication. The peak realized gain is ~−31 dBic at 2.4 GHz and the XPD is ~14.6 dB. A little gain enhancement was caused by the PEC material at the bottom of the ground plane. The results mean that the electrical components have a slight effect on the antenna's performance and the antenna can be further re-optimized with better performance.

In the previous study, the design and optimization of the proposed helical antenna and integrated capsule antenna were carried out using a one-layer muscle phantom. The use of this phantom allowed for efficient optimization and measurement, and the measured results were then compared with the simulated results to evaluate the design's performance. However, it is important to note that the human body is composed of various organs and tissues, and evaluating the proposed antenna in a more realistic human body model would provide more accurate results.

To study the proposed antenna design in a more realistic environment, a three-dimensional Gustav voxel human body model (Figure 3.49) is utilized. Numerical analyses are conducted using the CST Microwave Suite [24]. Three different implant positions in the human body, namely the stomach, small intestine, and colon, are considered for evaluating the performance of the capsule antenna (Figure 3.50). The capsule system, including the capsule antenna, is embedded in these three organs. The coordinate system's origin is shown in Figure 3.49, where the capsule antenna is positioned along the $+z$ axis.

The specific implant positions in the stomach, small intestine, and colon are located at (60 mm, 360 mm, 50 mm), (−38 mm, 263 mm, 50 mm), and (20 mm, 290 mm, 70 mm), respectively. As the implant depths vary with different organs, the simulated reflection coefficients of the optimized capsule antenna, with electrical components, are analyzed in different simulation environments – the simplified muscle phantom (Figure 3.42) and the human body model with the three implant positions (Figure 3.50).

Figure 3.51 shows the simulation results, revealing that the proposed capsule antenna covers the entire 2.4–2.48 GHz ISM-band in both simulation models. Notably, the reflection coefficient of the helical antenna is not significantly affected by the different implant positions. These results also demonstrate the accuracy and validity of using the simplified muscle phantom model (Figure 3.52) for capsule antenna design, especially in terms of evaluating the reflection coefficient performance.

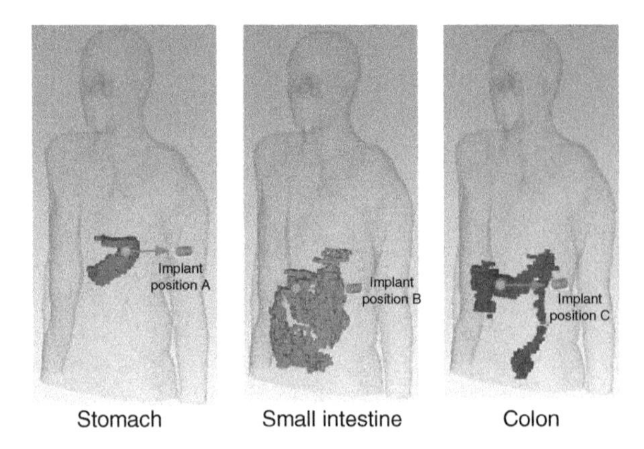

Figure 3.50 Three implant positions in the human body for the capsule antenna design.

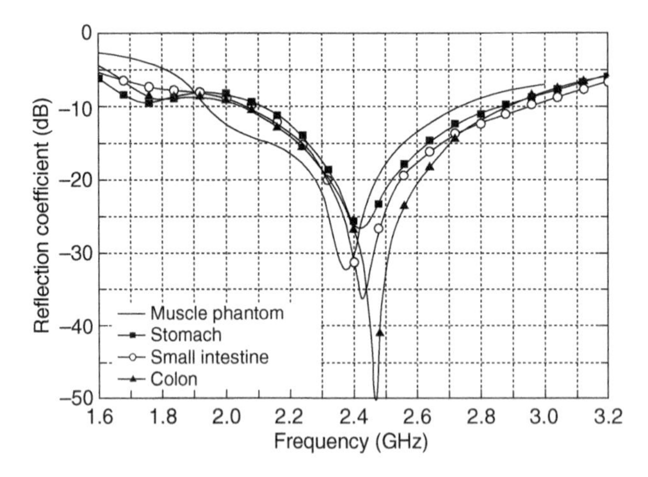

Figure 3.51 Simulated reflection coefficients of the capsule antenna with electric components in different simulation environments: simplified phantom in Figure 3.42 and CST human body with three different implant positions in Figure 3.50.

Figure 3.52 shows the simulated radiation patterns and AR of the antenna in the human body at yoz-plane. The main radiation directions with best AR performance of the capsule antenna in the three positions are −62°, −27°, and −18° in the yoz-plane, respectively. The radiation patterns in the small intestine and colon can achieve the 6-dB beam-width over 120°. Table 3.4 lists the detailed performance comparisons in three implant positions at 2.45 GHz. Figure 3.53 describes the simulated peak realized gain and AR values versus frequency at the main radiation directions. We can conclude that the peak realized gain in the small intestine and colon is much higher than that in the stomach. The reason is that the implant depth in the stomach is 20 mm deeper than that in the small intestine and colon. Meanwhile, the AR performance is affected by the different implant depths. The CP purity is good in the stomach but would be degenerated in the small intestine and colon. The AR is sensitive to the different implant depths and could be optimized in the desperate depth for good performance.

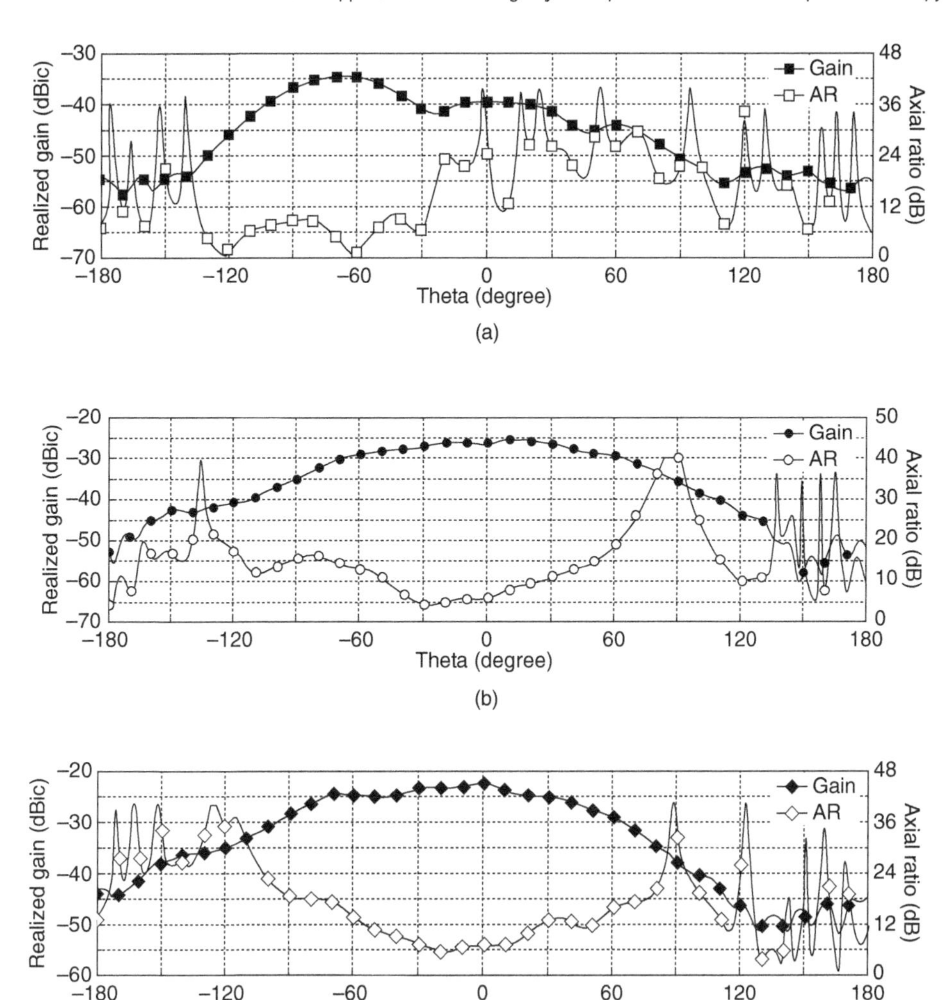

Figure 3.52 Simulated radiation patterns and AR of the capsule antenna in the human body with three different implant positions at 2.45 GHz: (a) the capsule antenna embedded in the stomach of Figure 3.50, (b) the capsule antenna embedded in the small intestine of Figure 3.50, and (c) the capsule antenna embedded in the colon of Figure 3.50.

In addition to the studies on reflection coefficients and radiation performance, safety considerations are crucial since the antenna will be swallowed inside the human body. Table 3.5 presents the specific absorption rate (SAR) distribution of the proposed antenna when embedded in the three different organs. The simulated maximum 1-g and 10-g average SAR values are as follows: 126.4 and 26.8 W/kg in the stomach, 180.7 and 35.7 W/kg in the small intestine, and 112.2 and 26.1 W/kg in the colon, respectively.

To comply with the SAR regulations [42, 43], the allowed transmitter power values are 8.85 mW for the 1-g SAR limit and 56.0 mW for the 10-g SAR limit. As the output power of a transmitter for a capsule endoscope system is much lower than these allowed transmitter

Table 3.4 Performance comparisons in three implant positions at 2.45 GHz.

Implant position	Gain (dBic)	AR (dB)	XPD (dB)
Stomach	−34.6	0.26	36.3
Small intestine	−26.7	4.16	12.6
Colon	−23.4	6.1	9.8

Table 3.5 Parameters of the link budget simulated maximum SAR (input power = 1 W), and maximum power for satisfying the SAR standard in the human body at 2.45 GHz.

Implant position	Max 1 g-avg SAR (W/kg)	Max input power (mW)	Max 10 g-avg SAR (W/kg)	Max input power (mW)
Stomach	126.4	12.7	26.8	74.6
Small intestine	180.7	8.85	35.7	56.0
Colon	112.2	14.3	26.1	76.6

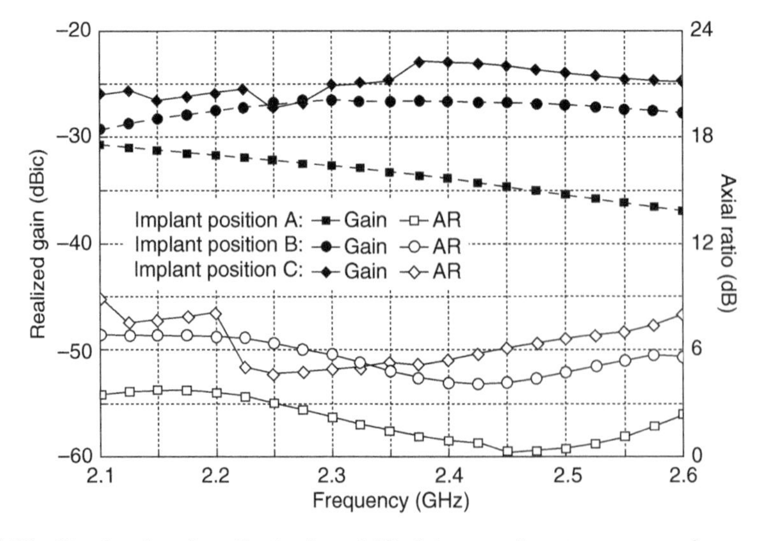

Figure 3.53 Simulated peak realized gain and AR of the capsule antenna versus frequency in the human body with three implant positions. The directions of AR values for three implant positions are −620, −270, and −180 in the yoz-plane of Figure 3.49, respectively.

power values, SAR should not be a concern in this scenario. The proposed capsule antenna's SAR values are within safe limits, ensuring its safe usage inside the human body.

3.4.1.4 In Vitro Testing and Discussions

To establish a data telemetry link with a capsule endoscope system, it is preferable to use a CP antenna for the base station to avoid polarization mismatching losses and ensure better communication quality. However, achieving an omnidirectional CP capsule antenna within different implant depths and various organs is challenging due to the varying directions of the capsule antenna in different positions inside the human body.

One approach to address this challenge is to use an omnidirectional CP receiver antenna. The receiver antenna should meet requirements such as having an omnidirectional radiation pattern, low-profile design, wide bandwidth, and good radiation efficiency.

One method to achieve an omnidirectional CP antenna is to use a conformal CP antenna array with a cylindrical carrier. While this method can be effective, it often comes with high costs and complicated fabrication processes. Another approach is to combine two wide-beam CP patch antennas to achieve an omnidirectional CP antenna. However, this method may result in an unstable radiation pattern.

A different approach, as proposed in [44], involves using the Zero-Order Resonance (ZOR) mode of epsilon-negative (ENG) transmission lines to achieve a low-profile omnidirectional CP antenna. Nevertheless, this method has a narrow AR bandwidth, which might not cover the entire 2.4 GHz ISM-band.

An alternative and simple way to design an omnidirectional CP antenna with a stable azimuthal plane radiation pattern is to combine vertical and horizontal polarization radiators, as discussed in [45, 46]. This method can provide an effective solution to achieve the desired omnidirectional CP performance for the receiver antenna.

Figure 3.54 illustrates the configuration of the omnidirectional CP antenna. The antenna comprises a top-loaded monopole for omnidirectional vertical polarization radiation and four arc-shaped dipoles for horizontal polarization radiation. The ground plane and bottom substrate have radii of 23 mm and 30 mm, respectively, while the top-loaded circular patch and top substrate have radii of 10 mm and 15 mm, respectively. Rogers RO4003 is used as the material for the antenna design, with a thickness of 32 mils for both substrates. The antenna is fed by a coaxial cable that directly connects the dipoles and the top-loaded patch. The final geometry parameters are marked in Figure 3.54. The main radiation of the antenna in the azimuthal plane is RHCP. The simulated average gain in the azimuthal plane exceeds 1.0 dBic across the entire 2.4-GHz ISM-band. The simulated results demonstrate that the proposed antenna achieves good omnidirectional CP performance with a low-profile, simple structure, and wide bandwidth, making it suitable for a base station.

Figure 3.55 showcases the fabricated multilayer helical antenna, which was measured using a homogeneous mixture of liquid muscle phantom. The recipe for the phantom

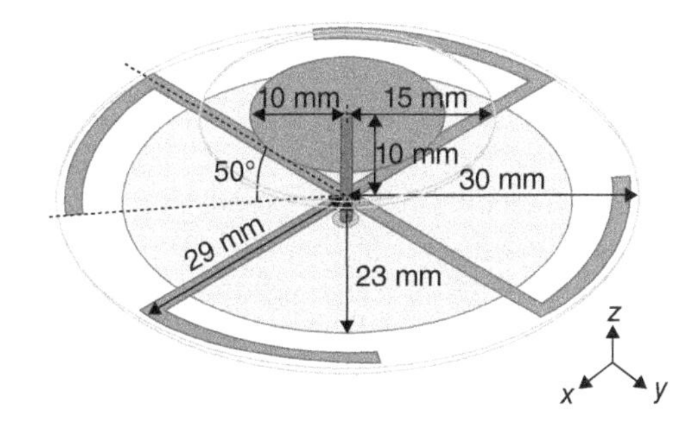

Figure 3.54 Configuration of the omnidirectional CP antenna.

Figure 3.55 Photograph of the fabricated multilayer helical antenna.

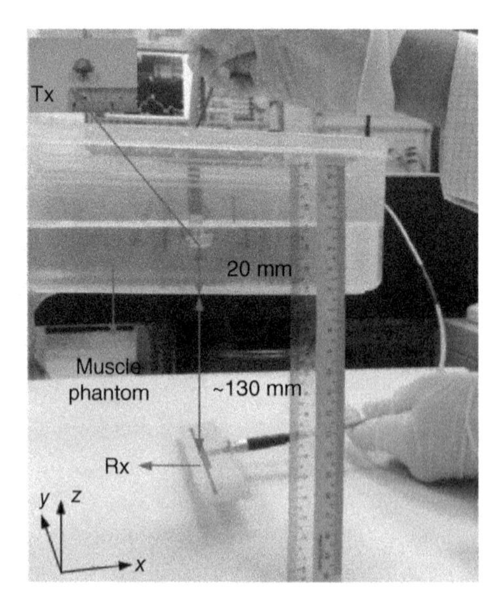

Figure 3.56 Illustration of the S-parameters measurement setup.

consists of 73.2% deionized water, 0.04% salt (NaCl), and 26.7% DGBE by weight. In-vitro measurements were conducted using a Rohde & Schwarz ZVA50 VNA, as depicted in Figure 3.56. This figure also illustrates the communication link between the transmit (Tx) and receive (Rx) antennas. For this study, a dipole antenna is employed as an external antenna to demonstrate the polarization property of the proposed antenna. The simulated and measured reflection coefficients of the proposed antenna in the one-layer muscle phantom are presented in Figure 3.57.

The dipole antenna has an impedance bandwidth of ~8.06% (2.38–2.58 GHz). The proposed helical antenna has a wide bandwidth over 2.3–3 GHz. A little frequency shift and mismatching of the helical antenna could be due to very thin air gap between layers as each layer of the helical antenna was fabricated separately and assembled by our fabrication laboratory. It can be overcome by introducing multilayer fabrication or assembly processes

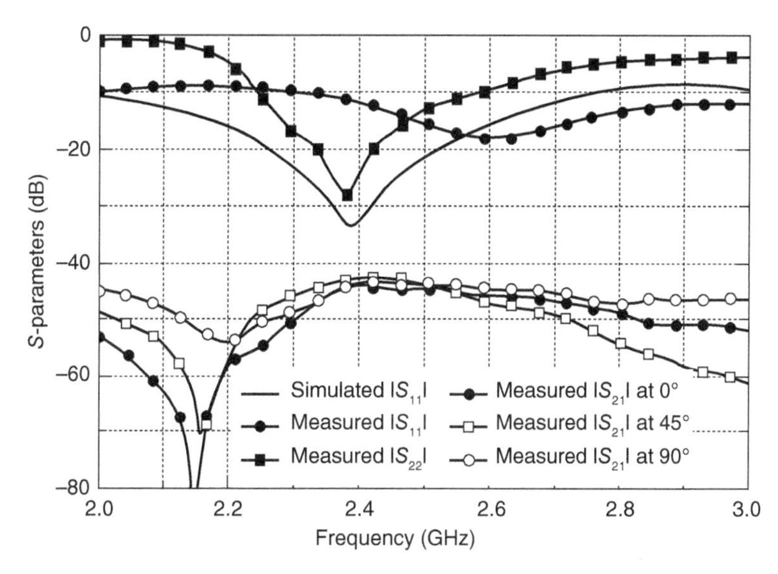

Figure 3.57 Simulated and measured S-parameters of the proposed antenna and an external dipole antenna when the dipole antenna is placed at phi = 0°, 45°, and 90° at xoy-plane.

such as multilayer Printed circuit boards (PCBs) or Low-temperature Co-fired Ceramics (LTCC). The communication link of the proposed antenna and the dipole antenna was measured when the dipole was placed at phi = 0°, 45°, and 90°, respectively. Note that the distance between the Tx and Rx is ~130 mm. The CP purity of the proposed antenna can be calculated by comparing the communication link levels for two orthogonal polarizations. We can get good polarization purity within a wide bandwidth ($|S_{21}|$ [phi = 0°] ≅ $|S_{21}|$ [phi = 90°]). It is good to see that the $|S_{21}|$ values would have no significant difference when the dipole antenna is placed at any degree in the xoz-plane. The measured results are well matched with the simulated ones as shown in Figure 3.43.

After demonstrating the CP purity using the LP dipole antenna, we proceeded to fabricate and measure the omnidirectional CP antenna to verify its communication link with the implantable helical antenna. Figure 3.58 shows the measured S-parameters of the helical antenna (Tx) and the omnidirectional CP antenna (Rx), along with a photograph of the omnidirectional CP antenna. The impedance bandwidth of the omnidirectional CP antenna ranges from 2.436 to 2.498 GHz for $|S_{11}| < -10$ dB. The implant depth of the helical antenna and the distance between the helical antenna and the omnidirectional CP antenna remained the same as shown in Figure 3.56. Additionally, the coupling strength remained consistent when we separately rotated the helical antenna and the omnidirectional CP antenna. The measured results confirm the CP purity of both CP antennas.

Comparing Figures 3.57 and 3.58, we observe that the $|S_{21}|$ values are similar at 2.45 GHz when using an external antenna, whether it is a linearly polarized dipole antenna or an omnidirectional CP antenna. However, the omnidirectional CP antenna exhibits less polarization loss. Several reasons contribute to this difference: First, the realized gain of the omnidirectional CP antenna (~1 dBic at 2.45 GHz) is smaller than that of the linearly polarized dipole antenna (~2.15 dBi at 2.45 GHz). Second, the frequency shift of the capsule

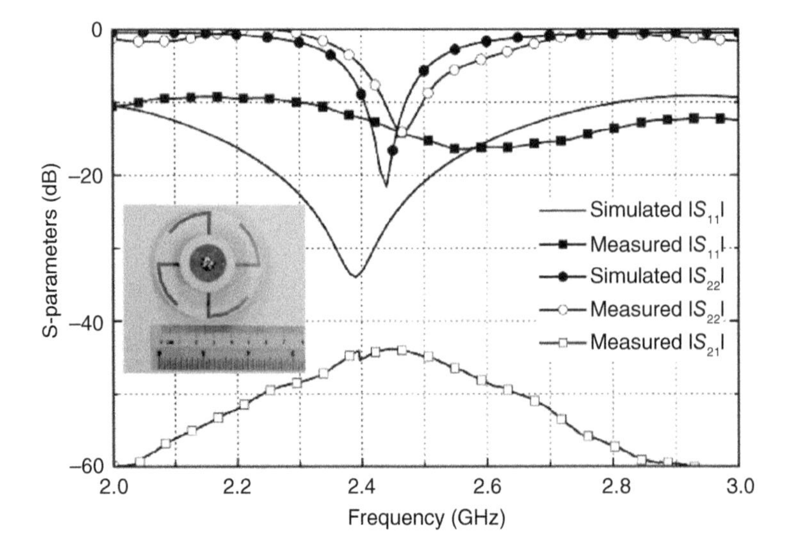

Figure 3.58 Simulated and measured S-parameters of the proposed antenna and an external omnidirectional CP antenna when the azimuthal plane of the external antenna is right against the main radiation direction of the helical antenna.

antenna and the omnidirectional CP antenna can cause a slight mismatching loss and polarization loss. Third, the gain versus frequency bandwidth of the omnidirectional CP antenna is narrower than that of the LP dipole antenna, which accentuates the effect of the frequency shift. Lastly, the misalignment of the implant and external antennas may introduce measurement errors.

3.4.2 Conformal CP Antenna for Wireless Capsule Endoscope Systems

According to the arrangement of the antenna within the capsule, we can classify capsule antennas into two types: embedded structures and conformal structures. Embedded antennas are placed inside the capsule cavity, occupying a significant amount of space, which can be a limitation when integrating other devices. For instance, in Section 3.4.1, the multilayer CP helical antenna, consisting of vertically connected three open loops, is an example of an embedded structure with a complex 3D design that occupies substantial internal space.

On the other hand, conformal capsule antennas are designed to adhere to the inner or outer surface of the capsule shell in a 2D configuration, minimizing the antenna's footprint within the capsule. In this section, we present a conformal CP capsule antenna that offers several advantages, including a reduced in-capsule footprint, ease of fabrication, and a broad overlapped impedance and ARBW.

3.4.2.1 Antenna Layout and Simulation Phantom

Figure 3.59 shows the planar radiation layer of the designed capsule antenna, which is printed on the top of a 0.254 mm thick Rogers RO6010 substrate (relative permittivity ε_r = 10.2 and loss tangent σ = 0.0023). The shaded conductor in Figure 1 consists of

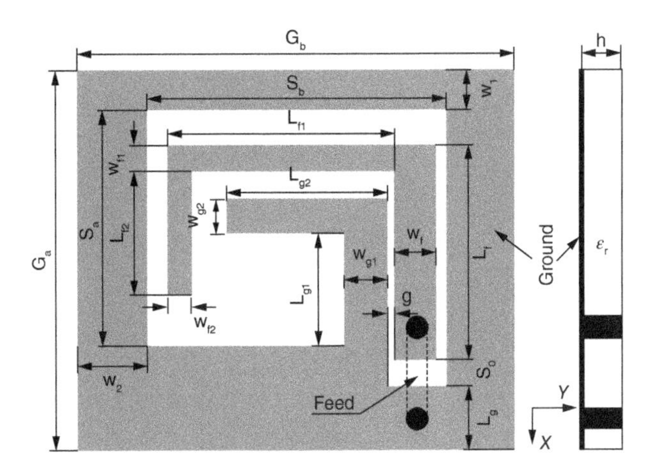

Figure 3.59 Planar structure of the capsule antenna.

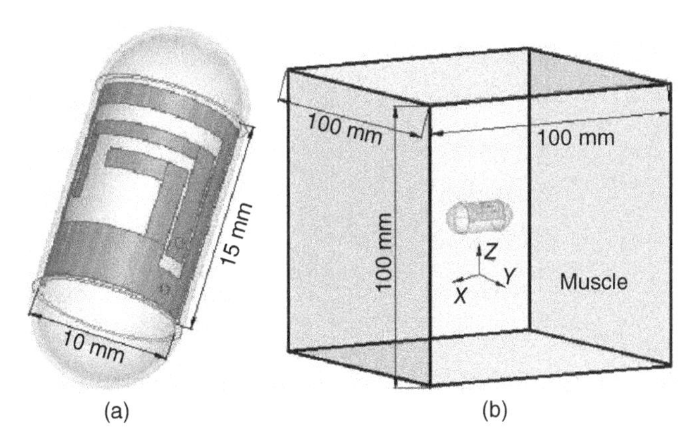

Figure 3.60 Overview of conformal capsule antenna and homogeneous muscle phantom: (a) capsule antenna, (b) muscle phantom.

a rectangular loop, an asymmetric U-shaped strip, and a protruding L-shaped stub. The antenna is corner-fed by a 50 Ω CPW, where the widths of signal strip and two gaps are Wf and g, respectively. For tight adhesion of the metal layer to the inner wall of the capsule shell, the proposed antenna was fed through two metallized vias from the back of the substrate, as shown in the dashed box in Figure 3.59. The planar structure in Figure 3.59 was wrapped around a cylinder to fit into the capsule of 10 mm inner diameter, as shown in Figure 3.60(a). The capsule shell is made of 0.5 mm thick acrylic material with a relative permittivity of 3.0 and a loss tangent of 0.001. Figure 3.60(b) shows a one-layer homogeneous muscle phantom model. The dimension of this model is 100 mm × 100 mm × 100 mm, and the capsule antenna was placed in the center of the muscle model. The dielectric properties of the muscle tissue ($\varepsilon_r = 52.73$ and $\sigma = 1.74$ S/m) at 2.45 GHz were used in the simulation. The final antenna dimensions in Figure 3.59 are listed in Table 3.6, which were optimized through the parametric study.

Table 3.6 Dimensions of the designed antenna (unit: mm).

Symbol	Value	Symbol	Value	Symbol	Value
G_a	14.2	L_{f1}	8.76	L_{g2}	6.09
G_b	16.64	W_{f1}	1.0	W_{g2}	1.35
S_a	8.8	L_{f2}	4.6	W_1	1.5
S_b	11.38	W_{f2}	0.88	W_2	2.63
L_f	8	L_{g1}	4.15	S_0	1.0
W_f	1.7	W_{g1}	1.62	L_g	2.4
g	0.3	ε_r	10.2	h	0.254

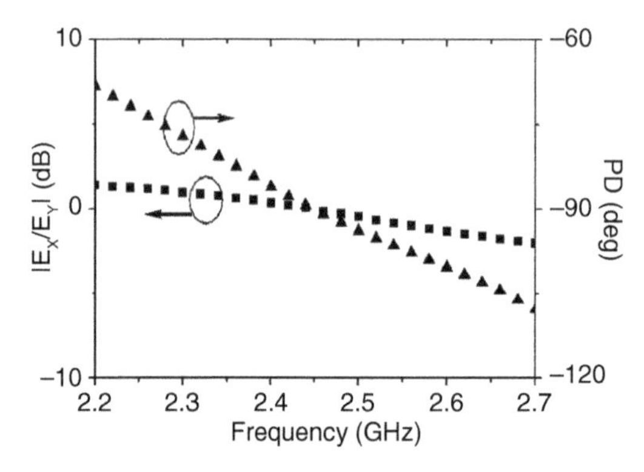

Figure 3.61 Amplitude ratio and phase difference of the far-field E_X and E_Y of the proposed antenna.

3.4.2.2 Mechanism of CP Operation

Based on [47], by using asymmetric CPW feed and properly introducing perturbation into a slot antenna, the CP wave can be generated. Here, we use the asymmetric U-shaped strip as a perturbation. By properly adjusting L_f, L_{f1}, and L_{f2} of the asymmetric U-shaped strip, the two equal-amplitude orthogonal electric fields ($|E_X/E_Y| = 0$ dB) with 90° phase difference can generate an ideal CP wave in the $+z$ direction. If phase difference is −90°, then LHCP wave will be produced. Figure 3.61 gives the simulated amplitude ratio and phase difference of the far field E_X and E_Y for the proposed antenna. At the center frequency of 2.45 GHz, they just have equal amplitude and −90° phase difference. The variation of amplitude imbalance and phase difference stays stable from 2.2 to 2.7 GHz, which explains its broad ARBW.

To further illustrate the operation mechanism of CP waves, the surface current density distributions varying with time are investigated. Figure 3.62 shows the current distributions at phase of 0°, 90°, 180°, and 270° at a CP frequency of 2.45 GHz. At $\omega t = 0$°, the upward predominant surface current is observed. At $\omega t = 90°$, the predominant surface current is rightward. It can be observed that the predominant surface current at $\omega t = 180°$ (270°) has equal magnitude and opposite phase with that of 0° (90°). As ωt increases, the

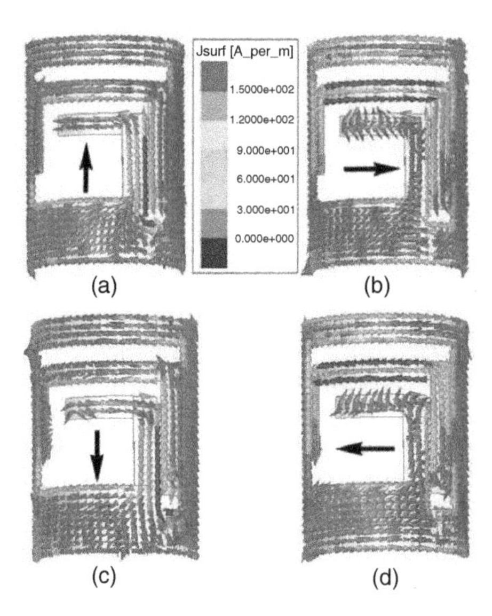

Figure 3.62 Surface current distributions of the designed antenna at 2.45 GHz for four different time phases: (a) 0°, (b) 90°, (c) 180°, and (d) 270°.

surface current direction rotates clockwise. Hence, the polarization sense is the LHCP wave in the $+z$.

In this design, the dimensions of the asymmetric U-shaped strip play a crucial role in determining the AR and impedance bandwidths. Additionally, the protruding L-shaped stub is used to fine-tune the center frequency of the AR and impedance bands, achieving a broad overlapped bandwidth. Through a parametric study, the optimal size of the antenna topology can be determined.

The optimized 3 dB ARBW of the proposed antenna is 13.11% (ranging from 2.28 GHz to 2.6 GHz), and its AR reaches an impressively low value of only 0.07 dB at 2.45 GHz. These results indicate that the antenna exhibits excellent performance with a wide frequency range and minimal AR distortion at the center frequency.

3.4.2.3 Results and Discussion

The planar geometry shown in Figure 3.59 was initially fabricated to validate the design method developed for the antenna. Subsequently, the planar antenna was transformed into a conformal structure and securely attached to the inner surface of a capsule shell with an inner diameter of 10 mm. The fabricated capsule antenna and the setup for measuring the reflection coefficient are depicted in Figure 3.63. For the measurement, a homogeneous mixture of liquid muscle phantom was prepared, consisting of 72% deionized water, 23% DGBE, and 5% Triton X-100 (polyethylene glycol mono phenyl ether).

Figure 3.64 displays the comparison of simulated and measured $|S_{11}|$ results obtained using the Rohde & Schwarz ZVA50 VNA. The simulated 10 dB impedance bandwidth is 31.58% (ranging from 2.08 GHz to 2.86 GHz), while the measured 10 dB impedance bandwidth is slightly broader at 39.21% (ranging from 2.05 GHz to 3.05 GHz). The slight discrepancies between the measured and simulated data can primarily be attributed to two

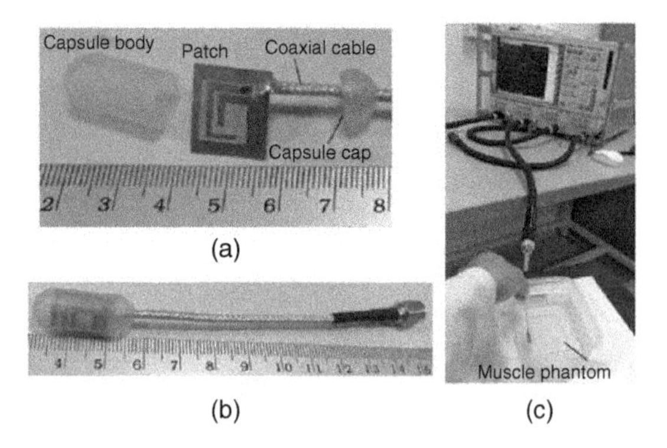

Figure 3.63 Fabricated capsule antenna: (a) decomposed capsule antenna, (b) complete conformal capsule antenna, and (c) measurement setup.

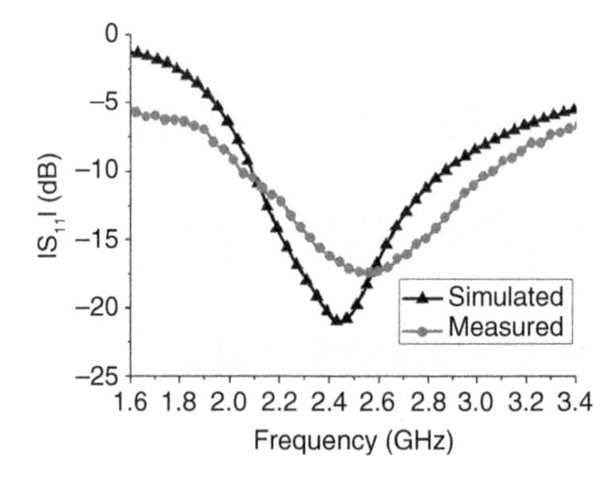

Figure 3.64 Simulated and measured $|S_{11}|$ of the proposed antenna.

factors: (1) the thin air gap between the radiation layer and the capsule shell and (2) minor fabrication variations.

The radiation patterns of the antenna were measured inside the SATIMO Starlab anechoic chamber, as shown in Figure 3.65. To avoid any liquid spillage on the rotating platform during measurements, chopped pork was used as a substitute for the liquid muscle platform depicted in Figure 3.63(c).

Figure 3.65 presents a comparison of the simulated and measured radiation patterns in the xz and yz planes at a frequency of 2.45 GHz. The measured radiation patterns exhibit shapes that are similar to the simulated ones in both the xz and yz planes. The simulated peak gain of the antenna in homogeneous muscle tissue is −29.1 dBi, while the measured maximum gain value of −31.6 dB was obtained in the chopped pork. The decrease in gain could be attributed to the utilization of a slightly larger pork container compared to the simulation platform.

As depicted in Figure 3.65, the radiation power of the LHCP is notably higher than that of the RHCP in the $+z$-axis for both the xz and yz planes. Consequently, the LHCP wave is predominantly excited in the $+z$ direction.

For safety considerations regarding the human body, the 1-g average SAR should not exceed 1.6 W/kg, as specified by the IEEE C95.1-1999 standard [42]. Given that the input

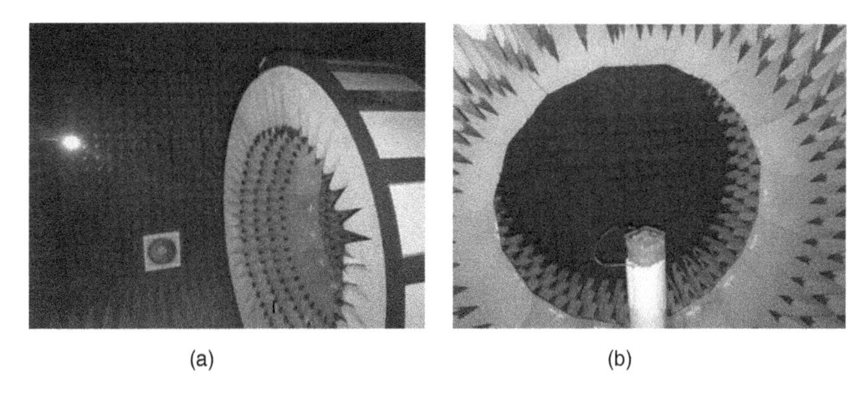

Figure 3.65 Radiation patterns measurement setup: (a) whole anechoic chamber, (b) local measurement phantom using chopped pork.

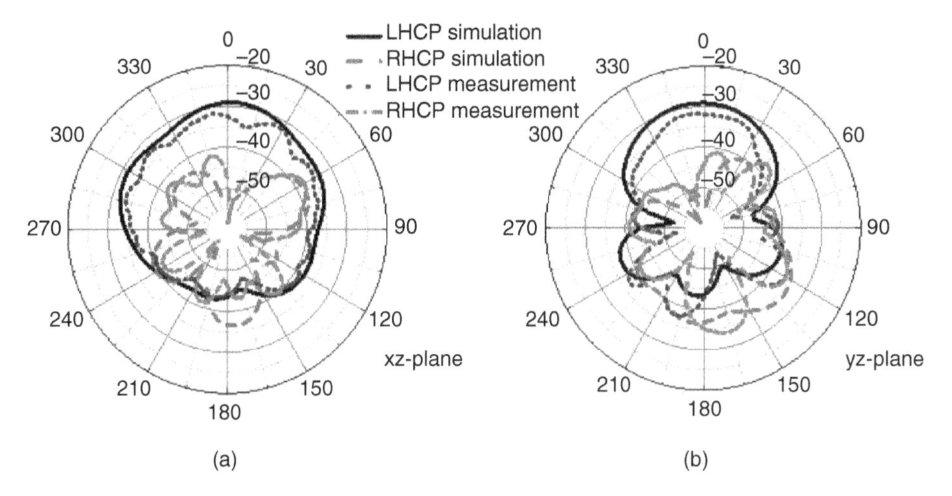

Figure 3.66 Simulated and measured radiation patterns at 2.45 GHz: (a) *xz* plane, (b) *yz* plane.

power of the antenna is 1 W, the peak SAR value at 2.45 GHz reaches 368.7 W/kg. Therefore, in order to comply with the regulated IEEE SAR standard, the delivered power must be limited to less than 4.34 mW at 2.45 GHz. This precaution ensures that the antenna's radiation remains within safe levels for the human body [42] (Figure 3.66).

As demonstrated above, integrating implantable CP antennas with capsule endoscopes necessitates a thoughtful and comprehensive approach considering the specific requirements and limitations of the implant's actual structure. This approach involves careful design considerations that take into account the physical constraints and demands of the implant environment. By factoring in these elements, the CP antenna can be fine-tuned and optimized to establish a reliable and efficient communication link with the capsule endoscope, ensuring superior medical imaging and seamless data transmission.

Considering the practical implications of the implant structure is of utmost importance during the design process of an implantable CP antenna for capsule endoscopy applications. By doing so, we can ensure the successful integration and functionality of the antenna within the capsule endoscope, leading to improved diagnostic capabilities and enhanced medical treatments.

3.5 In Vivo Testing of Circularly Polarized-implantable Antennas

Typically, the initial design of implantable antennas involves the use of a one-layer phantom for evaluation. The antennas are then in-vitro measured in the one-layer liquid/solid phantom and compared to the simulated results to assess the design concept. However, it is important to note that the one-layer phantom model does not fully represent the complex multilayer tissue environment found in real-life scenarios, as live tissues' electric properties can vary based on factors such as temperature, age, size, and sex of the subject.

Conducting in vivo testing of implantable CP antennas in live animals offers several advantages over using tissue phantoms. It allows for a more accurate assessment of the antenna's performance in a real-world environment, considering the influence of various tissues, fluids, and other physical parameters. Live animal testing enables researchers to better evaluate the antenna's safety, biocompatibility, and long-term performance, providing valuable insights for real-life medical applications. Moreover, testing in live animals provides a more realistic and representative picture of the antenna's behavior, particularly concerning implantable medical devices, where human body properties can significantly impact performance.

In this section, we present the in vivo testing of two CP-implantable antennas inside a rat [48]. Specifically, we will discuss the capacitively loaded CP-implantable patch antenna from Section 3.2.1 and the axial-mode multilayer helical antenna from Section 3.4.1. The in vivo testing took place at the Singapore Institute for Neurotechnology, National University of Singapore.

3.5.1 In Vivo Testing Configuration

To further investigate the sensitivity of the two antennas in a live multi-tissue environment, we conducted an experiment involving the embedding of the antennas inside a rat, which was approximately 18 weeks old. The detailed in vivo experiment steps are outlined in Figure 3.67 as follows:

(1) All experiment devices, including the antennas under test, coaxial cables, and SubMiniature version A (SMA) connectors, were thoroughly disinfected using medical-grade disinfectant.
(2) The hair in the areas where the antennas would be implanted was carefully trimmed, as depicted in Figure 3.67 (1).
(3) An incision was made through the skin, followed by another incision through the muscle layer, as shown in Figure 3.67 (2). In this experiment, we first implanted the antennas (as shown in Figures 3.6 and 3.55) under the muscle layer, as the proposed helical antenna is designed for a capsule system. Subsequently, the implantable patch antenna was implanted under the skin layer to observe the effect of different implant positions.

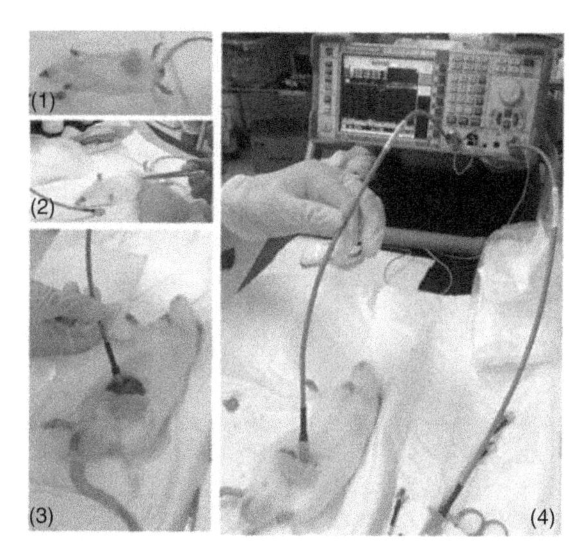

Figure 3.67 Surgical implantation of the implantable antennas into the rat.

(4) The implantable antennas were carefully placed under the muscle and skin layers, as shown in Figure 3.67 (3).
(5) After ensuring proper placement, the incisions were sutured, and the reflection coefficient measurement was performed using a Rohde & Schwarz R&S ZVL13 VNA (VNA), as shown in Figure 3.67 (4). The measured data were recorded under stable conditions.

Once the measurements were recorded, the experiment devices were cleaned, and the rat was humanely sacrificed without experiencing any pain.

By conducting this in vivo experiment, we aimed to gain valuable insights into the performance and behavior of the antennas in a real-life, multi-tissue environment, providing a more accurate assessment of their potential for medical applications.

3.5.2 Measured Reflection Coefficient

Figure 3.68(a) displays a comparison of the simulated and measured reflection coefficients of the multilayer helical antenna in different environments, including inside the rat. In comparison to the simulated results shown in Figure 3.43, the measured impedance bandwidth inside the rat covers approximately 2.024–2.379 GHz (16.1%) for $|S_{11}| < -10$ dB. The frequency point with good matching is shifted from 2.4 to 2.1 GHz. While the impedance matching of the implantable helical antenna in the rat is within acceptable performance, the resonant frequency has shifted to a lower frequency. However, in this in vivo testing, we did not focus on the polarization property of the antenna, as the antennas were placed facing the rat body directly. In future studies and experiments, we plan to use the complete implant system, including the battery and the transmitter, to excite the implantable antenna and study its polarization properties more comprehensively.

The measured reflection coefficient of the patch antenna is depicted in Figure 3.68(b). The antenna was implanted twice, one under the muscle layer and the other under the skin

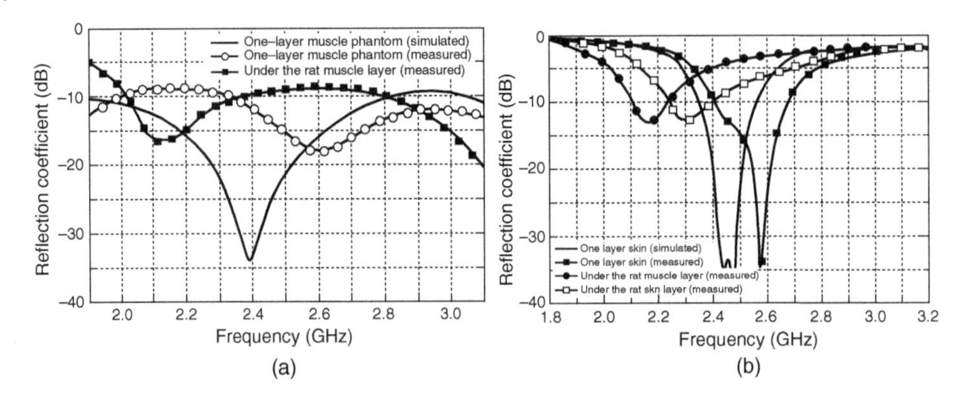

Figure 3.68 Comparisons of the simulated and measured reflection coefficients: (a) helical antenna, (b) patch antenna.

layer, to observe the effects of different implant positions. In comparison to the simulated results shown in Figure 3.5, the measured impedance bandwidth for the patch antenna is found to be in the frequency range of 2.111–2.232 GHz when implanted under the muscle layer and 2.242–2.392 GHz when implanted under the skin layer. Both cases show a shift in the resonant frequency of the patch antenna to lower frequencies compared to the simulation results.

3.5.3 Analysis of the Results and Discussions

Table 3.7 depicts the simulated and measured performances of the two implantable antennas. The frequency shift percent is defined as $(f_{\text{measured}} - f_{\text{simulated}})/f_{\text{simulated}}$. Note that the resonant frequencies of both antennas shifted to lower frequencies when implanted in the rat. The reasons why both the resonant frequencies of the two antennas shift could be:

(1) **Air gap between the antenna and the tissue**: The incision made in Figure 3.67 was large to cover the implant. Air gaps between the antenna and the tissue are inevitable in the testing. This point was mentioned in [29].
(2) **Implant positions in the in vivo testing**: In the previous antenna design, the antennas were designed directly to the free space. In the in vivo testing, the antennas were placed facing the rat body. When the antennas were under the muscle layer, the relative dielectric constant of relevant live tissues under the muscle may be higher than the muscle and skin. The higher relative dielectric constant of the tissues would cause the frequency to shift to lower frequencies.
(3) **Inexact simulation model**: The one-layer skin/muscle phantom is utilized for initial easy optimization and demonstration of the design concept. However, the rat consists of various organs. The simulation model could be updated with various organs.
(4) **Attached rat body fluid or blood**: The environment is very complicated under the muscle layer. The attached rat body fluid or blood that covered over the top surface of the antennas under test could affect the measured results.

Based on above possible reasons, we rebuild the simulation model to demonstrate the effect of the dielectric properties of rat tissues. Table 3.8 lists the dielectric properties

Table 3.7 Performance comparisons for two types of CP antennas.

	Antenna type	Helical	Patch
Simulation	Volume (mm^3)	$\pi \times (5.5)^2 \times 3.81$	$10 \times 10 \times 1.27$
	Phantom	Muscle	Skin
	Impedance bandwidth ($\lvert S_{11}\rvert < -10$ dB)	1.85–2.8 GHz	2.36–2.55 GHz
	AR bandwidth (AR < 3 dB)	2.0–2.8 GHz	2.44–2.48 GHz
Measurement (in vivo)	Implant position	Under the muscle	Under the muscle/skin
	Impedance bandwidth ($\lvert S_{11}\rvert < -10$ dB)	2.024–2.379 GHz	2.111–2.232/ 2.242–2.392 GHz
	Center frequency	2.1 GHz	2.18/2.3 GHz
	Frequency shift percent	−12.5%	−11%/−6.1%

Table 3.8 Dielectric properties of rat tissues at 2.45 GHz.

Biological tissues	Relative permittivity	Conductivity (S/m)
Rat skin [49]	~29.3	1.07
Rat muscle [50]	~56.0	~2.3
Rat spleen [50]	~52.35	~2.05

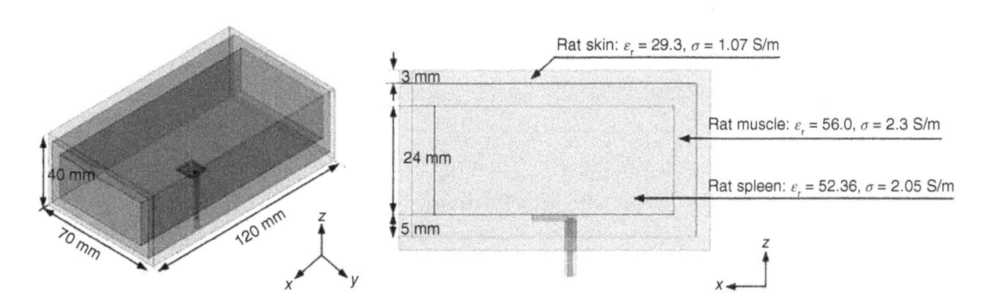

Figure 3.69 Rectangular phantom with three rat tissues.

of different rat tissues at 2.45 GHz. The simulation rectangular phantom is shown in Figure 3.69, where the phantom size is similar to the rat for testing. Two antennas were placed in the center of x-axis. Three rat tissues were chosen for the simulation as the antennas were embedded near the digestive organs.

Figure 3.70 shows the simulated results with different phantoms. The resonant frequency of the helical antenna was shifted from 2.42 GHz in the one-layer human muscle tissue to 2.37 GHz in the rectangular rat tissues. The resonant frequency of the patch antenna remains the same when the antenna is placed under the rat skin layer and in the one-layer

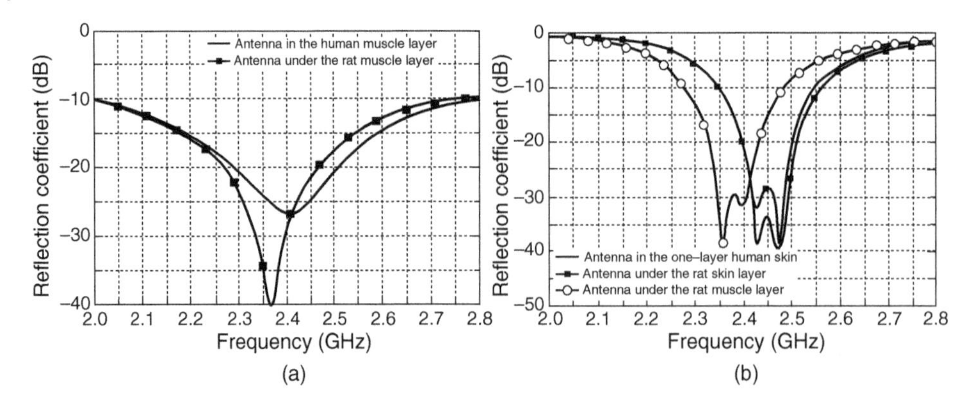

Figure 3.70 Simulated results of the two antennas with different phantoms: (a) helical antenna, (b) patch antenna.

human skin layer. The resonant frequency would shift to 2.37 GHz when the proposed antenna was placed under the rat muscle layer. Compared to the measured results in the rat, the simulated results in the simple rectangular model also show the shift toward lower frequencies, while the frequency shift rate $[(f_{\text{underratmuscle}} - f_{\text{underratskin}})/f_{\text{underratskin}}]$ for simulation in Figure 3.69 is smaller than that in the in vivo measurement.

The simulated results mean that the rat model would have some effect on the shift toward lower frequencies. Note that the dielectric constants of rat tissues in [50] were lower than the values reported by Burdette et al. [51]. After the patch antenna was removed from the rat and placed in free-space, we noticed that the proposed patch had a good impedance matching at around 2.45 GHz, which is much better than the measured results when the patch antenna was embedded under the rat muscle/skin layer. This means that the attached rat body fluid or blood covers the top surface of the antenna causing the further shift to lower frequencies.

3.6 Conclusions

To ensure effective communication with implantable devices, the design of implantable antennas must take into account various environments and conditions. One crucial consideration in antenna design is polarization, which refers to the orientation of the electromagnetic field generated by the antenna. CP antennas offer several advantages over other types, making them particularly well-suited for implantable medical devices. These benefits include consistent signal strength and reduced susceptibility to interference, making them highly suitable for use in challenging medical settings.

In our research, we have delved into the design and optimization of CP antennas for implantable medical devices. As discussed earlier, designing efficient CP antennas for implants involves considering multiple factors, such as device size and shape, the surrounding tissue and fluid environment, and the specific frequency range of interest. This necessitates a careful balance between performance and size, as well as a thoughtful consideration of the practical limitations and requirements of the implant environment.

Indeed, CP antenna designs can help solve the wireless communication link issues caused by polarization. However, another effective approach to address polarization-related problems is the use of dual-polarization antenna designs [52, 53]. Dual-polarization antennas have the capability to simultaneously transmit and receive two different polarization modes, effectively mitigating signal attenuation, and transmission instability caused by polarization mismatch.

Switching the polarization of the antenna is a common design method to enhance the communication link for implantable devices. For instance, in our previous work [52], we proposed a dual-polarized ingestible antenna operating at 434 MHz in the Industrial Scientific Medical (ISM) band. This compact conformal antenna exhibited a broad impedance bandwidth to withstand the dynamic environment of the GI tract. Additionally, due to the highly aligned current distributions, the omnidirectional loop antenna demonstrated high port isolation and low cross-polarization, enabling a robust communication link with external receiving antennas regardless of orientation and direction.

As illustrated in Figure 3.71(a), the dual-polarized antenna comprises two perpendicular counterparts printed on both sides of the substrate (RO3010). The current distribution, as shown in Figure 3.71(b), indicated that the proposed antenna operated on the fundamental mode of one-wavelength resonance. It had two peaks and two nulls along this path. The maximum currents occurred around the feeding point and at a position half the

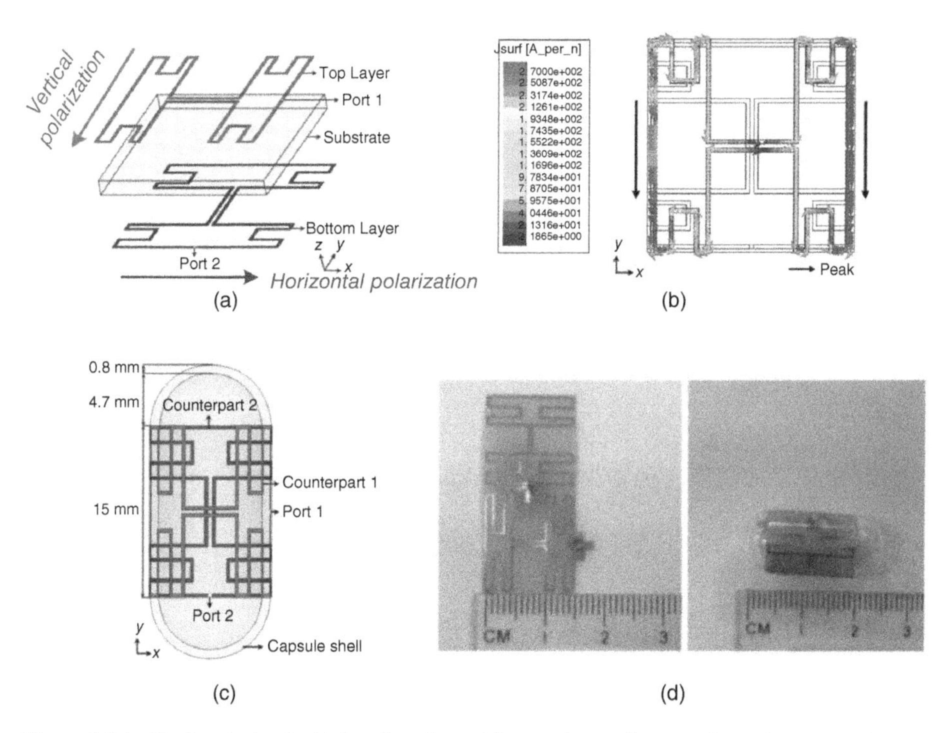

Figure 3.71 Dual-polarized wideband conformal loop antenna for capsule endoscopy systems: (a) perspective view, (b) current distributions of the planar dual-polarized antenna (top layer), (c) conformal configuration, and (d) fabricated prototype of the proposed conformal antenna. Source: Lei et al. [52]/IEEE.

wavelength distance away, which was located parallel at the exact opposite side. This led to the generation of linearly polarized radiation along the *y*-axis. Concurrently, within a period, the current distributions along the *x*-axis were weak and out of phase. Due to the symmetrical structure, most of them canceled each other out and contributed little to the radiation. For the lower counterpart, with the structure rotated 90° with respect to the *z*-axis, another linear polarization along the *x*-axis could be excited. Consequently, the antenna radiated with two orthogonal polarizations produced by two pure linear *E*-fields, achieving good isolation and omnidirectional patterns, thanks to the highly in-phase current vectors introduced.

To better integrate with capsule endoscopy systems and enhance radiation efficiency, the initial structure of the planar antenna was modified to a corresponding conformal one, as shown in Figure 3.71(c) and (d). As a result, the proposed dual-polarized antenna achieved a broad impedance bandwidth of 271 and 247 MHz for each port, covering the entire Medical Device Radiocommunications Service band as well as the 433.05–434.79 MHz ISM band. The isolation between two ports was measured to be around 42 dB at 434 MHz and remained higher than 39 dB within the targeted frequency range.

Additionally, polarization diversity designs have been considered to address the direction sensitivity problem for implantable devices. Antennas with polarization diversity offer a clever solution to counter the multipath fading effect in complex propagation environments. For instance, in [54], a diverse polarization antenna was designed for a 2.4-GHz ISM band capsule endoscopy system. By bending a normal dipole, three orthogonal currents were obtained, providing polarization diversity characteristics that realized orientation-insensitive performance.

Achieving multi-polarization in implant antennas can be challenging due to size constraints and specific requirements of implantable devices. Therefore, in some cases, it may be more practical to employ multi-polarization designs for external receiving antennas [55]. This approach involves using multiple antennas oriented in different directions to receive signals from the implant antenna. By combining the signals from these antennas, the overall signal strength and quality can be improved, resulting in more reliable wireless communication. However, this approach also requires careful design and testing to ensure optimal performance and minimize interference.

In conclusion, designing and optimizing antenna polarization for implantable medical devices is a complex and multifaceted process that demands careful consideration of various factors. Additionally, researching the wireless communication link characteristics from in-body to off-body caused by polarization is of particular importance.

References

1 Wong, H., Lin, W., Huitema, L., and Arnaud, E. (2017). Multi-polarization reconfigurable antenna for wireless biomedical system. *IEEE Trans. Biomed. Circuits Syst.* 11 (3): 652–660.

2 Arx, J. A. V., Rawat, P., Mass, W. R. & Vallapureddy, V. (2013). *Diversity Antenna System for Communication with an Implantable Medical Device.* United States Patent, Patent No.: US 8,353,040 B2.

3 Gao, S., Luo, Q., and Zhu, F. (2014). *Circularly Polarized Antennas*. John Wiley & Sons.
4 Wong, H., So, K.K., Ng, K.B. et al. (2010). Virtually shorted patch antenna for circular polarization. *IEEE Antennas Wireless Propag. Lett.* 9: 1213–1216.
5 Chen, W.S., Wu, C.K., and Wong, K.L. (2001). Novel compact circularly polarized square microstrip antenna. *IEEE Trans. Antennas Propag.* 49 (3): 340–342.
6 Lam, K.Y., Luk, K.-M., Lee, K.F. et al. (2011). Small circularly polarized U-slot wideband patch antenna. *IEEE Antennas Wireless Propag. Lett.* 10: 87–90.
7 Iwasaki, H. (1996). A circularly polarized small-size microstrip antenna with a cross slot. *IEEE Trans. Antennas Propag.* 44 (10): 1399–1401.
8 Elsdon, M., Sambell, A., Gao, S.C., and Qin, Y. (2003). Compact circular polarised patch antenna with relaxed manufacturing tolerance and improved axial ratio bandwidth. *Electron. Lett.* 39 (18): 1296–1298.
9 Nasimuddin, Qing, X., and Chen, Z.N. (2011). Compact asymmetric-slit microstrip antennas for circular polarization. *IEEE Trans. Antennas Propag.* 59 (1): 285–288.
10 Wong, K.L. and Lin, Y.F. (1998). Circularly polarised microstrip antenna with a tuning stub. *Electron. Lett.* 34 (9): 831–832.
11 Chen, H.-M. and Wong, K.-L. (1999). On the circular polarization operation of annular-ring microstrip antennas. *IEEE Trans. Antennas Propag.* 47 (8): 1289–1292.
12 Liu, C., Xiao, S., Guo, Y.X. et al. (2012). Compact circularly-polarized microstrip antenna with symmetric-slit. *Electron. Lett.* 48 (4): 195–196.
13 Row, J.-S. and Ai, C.-Y. (2004). Compact design of single-feed circularly polarised microstrip antenna. *Electron. Lett.* 40 (18): 1093–1094.
14 Chen, H.-D. (2002). Compact circularly polarized microstrip antenna with slotted ground plane. *Electron. Lett.* 38 (13): 616–617.
15 Bao, X.L. and Ammann, M.J. (2007). Dual-frequency circularly-polarized patch antenna with compact size and small frequency ratio. *IEEE Trans. Antennas Propag.* 55 (7): 2104–2107.
16 Tang, X., Lau, K.-L., Xue, Q., and Long, Y. (2010). Design of small circularly polarized patch antenna. *IEEE Antennas Wireless Propag. Lett.* 9: 728–731.
17 Dong, Y., Toyao, H., and Itoh, T. (2012). Compact circularly-polarized patch antenna loaded with metamaterial structures. *IEEE Trans. Antennas Propag.* 59 (11): 4329–4333.
18 Oh, J. and Sarabandi, K. (2013). A topology-based miniaturization of circularly polarized patch antennas. *IEEE Trans. Antennas Propag.* 55 (7): 1422–1426.
19 Song, J.H. and Lancaster, M.J. (1994). Capactively loaded microstrip loop resonator. *Electron. Lett.* 30 (18): 1494–1495.
20 Oh, J. and Sarabandi, K. (2012). Low profile, miniaturized, inductively coupled capacitively loaded monopole antenna. *IEEE Trans. Antennas Propag.* 60 (3): 1206–1213.
21 Gabriel, S., Lau, R.W., and Gabriel, C. (1996). The dielectric properties of biological tissues. *Phys. Med. Biol.* 2231–2293. Available: http://niremf.ifac.cnr.it/tissprop/ (Oct. 2004).
22 Liu, C., Guo, Y.X., and Xiao, S. (2014). Capacitively loaded circularly polarized implantable patch antenna for ISM-band biomedical applications. *IEEE Trans. Antennas Propag.* 62 (5): 2407–2417.

23 Yilmaz, T., Karacolak, T., and Topsakal, E. (2008). Characterization and testing of a skin mimicking material for implantable antennas operation at ISM band (2.4 GHz-2.48 GHz). *IEEE Antennas Wireless Propag. Lett.* 7: 418–420.

24 CST Microwave Suite Version. 2012.*08, CST Voxel Human Body Model.* Available: http://www.cst.com.

25 Xia, W., Saito, K., Takahashi, M., and Ito, K. (2009). Performances of an implanted cavity slot antenna embedded in the human arm. *IEEE Trans. Antennas Propag.* 57 (4): 894–899.

26 Mizuno, H., Ito, K., Takahashi, M., and Saito, K. (2010). A helical folded dipole antenna for implantable communication devices. In: *Proceedings of the IEEE Antennas and Propagation Society International Symposium (APSURSI)*, 1–4.

27 Merli, F., Bolomey, L., Zürcher, J.-F. et al. (2011). Design, realization and measurements of a miniature antenna for implantable wireless communication systems. *IEEE Trans. Antennas Propag.* 59 (10): 3544–3555.

28 Warty, R., Tofighi, M.R., Kawoos, U., and Rosen, A. (2008). Characterization of implantable antennas for intracranial pressure monitoring: Reflection by and transmission through a scalp phantom. *IEEE Trans. Microwave Theory Tech.* 56 (10): 2366–2376.

29 Karacolak, T., Cooper, R., Butler, J. et al. (2010). In vivo verification of implantable antennas using rats as model animal. *IEEE Antennas Wireless Propag. Lett.* 9: 334–337.

30 Soontornpipit, P., Furse, C.M., and Chung, Y.C. (2004). Design of implantable microstrip antennas for communication with medical implants. *IEEE Trans. Microwave Theory Tech.* 52 (8): 1944–1951.

31 Merli, F., Fuchs, B., Mosig, J.R., and Skrivervik, A.K. (2011). The effect of insulating layers on the performance of implanted antennas. *IEEE Trans. Antennas Propag.* 59 (1): 21–31.

32 Lin, Y.F., Chen, H.M., and Lin, S.C. (2008). A new coupling mechanism for circularly annular-ring patch antenna. *IEEE Trans. Antennas Propag.* 56 (1): 11–16.

33 Xu, L.J., Guo, Y.-X., and Wu, W. (2015). Miniaturized circularly polarized loop antenna for biomedical applications. *IEEE Trans. Antennas Propag.* 63 (3): 922–930.

34 Karacolak, T., Hood, A.Z., and Topsakal, E. (2008). Design of a dual-band implantable antenna and development of skin mimicking gels for continuous glucose monitoring. *IEEE Trans. Microwave Theory Tech.* 56 (4): 1001–1008.

35 Duan, Z., Guo, Y.X., Xue, R.F. et al. (2012). Differentially-fed dual-band implantable antenna for biomedical applications. *IEEE Trans. Antennas Propag.* 60 (12): 5587–5595.

36 Chen, Z.M., Cheng, K.W., Zheng, Y.J., and Je, M. (2011). A 3.4 mW 54.24 Mbps burst-mode injection-locked CMOS FSK transmitter. *IEEE Asian Solid State Circuits Conference* 289–292.

37 Lei, W., Chu, H., and Guo, Y.X. (2016). Design of a circularly polarized ground radiation antenna for biomedical applications. *IEEE Trans. Antennas Propag.* 64 (6): 2535–2540.

38 Merli, F. and Skrivervik, A.K. (2010). Design and measurement considerations for implantable antennas for telemetry applications. In: *Proceedings of the 4th European Conference on Antennas and Propagation (EuCAP)*, 1–5.

39 Izdebski, P.M., Rajagopalan, H., and Rahmat-Samii, Y. (2009). Conformal ingestible capsule antenna: a novel chandelier meandered design. *IEEE Trans. Antennas Propag.* 57 (4): 900–909.

40 Rajagopalan, H. and Rahmat-Samii, Y. (2012). Wireless medical telemetry characterization for ingestible capsule antenna designs. *IEEE Antennas Wireless Propag. Lett.* 11: 1679–1682.

41 Liu, C., Guo, Y.-X., and Xiao, S. (2014). Circularly polarized helical antenna for ISM-band ingestible capsule endoscope systems. *IEEE Trans. Antennas Propag.* 62 (12): 6027–6039.

42 IEEE (1999). *Standard for Safety Levels With Respect to Human Exposure to Radio Frequency Electromagnetic Fields, 3 kHz to 300 GHz*. IEEE Standard C95.1-1999.

43 IEEE (2005). *Standard for Safety Levels With Respect to Human Exposure to Radiofrequency Electromagnetic Fields, 3kHz to 300 GHz*. IEEE Standard C95.1-2005.

44 Park, B.-C. and Lee, J.-H. (2011). Omnidirectional circularly polarized antenna utilizing zeroth-order resonance of epsilon negative transmission line. *IEEE Trans. Antennas Propag.* 59 (7): 2717–2720.

45 Hsiao, F.-R. and Wong, K.-L. (2005). Low-profile omnidirectional circularly polarized antenna for WLAN access point. *Microwave Opt. Tech. Lett.* 46 (3): 227–231.

46 Yu, Y., Shen, Z., and He, S. (2012). Compact omnidirectional antenna of circular polarization. *IEEE Antennas Wireless Propag. Lett.* 11: 1466–1469.

47 Sze, J.Y., Wang, J.C., and Chang, C.C. (2008). Axial-ratio bandwidth enhancement of asymmetric-CPW-fed circularly-polarised square slot antenna. *Electron. Lett.* 44 (18): 1048–1049.

48 Liu, C., Guo, Y.-X., Jegadeesan, R., and Xiao, S. (2015). In vivo testing of circularly polarized implantable antennas in rats. *IEEE Antennas Wirel. Propag. Lett.* 14: 783–786.

49 Karacolak, T., Cooper, R., and Topsakal, E. (2009). Electrical properties of rat skin and design of implantable antennas for medical wireless telemetry. *IEEE Trans. Antennas Propag.* 57 (9): 2806–2812.

50 Kraszewski, A., Stuchly, M.A., Stuchly, S.S., and Smith, A.M. (1982). In vivo and in vitro dielectric properties of animal tissues at radio frequencies. *Bioelectromagnetics* 3: 421–432.

51 Burdette, E.C., Cain, F.L., and Seals, J. (1980). In vivo probe measurement technique for determining dielectric properties at VHF through microwave frequencies. *IEEE Trans. Microwave Theory Tech.* 28 (4): 414–427.

52 Lei, W. and Guo, Y.-X. (2018). Design of a dual-polarized wideband conformal loop antenna for capsule endoscopy systems. *IEEE Trans. Antennas Propag.* 66 (11): 5706–5715.

53 Li, R. and Guo, Y. (2021). A conformal UWB dual-polarized antenna for wireless capsule endoscope systems. *IEEE Antennas Wirel. Propag. Lett.* 20 (4): 483–487.

54 Li, Y., Guo, Y.X., and Xiao, S.Q. (2017). Orientation insensitive antenna with polarization diversity for wireless capsule endoscope system. *IEEE Trans. Antennas Propag.* 65 (7): 3738–3743.

55 Wang, H., Feng, Y., Hu, F., and Guo, Y.-X. (2022). A wideband dual-polarized ring-loaded cross bowtie antenna for wireless capsule endoscopes: design and link analysis. *IEEE Trans. Antennas Propag.* 70 (9): 7843–7852.

4

Differential-fed Implantable Antennas

4.1 Introduction

In the previous chapters, we explored a range of multipurpose implantable antennas, focusing on miniaturization, bandwidth enhancement (broadband and multiband), and polarization diversity (circular polarization and dual polarization). Many of these antennas have a single port feed design, and numerous single-fed implantable antennas have been proposed in earlier studies [1–12]. However, integrating single-fed implantable antennas with circuit systems presents challenges.

Single-fed implantable antennas often require baluns and other matching networks to integrate with radio frequency (RF) chipsets, which are commonly designed with differential ports to suppress the interferences, to achieve the desired performance. However, implementing these baluns and matching networks in a small form factor can be difficult and can also be sensitive to changes in the implant environment. Consequently, when used in differential systems, single-fed implantable antennas may face losses in baluns and matching circuits, resulting in lower efficiency. In contrast, differential-fed implantable antennas are better suited for biomedical differential systems.

Differential-fed implantable antennas utilize a differential feed technique, where two opposite but equal electrical signals are fed to the antenna. This design enhances signal quality and reduces noise levels in the communication link between the implanted device and an external device. Differential-fed implantable antennas also exhibit advantages in being less sensitive to changes in the surrounding tissue and body fluids, which can affect antenna performance.

To summarize, differentially fed antennas offer several inherent advantages, including common mode rejection, reduced mutual coupling effect, low noise, and high linearity. Additionally, some commercial RF chips are designed with complex impedance for RF signal output. In such cases, antenna designs with differential-fed complex impedance are worth considering to avoid losses and volume increase caused by additional impedance matching circuits for implants.

In this chapter, we will delve into the design considerations of differential-fed implantable antennas, covering miniaturization design, multi-frequency design, chip-integrated design, and complex input impedance design. We will also highlight the advantages and prospects of differential-fed implantable antennas in various implantable applications.

Antennas and Wireless Power Transfer Methods for Biomedical Applications, First Edition.
Yongxin Guo, Yuan Feng and Changrong Liu.

4.2 Dual-band Implantable Antenna for Neural Recording

In this section, we will begin with an analysis of the characterization of the differential reflection coefficient. Following that, we will proceed with the design of a dual-band implantable antenna with differential feeding, specifically tailored for neural recording applications. The antenna is intended to be connected to a dual-band complementary metal-oxide-semiconductor (CMOS) transmitter, which operates at frequencies of 542.4 and 433.92 MHz for the purpose of neural signal recording [13].

4.2.1 Differential Reflection Coefficient Characterization

The conventional S-parameters of Figure 4.1 can be expressed as follows:

$$\begin{bmatrix} b_1 \\ b_2 \end{bmatrix} = \begin{bmatrix} S_{11} & S_{12} \\ S_{21} & S_{22} \end{bmatrix} \begin{bmatrix} a_1 \\ b_1 \end{bmatrix} = S_{\text{std}} \begin{bmatrix} a_1 \\ b_1 \end{bmatrix} \tag{4.1}$$

where a_i and b_i represent the incident and reflected waves of port i, and S_{std} is the standard S-parameters matrix.

In contrast, when the ports 1 and 2 are viewed as a single differential port 1, the mixed-mode S-parameters of Figure 4.1 can be expressed as follows:

$$\begin{bmatrix} b_{d1} \\ b_{c1} \end{bmatrix} = \begin{bmatrix} S_{dd11} & S_{dc11} \\ S_{cd11} & S_{cc11} \end{bmatrix} \begin{bmatrix} a_{d1} \\ a_{c1} \end{bmatrix} = \begin{bmatrix} S_{\text{mm}} \end{bmatrix} \begin{bmatrix} a_{d1} \\ a_{c1} \end{bmatrix} \tag{4.2}$$

where a_{d1} and a_{c1} are the differential mode and common mode incident waves, and b_{d1} and b_{c1} are the differential mode and common mode reflected waves. And S_{mm} represents the mix-mode S-parameters matrix. S_{dd11} is the differential-mode S-parameter, S_{cc11} is the common-mode S-parameter, S_{dc11} and S_{cd11} is the mode-conversion or cross-mode S-parameter.

From [14, 15], we know that:

$$\begin{bmatrix} a_{d1} \\ a_{c1} \end{bmatrix} = \frac{1}{\sqrt{2}} \begin{bmatrix} 1 & -1 \\ 1 & 1 \end{bmatrix} \begin{bmatrix} a_1 \\ a_2 \end{bmatrix} = M \begin{bmatrix} a_1 \\ a_2 \end{bmatrix} \tag{4.3}$$

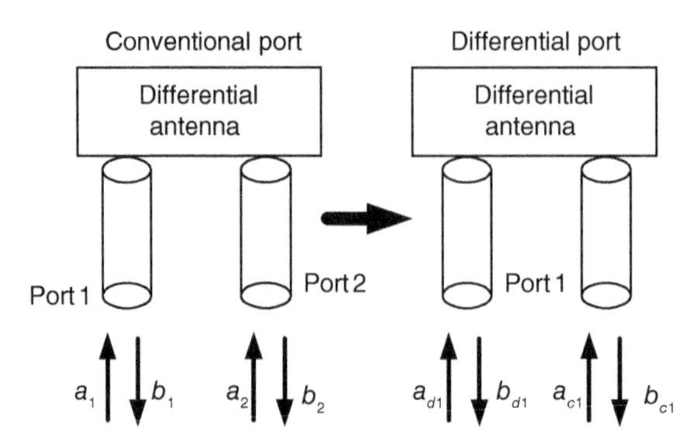

Figure 4.1 Simple schematic for conventional single-ended port to differential port conversion.

$$\begin{bmatrix} b_{d1} \\ b_{c1} \end{bmatrix} = \frac{1}{\sqrt{2}} \begin{bmatrix} 1 & -1 \\ 1 & 1 \end{bmatrix} \begin{bmatrix} b_1 \\ b_2 \end{bmatrix} = M \begin{bmatrix} b_1 \\ b_2 \end{bmatrix} \tag{4.4}$$

Therefore from (4.1) to (4.4), we can get the following equations:

$$S_{\text{mm}} = MS_{\text{std}}M^{-1} \tag{4.5}$$

$$\left.\begin{aligned} S_{dd11} &= \frac{1}{2}(S_{11} - S_{21} - S_{12} + S_{22}) \\ S_{dc11} &= \frac{1}{2}(S_{11} - S_{21} + S_{12} - S_{22}) \\ S_{cd11} &= \frac{1}{2}(S_{11} + S_{21} - S_{12} - S_{22}) \\ S_{cc11} &= \frac{1}{2}(S_{11} + S_{21} + S_{12} + S_{22}) \end{aligned}\right\} \tag{4.6}$$

For a symmetrically balanced antenna, as in the case of our proposed implantable antenna, $S_{11} = S_{22}$ and $S_{12} = S_{21}$, (4.6) can be further reduced to:

$$\left.\begin{aligned} S_{dd11} &= S_{11} - S_{12} \\ S_{dc11} &= 0 \\ S_{cd11} &= 0 \\ S_{cc11} &= S_{11} + S_{12} \end{aligned}\right\} \tag{4.7}$$

The result means for symmetrically balanced antenna, there is no conversion between differential mode and common mode. In other words, when the excitations of two ports are equal in amplitude and 180° out of phase, the key parameter of symmetrically balanced implantable antenna is the differential reflection coefficient or odd mode reflection coefficient, which is:

$$\Gamma_{\text{odd}} = S_{11} - S_{12} \tag{4.8}$$

The mixed-mode theory is applicable to general applications, not limited by coupled transmission lines and shielded balanced transmission lines. And the difference between the balanced cases and the unbalanced case is that when the structure is not balanced, mode-conversion will exist. For the following section, Γ_{odd} will be used to evaluate the reflection coefficient of the antenna.

4.2.2 Antenna Design and Operating Principle

Figure 4.2 shows the geometry of the dual-band antenna. The antenna is fabricated on the substrate of Rogers 6010 ($\varepsilon_r = 10.2$, $\tan\delta = 0.0023$), covered by a superstrate with the same material, each with a thickness of 25 mil (0.635 mm). The superstrate layer is used to protect the antenna from direct contact with the semi-conducting tissue. Also, the superstrate acts as a buffer between the metal radiator and human tissues by reducing RF power at the locations of lossy human tissues [16]. The total size of the antenna is 27 mm × 14 mm × 1.27 mm (480.06 mm^3). Because the antenna is a differentially fed one, we proposed a structurally symmetrical antenna where a spiral-shaped branch is connected to the main path to create the second resonance. The feeding location is on the same side of the antenna for better

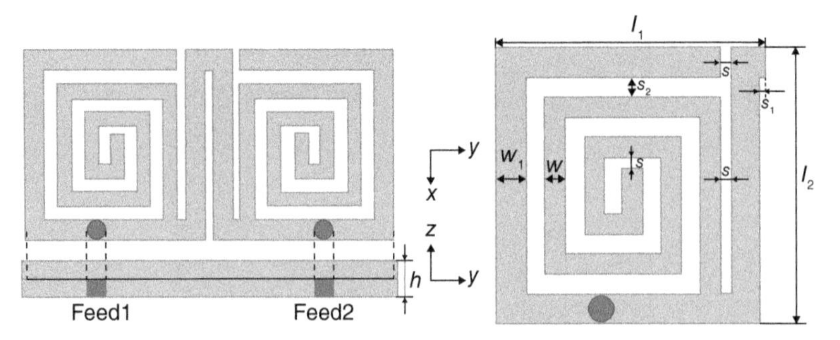

Figure 4.2 Geometry of the proposed dual-band differentially fed antenna.

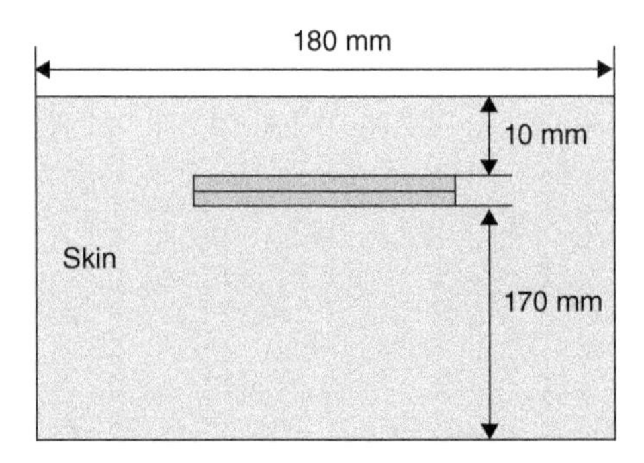

Figure 4.3 Simplified geometries for the one-layer tissue model (not in scale).

connection with the differential circuitry. Therefore, in Figure 4.2, the geometrical dimension only for one side is labeled. The differential input impedance of the proposed antenna is 100 Ω, meaning an impedance of 50 Ω for each port. The detailed values of geometrical dimension are summarized as follows: $l_1 = 13.1$ mm, $l_2 = 13.2$ mm, $s = 0.5$ mm, $s_1 = 0.3$ mm, $s_2 = 0.9$ mm, $w = 1.0$ mm, $w_1 = 1.4$ mm, $h = 1.27$ mm.

In Figure 4.3, we present the simplified geometry of a one-layer tissue model. The antenna under consideration is intended for neural recording applications, and we assume the tissue size to be 180 mm × 180 mm × 180 mm, with the maximum dimension matching that of a human head model [2]. For the initial evaluation, we consider a simple cubic tissue box with 10 mm thickness of tissue above the antenna and 170 mm thickness of tissue below the antenna. However, for the actual implantation, where the curvature of the human head and different implanting positions come into play, some adjustments to the geometrical parameters of the proposed antenna will be necessary.

Regarding the dielectric properties of the tissue, we use the values corresponding to dry skin for the one-layer tissue model, as the permittivity and conductivity of tissues can vary with frequency. Table 4.1 contains the numerical values of permittivity and conductivity for all tissues used in the models at the operating frequency, which can be found in [17]. Since the antenna operates at two closely spaced frequencies, we calculate the arithmetic average of the two frequencies to determine the dielectric properties used in the simulation model. For instance, for the dry skin used in the one-layer model, we assume a permittivity of 45.2 and a conductivity of 0.72 S/m.

Table 4.1 Dielectric properties of tissue.

Biological tissues	433.9 MHz		542.4 MHz		Arithmetic average	
	ε_r	σ (S/m)	ε_r	σ (S/m)	ε_r	σ (S/m)
Fat	5.57	0.04	5.53	0.04	5.55	0.04
Muscle	56.9	0.81	56.2	0.83	56.6	0.82
Skin	46.1	0.70	44.3	0.74	45.2	0.72

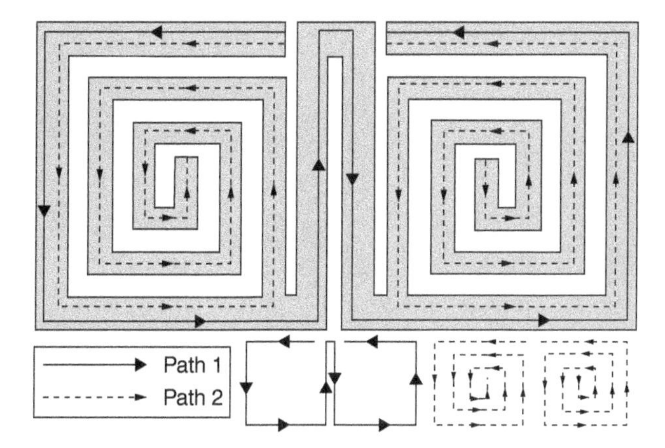

Figure 4.4 Electric current paths of the proposed dual-band differentially fed antenna.

In order to explain the operating principle of the proposed antenna, a simplified schematic of the antenna is given in Figure 4.4, and two electric current paths are indicated in this figure. Path 1 is with the same conductor track width (w_1). For spiral-shaped Path 2, from the turning point between the third and fourth segments (counting from the outermost to the innermost), the conductor track width changes from w_1 to w, and the spacing between each turn is fixed at s_2. It is observed that the first resonance around 433.9 MHz is controlled by both Path 1 and Path 2, while the second resonance around 542.4 MHz is mainly controlled by Path 2.

To validate this assumption, three antenna designs based on the original one are presented in Figure 4.5. Initially, we gradually removed inner segments of the spiral-shaped path, resulting in two antenna designs designated design 1 and design 2. Next, we removed the part that connects the two spirals of the original antenna, designated as design 3. The odd mode reflection coefficients of these three modified designs are compared with the original antenna in Figure 4.6.

From Figure 4.6, we can observe that as the spiral segments are gradually removed, the second resonant frequency shifts up and the differential S_{11} becomes worse, while the first frequency also shifts up slightly. When the spiral part is completely removed, as in the case of design 2, the second resonance disappears and the first resonance remains with a certain frequency shift and reduction of the reflection coefficient. For further verification, we removed the part connecting the two spirals of the original antenna, designated as design 3.

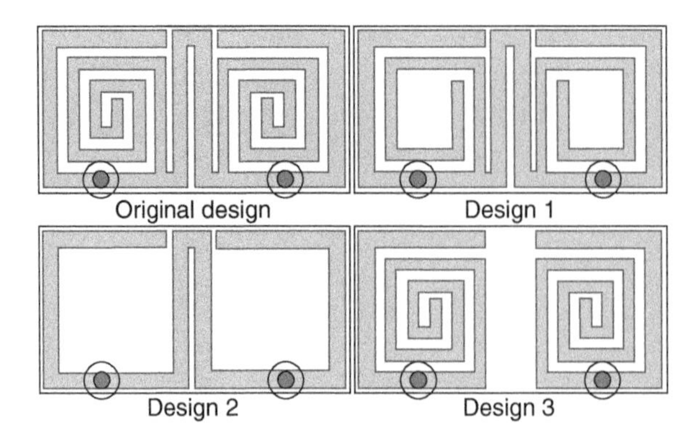

Figure 4.5 Antenna design variations to validate the operating principle.

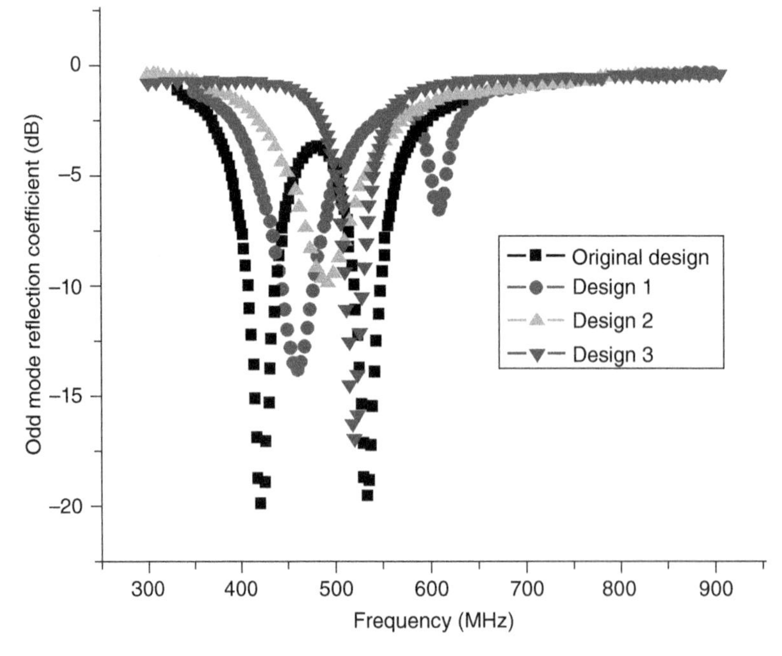

Figure 4.6 Odd mode reflection coefficient comparison of the original design and three modified designs for validating the operating principle.

From the comparison in Figure 4.6, it can be seen that the second resonance is almost unaltered while the first resonance disappears, which indicates that the second resonance is mainly controlled by the spiral part.

4.2.3 Measurement and Discussions

Figure 4.7 shows the measurement setup of the antenna, including a photograph of the fabricated antenna. The antenna is immersed in a tissue-mimicking solution filled in a plastic box, and we used a skin-mimicking gel for the solution, following the recipe proposed by

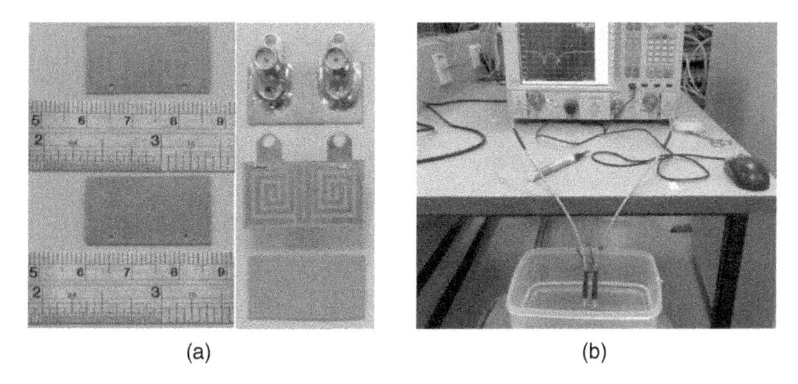

Figure 4.7 Measurement setup: (a) photograph of the fabricated implantable antenna, (b) measurement setup.

Karacolak et al. [6]. For the MICS band, the recipe consists of 56.18% sugar, 2.33% salt, and 41.49% deionized water.

To accurately measure the differential antennas, a four-port network analyzer with direct mix-mode measurement capability is preferred, as it avoids the need for extensive post-processing of data and ensures better accuracy. Thus, we used the Agilent N5222A four-port PNA Microwave Network Analyzer for the measurements, as shown in Figure 4.7(b).

The comparison of simulation and measurement results in both air and liquid tissue environments is presented in Figure 4.8. It can be observed that the tissue acts as a superstrate with a high dielectric constant and high loss tangent, resulting in a decrease in the resonant frequency, a widening of the bandwidth, and a reduction in the gain. The simulated and

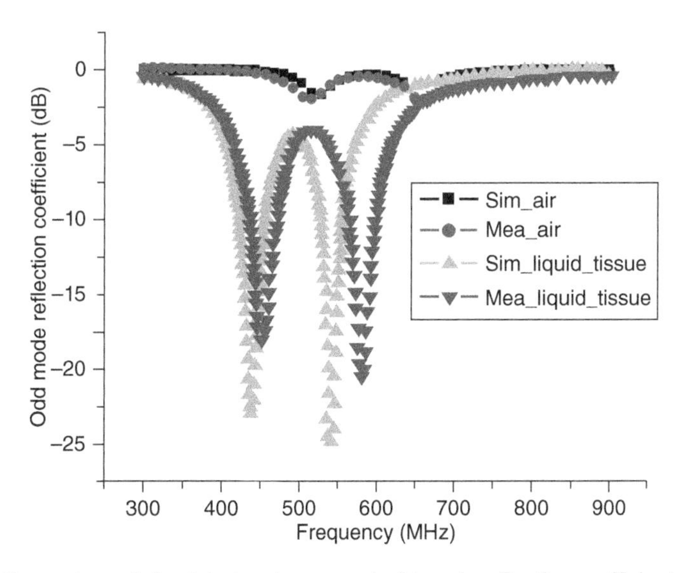

Figure 4.8 Comparison of simulated and measured odd mode reflection coefficients in air/liquid tissue.

measured results exhibit good agreement. The frequency bands obtained from simulations and measurements are 408–439 MHz (7.3% bandwidth) at the lower band and 435–471 MHz (7.9% bandwidth) at the upper band, respectively.

4.2.4 Communication Link Study

It is known that the above dual-band implantable antenna is designed as a transmitting antenna. In this connection, an external antenna with dual-band configuration should be designed for communication links. Figure 4.9 depicts the geometry of the external antenna with detailed geometrical parameters. The antenna is a typical dipole antenna with two u-shaped slots of the same width along the slot path in each arm to create another resonance. The antenna is printed on FR4 substrate. And the input impedance of the external dipole antenna is 100 Ω.

For the characterization of the communication link, a one-layer tissue model for the implanted antenna is used and the two antennas are spaced 130 mm apart to evaluate the coupling between them. The simulation setup is given in Figure 4.10(a), and the simulated S parameters of the two antennas are given in Figure 4.10(b). Assume the differential port of the external antenna as port 1 and the two bottom-feeding ports of the internal implanted antenna as port 2 and port 3. As port 2 and port 3 are symmetric, either S_{21} or S_{31} can be

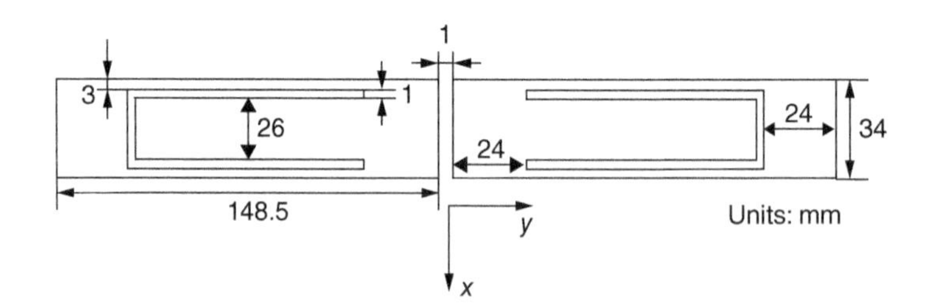

Figure 4.9 Geometry of the external antenna.

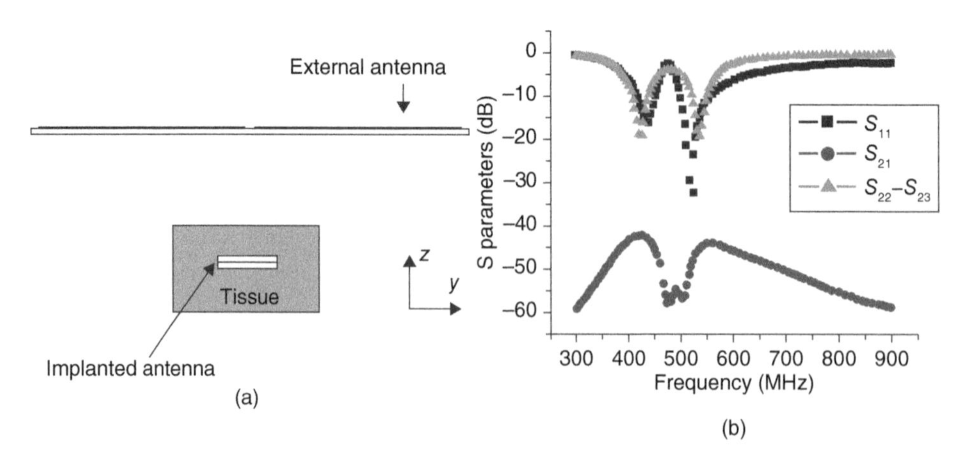

Figure 4.10 (a) Simulation setup for characterizing the communication link (length of antenna is not in scale), (b) S parameters of the antenna pair.

Table 4.2 Parameters of the link budget.

Operating frequency	433.9 MHz	542.4 MHz
Tx power	−19 dBm	
Distance	130 mm	
Path loss	42.4 dB	44.6 dB
Implementation loss	6 dB	
Receiver noise floor	−101 dBm	
SNR (BER = 1E-5)	14 dB	

considered as the coupling strength between the two antennas. Hence, $|S_{11}|$ is the reflection coefficient of the external antenna, and $|S_{22} - S_{23}|$ (Γ_{odd}) is the odd mode reflection coefficient of the implanted antenna. From Figure 4.10(b), it can be seen that a maximum coupling strength of −42.4 and −44.6 dB at 433.92 and 542.4 MHz can be achieved.

When the two antennas are perfectly matched, the S-parameter $|S_{21}|$ quantifies the power transmission ($|S_{21}|^2 = P_r/P_t$), and therefore it also gives the value of path loss. P_r is the received power and P_t is the transmitted power. When expressed in decibels, $|S_{21}|$ should be the same value of P_r/P_t and the negative value of path loss. Various parameters of the link budget are listed in Table 4.2.

The link margin (LM) can be expressed as:

$$\begin{aligned} \text{LM} &= P_T - \text{PL} - \text{RNF} - \text{IL} - \text{SNR} \\ &= -19\ \text{dBm} - 44.6\ \text{dB} + 101\ \text{dBm} - 6\ \text{dB} - 14\ \text{dB} \\ &= 17.4\ \text{dB} \end{aligned} \tag{4.9}$$

where PL is the path loss and RNF is the receiver noise floor and IL is the implementation loss and SNR is the signal-to-noise ratio of the receiver. For the value of PL in the calculation of link margin, we take the larger one for the path loss at the two resonant frequencies. The values of receiver noise floor and the SNR are from [18].

Additionally, the distance along the z-axis between the two antennas is increased from 100 to 500 mm to evaluate the coupling strength with distance. The simulation results of $|S_{21}|$ and the resulting LM at two resonant frequencies are given in Figure 4.11. After the distance along z-axis between two antennas is increased from 100 to 500 mm, a decrease of 14.2 dB for LM can be seen, and even at a large distance of 500 mm between the two antennas, a minimum LM of 6.3 dB can be obtained.

4.3 Integrated On-chip Antenna in 0.18 μm CMOS Technology

In Section 4.2, a dual-band implantable antenna was introduced for neuro applications, but its size may be too large for some specific applications, such as embedding devices in eyes, heads, blood vessels, or organs. In such cases, CMOS technology becomes essential to further reduce the antenna size and achieve high integration of the entire implantable system, including the implantable antenna and RF circuits on the same chip.

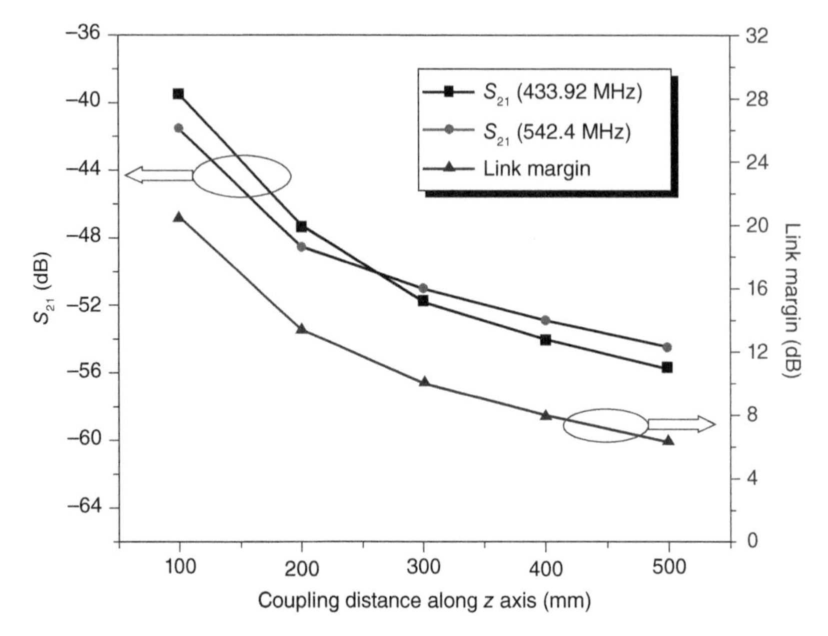

Figure 4.11 Variation of $|S_{21}|$ with respect to the different distances between two antennas.

In the neuro system application [19], two UWB antennas were designed, and biological channels were modeled for neuro recording systems. However, the UWB antenna size was still in the region of centimeters. To miniaturize further, a minimally invasive 64-channel wireless electrocorticography (ECoG) microsystem was developed [20], where a loop antenna with a side length of 6.5 mm was placed around the electrode array. In another work [21], a $1 \times 1 \times 1\,\text{mm}^3$ cubic loop antenna was designed for Wireless Brain–Machine Interface Systems.

For even smaller wireless biomedical devices, which could be placed in blood vessels, human heads, organs, and other tight spaces, it is necessary to further shrink the antenna size. Previous projects, such as Smartdust [22], aimed to integrate a sensor, communication channel, and a small microprocessor into a cubic millimeter package, but the actual devices were around 7 mm in size. However, recent advancements have shown promising progress toward cubic millimeter systems. For example, the Fraunhofer Institute demonstrated a cubic millimeter camera for endoscopy applications [23], and MIT researchers developed parts for cell-sized batteries [24]. These developments indicate that building cell-sized (less than $1\,\text{mm}^3$) implantable devices for medical applications in the future is possible.

In this section, we introduce a miniaturized differential-fed implantable on-chip antenna for the 902–928 MHz Industrial, Scientific, and Medical (ISM) band. Compared to single-fed implantable antennas, the differentially driven implantable antenna can act as an output balun, enabling a more compact and efficient integrated system. Furthermore, the demand for state-of-the-art single-chip and single-package radio systems calls for differential antennas to reduce the bill of materials and improve receiver noise performance and transmitter power efficiency [25].

4.3.1 System Requirement and Antenna Design

Table 4.3 lists the detailed system requirements. The output of the transmit (Tx) chip is differential and the Tx chip size is 1.5 mm × 1.8 mm using 0.18 μm CMOS technology. The antenna should be implanted with a depth of 20 mm. The distance between an exterior antenna and the top of the skin phantom is not strictly limited. The communication link $|S_{31}|$ value should >−70 dB (−60 dB is preferred) for the link budget, where ports 1 and 2 are the two ports of proposed differential-fed on-chip antenna, port 3 is the excitation port of an exterior dipole antenna. Therefore, we need to design a compact differential-fed on-chip antenna.

Considering designing a differentially driven implantable antenna, a dipole antenna is a good solution. Figure 4.12 shows the layout of the proposed dipole antenna. Figure 4.12(a) shows the detailed 0.18 μm CMOS process stackup and the proposed antenna is designed in M6 layer. The antenna structure is shown in Figure 4.12(b). Two spiral lines with each length of 14.45 and 3.94 mm are connected as one of the arms of the dipole antenna. The width of all lines is 20 μm. Ground-Signal-Ground-Single-Ground (GSGSG) pads are used

Table 4.3 Parameters of wireless data transfer system.

Transmitting antenna	
Operating frequency [MHz]	902–928
Tx chip size [mm^2]	1.5 × 1.8
Tx chip output	Differential
Tx antenna size	≤1.5 × 1.8
Process	0.18 μm CMOS
Receiving antenna	
Rx antenna	Unlimited
Communication link	
S_{31} [dB]	>−70, −60 preferred
d	Unlimited

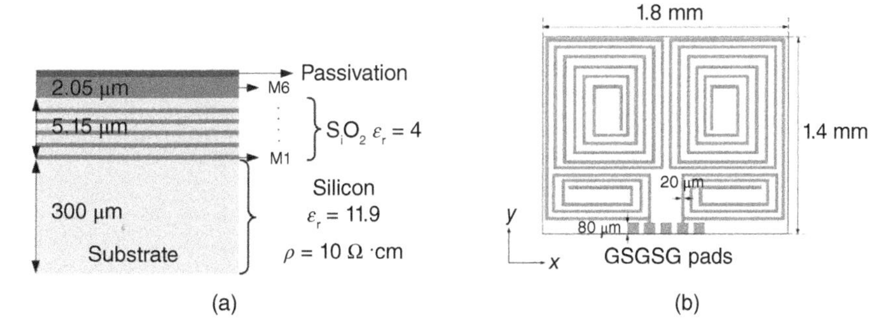

Figure 4.12 Layout of proposed on-chip antenna: (a) 0.18 μm CMOS process stackup, (b) the layout of M6.

for simulation. The excitations of two ports (ports 1 and 2 are with 50 Ω) are equal in amplitude and 180° out of phase.

The electrically small dipole antenna consists of two modified spiral arms. The reason why choose spiral arms is that the current vectors in the wire can be arranged in a matter so as to lower the resonant frequency for a given wire length and size, when the current vectors in directly adjacent wires are always in reinforcing directions [26]. In practice, a number of factors, such as the current vector alignment of each dipole arm, the capacitive coupling between dipole arms, and the feeding location, will determine the resonant frequency of the dipole antenna [27]. The on-chip antenna is optimized based on these factors, fixed area and feeding location.

Figure 4.13 shows the simulation scenario for coupling strength evaluation. The antenna is embedded at the depth of 20 mm in the skin tissue. The total volume of the skin phantom is $180 \times 200 \times 200$ mm^3. The skin electrical properties ($\varepsilon_r = 41.395$, $\sigma = 0.8675$ S/m) at 902 MHz were used in the one-layer skin simulation model. A simple dipole antenna as an exterior antenna is placed over the distance of d from the top of the skin tissue. The exterior antenna is designed using substrate FR-4 ($\varepsilon_r = 4.4$, $\tan\delta = 0.02$). Figure 4.14 reveals the simulated reflection coefficients and coupling values with different distances (d). The on-chip antenna has a wide impedance bandwidth that can cover over 0.7–1.1 GHz for $|S_{11}-S_{12}|$ less than −10 dB. And the coupling strength $|S_{31}|$ can meet the system requirement ($|S_{31}| > -60$ dB) when d is less than 20 mm.

To study the design in a realistic environment, the proposed antenna is evaluated within the Gustav voxel human body. Figure 4.15 shows the three-dimensional Gustav voxel human body used for the implantable design in a human head. Note that the Gustav human body model belongs to CST (CST Studio Suite) voxel family [28]. The so-called Voxel Man is based on a dissected male corpse sliced into several thousand layers. The relevant characteristics of Gustav human body model are as follows: Age/Sex: 38y/male, size: 176 cm, mass: 69 kg, resolution: $2.08 \times 2.08 \times 8.0$ mm^3. The total volume of the human head is $22 \times 17.5 \times 25$ cm^3. Numerical analyses are performed using the CST Microwave Suite [28]. The embedded depth is the same as that in the one-layer skin model. The distance between the implantable antenna and the exterior antenna is 40 mm.

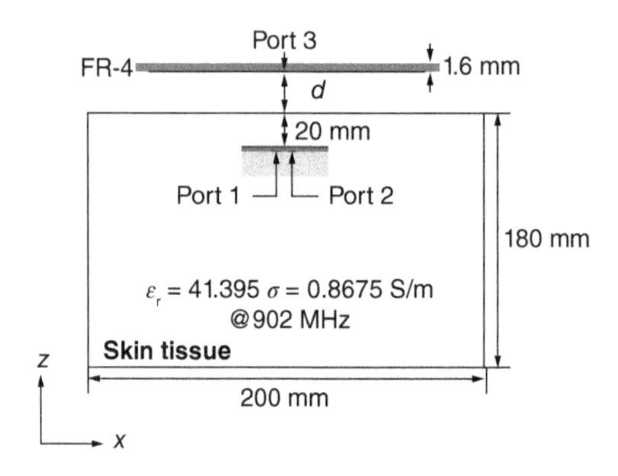

Figure 4.13 Simulation scenario for coupling strength evaluation.

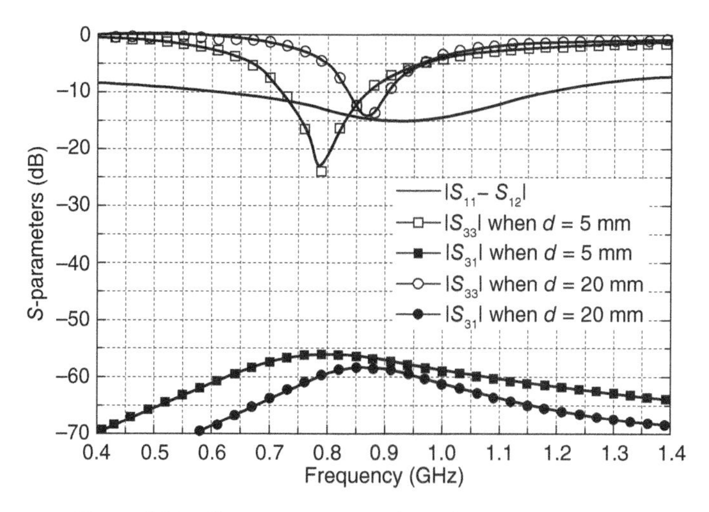

Figure 4.14 Simulated results with different distances (d).

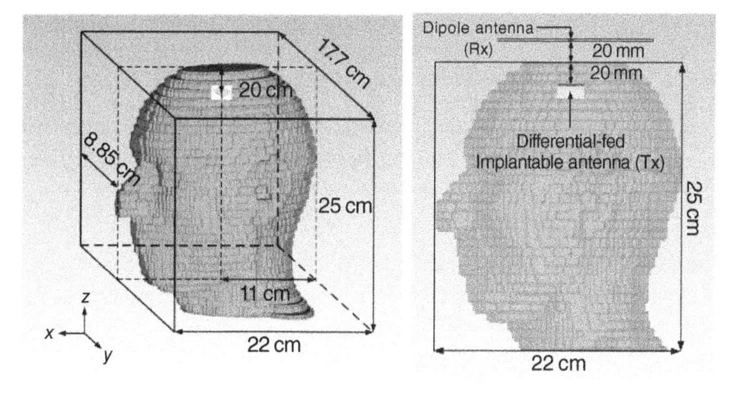

Figure 4.15 Three-dimensional Voxel human body used for the implantable antenna design in a human head.

Figure 4.16 presents the simulated S-parameters results. The reflection coefficient of the on-chip antenna is in the frequency range of over 0.4–1.05 GHz for $|S_{11}-S_{12}|$ less than −10 dB. Also, the impedance matching of the antenna in the human head is much better than that in the skin phantom. The coupling strength is around −56 dB at 902 MHz, which is similar to the value of ∼ −58 dB in the skin phantom at the same frequency. The simulated results demonstrate that the proposed antenna can achieve almost the same performance in the human head model.

As the antenna is embedded in the human head, specific absorption rate (SAR) should be studied to meet the 1-g and 10-g SAR regulations. When the proposed antenna is assumed to deliver 1 W, the simulated maximum 1-g and 10-g average SAR values are 134.15 and 13.43 W/kg, respectively. Therefore, the allowed transmitter power values are 11.9 mW (10.7 dBm) and 148.9 mW (21.7 dBm) to satisfy the 1-g and 10-g SAR regulation, respectively [29, 30]. SAR issue should not be a concern as the transmit power of commercial transmitters is much lower than the allowed transmit power.

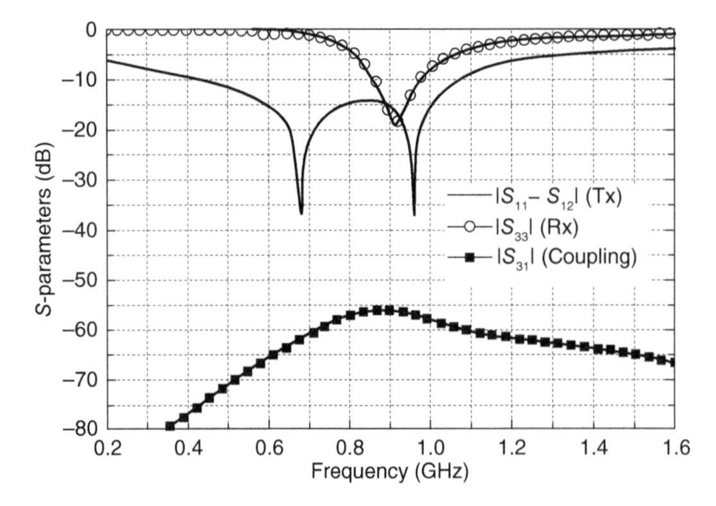

Figure 4.16 Simulated s-parameters of the proposed antenna in the human head when $d = 20$ mm.

4.3.2 Chip-to-SMA Transition Design and Measurement

The above-mentioned antenna design employed GSGSG pads, but conducting measurements using a probe station is impractical because the antenna should be embedded in the skin phantom. To address this, the on-chip antenna is connected to a test structure through a chip-to-SMA (SubMiniature version A) transition using wire bonding, as shown in Figure 4.17. The GSGSG pads are connected to the transition designed on Rogers 3010 ($\varepsilon_r = 10.2$, $\tan\delta = 0.0035$, 75 mil) using wire bonding. The simulated reflection coefficients of the on-chip antenna with and without wire bonding are compared in Figure 4.18, demonstrating that there is no significant difference between the two configurations.

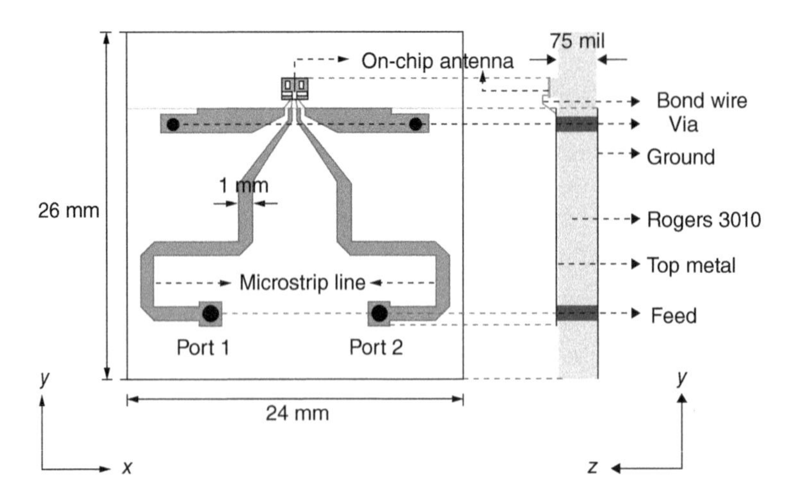

Figure 4.17 Top and side views of the chip-to-SMA transition.

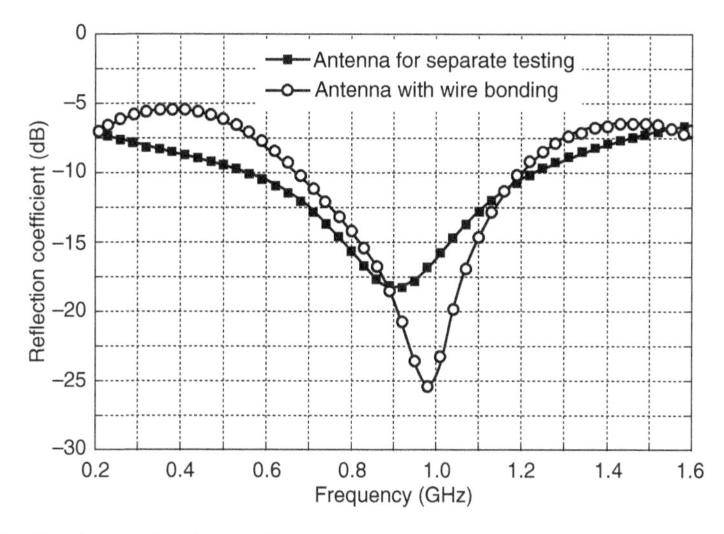

Figure 4.18 Simulated reflection coefficients of the proposed antenna with/without wire bonding (Legend not consistent antenna without wire bonding).

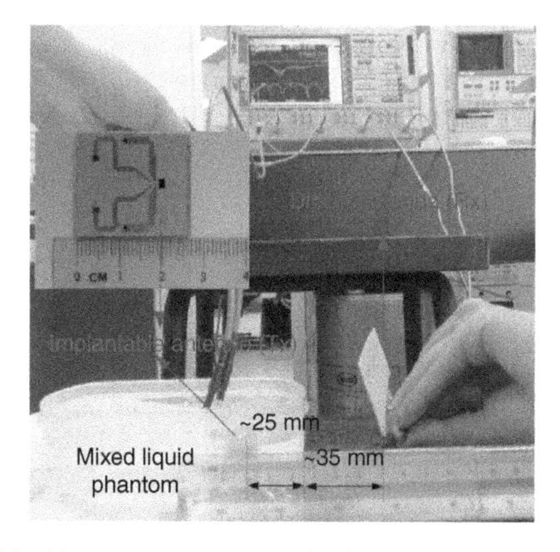

Figure 4.19 Measurement set up and fabricated antenna with wire bond.

To validate the design, the antenna is fabricated and tested. Figure 4.19 shows the measurement setup, including the photograph of the fabricated on-chip antenna with wire bonding. The on-chip antenna is embedded in the skin phantom at a depth of around 25 mm, and an exterior dipole antenna is placed at a distance of around 35 mm from the top surface of the skin phantom. For consistency, the simulated and measured results are compared in the same environment, including the implantable depth and the distance between the implantable antenna and the exterior antenna.

Comparison of simulated and measured results of the proposed antenna with wire bonding is shown in Figure 4.20. The simulated and measured results show good agreement,

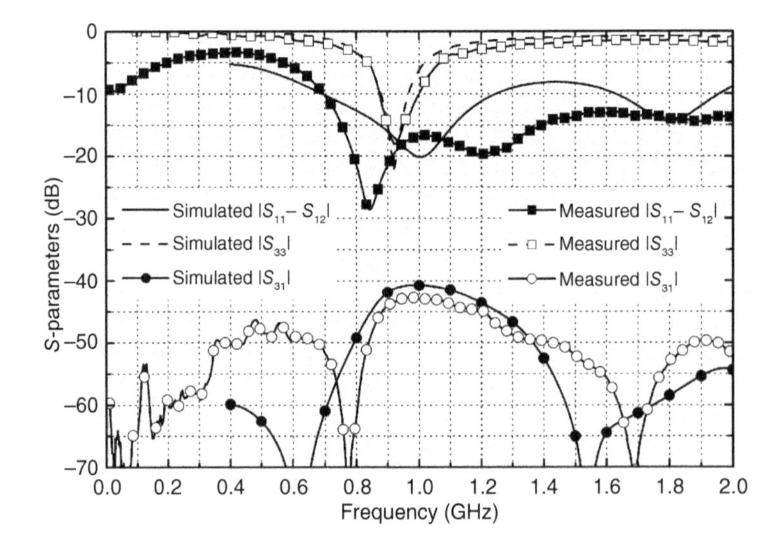

Figure 4.20 Comparison of simulated and measured results of the proposed on-chip implantable antenna with chip-to-SMA transition.

except for the matching of the on-chip antenna, which may be attributed to the differences in the bonding wire. The measured impedance bandwidth of the differential-fed on-chip antenna is in the frequency range of over 0.6–1.4 GHz. Additionally, the measured communication link between the implantable antenna and the exterior dipole antenna achieves ~ −42 dB at the ISM band when the distance between the top of the skin phantom and Rx is ~35 mm. The measured results indicate that the on-chip antenna can achieve a similar communication link without the transition.

Upon comparing the coupling coefficients in Figures 4.20 and 4.14, it is observed that the on-chip antenna with the transition structure exhibits a coupling coefficient more than 10 dB higher than the same on-chip antenna without the transition in the ISM band. This indicates that the transition structure also contributes to the radiation. To further confirm the radiation contribution from the test transition structure, Figure 4.21 shows the E-field distribution of the proposed antenna, including the transition, at 915 MHz. It is important to note that the large test structure was used solely for testing the antenna performance. In the actual circuit, the on-chip antenna will be integrated with the chips without the large test transition structure.

After designing and testing the on-chip antenna, the feeding part needs to be revised for integration, as the output pads of the Tx chip were not in the center of the axis. The size of the output pads is 70 μm × 70 μm, with a pitch of 100 μm. The initial antenna design is referred to as case 1, and the antenna with the revised feeding structure is referred to as case 2. Case 1 is used for separate testing to demonstrate the antenna's performance, while case 2 is intended for final integration with the Tx chip. Figure 4.22 shows the simulated reflection coefficients of the two cases, indicating that both cases exhibit similar results and coupling strengths with the external antenna.

Finally, the proposed antenna was fabricated, and the photograph of the integrated on-chip antenna is shown in Figure 4.23. The functionality of the circuitry is as follows:

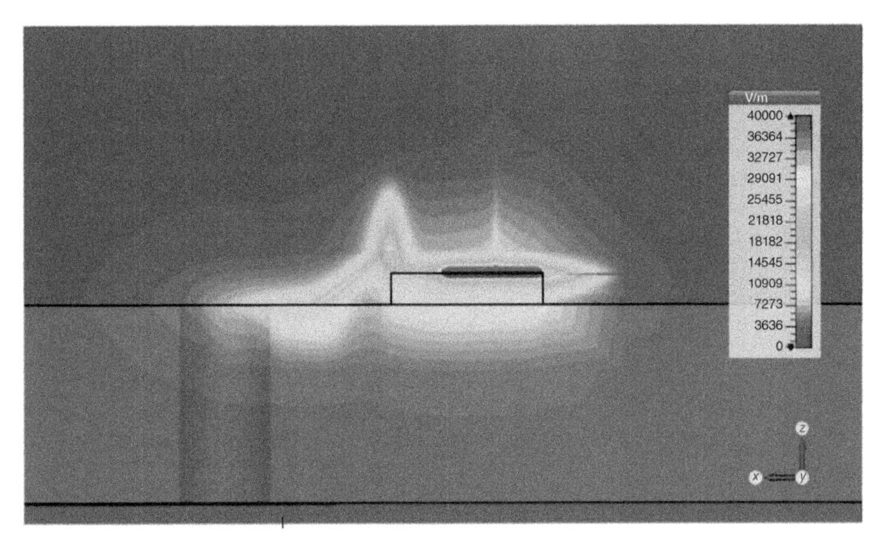

Figure 4.21 E-field distribution of the proposed antenna with transition in the voxel human head model.

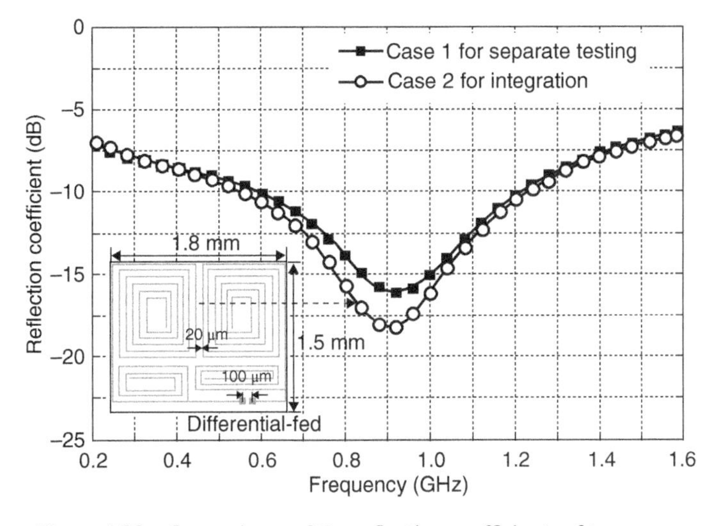

Figure 4.22 Comparisons of the reflection coefficients of two cases.

the transmitter converts the biosignals collected by implantable micro-electromechanical systems (MEMS) sensors or probes into electrical signals, which are then processed and transmitted to external devices for further analysis and processing.

Table 4.4 summarizes the recent research on antennas embedded in the human head. It can be seen that different kinds of antennas can be utilized for antenna design in the human body. And Table 4.4 indicates that the introduced on-chip antenna [32] would have small volumes and acceptable link gain for biomedical applications.

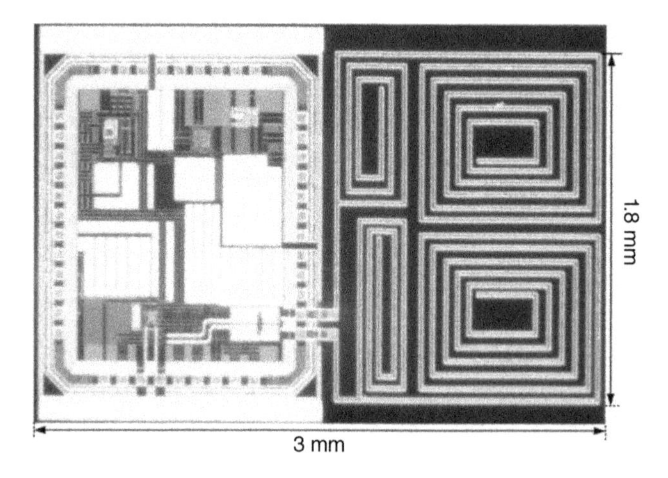

Figure 4.23 Photo of the integrated on-chip antenna.

Table 4.4 Performance Comparisons of Antennas in the Human Head Models.

References	[31]	[20]	[21]	[32]
Antenna type	Monopole	Loop	Loop	Dipole
Frequency	3.1–10.6 GHz	300 MHz	907.5 MHz	915 MHz
Size [mm^3]	12 × 12 × 2.8	6.5 × 6.5	1 × 1 × 1	1.4 × 1.8 × 0.308
Tx–Rx distance [mm] Head + Air	13 + 2	11 + 5	27 + 5	20 + 20
Link Gain [dB]	∼ − 20 at 8GHz (Mea.)	−16.5 (Sim.)	−57.19 (Mea.)	−56 (Sim.)
Miniaturization technique	Truncated ground	IC inside the loop	Cubic loop	Spiral arm

4.4 Dual-band Implantable Antenna for Capsule Systems

In this section, a differentially fed dual-band implantable antenna operating at both the MICS and ISM bands is presented [33]. This antenna's dual-band capability allows it to be utilized in a system with two modes: a sleep mode and a wake-up mode, effectively extending the lifetime of the implant.

4.4.1 Planar-implantable Antenna Design

A frequency-dependent skin tissue model with size of 100 mm × 100 mm × 100 mm is utilized, and the antenna is located 3 mm below the surface of the skin. The permittivity and conductivity of the skin at different frequencies can be obtained from [17].

Figure 4.24 depicts the geometry of the dual-band planar implantable antenna, which is symmetrical with respect to x-axis. The substrate supporting the antenna is Rogers RO3010 (ε_r = 10.2, tanδ = 0.0035) with thickness of 25 mil (0.635 mm). Via holes with

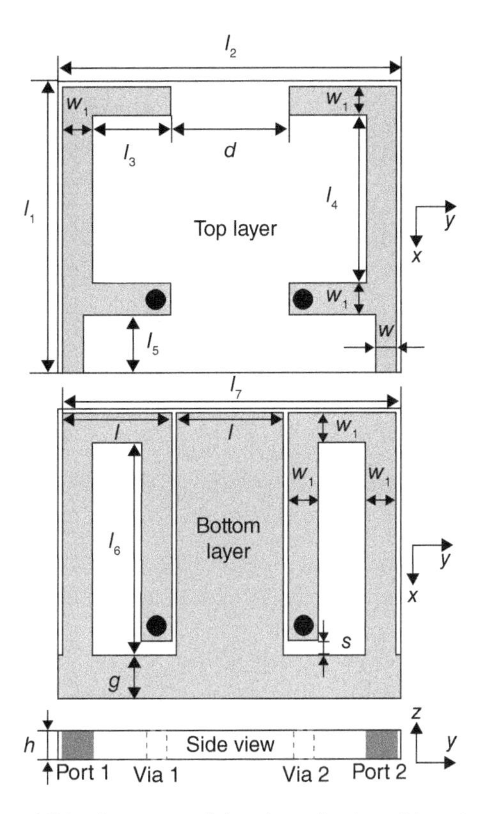

Figure 4.24 Geometry of the planar implantable antenna.

diameter of 0.9 mm are located in the center of the copper trace. The antenna is wrapped by a biocompatible material named parylene-C with thickness of 0.1 mm ($\varepsilon_r = 2.95$, $\tan\delta = 0.013$). The total size of the antenna including the biocompatible encapsulation is 13.4 mm × 16 mm × 0.835 mm. The optimized parameters of the antenna are as follows: $w_1 = 1.4$ mm, $w = 1$ mm, $l = 5$ mm, $l_1 = 13.2$ mm, $l_2 = 15.8$ mm, $l_3 = 3.6$ mm, $l_4 = 7.6$ mm, $l_5 = 2.6$ mm, $l_6 = 9.6$ mm, $l_7 = 15.4$ mm, $g = 2$ mm, $s = 0.6$ mm, $d = 5.4$ mm, and $h = 0.635$ mm.

For comparison, several different settings were used in simulation. For High Frequency Structural Simulator (HFSS) simulation, two lumped ports were first used, as shown in Figure 4.24. However, considering the ground size is very small, coaxial cable along the feeding line in testing will have effect on the performance of the antenna. Hence, two wave ports are applied to the two coaxial feeds. Also, for further validation, the antenna was placed in the human model of Computer Simulation Technology (CST) Microwave Studio. In CST, two discrete ports were utilized to feed the antenna. The simplified simulation setup in HFSS and CST for one layer model is shown in Figure 4.25(a).

For the simulation setup in the human model of CST, a male model Gustav is used. Also, for reducing the simulation time, only the upper part of the human torso including the head is imported. The front and side views of the human model are shown in Figure 4.25(b). To evaluate the performance of the implantable antenna, the antenna was

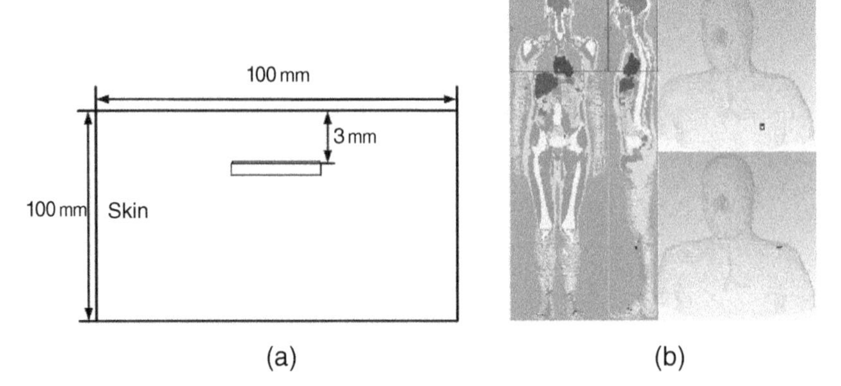

Figure 4.25 (a) Simulation setup in HFSS and CST for one skin layer model (not in scale), (b) simulation setup in CST for chest and shoulder implantation.

placed in two different locations: the shoulder and the chest as shown in Figure 4.25(b). The total volume for the Gustav human body is (0, 0, 0) – (256, 127, 882), and the selected part is (0, 0, 0) – (256, 127, 224). The resolution of the human voxel data is $2.08 \times 2.08 \times 2\ \text{mm}^3$, which means a total human size of $532.48 \times 264.16 \times 1764\ \text{mm}^3$.

Figure 4.26 shows the comparisons of differential reflection coefficients of the planar antenna for different simulation setups. From Figure 4.26, it can be indicated that the curves for CST and HFSS simulation results in one layer skin model almost coincide with each other. The resonant frequencies for the HFSS lumped model simulation are 413 MHz and 2.44 GHz. The bandwidth ($|S_{11}| < -10$ dB) is 370–466 MHz (23.0%) for the lower band and 2.22–2.75 GHz (21.3%) for the upper band. For the actual implantation in chest and shoulder in CST Gustav human model, certain off-resonance can be seen. However, two resonant frequencies around 400 MHz MICS band and 2.45 GHz ISM can still be clearly noticed.

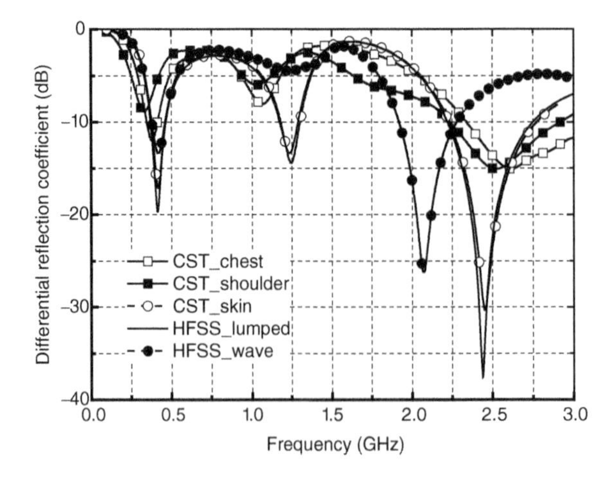

Figure 4.26 Comparison of differential reflection coefficient of planar antenna for different simulation setups.

When the coaxial cables are applied to the feeding position as indicated in Figure 4.26, the third resonance around 2.44 GHz shifted downwards to around 2.07 GHz, while the first resonance around 400 MHz remains more or less unaffected. This means that the antenna is more easily affected by the size of ground plane at higher frequency due to the fact that the SMA is electrically larger, the effect of which is neglected in lumped model simulation. The way to mitigate the effect is to increase the size of ground plane. This will not only change the resonance properties but will also increase the total size of the implantable antenna, which would be a disadvantage for actual implantation. However, for system integration such as conformal capsule antenna case, the ground part can be intended for the position where battery is placed.

4.4.2 Conformal Capsule Design

When the antenna is designed on a flexible substrate, it can be wrapped around a cylinder-shaped device and be used in a biotelemetric capsule system for medical purposes. For actual implantation, physical size constraint has to be considered. The size of a capsule is typically 11 mm (diameter of the capsule) × 26 mm (length of the capsule).

The designed planar antenna is modified and a conformal one on a flexible substrate named polyimide ($\varepsilon_r = 3.5$, $\tan\delta = 0.008$) with thickness of 0.15 mm is optimized. The outer and inner radius for the cylinder is 5 and 4.85 mm respectively. Also, the antenna is wrapped by parylene-C with thickness of 0.1 mm and the whole size of the antenna including the encapsulation is 186.3 mm^3 ($\pi \times [5.1^2–4.75^2] \times 17.2$). The diameters for via holes are 0.9 mm. The whole antenna (10.2 mm × 17.2 mm) is just slightly larger than a capsule's half-size. Therefore, for actual application, the size of the ground plane can be extended to save this part of the capsule for implanting the battery and the necessary electronic circuits. The conceptual application in a capsule case is shown in Figure 4.27.

The geometry of the flexible antenna is shown in Figure 4.28. The simulation environment in HFSS one-layer skin model for the flexible antenna is the same as the planar one except that the center of the capsule cylinder is now 8 mm from the surface of the skin tissue. For HFSS simulation, coaxial cables are added along the feeding line to represent the effect of cables during the actual measurement. In this case, wave port excitations are applied to the coaxial feeds. For CST simulation, we put the antenna in the stomach contents in Gustav human model. The selected volume for stomach implantation is (54, 0, 224) – (198, 127, 365). The simulation environment for CST stomach case is shown in Figure 4.28.

For a clearer presentation of the conformal design, a cut was made at the bottom center of the antenna and the flexible substrate was spread out as shown in Figure 4.29, which is also the layout for fabrication. The geometrical dimensions of the design shown in

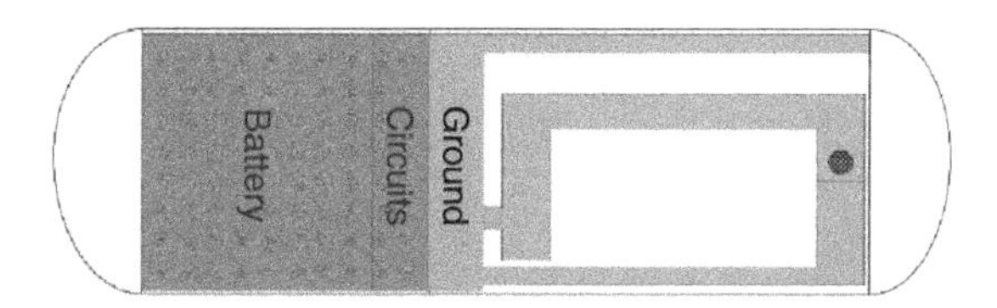

Figure 4.27 The conceptual application of the flexible antenna in a capsule.

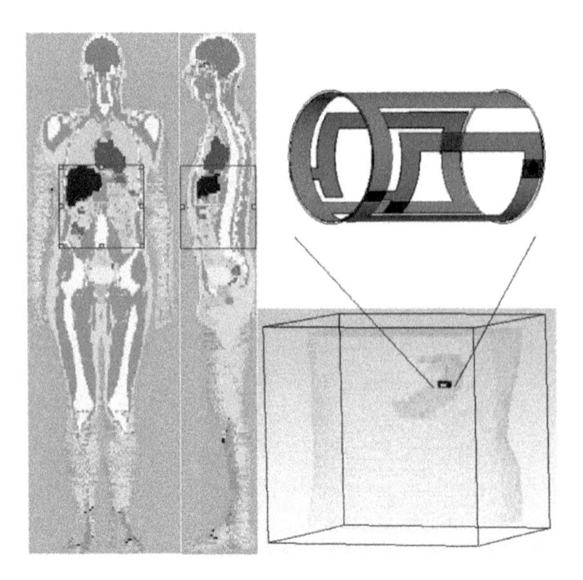

Figure 4.28 Geometry of the introduced flexible antenna and the simulation setup for CST stomach implantation.

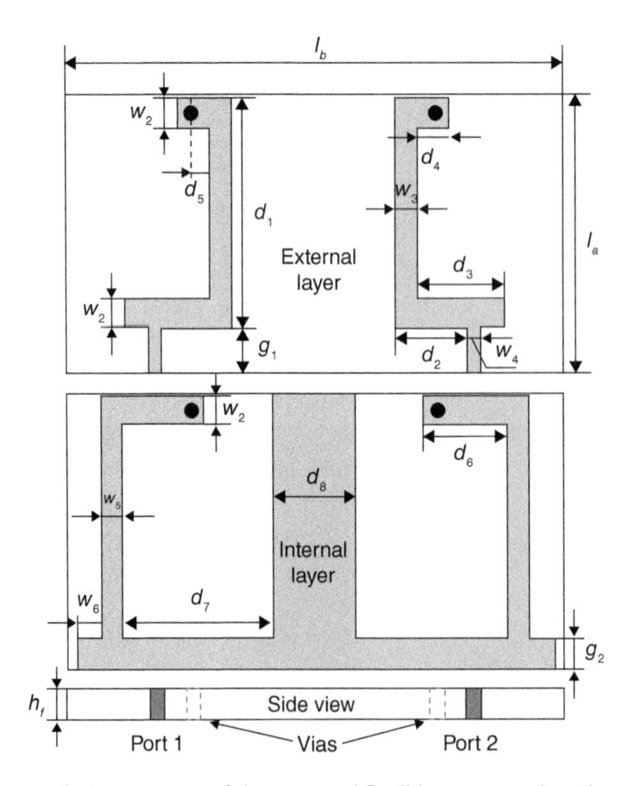

Figure 4.29 Geometrical parameters of the proposed flexible antenna when the conformal design is spread out.

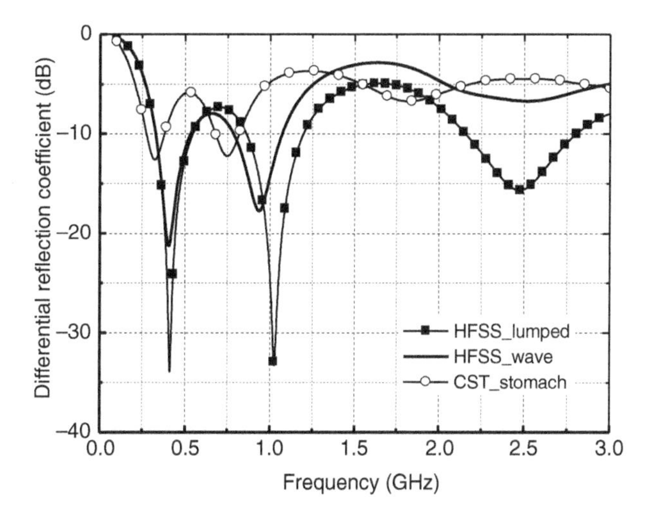

Figure 4.30 Comparison of differential reflection coefficient of flexible antenna for different simulation setups.

Figure 4.29 are: l_a = 17 mm, l_b = 31.4 mm, w_2 = 1.8 mm, w_3 = 1.4 mm, w_4 = 0.87 mm, w_5 = 1.34 mm, w_6 = 1.59 mm, g_1 = 2.8 mm, g_2 = 2 mm, d_1 = 14 mm, d_2 = 4.45 mm, d_3 = 5.41 mm, d_4 = 2.01 mm, d_5 = 1.15 mm, d_6 = 5.28 mm, d_7 = 9.53 mm, d_8 = 5.24 mm.

The comparison of simulation results from HFSS and CST for different implantation scenarios is shown in Figure 4.30. The bandwidths for the HFSS lumped case are 321–532 MHz (49.5%) for the MICS band and 2.15–2.74 GHz (24.1%) for the ISM band. For the actual implant application in different scenarios, further tuning of the geometrical parameters of the proposed antenna is necessary.

In consideration of the fact that the battery and the electronics may have effect on the performance of the antenna, four cases are shown in Figure 4.31.

Case 1: A PEC cylinder was placed with a size of 9.5 mm × 10 mm at the side of the antenna to test the robustness of matching performance. The inner part of the perfect electrical conductor (PEC) coincides with the ground for a length of 2 mm. The total size of the antenna in case 1 is 10.2 mm × 25.2 mm, which is approximately the size of a capsule.

Case 2: The size of ground plane (g_2 = 10 mm) was extended. Besides, the inner part of the capsule is filled with tissue material for the original simulation model.

Case 3: The inner tissue portion of the cylinder with a 9.5 mm × 10 mm vacuum layer.

Note that the feeding ports remain unchanged in all these cases. Figure 4.32 depicts the comparison of differential reflection coefficients of a flexible antenna for different simulation setups. It can be seen that compared with the original case, the first resonance remains unchanged with the PEC added (case 1), while the second resonance becomes weak; however, the frequency shift is not significant. The curve of case 2 almost coincides with that of case 1, meaning the extended ground plane would not affect the performance of the antenna as long as the positions of the feeding ports remain unchanged. With the inner portion of the

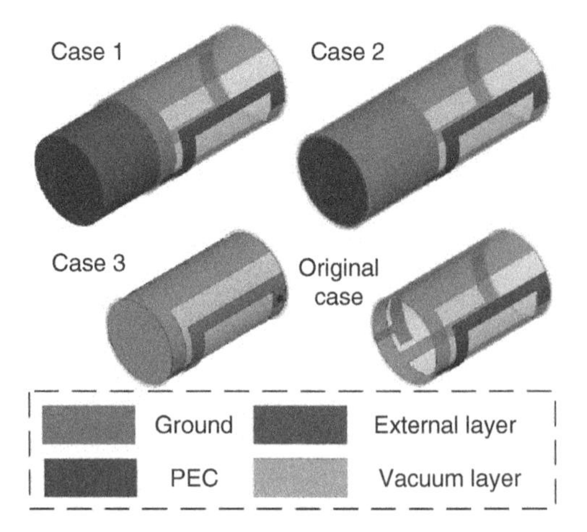

Figure 4.31 Simplified schematic for different simulation cases as in the real implantation scenarios.

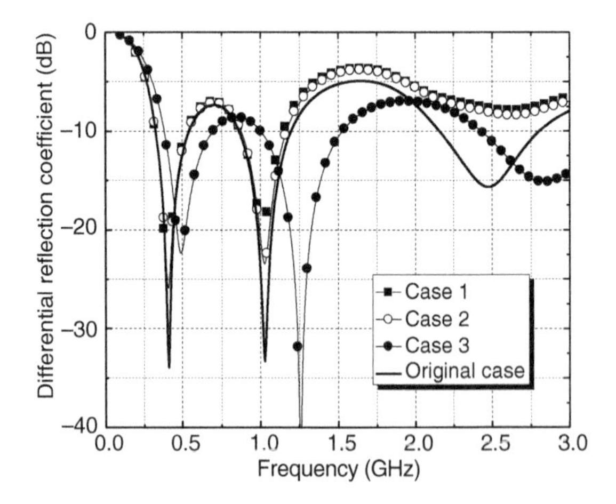

Figure 4.32 Comparison of differential reflection coefficients of flexible antenna for different simulation setups.

capsule substituted by vacuum layer as in case 3, all resonance frequencies shift upwards. However, the matching is still decent for all resonance frequencies.

Besides the reflection coefficients, radiation performance of the antenna should be studied. The realized gains of the planar antenna are −30.6 dB for the MICS band and − 19.1 dB for the ISM band.

The coupling strength between the external half-wavelength dipole and the implanted antenna is shown in Figure 4.33. The external dipole is constructed on Roger RO4003 ($\varepsilon_r = 3.55$, $\tan\delta = 0.0027$) with thickness of 32 mil (0.813 mm). For each frequency band, a different dipole is adopted, and the total length of the dipole is 336.5 mm for MICS band and 51.5 mm for the ISM band. Figure 4.33 indicates that for extremely close distances at 20 mm, for instance, the coupling at ISM band is 6.7 dB higher than MICS band. Because of this distance, antennas are both within the near-field and the gain for the ISM band

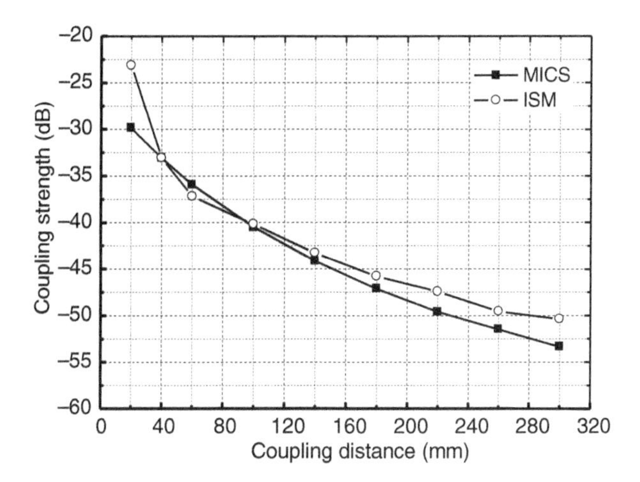

Figure 4.33 Coupling strength of external half-wavelength dipole with planar antenna for MICS band and ISM band.

is larger. As the coupling distance increases, for the ISM band it gradually enters the intermediate region, while for the MICS band, it is still within near-field region. Therefore, the coupling strength for the MICS band takes over between a coupling distance of 40 and 100 mm. For coupling distance larger than 100 mm, the coupling at ISM band dominates again due to larger gain values.

For the flexible antenna, as the gain for the stomach implantation varied significantly when the antenna was moved from one position to another, only the gain of HFSS one-layer implantation was given, which is −30.5 dB for MICS band, and − 22.2 dB for ISM band. The gain values are similar to that of the planar case, so the coupling strength evaluation is not given for the flexible case.

4.4.3 Coating and In Vitro Measurement

After fabrication, both the planar antenna and the flexible antenna are coated by parylene-C with thickness of 0.1 mm. For the parylene to be coated on both sides of the antenna, some clips were used to hold the antennas and stick them up, and then the clips were pasted to the coating platform. For every operating process of 8 hours, 25 μm of parylene coating can be deposited, and 25 g of parylene is needed according to our test samples. Therefore, four iterations are performed for a total thickness of 100 μm.

The fabricated antennas after parylene coating are shown in Figure 4.34. The in vitro measurement is performed in minced pork with a four-port network analyzer Rohde & Schwarz R&S ZVA 50, which can perform differential return loss (RL) measurement directly, as shown in Figure 4.34. The differential reflection coefficient of the planar antenna is shown in Figure 4.35. It means that the results are generally the same at MICS band, while at higher frequencies, the curve of HFSS_lumped differs with the measurement result to a greater extent. This is caused by the effect of SMA connectors and cables being connected to the ports, which is neglected in lumped model. In wave port model, with added cables alongside the ports, the curve of HFSS_wave is closer to the measurement result.

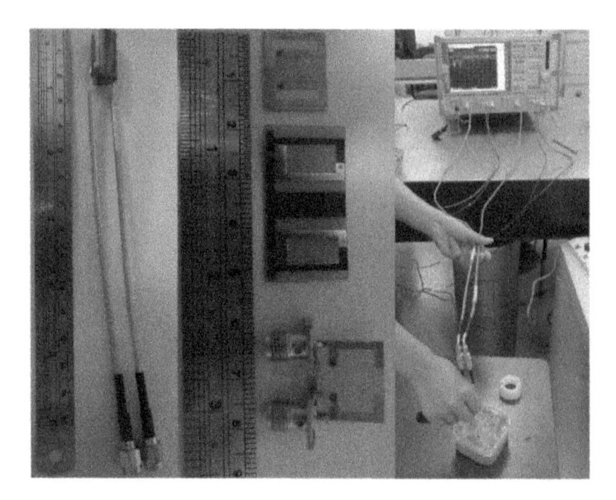

Figure 4.34 Fabricated implantable antenna and measurement setup for the implantable antenna.

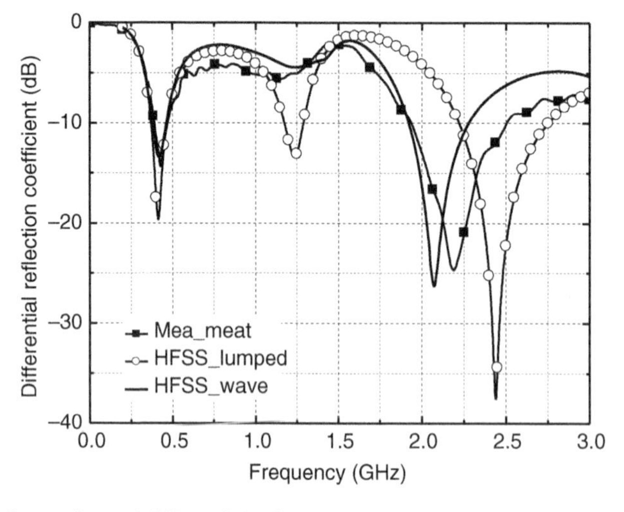

Figure 4.35 Comparison of differential reflection coefficient of measurement and simulation results for the planar antenna.

The comparison of differential reflection coefficients for flexible capsule antenna is shown in Figure 4.36. It indicates that for the flexible antenna, the ISM is even more severely affected. Also, the measurement result is closer to the curve of HFSS_wave due to the effect brought by the small ground and feeding cable.

4.5 Miniaturized Differentially Fed Dual-band Implantable Antenna

In a recently published paper [34], a differentially fed dual-band implantable antenna has been presented and discussed. The antenna's two center working frequencies are

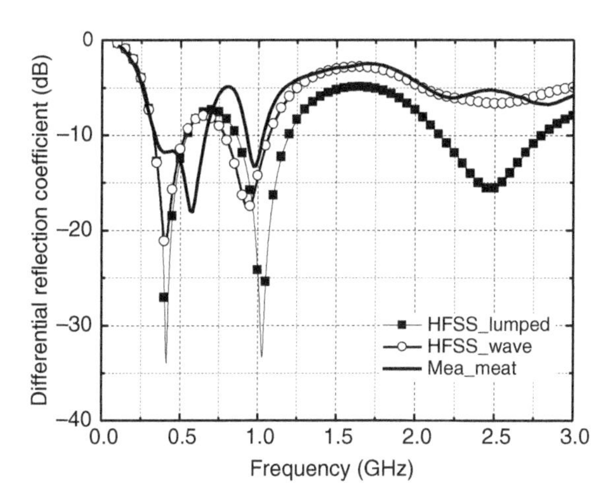

Figure 4.36 Comparison of differential reflection coefficient of measurement and simulation results for the flexible antenna.

both around the 401–406 MHz Medical Device Radiocommunications Service (MedRadio) band, and they exhibit narrow bandwidths. However, for practical applications, it is preferable to have dual-band implantable antennas that operate at both the MedRadio and ISM bands. This way, the two bands can be separately used for a sleep mode and a wake-up mode, offering more versatility in implantable device functionality.

Also, the size of the antenna in [34] is 27 mm × 14 mm × 1.27 mm (480.06 mm^3), which is a little too large than the single-fed implantable planar inverted-F antenna. This section will focus on a miniaturized differentially fed dual-band implantable antenna [35]. A shorting strip is implemented in this design to achieve the size reduction. Working principle of the proposed antenna will be analyzed.

4.5.1 Miniaturized Dual-band Antenna Design

Figure 4.37 shows the configuration of the differentially fed antenna with a relatively small size of 308.6 mm^3 (27 mm × 9 mm × 1.27 mm). The antenna is designed by using 0.635 mm thick Rogers 3010 for both substrate and superstrate. The antenna is fed by two 50 Ω input coaxial cables operating in the differential mode. They are separately located at ports 1 and 1′, which are 3.2 mm away from the center of y-axis to offer sufficient room to solder the SMA connectors in measurement. The detailed dimensions are l_1 = 12.6 mm, l_2 = 8.6 mm, l_3 = 1.7 mm, l_4 = 4.5 mm, l_5 = 1.5 mm, l_6 = 4.6 mm, l_7 = 10 mm, l_8 = 0.9 mm, l_s = 22 mm, w_1 = 1.3 mm, w_2 = 0.6 mm, w_3 = 0.5 mm, s = 1.4 mm, and h = 1.27 mm.

The geometry of the antenna is symmetrical with respect to the x-axis with two arms applied at either side to achieve the dual-band operation. The meandered arm with longer length is mainly resonant at the 401 MHz MedRadio band; the shorter L-shaped one primarily contributes to the 2.4 GHz ISM band. For the MedRadio frequency band, a shorting strip between two sides acts as an inductive loading and compensates the capacitive effect of the longer meandered arms to achieve impedance matching. By making full use of the

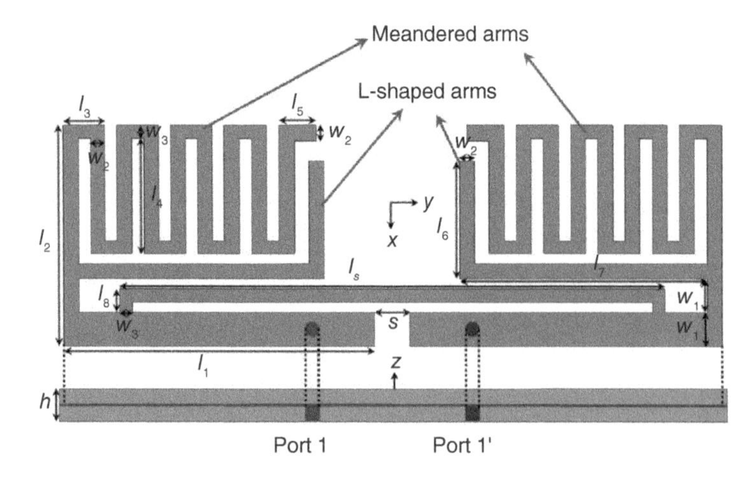

Figure 4.37 Geometry of the miniaturized differentially fed antenna.

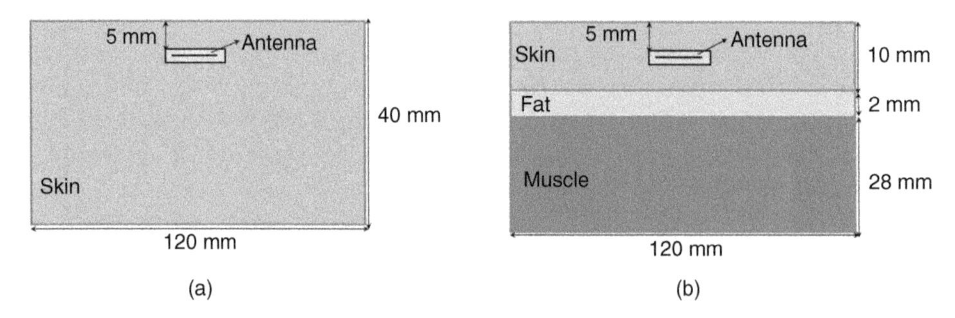

Figure 4.38 Simplified geometries of (a) one-layer tissue model, (b) multilayer tissue model (not in scale).

space between the shorting strip and the longer arms, shorter arms resonant at 2.4 GHz ISM band can be established.

Initially, one-layer and multilayer tissue models, which are with the same electrical properties of human skin, fat, and muscle at 401 MHz and 2.4 GHz, are used in simulation (as presented in Figure 4.38). The numerical values of permittivity and conductivity for biological tissue models at the operating frequencies are from [17]. The dimensions of the one-layer skin model are 120 mm × 120 mm × 40 mm and the antenna is put in the middle of the tissue box, which is 5 mm from the top surface and 35 mm from the bottom surface. Performance of the proposed antenna inside the multilayer tissue model has also been simulated and compared with the one-layer model. To keep consistence, the antenna is located at the same position as the skin layer, and the dimensions of the tissue model are kept the same as the one-layer model.

A shorting strip is introduced as an inductive loading, which is used to compensate for the capacitive effect of the meandered arms. To verify this working mechanism, the dual-band antenna was modified to a single-fed one in Figure 4.39. As can be seen, only half side of the differentially fed antenna is used and the end of the shorting strip is directly

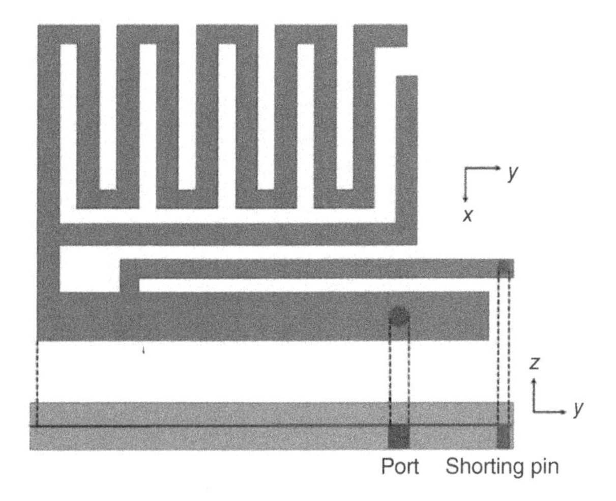

Figure 4.39 Configuration of the corresponding single-fed antenna.

connected to the ground with a shorting pin located 1 mm from the strip end. The input impedance of the single-fed antenna is 50 Ω. Figure 4.40 shows the comparison between the curves of reflection coefficients under different feeding methods, inside both one-layer and multilayer tissue boxes. It is noticed that, the working frequencies and bandwidths of the curves with multilayer tissue model agree with the results inside one-layer skin model. Meanwhile, the curves of the single-fed antenna and differentially fed one match well at desired bands. Based on the above simulation results, the differentially fed antenna can resonate at the desired frequencies with sufficient bandwidths for reflection coefficients less than −10 dB. For the single-layer case, the bandwidths are 30 MHz at MedRadio band and 119 MHz at ISM band. While for the multilayer case, the bandwidths are 32 and 112 MHz, respectively.

A CST human voxel model considering the anatomy of a human body is used in simulation for better assess of the proposed antenna. Simulated results of the proposed

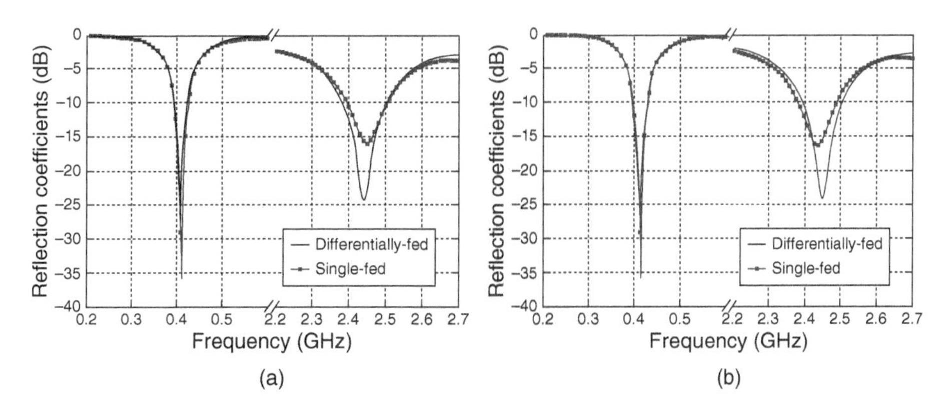

Figure 4.40 Reflection coefficients comparison of the differentially fed antenna and the single-fed antenna inside (a) one-layer tissue model, (b) multilayer tissue model.

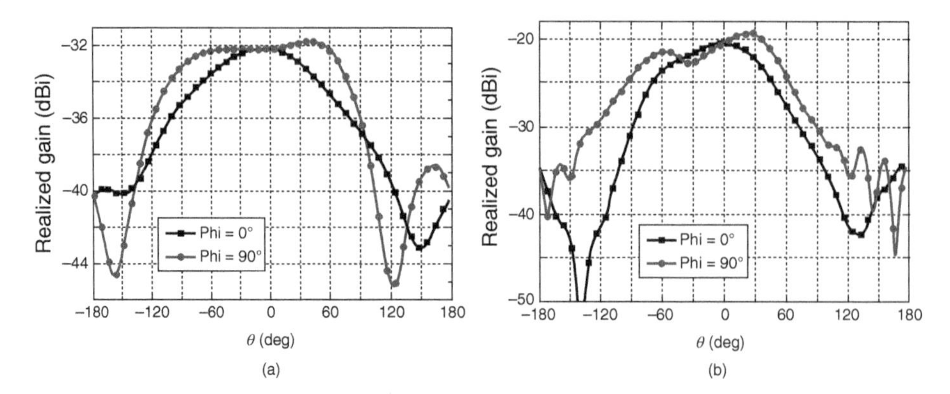

Figure 4.41 Simulated radiation patterns at (a) 401 MHz, (b) 2.45 GHz.

antenna (beneath the muscle layer of the upper arm) with CST human body model demonstrate that the maximum realized gains of the proposed antenna are −31.85 dBi at 401 MHz and − 19.34 dBi at 2.45 GHz. The radiation patterns at the above two frequencies are illustrated in Figure 4.41. It can be seen that both of the radiations are almost toward the boresight. Though the back-lobe is only around 8 dB lower than the boresight gain at 401 MHz, the value in the main direction is sufficient for biomedical applications.

For completeness, realized gains of the corresponding single-fed antenna have also been simulated, which are −39.46 dBi at 401 MHz and − 20 dBi at 2.45 GHz, respectively. Although the single-fed and differentially fed antennas would have similar impedance bandwidths, the realized gain of the differentially fed antenna is much higher than that of the single-fed antenna. Note that the radiation area of the differentially fed antenna is double that of the single-fed antenna.

4.5.2 Parametric Analysis and Measurement

4.5.2.1 The Effect of the Shorting Strip

Figure 4.42 presents the simulated odd-mode reflection coefficients with different lengths of the shorting strip. As seen in Figure 4.42(a), this shorting structure helps establish the current path for the resonance at the 401 MHz MedRadio frequency band. In addition, as plotted in Figure 4.42(b), the existence of the shorting strip has no significant influence on the 2.4 GHz ISM band resonance. The length of the shorting strip plays a key role in the impedance matching of this resonance within the MedRadio band, since its inductive characteristic helps to compensate the capacitive loadings. This is proved by the significantly varied resonance depths with different lengths of the shorting strip.

4.5.2.2 The Effect of the Length of L-shaped Arms

The effects of the length of the L-shaped arms were studied by changing the value of l_6 as illustrated in Figure 4.43, for the proposed differentially fed antenna. It is observed that the curves at ISM band change substantially along with the variation of this parameter. When the length increases from 1.6 to 4.6 mm, the center resonance shifts toward lower frequency

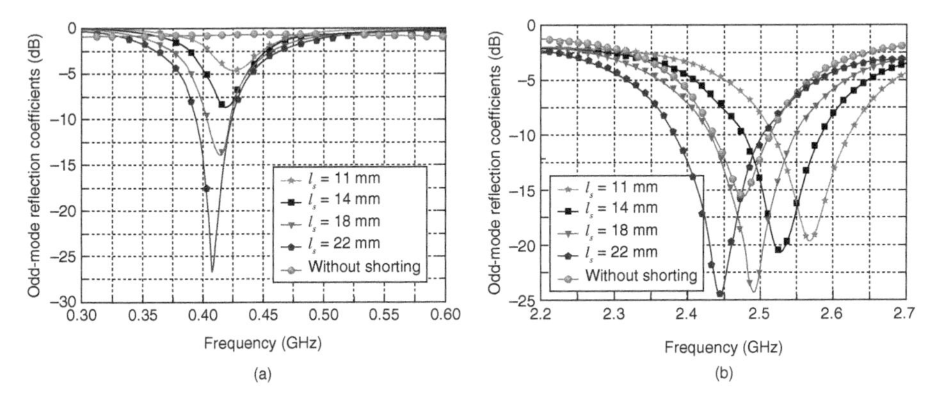

Figure 4.42 Odd-mode reflection coefficients for different conditions of the shorting strip at (a) 401 MHz MedRadio band, (b) 2.4 GHz ISM band.

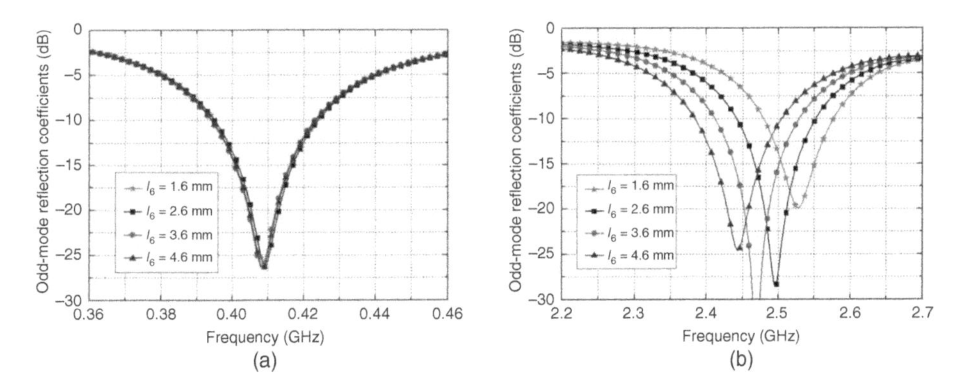

Figure 4.43 Odd-mode reflection coefficients for different lengths of the shorter L-shaped arms at (a) 401 MHz MedRadio band, (b) 2.4 GHz ISM band.

from 2.53 to 2.44 GHz. Therefore, the proposed antenna can operate at the desired ISM frequency band properly. Besides, the simulation results at MedRadio band demonstrate that the performances of the implantable antenna are kept almost unchanged with different l_6. Also as shown, the length of the L-shaped arms mainly affects the resonant frequency, while having little effect on the impedance matching.

4.5.2.3 Measurement

To validate our design, experimental measurements over a prototype were performed. The fabricated antenna with experimental setups of in vitro tests is shown in Figure 4.44. Measurement of the antenna was implemented in a rectangular container, which was filled with skin-mimicking liquids with the same properties of human tissue at the 401 MHz MedRadio band and 2.45 GHz ISM band.

Figure 4.45 depicts the comparison of simulation and measurement for odd-mode reflection coefficients at MedRadio and 2.4 GHz ISM bands. The experimental bandwidths of the differentially fed antenna can cover the 401–406 MedRadio band and the 2.4–2.48 GHz ISM band.

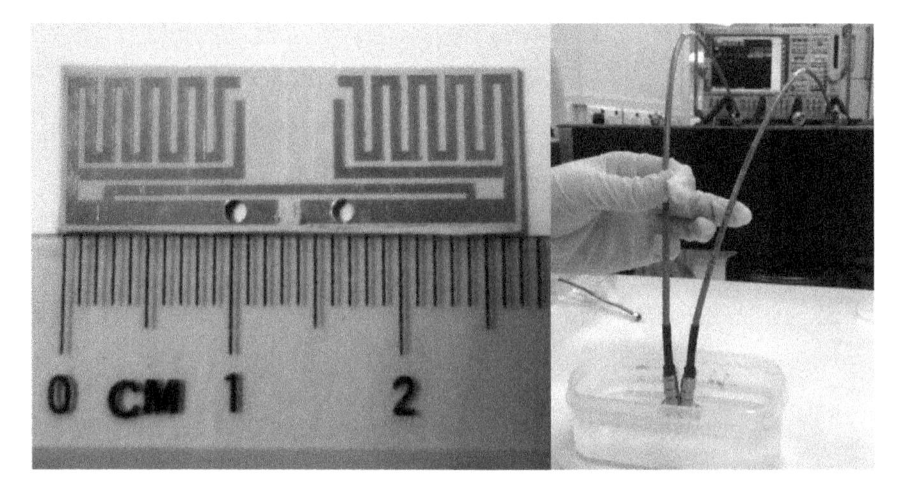

Figure 4.44 Manufacture and experimental setup for in vitro test of the proposed differentially fed antenna.

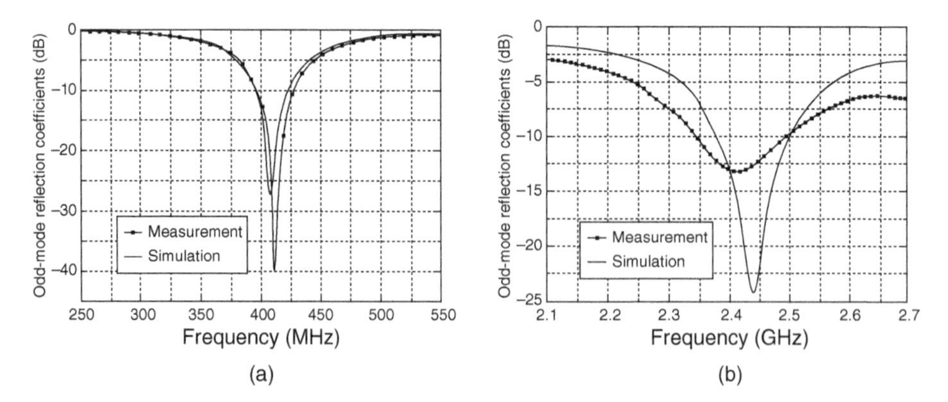

Figure 4.45 Depicts the comparison of simulation and measurement for odd-mode reflection coefficients at MedRadio and 2.4 GHz ISM bands. The experimental bandwidths of the differentially fed antenna can cover the 401–406 MedRadio band and the 2.4–2.48 GHz ISM band.

4.6 Differentially Fed Antenna With Complex Input Impedance for Capsule Systems

The differentially fed antennas described earlier are all designed to match a 50 Ω load. On one hand, commercial RF and microwave chips, such as the one used in [36], or RF chips such as Higgs-4 SOT [37], are usually of a complex input impedance. An additional impedance-matching network is always needed for 50 Ω antennas in these systems. Therefore, it is necessary to use differential-fed implantable antennas with complex impedance designs to achieve full miniaturization for implantable devices. In the previous chapters, we discussed how wideband antenna designs can help ensure the stability of implantable antennas in the human body. For complex impedance design, it is also challenging to tune both the real and imaginary parts of the complex input impedance independently in a wide bandwidth range. In response to these challenges, it is necessary to study differential-fed implantable antennas with complex impedance [38, 39].

In this section, we want to introduce a novel miniaturized antenna with a complex input impedance that can be tuned flexibly [38] as a reference to guide engineering practice. The differential-fed antenna is designed for capsule endoscope systems, achieving significant miniaturization while simplifying the circuit system design. Moreover, the antenna is designed using printed circuit board (PCB) technology, making it more suitable for integration with a capsule endoscope system compared to conformal antennas, which are conformal to the capsule shell. The antenna is circular in shape with a diameter of 10 mm and works at the ISM 2.45 GHz band. Its size is reduced significantly using the meandered-line structure. More importantly, using a novel Π-match together with the coupling between the antenna end and the feed point, the input impedance of the proposed antenna (both the real and imaginary parts) can be tuned flexibly in a very wide bandwidth range. In addition, this input impedance varies smoothly with frequency, which implies a wide bandwidth. Hence, the antenna can be conjugately matched to a transceiver directly without using an additional impedance-matching circuit.

4.6.1 Antenna Geometry

Figure 4 46(a) shows the antenna placed in a capsule shell where all other components are ignored. The capsule shell is made of a photosensitive resin and comprises a cylindrical body and two hemispherical caps located on the top and bottom of the cylindrical body, respectively. It has an inner diameter of 10 mm and an inner length of 25 mm. Limited by the fabrication facility and meanwhile to ensure mechanical strength, the thickness of the capsule shell is chosen to be 0.5 mm. The antenna also has a diameter of 10 mm and is placed at the top portion of the cylindrical body of the capsule shell. It consists of two 10 mil-thick Roger 5880 substrates which are separated by a 1.0 mm-thick air gap. The radiating structure is printed on the top surface of the upper substrate; meanwhile, the bottom surface of the lower substrate is covered by a copper layer (ground). The lower substrate with the ground is used to emulate the PCB existing in the practical capsule system. The presence of the air gap is because there are components mounted on the PCB, and hence, the antenna cannot be attached to the PCB directly. The PCB also helps partially block the influences of other components on the antenna in actual scenarios when those components are added inside the capsule shell. The bottom surface of the upper substrate and the top surface of the lower substrate are empty.

Figure 4.46(b) shows the top surface of the upper layer, that is, the radiating structure and a Π-match of the antenna, in detail. Basically, it is a meandered dipole fed with a Π-match – the Π-match is located in the middle of the substrate, which is enclosed by the dashed line, while the two meandered arms of the dipole are located on the two flanks of the Π-match. The antenna geometry is symmetrical with respect to the central line. The Π-match comprises a section of coplanar strip (CPS) with a shunt copper strip at the upper end. Referring to Figure 4.46(b), the two meandered arms of the dipole are also connected to the upper end of the CPS. The widths of the shunt strip and the meandered arms are w_2 and w_3, respectively. The feed points of the antenna are located on the lower end of the CPS. Two probes are used to feed the antenna – connecting the feed points to the balanced

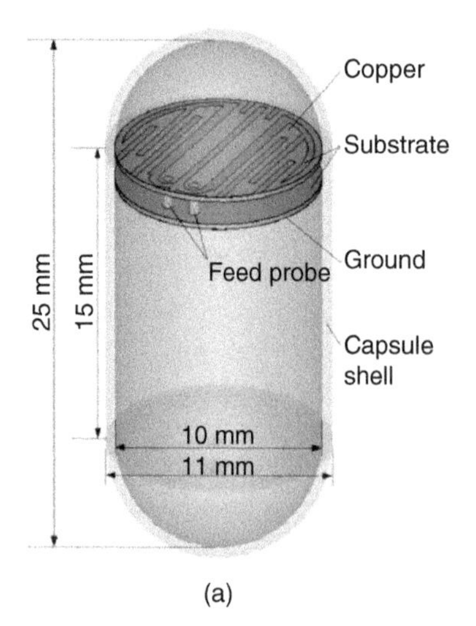

(a)

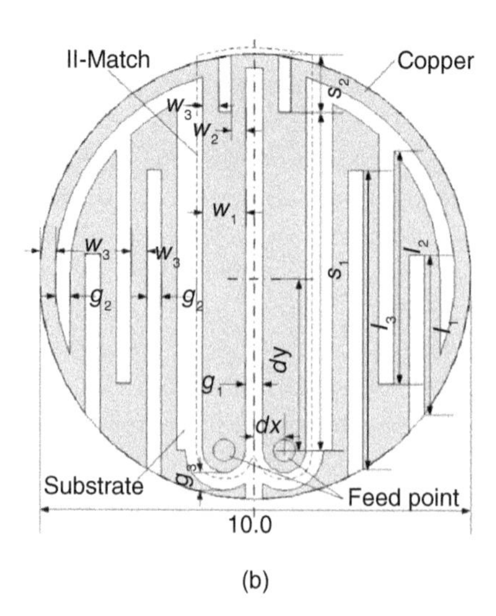

(b)

Figure 4.46 (a) The proposed antenna placed inside a capsule shell. The lower substrate with ground is used to emulate the PCB existing in practical capsule system, (b) the top surface of the upper substrate. The structure enclosed by dash line is the Π-match. The antenna geometry is symmetrical with respect to the center line. $dx = 0.7$, $dy = 3.9$, $g_1 = 0.4$, $g_2 = 0.35$, $g_3 = 0.4$, $s_1 = 7.6$, $s_2 = 1.25$, $w_1 = 1$, $w_2 = 0.35$, $w_3 = 0.36$, $l_1 = 3.6$, $l_2 = 5.2$, $l_3 = 6.7$ (Units: mm).

RF output pins of the transceiver chip on the lower substrate. Two holes are drilled on the upper substrate to let the probes go through.

As shown in Figure 4.46(b), the meandered arms of the dipole are not of uniform width. They are formed simply by carving slots on the copper layer for the sake of fabrication convenience. This is because the substrate is so thin that carving on the substrate should be as less as possible. The lengths of the meandered arms can be changed simply by changing the lengths of these slots l_1, l_2, and l_3. The parallel sections of the meandered arms are of the same width w_3 and the gaps between these parallel sections are of the same width g2. The ends of the meandered arms are very close to the feed points and the mutual coupling between them can be tuned by changing the gap width g3.

The simulation setup of the proposed antenna is shown in Figure 4.47(a). The capsule shell with the antenna inside is put at the center of a muscle-mimicking phantom of size 100 mm × 100 mm × 100 mm. The snapshot of the simulated currents on the proposed antenna is shown in Figure 4.47(b), from which it is found that the proposed antenna indeed is a meandered dipole fed with a Π-match.

4.6.2 Operating Principle

The proposed antenna is designed for the 2.4 GHz transceiver nRF24L01+ [37] and is desired to have an input impedance of $15 + j88\,\Omega$ in the frequency range from 2.4 to 2.525 GHz. As shown in Figure 4.46(b), the antenna size is reduced significantly by meandering the two arms of the dipole. Whereas the input impedance of the obtained meandered dipole is far from the desired value. Therefore, the Π-match and the mutual

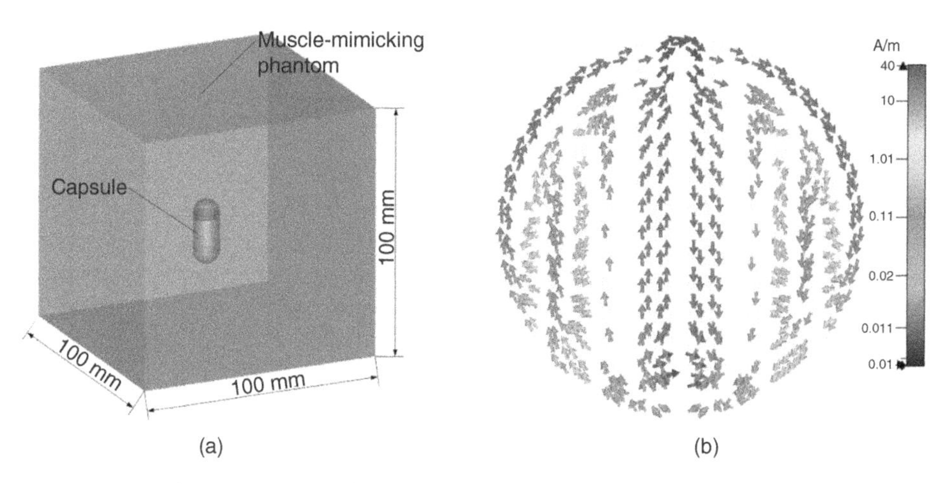

Figure 4.47 (a) Simulation setup of the proposed antenna, (b) snapshot of the simulated currents on the proposed antenna.

coupling between the ends of the meandered dipole and the feed points are used to tune the input impedance of the antenna in a wide frequency range.

4.6.2.1 Equivalent Circuit

The equivalent circuit of the proposed antenna is derived to gain a deep understanding of its working principle. The equivalent circuits of each section of the antenna are shown correspondingly in Figure 4.48(a). Among them, the equivalent circuit of the meandered dipole is adopted from [40] and [41]. To make the equivalent circuit balanced, all the components of the equivalent circuit of the meandered dipole, that is, L_0, C_0, L_1, C_1, and R_1, are divided into two parts on purpose. C_2 denotes the parasitic capacitance between the two strips of the CPS; C_3 denotes the parasitic capacitance between the CPS and the ground; C_4 denotes the parasitic capacitance between the meandered arms and the ground; and C_5 denotes the capacitance between the antenna end and the feed point. L_2 and L_3 denote the inductances

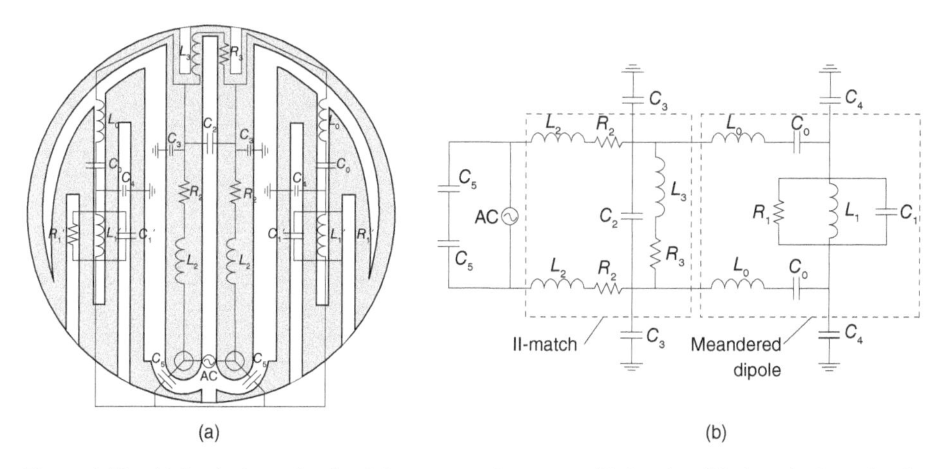

Figure 4.48 (a) Equivalent circuit of the proposed antenna, (b) the simplified equivalent circuit.

of the CPS and the copper strip, respectively. R_2 and R_3 denote the resistances of the CPS and the copper strip, respectively, among which R_2 represents a negligible resistance caused by the conduction loss and R_3 is a radiation resistance.

Figure 4.48(b) shows the equivalent circuit of the proposed antenna after combining components of the meandered dipole so that its equivalent circuit is similar to that of the classic one [41]. The parts corresponding to the meandered dipole and the Π-match have been indicated, respectively, by dash lines in Figure 4.48(b). It can be observed that the structure of Π-match is similar to the symbol Π. The initial values of these components in Figure 4.48(b) are obtained as follows. First, the Π-match is cut away from the antenna, and a lumped port is added directly between the two arms of the dipole to simulate its resistance and reactance over frequency from 0.5 to 5 GHz. L_0, C_0, L_1, C_1, and R_1 are then obtained by following the method proposed in [40, 41] with an assumption that the antenna resonates at 2.45 GHz. Second, the Π-match is simulated separately with the meandered dipole being removed, after which the shunt copper strip of the Π-match is simulated separately again. The initial values of L_2, R_2, C_2, L_3, and R_3 are hereby obtained. Third, the equivalent circuit of the antenna is then built in the Advanced Design System (ADS). The values of C_3, C_4, and C_5 as well as the aforementioned components are tuned to make the obtained input impedance match with the one obtained using CST STUDIO SUITE (CST) as closely as possible. The final values of these components in Figure 4.48(b) are shown in Table 4.5.

The simulated input impedances of the proposed antenna by CST and the equivalent circuit are given in Figure 4.49(a). It is observed that they agree very well across the operating frequency range from 2.4 to 2.525 GHz. Meanwhile, the discrepancy increases slightly as the frequency goes far from this range. This is mainly caused by the fact that in full-wave simulations, the values of the distributed components change with frequency and the muscle is dispersive, whereas these lumped elements of the equivalent circuit are frequency-independent. Figure 4.49(b) shows the simulated input impedances of the proposed antenna placed in muscle, stomach, small intestine, and colon. The input impedances remain almost unchanged in these diverse tissues.

4.6.2.2 Parametric Study

A parametric study is carried out in this section to demonstrate that the proposed antenna is able to realize a wide tuning range of input impedance. As we all know, the lengths of the

Table 4.5 Values of the components of the equivalent circuits shown in Figure 4.48(b).

Capacitor	pF	Inductor	nH	Resister	Ω
C_0	1.21	L_0	0.81		
C_1	0.24	L_1	3.5	R_1	405
C_2	1.6	L_2	2.16	R_2	0
C_3	0.1	L_3	0.59	R_3	0.8
C_4	0.08				
C_5	0.2				

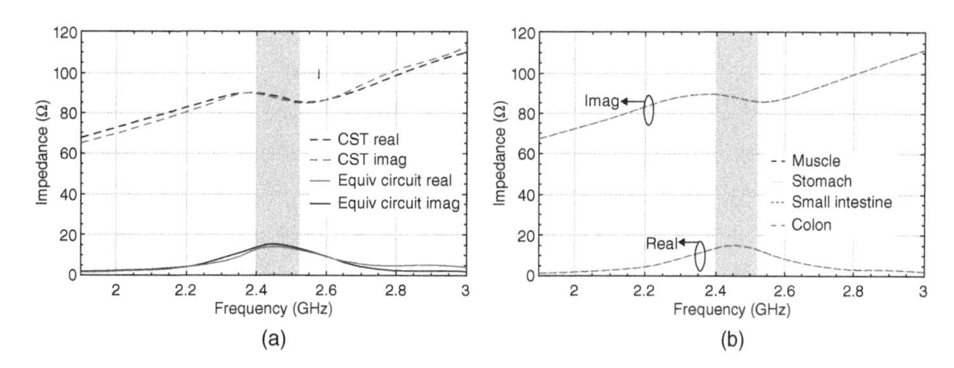

Figure 4.49 (a) Simulated input impedance of the proposed antenna placed in the muscle-mimicking phantom using CST and the equivalent circuit, (b) simulated input impedance of the antenna in diverse tissues.

meandered arms of the dipole directly affect the resonant frequency of the antenna. Hence, it has been skipped here. Then, it is s_1, the length of the CPS, which determines the value of components L_1 and R_1 in the equivalent circuit. The initial value of s_1 is 7.6 mm. The effects of s_1 on the input impedance of the proposed antenna are shown in Figure 4.50. It is observed that with the decrease in s_1, first, the reactance of the antenna decreases in a monotonic manner; second, the resistance near the resonate frequency decreases; third, the resonant frequency shifts downward which is because the arm lengths of the meandered dipole increase as s_1 decreases. It should be noted that the resistance and reactance of the antenna at 2.45 GHz change from 2.0 to 14.1 Ω and from 57.7 to 87.0 Ω, respectively when s_1 increases from 4.6 to 7.6 mm.

As mentioned above, the lengths of the meandered arms change accordingly with s_1; hence, besides L_1 and R_1, the values of nearly all other components of the equivalent circuit have to be calculated and tuned again for every single value of s_1. Thus, the simulated input impedances of the proposed antenna using equivalent circuit have been omitted.

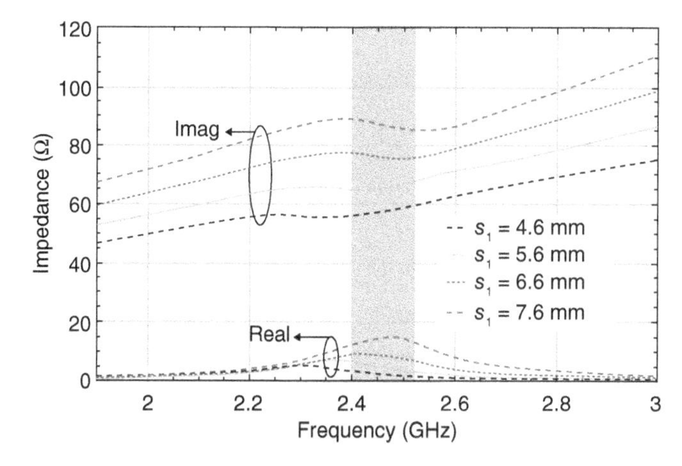

Figure 4.50 Simulated input impedance using CST of the proposed antenna versus s_1 – the length of the CPS. s_1 mainly affects components L_1 and R_1 in the equivalent circuit most significantly.

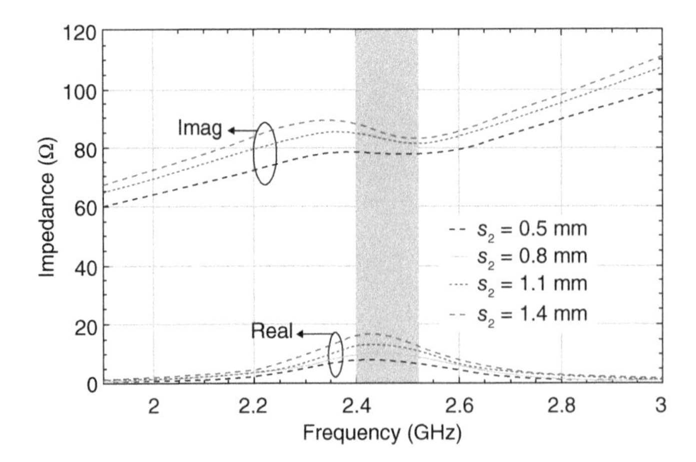

Figure 4.51 Simulated input impedance using CST of the proposed antenna versus s_2 – the length of the shunt copper strip on the CPS. s_1 is shortened slightly to 7.1 mm to avoid the copper strip going beyond the substrate as s_2 is increasing.

Figure 4.51 shows the effects of s_2 on the input impedance of the proposed antenna. Referring to Figure 4.46(b), s_2 is the length of the shunt copper strip on the CPS and it determines components L_3 and R_3 of the equivalent circuit. It is observed that in contrast to s_1, s_2 has equivalent effects on the resistance and reactance of the antenna in terms of the variation in absolute value – the resistance and reactance of the proposed antenna at 2.45 GHz increase from 8.4 to 16.3 Ω and from 77.5 to 84.3 Ω, respectively, when s_2 increases from 0.5 to 1.5 mm. The width of the shunt copper strip on the CPS, w_2, also affects the input impedance considerably. It affects the resistance more significantly than it does on the reactance of the antenna. Figure 4.51 shows the effect of w_2.

By changing g3, the gap width between the arm end and the feed point of the antenna, and the intensity of the mutual coupling between them, that is, *C*5, can be controlled. The effect of g_3 on the input impedance of the antenna is shown in Figure 4.52. It is observed that g_3 has a more substantial effect on resistance than on reactance, which is similar to w_2. The resistance increases from 9.3 to 15.1 Ω, while the reactance changes from 83.9 to 87.0 Ω as g_3 increases from 0.2 to 0.5 mm.

Now the key parameters that affect the input impedance of the proposed antenna considerably have been discussed one by one. By tuning these parameters, that is, s_1, s_2, w_2, g_3, and the arm length, the resistance and reactance of the proposed antenna can be tuned almost independently, and hence a wide tuning range of input impedance can be achieved.

4.6.2.3 Comparison With T-Match

To further demonstrate the advantage of the proposed antenna, a comparison between the proposed antenna and a typical miniaturized antenna with a complex input impedance is carried out in this section. A dipole antenna with T-match is chosen as the reference antenna, since it has been widely used as tag antennas for the applications of radio frequency identification (RFID), where, similar to the applications of the proposed antenna, a complex input impedance is demanded. All geometries/settings of the reference antenna are the same as those of the proposed antenna, except for the geometries on

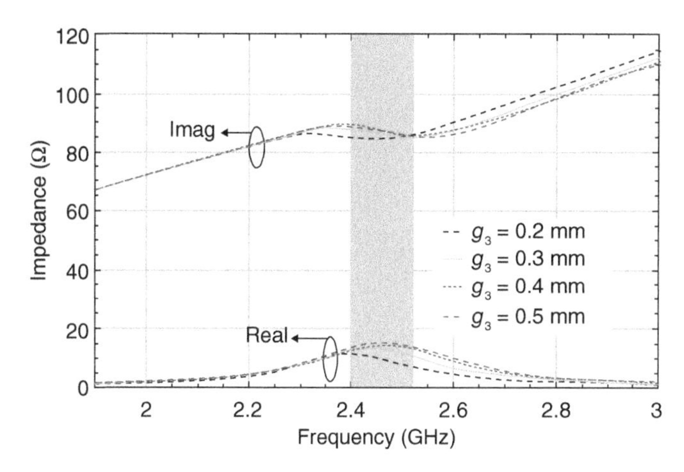

Figure 4.52 Simulated input impedance of the proposed antenna versus g_3 the gap width between the antenna end and the feed point.

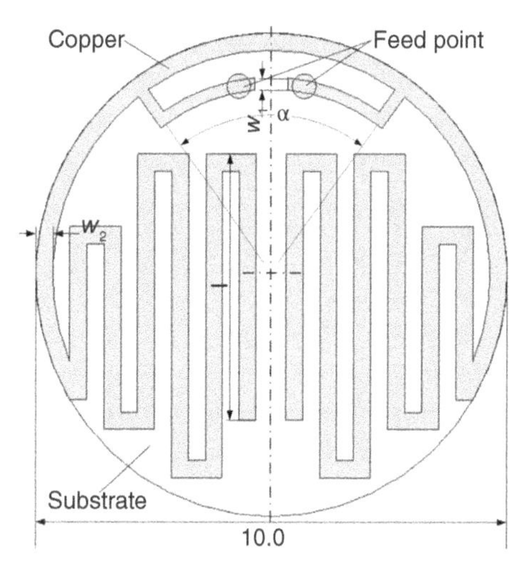

Figure 4.53 The top surface of the upper substrate of the reference antenna – a meandered dipole with a T-match. The geometries are symmetrical with respect to the center line. All other geometries/settings of the reference antenna are the same as those of proposed antenna. The key parameters to tune the input impedance of this reference antenna are α, ratio w_2/w_1, and arm length. $\alpha = 35.5°$, $w_1 = 0.25$, $w_2 = 0.36$, $l = 5.5$ (Units: mm).

the top surface of the upper substrate. Referring to Figure 4.53, the reference antenna is a meandered dipole with T-match, which is fit into a circular area of diameter 10 mm. Besides the lengths of the dipole arms, the length of the copper strip in shunt to the dipole (angle α in this embodiment) and the ratio of w_2/w_1 affect the input impedance of the reference antenna most considerably [42], where w_1 and w_2 are the widths of the strip in shunt and the dipole arms, respectively.

Figure 4.54(a) and (b) shows the effects of angle α and width w_1 on the input impedance of the reference antenna, respectively. Referring to Figure 4.54(a), both the resistance and

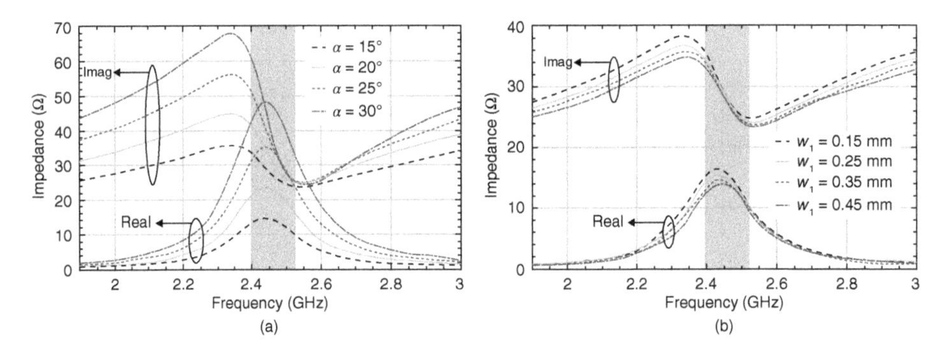

Figure 4.54 Simulated input impedance of the reference antenna versus key parameters (a) α when $w_1 = 0.25$ mm and (b) w_1 when $\alpha = 15°$. The resistance at 2.45 GHz is close to 15 Ω when $\alpha = 15°$.

reactance of this antenna increase in a monotonic manner as α increases from 15° to 30°. But around the resonant frequency of 2.45 GHz, the resistance increases from 14.5 to 48.3 Ω, which is much larger than the desired value of 15 Ω; while the reactance increases from 28.4 to 43.0 Ω, which is far below the desired value of 88 Ω. The resistance and reactance of this antenna cannot reach the desired values simultaneously. More importantly, the reactance around the resonant frequency changes steeply with frequency, which implies that the bandwidth of the reference antenna would be very narrow even if its reactance near 2.45 GHz could be tuned to 88 Ω.

The effect of w_2 on the input impedance of the reference antenna is shown in Figure 4.54(b). It is observed that w_2 has less effect compared with α. Obviously, the resistance and reactance of the reference antenna cannot be tuned to the desired values via changing w_2. In addition, the effect of the dipole arm lengths is revealed in Figure 4.55. By changing α to 35.5° and sweeping the length of the dipole arms, the reactance of the reference antenna can be tuned to the desired value of 88 Ω ($l = 4.3$ mm, black dashed line). But now it is the resistance that deviates from the desired value of 15 Ω. It is also observed that the resistance of the reference antenna changes rapidly with frequency within the desired frequency range.

The RLs of the reference antenna and the proposed antenna are compared in Figure 4.56. Two representative configurations of the reference antenna are chosen since their input impedances are closest to the desired value. They are: (i) $\alpha = 15°$ (black dashed line) in Figure 4.54(a) with an input impedance of $15.1 + j28.6$ Ω at 2.45 GHz and (ii) $l = 4.3$ mm (black dashed line) in Figure 4.55 with an input impedance of $29.1 + j87.1$ Ω at 2.45 GHz. Since the input impedance of the transceiver is complex, the RL is calculated using the reflection coefficient for power waves [43]:

$$\mathrm{RL(dB)} = -20\log_{10}\left|\frac{Z_L - Z_R^*}{Z_L + Z_R^*}\right| \tag{4.10}$$

where Z_L is the input impedance of the antenna, Z_R is the input impedance of the transceiver ($15 - j88\Omega$), and * denotes complex conjugation. It is observed in Figure 4.56 that the advantage of the proposed antenna with Π -match over the reference antenna is overwhelming – it has a wider bandwidth and much lower RL across the desired frequency range.

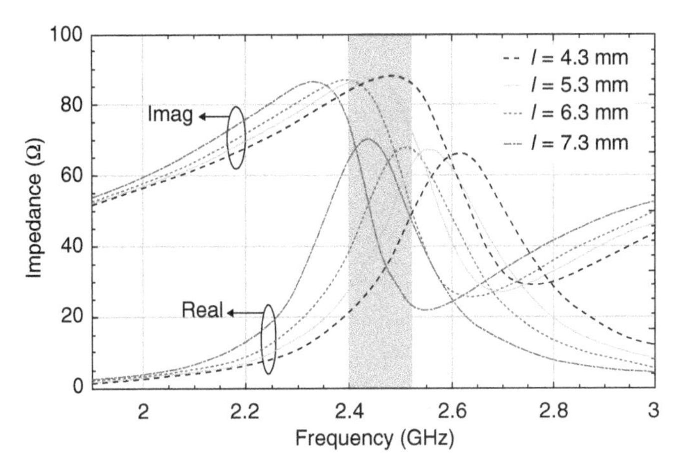

Figure 4.55 Simulated input impedance of the reference antenna versus 1 – the length of the last section of the meander arms, when α =35.5° and w_1 = 0.25 mm so that the reactance is close to the desired value of 88 Ω.

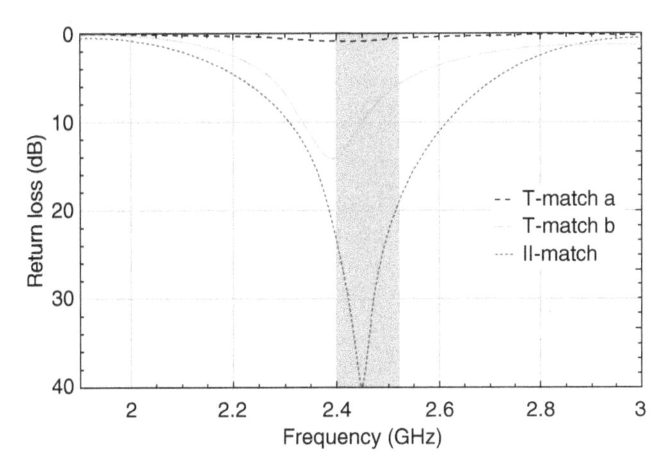

Figure 4.56 Simulated RLs of the reference antenna with T-match of two different configurations (a $\alpha = 15°$ in Figure 4.54 (a) with an input impedance of 15.1 + j28.6 Ω, Figure 4.54(b) corresponds to l = 4.3 mm in Figure 4.55 with an input impedance of 29.1 + j87.1 Ω) and the proposed antenna with Π -match with an input impedance of 15.2 + j88.0 Ω. All the values of input impedances are taken at 2.45 GHz.

Therefore, it can be concluded that the bandwidth of the reference antenna cannot be as wide as that of the proposed antenna, because, for one thing, its resistance and reactance cannot be tuned independently; for another thing, its input impedance near the resonate frequency changes much more rapidly with frequency than does that of the proposed antenna.

4.6.3 Experiment

Figure 4.57(a)–(d) shows the components of the antenna and the antenna after assembling. A Rohacell 31 HF [44] foam of 1 mm thickness is used to support the two substrates. It has

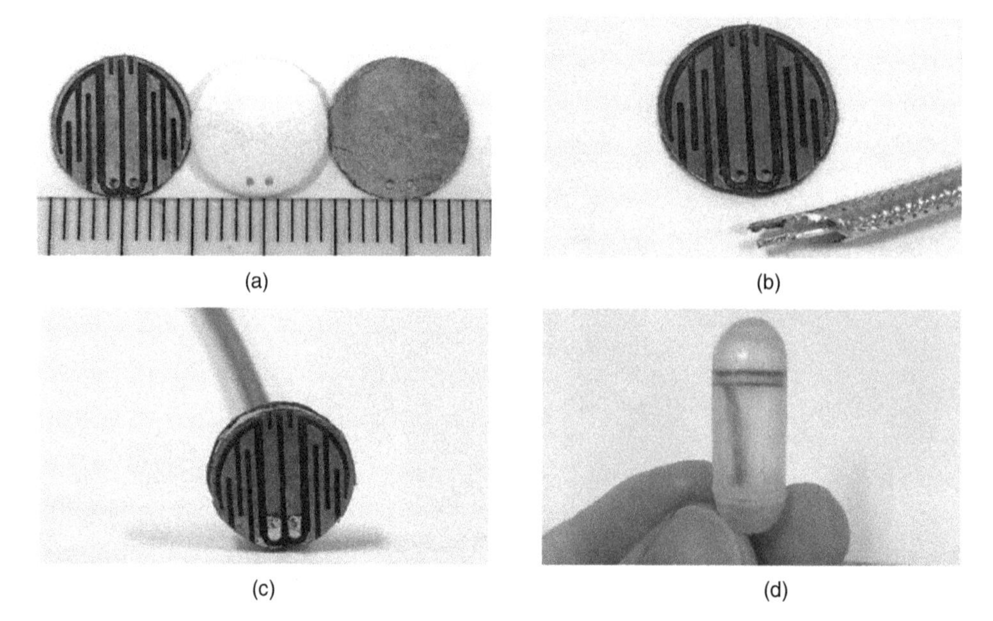

(a) (b) (c) (d)

Figure 4.57 Fabricated prototype of the proposed antenna: (a) two substrates and foam, (b) upper substrate and tapered coaxial cable, (c) fabricated antenna, and (d) fabricated antenna placed inside a capsule shell.

a dielectric constant of 1.05 and a loss tangent of less than 0.0002 at 2.5 GHz, the effects of which on the antenna are negligible. The substrates and foam are adhered together with glue. The antenna is fed by a semirigid coaxial cable. As shown in Figure 4.57(b), the coaxial cable is tapered gradually to a balanced two-conductor transmission line [45], which, after going through the holes on the substrates and foam, is soldered to the feed points on the antenna. The antenna is then inserted into a capsule shell, as shown in Figure 4.57(d), and a hole is drilled into the cap of the capsule shell to allow the coaxial cable to go through. The other end of the coaxial cable, that is, an SMA connector, is connected to a vector network analyzer (VNA). During the measurement, the capsule shell is immerged inside a muscle-mimicking gel, which is about 10 cm × 10 cm × 10 cm in size. The gel is made based on the method in [6, 46] and measured using an HP 85070A dielectric probe kit.

To measure the input impedance of the fabricated antenna precisely, the phase variations and insertion losses caused by the coaxial cable, SMA connector, and adapter have to be taken into account. Therefore, S_{21} (magnitude and phase) of the two coaxial cables of different lengths shown in Figure 4.58 are measured using a VNA. With these two measurements, the phase variations and insertion losses caused by the SMA connector together with the adapter, and the coaxial cable per unit length can be determined. Figure 4.59 shows the RL and phase variation introduced together by the coaxial cable, SMA connector, and adapter, which feed the fabricated antenna in the experiment. The input impedance and RL of the proposed antenna can then be calculated as follows.

It is observed in Figure 4.59 that the insertion loss caused together by the coaxial cable, SMA connector, and adapter is less than 0.14 dB across the desired frequency range, which is negligible. Hence, the reflection coefficient seen at the feed probes of the fabricated antenna

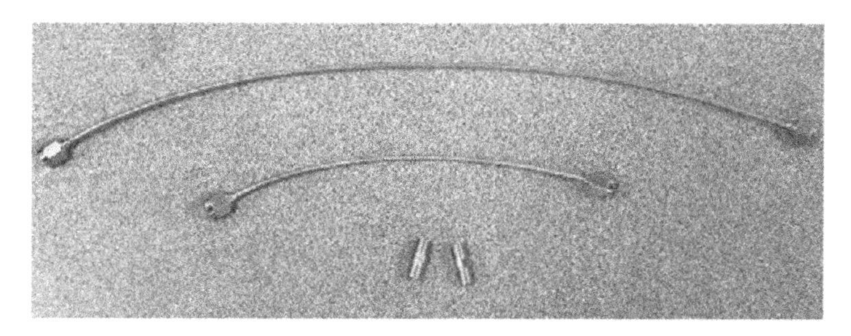

Figure 4.58 S_{21} (magnitude and phase) of the two coaxial cables of different lengths is measured. The phase variations and insertion losses introduced by the SMA connector together with the adapter and coaxial cable per unit length can hence be determined.

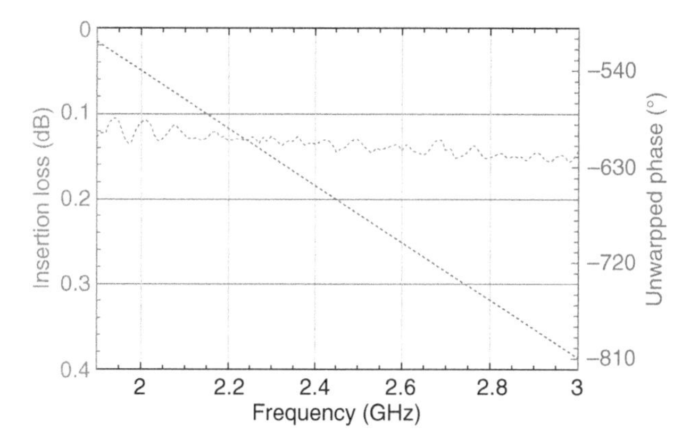

Figure 4.59 Measured insertion loss and phase variation introduced together by the coaxial cable, SMA connector, and adapter, which feed the proposed antenna in the experiment.

Γ can be obtained by [47]

$$\Gamma = e^{-j2\theta} S_{11} \tag{4.11}$$

where S_{11} is measured by the VNA, and θ is the phase variation introduced together by the coaxial cable, SMA connector, and adapter (Figure 4.59). Note that the term $(-j2\theta)$ is positive here since the reference plane is shifted inward the load, namely, the antenna. The input impedance seen at the feed probes of the fabricated antenna Z_{Ant} is then calculated as follows:

$$Z_{\text{Ant}} = \frac{1+\Gamma}{1-\Gamma} Z_0 \tag{4.12}$$

where Z_0 is the characteristic impedance of the coaxial cable, namely, 50 Ω.

Figure 4.60(a) shows the simulated and measured input impedances of the proposed antenna. The two sets of results agree quite well. The simulated and measured RLs of the proposed antenna are depicted in Figure 4.60(b). It is noted that the discrepancy between the simulated and measured RLs is more remarkable than that between the simulated and measured input impedances, which suggests it would be more difficult to match an antenna with complex input impedance than to match an antenna with pure resistance. The discrepancy could be caused by many factors, such as fabrication errors, the prototype being

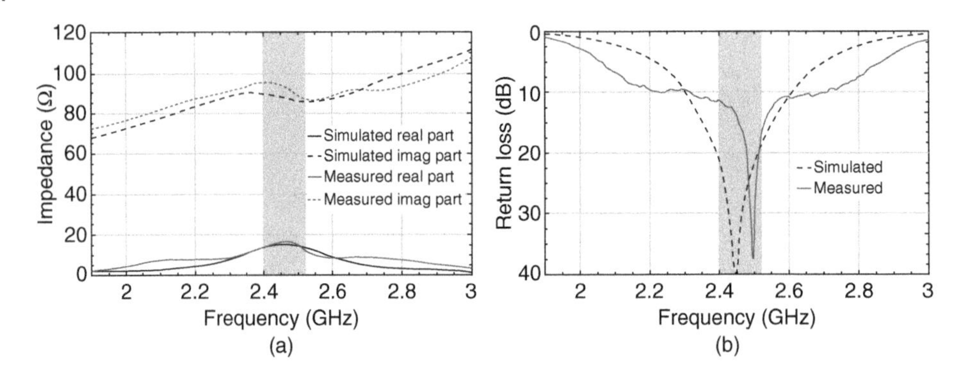

Figure 4.60 Simulated and measured (a) input impedance, (b) RL of the proposed antenna.

deformed, the muscle-mimicking gel deviating from desired properties, the antenna not being located at the position exactly the same as that in simulations, and air gap/bubbles between the capsule and the muscle-mimicking gel. Nevertheless, the measured result is acceptable considering the fact it is difficult to maintain the same environment in simulation and experiment for an antenna operating in body, and the desired frequency range is covered.

Given the fact that the gain and the radiation pattern of an antenna operating in the body depends highly on the location and the surrounding environment of the antenna [48], the novelty of the proposed antenna lies in its capability to flexibly adjust its complex input impedance. Simulations show that it has an omnidirectional pattern with a gain of −32.6 dBi when being placed at the center of the 100 mm × 100 mm × 100 mm muscle-mimicking phantom. As a comparison, the antenna with T-match shown in Figure 4.53 also has a gain of −32.6 dBi in the identical simulation setup.

4.7 Conclusions

The use of differential-fed antennas brings significant advantages, such as improved signal-to-noise ratio and reduced distortion caused by tissue loss and body movements. These benefits contribute to enhancing the overall performance of implantable medical devices relying on wireless communication. In the previous sections, various types of differential-fed implantable antennas were proposed, each with its own strengths and limitations. The selection of a particular antenna design will depend on the specific requirements of the medical device, the implantation location, and the surrounding tissue environment.

The adoption of differential-fed implantable antennas facilitates better integration between the antenna, circuitry, and chip system, eliminating the need for additional matching circuits and baluns. This integration is crucial for designing smaller implantable devices, which is an essential goal in medical applications. However, despite the advantages, there are still challenges that need to be addressed in the development of differential-fed implantable antennas.

One critical challenge is achieving miniaturization without compromising performance. The antennas should be small enough to be embedded in specific phantoms or body parts, such as eyes, heads, blood vessels, or organs, but still maintain efficient and reliable communication. Additionally, ensuring biocompatibility and safety is of utmost importance, as implantable antennas come into direct contact with living tissues.

Another crucial aspect is in vivo verification, which involves testing the antennas in live animals or human subjects to assess their performance in real-world environments. This step is essential to validate the antenna's functionality, safety, and effectiveness in practical medical applications.

Measuring differential-fed implantable antennas accurately remains a significant challenge. While some measurement methods have been presented in this chapter, effective measuring techniques need to be further developed to ensure precise evaluation and characterization of these antennas.

To advance the field, future research should focus on addressing these challenges, aiming to enhance the performance, reliability, and safety of implantable medical devices that utilize differential-fed antennas. By continuing to improve these antennas, we can pave the way for more advanced and efficient medical applications, ultimately benefiting patients and healthcare providers alike.

References

1 Kiourti, A. and Nikita, K.S. (2012). A review of implantable patch antennas for biomedical telemetry: challenges and solutions. *IEEE Antennas Propag. Mag.* 54 (3): 210–228.

2 Kim, J. and Rahmat-Samii, Y. (2004). Implanted antennas inside a human body: simulations, designs, and characterizations. *IEEE Trans. Microwave Theory Tech.* 52 (8): 1934–1943.

3 Soontornpipit, P., Furse, C.M., and Chung, Y.C. (2004). Design of implantable microstrip antennas for communication with medical implants. *IEEE Trans. Microwave Theory Tech.* 52 (8): 1944–1951.

4 Soontornpipit, P., Furse, C.M., and Chung, Y.C. (2005). Miniaturized biocompatible microstrip antenna using genetic algorithm. *IEEE Trans. Antennas Propag.* 53 (6): 1939–1945.

5 Lee, C.-M., Yo, T.-C., Huang, F.-J., and Luo, C.-H. (2008). Dual-resonant π-shape with double L-strips PIFA for implantable biotelemetry. *Electron. Lett.* 44 (14): 837–838.

6 Karacolak, T., Hood, A.Z., and Topsakal, E. (2008). Design of a dual-band implantable antenna and development of skin mimicking gels for continuous glucose monitoring. *IEEE Trans. Microwave Theory Tech.* 56: 1001–1008.

7 Liu, C., Guo, Y.X., and Xiao, S. (2012). Compact dual-band antenna for implantable devices. *IEEE Antennas Wirel. Propag. Lett.* 11: 1508–1511.

8 Xu, L.J., Guo, Y.X., and Wu, W. (2014). Miniaturized dual-band antenna for implantable wireless communications. *IEEE Antennas Wirel. Propag. Lett.* 13: 1160–1163.

9 Huang, F.-J., Lee, C.-M., Chang, C.-L. et al. (2011). Rectenna application of miniaturized implantable antenna design for triple-band biotelemetry communication. *IEEE Trans. Antennas Propag.* 59 (7): 2646–2653.

10 Liu, C., Guo, Y.X., and Xiao, S. (2014). Capacitively loaded circularly polarized implantable patch antenna for ISM-band biomedical applications. *IEEE Trans. Antennas Propag.* 62 (5): 2407–2417.

11 Xu, L.J., Guo, Y.X., and Wu, W. (2015). Miniaturized circularly polarized loop antenna for biomedical applications. *IEEE Trans. Antennas Propag.* 63 (3): 922–930.

12 Liu, C., Guo, Y.X., and Xiao, S. (2014). Circularly polarized helical antenna for ISM-band ingestible capsule endoscope systems. *IEEE Trans. Antennas Propag.* 62 (12): 6027–6039.

13 Chen, Z.M., Cheng, K.W., Zheng, Y.J., and Je, M. (2011). A 3.4 mW 54.24 Mbps burst-mode injection-locked CMOS FSK transmitter. In: *IEEE Asian Solid State Circuits Conference*, 289–292.

14 Bockelman, D.E. and Eisenstadt, W.R. (1995). Combined differential and common-mode scattering parameters: theory and simulation. *IEEE Trans. Microwave Theory Tech.* 43 (7): 1530–1539.

15 Bockelman, D.E. and Eisenstadt, W.R. (1997). Pure-mode network analyzer for on-wafer measurements of mixed-mode S-parameters of differential circuits. *IEEE Trans. Microwave Theory Tech.* 45 (7): 1071–1077.

16 Liu, W., Sivaprakasam, M., Wang, G. et al. (2005). Implantable biomimetic microelectronic systems design. *IEEE Eng. Med. Biol. Mag.* 24 (5): 66–74.

17 Gabriel, S., Lau, R.W., and Gabriel, C. (1996). The dielectric properties of biological tissues. *Phys. Med. Biol.* http://niremf.ifac.cnr.it/tissprop (Oct. 2004).

18 Recommendation ITU-R RS.1346 (1998). Int. Telecommun. Union, Geneva, Switzerland.

19 Wise, K.D., Anderson, D.J., Hetke, J.F. et al. (2004). Wireless implantable microsystems: high-density electronic interfaces to the nervous system. *Proc. IEEE* 92 (1): 76–97.

20 Muller, R., Le, H.-P., Li, W. et al. (2015). A minimally invasive 64-channel wireless μECoG implant. *IEEE J. Solid-State Circuits* 50 (1): 344–359.

21 Bahrami, H., Mirbozorgi, S.A., Rusch, L.A., and Gosselin, B. (2015). Backscattering neural tags for wireless brain-machine interface systems. *IEEE Trans. Antennas Propag.* 63 (2): 719–726.

22 K. S. J. Pister, J. M. Kahn and B. E. Boser (2001). SMART DUST: autonomous sensing and communication in a cubic millimeter. http://robotics.eecs.berkeley.edu/~pister/SmartDust.

23 Research news: cameras out of the salt shaker (2011). http://www.fraunhofer.de/en/press/research-news/2011/march/cameras-out-of-the-salt-haker.html.

24 Nam, K.T. et al. (2008). Stamped microbattery electrodes based on self-assembled M12 viruses. *PNAS* 105 (45): 17227–17231.

25 Zhang, Y.P. and Wang, J.J. (2006). Theory and analysis of differentially driven microstrip antennas. *IEEE Trans. Antennas Propag.* 54 (4): 1092–1099.

26 Fenwick, R. (1965). A new class of electrically small antennas. *IEEE Trans. Antennas Propag.* 13: 379–383.

27 Best, S.R. and Morrow, J.D. (2003). On the significance of current vector alignment in establishing the resonant frequency of small-space filling wire antennas. *IEEE Antennas Wirel. Propag. Lett.* 2: 201–204.

28 CST Microwave Suite Ver. 2012.08, CST Voxel human body model. http:// www.cst.com.

29 IEEE Standard for Safety Levels with Respect to Human Exposure to Radio Frequency Electromagnetic Fields, 3 KHz to 300 GHz (1999), IEEE Standard C95.1–1999.

30 IEEE Standard for Safety Levels with Respect to Human Exposure to Radiofrequency Electromagnetic Fields, 3kHz to 300 GHz (2005), IEEE Standard C95.1–2005.

31 Bahrami, H., Mirbozorgi, S.A., Rusch, L.A., and Gosselin, B. (2015). Biological channel modeling and implantable UWB antenna design for neural recording systems. *IEEE Trans. Biomed. Eng.* 62 (1): 88–98.

32 Liu, C., Guo, Y.-X., Liu, X., and Xiao, S. (2016). An integrated on-chip implantable antenna in 0.18-μm CMOS technology for biomedical applications. *IEEE Trans. Antennas Propag.* 64 (3): 1167–1172.

33 Duan, Z., Guo, Y.X., Je, M., and Kwong, D.L. (2014). Design and in vitro test of differentially fed dual-band implantable antenna operating at MICS and ISM bands. *IEEE Trans. Antennas Propag.* 62 (5): 2430–2439.

34 Duan, Z., Guo, Y.X., Xue, R.F. et al. (2012). Differentially-fed dual-band implantable antenna for biomedical applications. *IEEE Trans. Antennas Propag.* 60 (12): 5587–5595.

35 Lei, W. and Guo, Y.-X. (2015). Miniaturized differentially fed dual-band implantable antenna: design, realization, and in vitro test. *Radio Sci.* 50: https://doi.org/10.1002/2014RS005640.

36 Zhang, K., Liu, C., Liu, X. et al. (2018). A conformal differentially fed antenna for ingestible capsule system. *IEEE Trans. Antennas Propag.* 66 (4): 1695–1703.

37 nRF24L01+ single chip 2.4GHz transceiver (2008). https://www.nordicsemi.com/Products/Low-power-short-range-wireless/nRF24-series.

38 Bao, Z. and Guo, Y.X. (2020). Novel miniaturized antenna with a highly tunable complex input impedance for capsules. *IEEE Trans. Antennas Propag.* 69 (6): 3106–3114.

39 Wang, H., Feng, Y., and Guo, Y. (2021). A differentially fed antenna with complex impedance for ingestible wireless capsules. *IEEE Antennas Wirel. Propag. Lett.* 21 (1): 139–143.

40 Liao, Y., Hubing, T.H., and Su, D. (2012). Equivalent circuit for dipole antennas in a lossy medium. *IEEE Trans. Antennas Propag.* 60 (8): 3950–3953.

41 Hamid, M. and Hamid, R. (1997). Equivalent circuit of dipole antenna of arbitrary length. *IEEE Trans. Antennas Propag.* 45 (11): 1695–1696.

42 Balanis, C.A. (2005). *Antenna Theory Analysis and Design*, 3e, 523–541. Hoboken, New Jersey: Wiley.

43 Rahola, J. (2008). Power waves and conjugate matching. *IEEE Trans. Circuits Syst. II Express Briefs* 55 (1): 92–96.

44 ROHACELL HF product information (2019). https://www.rohacell.com/product/peek-industrial/downloads/rohacell%20hf%20product%20information.pdf.

45 Duncan, J.W. and Minerva, V.P. (1960). 100:1 Bandwidth balun transformer. *Proc. IRE* 48 (2): 156–164.

46 Yilmaz, T., Karacolak, T., and Topsakal, E. (2008). Characterization and testing of a skin mimicking material for implantable antennas operating at ISM band (2.4 GHz–2.48 GHz). *IEEE Antennas Wirel. Propag. Lett.* 7: 418–420.

47 Pozar, D.M. (2012). *Microwave Engineering*, 4e. Wiley.

48 Bao, Z. (2019). Comparative study of dual-polarized and circularly-polarized antennas at 2.45 GHz for ingestible capsules. *IEEE Trans. Antennas Propag.* 67 (3): 1488–1500.

5

Wearable Antennas for On-/Off-Body Communications

5.1 Introduction

The world of wearable medical devices and consumer electronics has experienced an unprecedented surge in growth in recent years. These multifaceted devices have become an increasingly common feature in our daily lives. From monitoring key health metrics such as heart rate, blood oxygenation, respiration rate, and physical activity, to providing a variety of information and entertainment services such as music, mobile communication, and navigational assistance, these devices form an integral part of the wireless body area network (WBAN).

Groundbreaking advancements in biomedical devices featuring wearable technology have revolutionized healthcare and wellness management. These devices, which ingeniously incorporate wearable antennas, provide continuous and real-time monitoring of vital health parameters, and offer invaluable insights into an individual health status. An innovative biomedical device, as depicted in Figure 5.1, was proposed by researchers [1]. This device leverages a wireless, noninvasive technology that not only meets the measurement accuracy of the current clinical standards for heart rate, respiration rate, temperature, and blood oxygenation but also provides a range of additional features. These include tracking movements and body orientation, quantifying the physiological benefits of skin-to-skin contact, recording the acoustic signatures of cardiac activity, identifying vocal biomarkers related to the tonality and temporal patterns of crying, and monitoring a surrogate measure for systolic blood pressure. Such pioneering devices equipped with wearable antennas have the potential to dramatically improve the standard of neonatal and pediatric critical care.

Furthermore, wearable technology, when equipped with wearable antennas, has found widespread application in a variety of fields, from security systems and sports fitness trackers to internet of things (IoT) devices and WBAN. These cutting-edge technologies, capitalizing on wearable antennas, drastically enhance data transfer efficiency, thereby augmenting their functionality and enriching user experiences. As such, the design and development of wearable antennas become a critical area of focus.

Antennas and Wireless Power Transfer Methods for Biomedical Applications, First Edition.
Yongxin Guo, Yuan Feng and Changrong Liu.

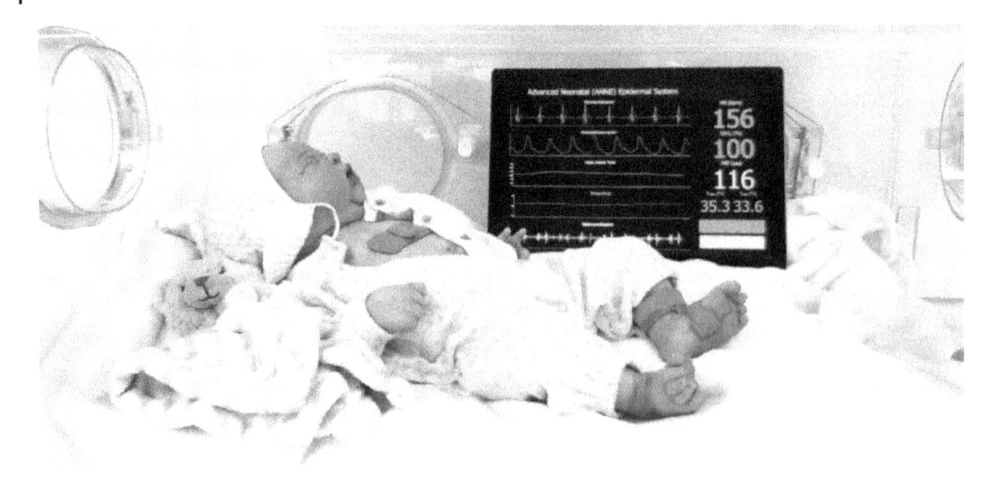

Figure 5.1 Advanced wireless physiological monitoring: wireless wearable devices on the chest and limb for neonatal and pediatric patients in the intensive care units. Source: Chung et al. [1]/Reproduced from Springer Nature.

Despite the significant advancements and substantial potential of wearable antennas, their design and integration remain intricate challenges that require comprehensive solutions for the sustainable progression of wearable technology.

(1) Material Selection: The selection of materials plays a crucial role in determining not only the performance of the antenna but also the comfort and safety of the wearer. The ideal material should be lightweight, flexible, skin-compatible, and capable of withstanding various environmental conditions, such as moisture, heat, and physical wear and tear.
(2) Manufacturing Process: Wearable antennas necessitate the employment of sophisticated manufacturing processes that respect the unique constraints of wearable technology. These considerations include the flexibility of materials, the durability of the final product, the possibility for miniaturization without compromising performance, and the seamless integration of the antenna with the rest of the device components.
(3) Interaction with the Human Body: Due to the close proximity of wearable antennas to the human body, their performance could be affected by signal absorption and reflection caused by the body. This challenge mandates the development of design strategies and shielding techniques that effectively mitigate these interferences, ensuring consistent and reliable performance.
(4) Impact of Bending and Flexibility: Given the dynamic and flexible nature of wearable devices' applications, the antennas incorporated within them need to withstand bending, twisting, and stretching while maintaining optimal performance. The design of these antennas must, therefore, incorporate structural resilience and functional stability, even under varying physical stresses.

In the following sections, we aim to delve deeper into these challenges, offering potential solutions and shedding light on the future direction of wearable antenna design. This approach ensures the continued evolution of wearable technology and its impactful contribution to sectors such as healthcare, fitness, security, and more.

5.2 Exploring Wearable Antennas: Design and Fabrication Techniques

5.2.1 Typical Designs of Wearable Antennas

An array of research and development efforts has been focused on harnessing the potential of conductive textiles for wearable antennas. These innovative materials are crafted primarily using two main approaches. The first involves the process of metallization, which entails coating yarns or fibers with conductive metal [2–7]. The second process involves weaving threads made of conductive metal or conductive polymers into conventional fabric [8, 9]. The suitability of these conductive textiles for wearable applications has been extensively studied and validated. Their flexibility, durability, and excellent electrical conductivity make them apt candidates for creating wearable antennas. Among various antenna designs developed using these materials, microstrip antennas and their arrays [2, 5–7, 10] have emerged as the preferred choice. This can be attributed to their compact size, conformability, and relatively lower sensitivity to the wearer's body.

The integration of antennas into wearable garments significantly enhances their utility by offering a larger coverage area and closer proximity to the human body. These advantages, however, could inadvertently influence the performance characteristics of the antennas. In comparison to other antenna types, such as dipole antennas, microstrip antennas are found to be less affected by the human body; thanks to the presence of a ground plane in their design. Figure 5.2 showcases a practical implementation of this concept with an eight-element microstrip antenna array. This array is fabricated using Nora, a conductive fabric with a sheet resistance of 0.03 Ω/sq. The antenna operates within the 2.45 GHz Industrial, Scientific, and Medical (ISM) band, demonstrating its relevance and effectiveness in real-world applications. Similarly, Figure 5.3 highlights a four-element microstrip antenna array, which incorporates a feed network and is created using silver-coated AmberStrand fibers.

The inherent flexibility of wearable antennas demands them to conform to the body-worn textiles they are integrated into. Consequently, the bending and flexing of these textiles are likely to affect the performance of the antennas. Nevertheless, preliminary studies and practical implementations, such as those illustrated in Figures 5.2 and 5.3, have shown that

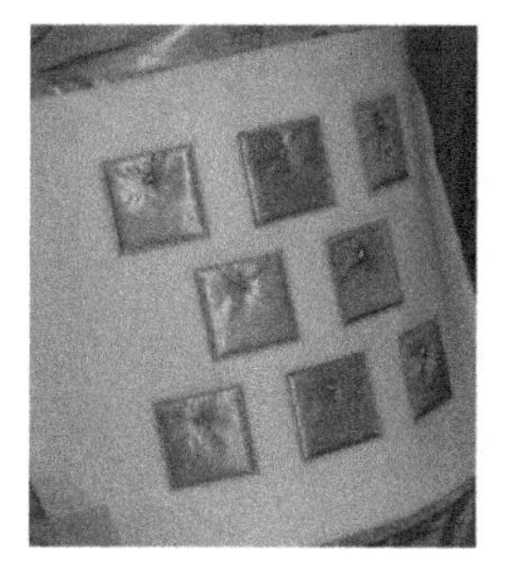
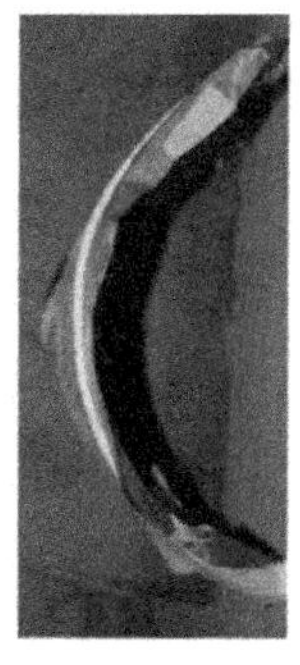

Figure 5.2 Eight-element wearable antenna array operating at 2.45 GHz ISM band being bent outward. (left) Front view and (right) side view. Source: Kennedy et al. [2]/from IEEE.

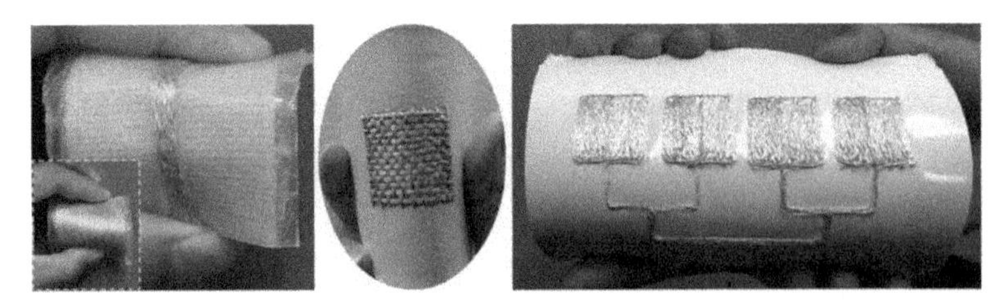

Figure 5.3 From the left: the microstrip transmission line, patch, and patch array with feed lines that are based on embroidered conductive metal–polymer fibers on polymer–ceramic composites. Source: Wang et al. [6]/from IEEE.

these antennas continue to function effectively, even when subject to bending and wearing on the body. This real-world resilience of wearable antennas underscores their potential in shaping the future of wearable technology, offering solutions that are flexible, robust, and efficient.

Wearable antennas' versatility and potential extend well beyond the domains of medical monitoring and consumer electronics within the WBAN. They have been envisaged and realized in an impressive variety of applications that cater to different functional and operational needs. For instance, wearable patch antennas have been proposed for ultra high frequency radio frequency identification (UHF RFID) applications [10]; wearable circularly polarized patch antennas have been proposed for personal satellite communication and navigation [11]; wearable dual-polarized patch antennas have been proposed for firefighter [12]; and a wearable two-antenna system on a life jacket has been proposed for Cospas–Sarsat personal locator beacons [13].

Electromagnetic band-gap (EBG) structures have been effectively incorporated into wearable antennas to mitigate body-bound radiation. Figure 5.4 illustrates a wearable coplanar waveguide (CPW)-fed dual-band antenna positioned on a band-gap array consisting of 3 × 3 elements [14]. The antenna operates around the frequency bands of 2.4 GHz and 5.8 GHz. Utilizing an Artificial Magnetic Conductor (AMC) ground substantially curtails body-bound radiation by 10 dB while enhancing the antenna gain by 3 dB. Additionally,

Figure 5.4 A wearable CPW-fed dual-band antenna on EBG plane with 3 × 3 elements. Source: Zhu et al. [14]/from IEEE.

Figure 5.5 A monopole antenna placed on AMC ground fabricated with flexible and transparent PDMS substrate. Source: Raad et al. [19]/from IEEE.

the antenna retains its performance efficiency when positioned on the thigh, thus demonstrating its effectiveness in wearable applications.

The AMC demonstrates a unique capability to act as ground for antennas [15–17]. Unlike perfect electric conductor (PEC) grounds, antennas on AMC grounds do not necessitate a quarter-wavelength separation, hence, affording a low-profile advantage. AMC structures have found practical implications in wearable applications too. For example, a flexible monopole antenna on an AMC ground, constituted by an EBG array over a copper sheet, has been proposed [18]. Further, a CPW-fed monopole antenna placed above an AMC ground has been reported for telemedicine applications [19]. As depicted in Figure 5.4, both the antenna and the AMC structure exhibit flexibility. The substrate material employed is the flexible and transparent polydimethylsiloxane (PDMS). The AMC ground implementation boosts the antenna front-to-back ratio and gain by 8 dB and 3.7 dB, respectively, thus enhancing the overall performance (Figure 5.5).

Wearable end-fire textile antenna for on-body communication at 60 GHz has also been reported [20]. Stretchable wearable antenna manufactured by using metal/polymer has been reported [21]. Wearable sensors have also been reported [5, 22]. Besides the flexible antennas or sensors that can be made conformal to the body, conventional microstrip antennas with miniaturized sizes are still playing an important role in WBAN applications. For instance, miniaturized dual-band microstrip antennas were proposed for the WBAN applications in [23, 24]. In addition, the switchable antennas for both on- and off-body are also of interest in some scenarios, e.g., the one for 2.45 GHz WBAN applications [25]. These various types of wearable antennas or sensors meet various demands of wearable electronics.

5.2.2 Variation of Antenna Characteristics and Design Considerations

The characteristics of wearable antennas typically vary due to two primary factors. First, the close proximity of a wearable antenna to the human body, which is characterized by a lossy medium with high dielectric constants, often leads to distortions in the radiation pattern and shifts in the resonance frequency. Several studies have analyzed the performance of wearable antennas under different states when placed on a body [2, 3, 11–14, 18, 26, 27]. As previously discussed, incorporating AMC structures with antennas that display boresight radiation [17, 19, 28] can reduce body radiation and make these antennas less susceptible to position changes on the body. This variability in antenna characteristics due to proximity to the human body must be continually considered throughout the design process of any wearable antenna.

Second, deformation, whether bending or crumpling, can lead to significant variations in the antenna characteristics. The double-edged nature of flexibility becomes evident here: while it allows wearable antennas to conform to the wearer's body shape, it also exposes them to the effects of external forces. Consequently, understanding the antenna behavior under various deformation states becomes a priority. A focused study examining the effects of such deformation, particularly bending and crumpling, on a dual-band wearable antenna is outlined in [26]. Typically, a thorough investigation into wearable antennas' characteristics under deformation, much like those demonstrated in Figures 5.2 and 5.3, is a prerequisite before these devices can be deemed fit for use. It is essential to understand how these characteristics might change with physical alterations to ensure the reliability and efficiency of wearable antenna technology in real-world applications.

The wearable antennas are desired to be flexible and they should maintain stable characteristics when being bent or crumpled. In addition, they are also required to comply with the Federal Communications Commission (FCC) regulations concerning human safety with respect to exposure to radio frequency (RF) electromagnetic fields [29, 30], which are the same as those applied to the implantable antennas discussed in previous chapters.

5.2.3 AMC-Backed Near-Endfire Wearable Antenna

As mentioned earlier, enhancing the front-to-back of a wearable antenna means reducing its radiation toward the human body. Meanwhile, the characteristic variation of the antenna due to the presence of the body can also become less significant. It is demonstrated in [28] that a significant improvement in front-to-back ratio for boresight-radiation antennas can be achieved by employing a multilayered AMC ground compared with the PEC or single-layered AMC (SAMC) grounds. Hence, in the following sections, we aim to use a flexible double-layer AMC (DAMC) [31] as an example to illustrate the key issues that need to be considered in the design of wearable antennas.

We have designed an AMC with a minimized unit cell size to achieve a 0° reflection phase at the analytical frequency of 2.44 GHz, and it is utilized as a reflector. We performed a comparative analysis using three different reflectors, PEC, SAMC surface, and DAMC surface, with a view to altering the radiation pattern. The analysis indicated that achieving impedance matching for low-profile antenna geometry is challenging with a PEC-backed Yagi antenna (Yagi_{PEC}). However, a SAMC-backed Yagi antenna ($\text{Yagi}_{\text{SAMC}}$) is capable of maintaining impedance matching while altering the radiation orientation. Moreover, when compared to $\text{Yagi}_{\text{SAMC}}$, a DAMC-backed Yagi antenna ($\text{Yagi}_{\text{DAMC}}$) further reduces radiation toward the human body, thereby improving antenna efficiency and minimizing peak specific absorption rate (SAR) values. An in-depth study of the characteristics of an optimally designed $\text{Yagi}_{\text{DAMC}}$ antenna under deformation has also been conducted.

5.3 Latex Substrate and Screen-Printing for Wearable Antennas Fabrication

The design of the Yagi radiator and AMC surface on the same substrate necessitates the use of thick, flexible materials to facilitate the realization of the periodic metamaterial surface. Fabric materials, inherently thin, demand layer stacking to reach the desired thickness. Coupled with uncertain result repeatability, this led to the selection of latex material from the category of nonwoven wearable materials for this study. To achieve the desired functionality of the wearable antenna, the electromagnetic properties of the material, including permittivity and loss tangent, must be known within the operating frequency band. Accordingly, these properties of latex were characterized using the T-resonator method, as detailed in [32, 33].

The T-resonator consists of a T-shaped pattern, which includes an open-end transmission stub and feed lines, as illustrated in Figure 5.6. It resonates at odd multiples of the quarter-wavelength frequency for which it is designed. This method's fundamental premise for determining dielectric constant values involves designing a resonator structure for a specific primary resonant frequency based on an estimated dielectric constant value. If there is a discrepancy between the estimated and the actual relative permittivity, the measured resonant frequencies will deviate from the designed frequencies.

The T-resonator's open end and T-junction, being discontinuities, affect the resonator's electrical length and induce radiation losses. The open end extends the stub's electrical length, while the T-junction reduces it. The precise stub length is calculated by factoring in corrections due to these discontinuities. The values of the effective dielectric constant can be deduced from the resonant frequencies and corresponding stub lengths. The material's actual dielectric constant values are then calculated from the measured effective dielectric constant values through the iterative process of the closed-form expressions described in [33]. The loss tangent is determined from the measured loaded quality factor values and the calculated unloaded quality factor of the resonator, as well as the quality factors attributable to the conductor and radiation.

Two sets of T-resonator samples were designed (as listed in Table 5.1) and measured using a vector network analyzer. MATLAB coding of the expressions from [33] facilitated the extraction of permittivity and loss tangent values for both samples. With the values obtained

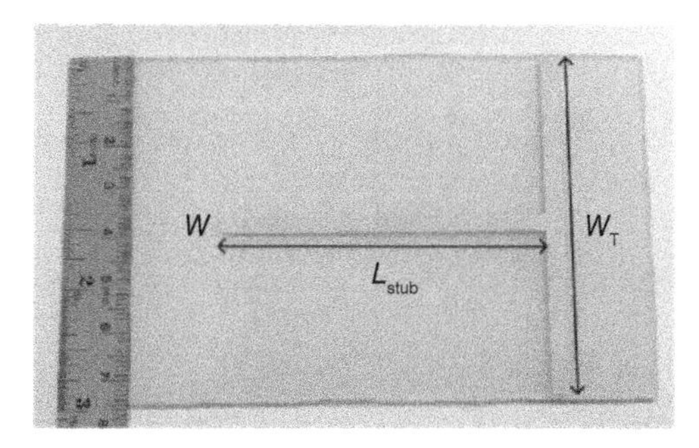

Figure 5.6 T-resonator prototype on latex material (with dimensions listed in Table 5.1 for both the samples in millimeters). Source: Agarwal et al. [31]/from IEEE.

Table 5.1 Dimensions of the *t*-resonators designed (millimeter).

Sample	Thickness	w	L_{stub}	W_T
T1	1	2.5	72.2	72.5
T2	1.5	3.7	72.2	72.5

from both samples being remarkably similar, the final permittivity and loss tangent values, 3.31 and 0.028 respectively, were derived by averaging these values.

Fabrication of the antenna and AMC reflector surfaces involves screen-printing silver ink onto latex substrates, followed by heat curing of the printed ink at 130 °C. This method outperforms the fabrication of textile material antennas in terms of repeatability and reduced manufacturing time. Furthermore, it offers enhanced precision for crafting multilayered periodic meta-surfaces.

In this process, the Yagi antenna is printed on a 1 mm-thick latex layer, while the SAMC is created using a 3 mm-thick grounded latex substrate. The DAMC, on the other hand, is crafted using two separate 1.5 mm-thick latex substrates for each layer, resulting in an overall thickness of 3 mm for both the SAMC and DAMC surfaces. Commercially available flexible Styrofoam (with a relative permittivity, ε_r, approximately equals 1) of 5.5 mm thickness is employed to create a separation between the AMC surfaces and the Yagi radiator for both the $Yagi_{SAMC}$ and $Yagi_{DAMC}$ antenna prototypes.

5.4 AMC-backed Endfire Antenna

All AMC-backed Yagi antennas discussed in this chapter are specifically designed to operate in the 2.4-GHz ISM band, exhibiting near-endfire off-axis radiation suitable for on-body communications. The cross-sectional and top views of a Yagi antenna situated over a DAMC reflector are depicted in Figure 5.7. As demonstrated in Figure 5.8, the feed mechanism utilizes a 50 Ohm coaxial probe linked to the driver element of the Yagi antenna via a surface mount device balun (as shown in the schematic in the inset of Figure 5.8(b)) to facilitate the conversion from unbalanced to balanced signals. The designs of bidirectional-endfire Yagi, unidirectional near-endfire $Yagi_{SAMC}$, and $Yagi_{DAMC}$ antennas are elaborated upon in the following sections.

5.4.1 Bidirectional Yagi Antenna for Endfire Radiation

A traditional planar Yagi antenna is deployed for endfire radiation at 2.4 GHz. The optimized dimensions for the driver, director, and reflector elements are illustrated in Figure 5.8(a). Figure 5.8(b) exhibits the in-house fabricated prototype using Rogers RO4003 material, validating the design concept along with the feeding structure.

5.4.2 Near-Endfire Yagi Antenna Backed by SAMC

The bidirectional-endfire radiation of the Yagi antenna is modified to off-axis near-endfire radiation using the AMC surface that introduces a 0° reflection phase. Known for imparting

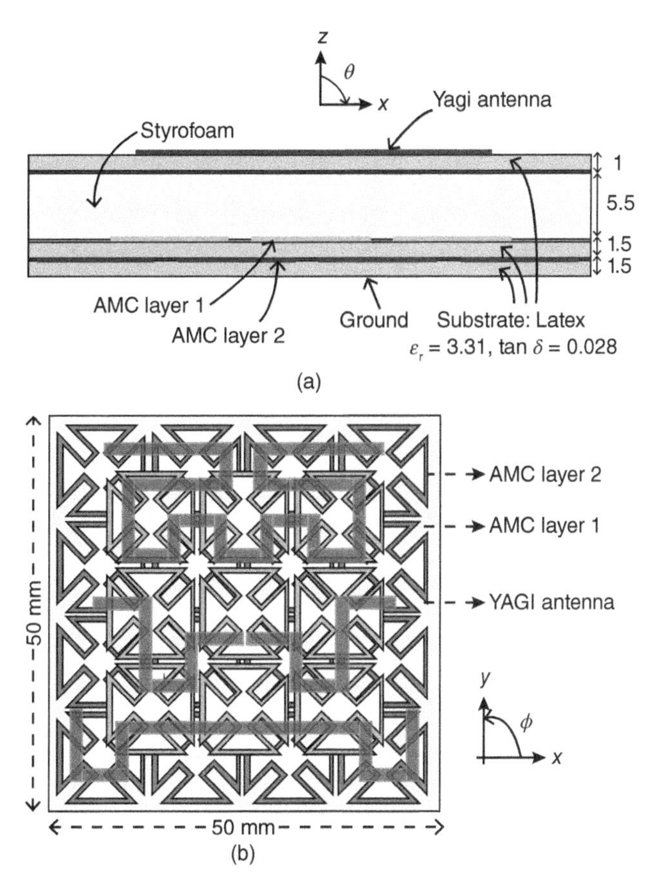

Figure 5.7 Proposed Yagi-antenna over double-layer AMC ground: YagiDAMC: (a) cross-sectional view, (b) top view of Yagi antenna over DAMC reflector (units: mm).

180° and 0° reflection phases respectively, the PEC and PMC surfaces inform the design of our SAMC unit cell. This unit cell is designed to manifest a 0° reflection phase at 2.44 GHz under floquet boundary conditions. A symmetrical SAMC geometry, as shown in the inset of Figure 5.9(a), is crafted to minimize the unit cell size and enhance the stability of the reflection phase response concerning the direction of the incident wave. The SAMC surface is realized using a 3×3 array of these units along the x- and y-axes.

The entire SAMC surface, initially optimized under floquet boundary conditions (considering an infinite ground plane size and infinite number of units), helps in determining the rough dimensions of the unit-cell geometry for the 0° reflection phase frequency. Subsequently, slight optimization of the SAMC surface (and unit-cell parameters) is conducted to account for the finiteness of the realized substrate (50 mm × 50 mm ground plane size) and the limited number of SAMC units. This is accomplished by changing the floquet boundary conditions to PEC and PMC walls (in the direction of E and H fields) for the finite SAMC surface in the simulation model.

Figure 5.9(a) exhibits the reflection phase response of the unit cell for a 0° reflection phase at 2.44 GHz, alongside the final optimized dimensions. The concept verification is

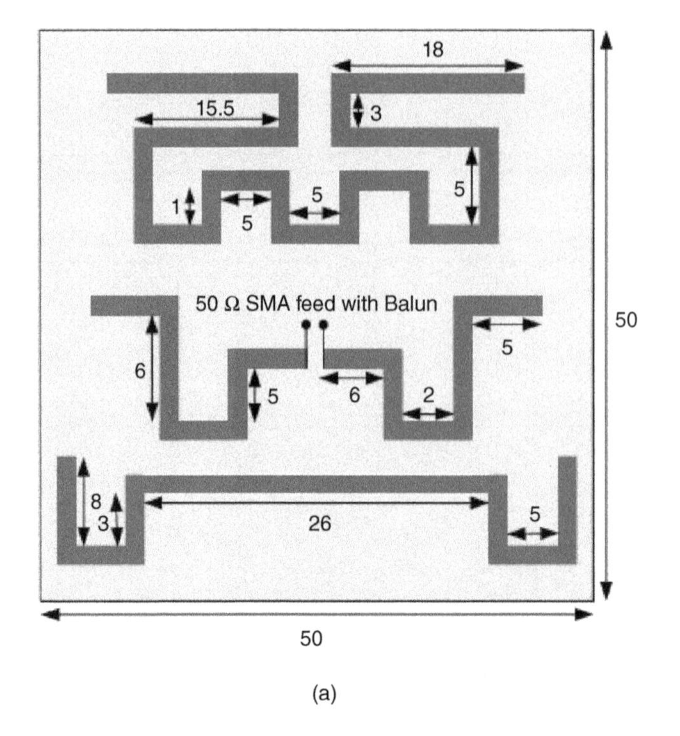

(a)

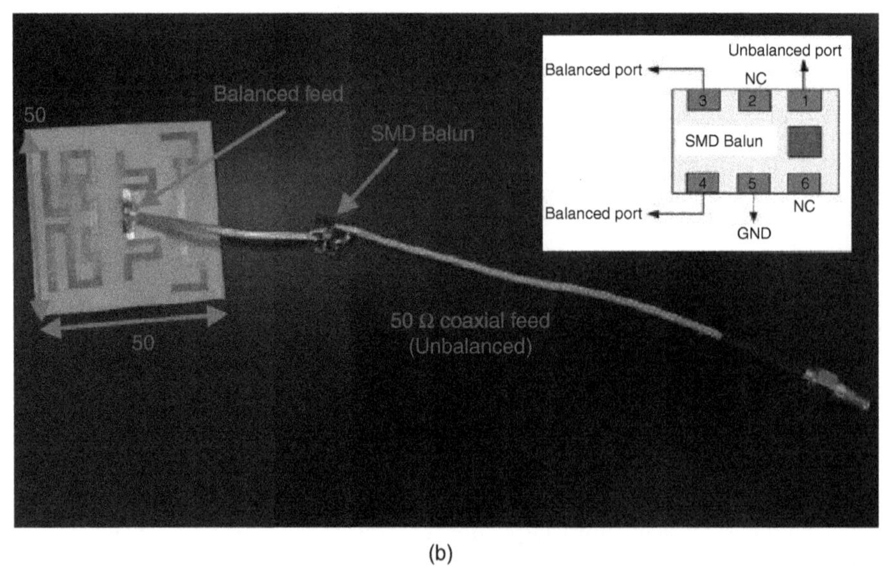

(b)

Figure 5.8 (a) Proposed endfire Yagi antenna with optimized dimensions, (b) fabricated Yagi antenna prototype on Rogers RO4003 substrate to verify the concept (units: mm).

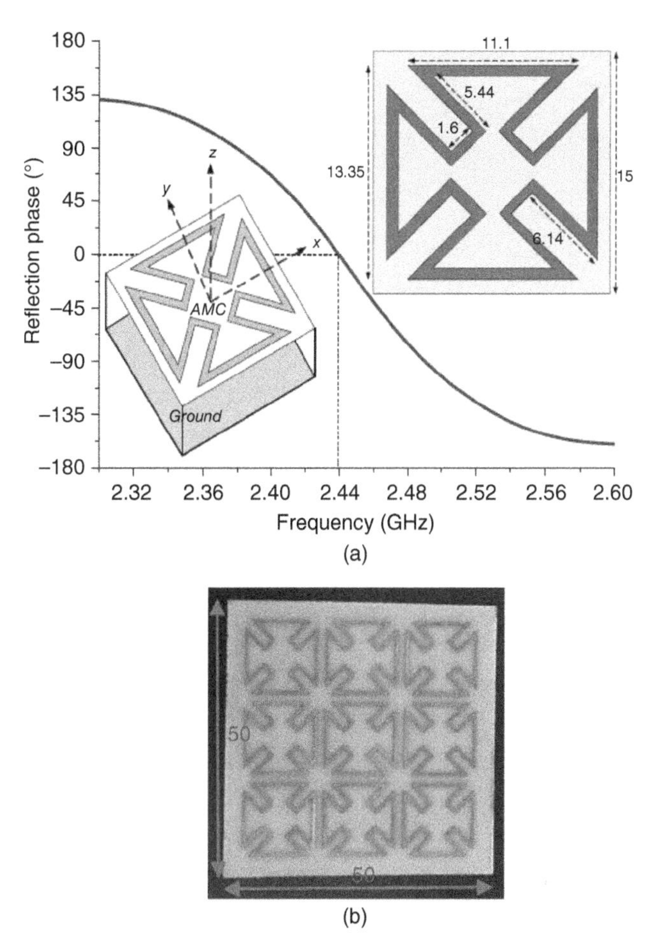

Figure 5.9 (a) Reflection phase of the SAMC unit cell (shown in the lower inset) with the SAMC unit cell being enclosed by the boundary conditions as x_{min}, x_{max}, y_{min}, and y_{max} = periodic boundary, z_{min} = electric ($Et = 0$), and z_{max} = floquet port. The upper inset shows the final optimized dimensions of the unit cell, (b) fabricated FR4 PCB prototype to realize the SAMC design for verifying the concept (unit: mm).

demonstrated in Figure 5.9(b) showing the in-house fabricated FR4 prototype of the SAMC metasurface.

5.4.3 Near-Endfire Yagi Antenna Backed by DAMC

This section delves into the design of the DAMC structure, an innovation aimed at enhancing the antenna's front-to-back ratio. The DAMC structure's units, incorporating both the upper and lower layers, are optimized. It is observed that a smaller unit cell size can achieve the 0° reflection phase response, compared to the SAMC unit size at an equivalent frequency. As illustrated in Figure 5.5(b), the DAMC surface's lower layer, placed beneath the Yagi radiator for an overall lateral dimension of 50 mm × 50 mm, consists of a 4 × 4 unit cell array. Meanwhile, the upper layer contains a 3 × 3 unit cell array.

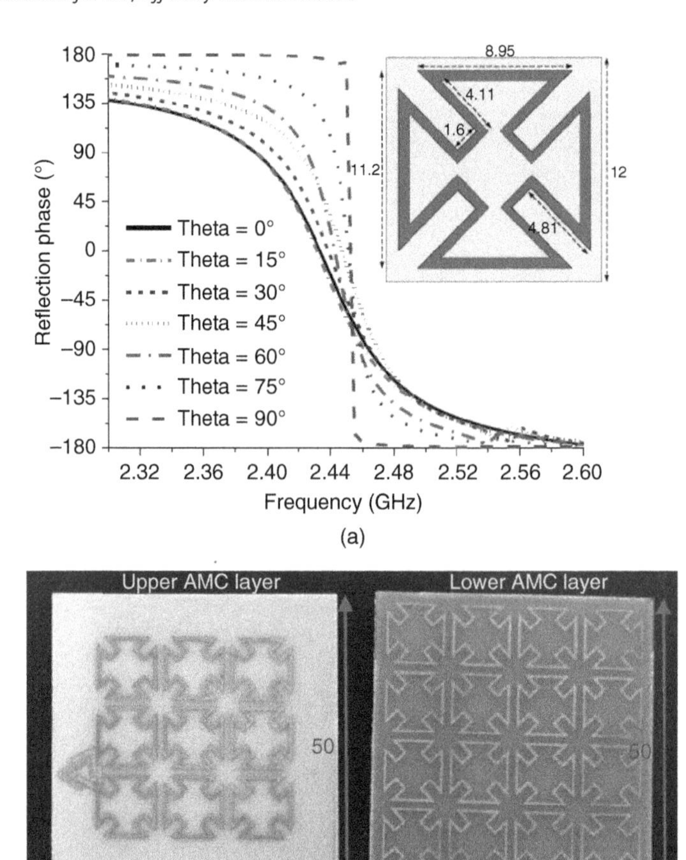

Figure 5.10 (a) Reflection phase of the DAMC surface versus the angle of incident waves. The inset shows the optimized dimensions of AMC unit, (b) fabricated prototype of the DAMC design using FR-4 substrate (unit: mm).

As both the SAMC and DAMC reflector surfaces are employed for endfire radiation, it is crucial to study the effect of the angle of incident waves striking the AMC surface. This aids in comprehending the utilization of AMC surfaces for endfire-radiation antennas. Figure 5.10(a) indicates that different incident angles create varying slopes around the 0° reflection phase frequency. However, no significant shift in the AMC frequency range is discerned (−45° to +45°).

Furthermore, considering the AMC surface is separated from the Yagi antenna by a distance of 5.5 mm, the waves striking the surface will have a maximum incident angle of around 60° or less. It is important to note that the incident waves will be perfectly parallel to the AMC surface at an incident angle of 90°, making the reflection phase graph continuous with respect to frequency.

The inset of Figure 5.10(a) showcases the final optimized AMC unit dimensions for the DAMC structure, which has a 0° reflection phase at 2.4 GHz. Concept verification is provided in Figure 5.10(b), depicting the in-house fabricated prototypes of the DAMC structure using an FR-4 substrate.

5.5 Simulations of the Antennas in Free Space

To facilitate comparison, the Yagi antenna is also situated on a PEC reflector ($Yagi_{PEC}$), maintaining the same Yagi-to-reflector separation distance of 5.5 mm as in the $Yagi_{SAMC}$ and $Yagi_{DAMC}$ antennas. This section examines and compares the free-space performances of all the antenna configurations in terms of return loss, radiation pattern, and gain.

5.5.1 Return Loss

The simulated return losses for these antennas in free space are shown in Figure 5.11. Note that it is hard to achieve impedance match for the $Yagi_{PEC}$ antenna in which the Yagi radiator is placed over the PEC ground with a separation distance of 5.5 mm, namely 0.044 λ_0 where λ_0 is the wavelength in free space at 2.4 GHz. It is also observed that the AMC-based Yagi antennas have narrower return-loss bandwidths compared with the Yagi antenna with no ground, which could be due to the frequency bandgap constraints of the AMC structures. Nevertheless, within the narrow band, the $Yagi_{SAMC}$ and the $Yagi_{DAMC}$ show better impedance match than the Yagi antenna does.

5.5.2 Radiation Patterns

The normalized simulated radiation patterns are shown in Figure 5.12(a) and (b) for *xy* ($\theta = 90°$) and *yz* ($\varphi = 90°$) planes, respectively. It is observed that the bidirectional-endfire radiation of the Yagi antenna (green dash curve) is changed to off-axis beam-tilt radiations of ~74° toward endfire direction by using the AMC structures in $Yagi_{SAMC}$ and $Yagi_{DAMC}$ antennas. This shift in the main beam direction is sensitive to the separation distance between the Yagi radiator and the upper AMC surface. As the separation is reduced below 5 mm, the direction of main lobe starts to shift toward the boresight ($\theta = 0°$) direction. In addition, note that $Yagi_{DAMC}$ enhances the back radiation by nearly 3 dB compared with $Yagi_{SAMC}$ and 6 dB compared with Yagi antenna with no AMC structure.

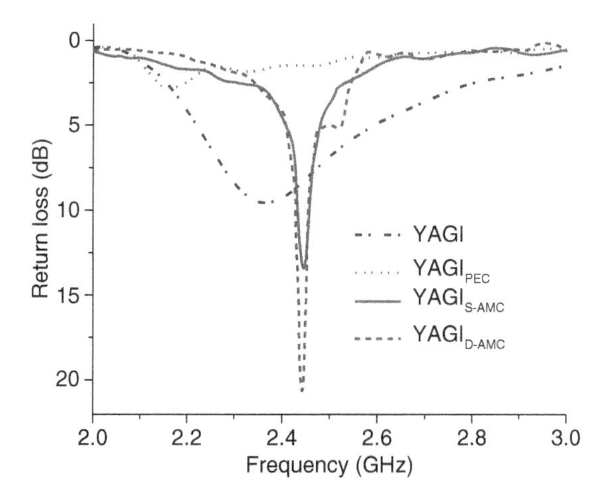

Figure 5.11 Simulated return losses of the Yagi antennas over PEC, SAMC, and DAMC reflectors, respectively, in free space.

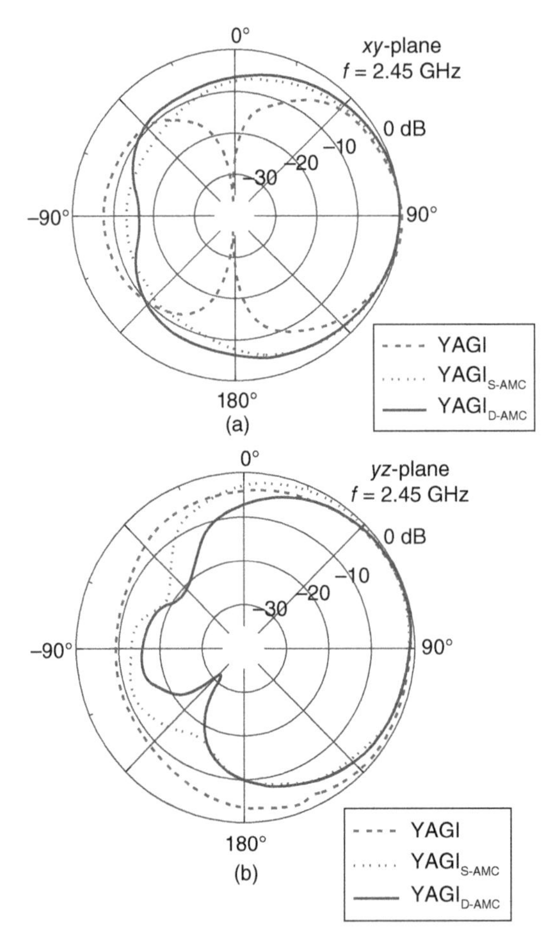

Figure 5.12 Normalized simulated radiation patterns of the Yagi antennas without and with the SAMC and DAMC structures at 2.45 GHz in (a) *xy* plane ($\theta = 90°$), (b) *yz* plane ($\varphi = 90°$).

5.5.3 Gain

Figure 5.13 shows the endfire gain (in the direction of $\theta = 90°$ and $\varphi = 90°$) of the Yagi antennas with and without AMC grounds. Note that the endfire gains of the Yagi antennas with AMC grounds are lower than the gain of the simple Yagi antenna with no ground due to the shift in the direction of main mean for the Yagi antennas with an AMC ground. For a proper comparison of gain reductions, the simulated gains in the main beam direction are also included for the $\text{Yagi}_{\text{SAMC}}$ and $\text{Yagi}_{\text{DAMC}}$ antennas in Figure 5.13 and are listed in Table 5.2. The Yagi antennas with an AMC ground have a broader beam compared with the simple Yagi antenna with no ground. Hence, the gain reductions are observed for the Yagi antennas with an AMC ground. It is worth mentioning that a broad beam is usually desired for wearable applications, because people usually keep moving and an on-body antenna with a broader beam would have a higher likelihood to cover the direction of the receiver. In addition, as emphasized previously, the Yagi antennas with an AMC ground yield higher front-to-back ratios, which are desired in wearable applications.

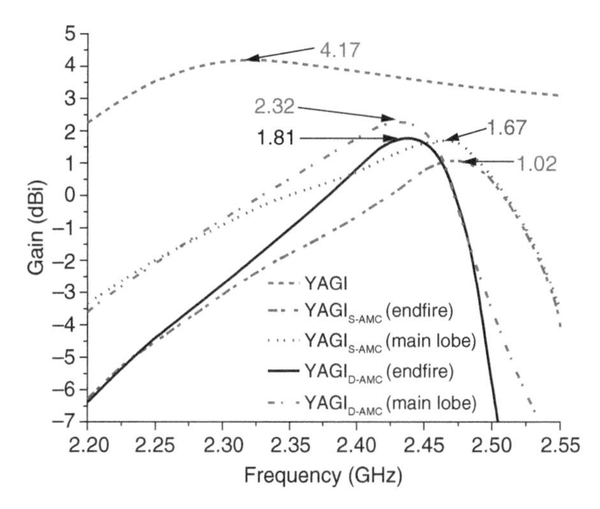

Figure 5.13 Simulated gain versus frequency of the Yagi antennas without and with SAMC and DAMC reflectors in the direction of endfire radiation and main lobe.

Table 5.2 Gain of the Yagi antennas without and with the SAMC or DAMC grounds.

Antenna design	Gain in endfire direction (dBi)	Gain in main beam (dBi)
Yagi	4.17	4.17
$\text{Yagi}_{\text{SAMC}}$	1.02	1.67
$\text{Yagi}_{\text{DAMC}}$	1.81	2.32

5.6 Simulations of the Antennas on Human Body

The on-body performances of all the antennas are simulated using the CST Microwave & RF Studio with the voxel model Gustav (male adult). As shown in Figure 5.14(a), a separation distance of 3 mm between the bottom surfaces of the AMC structures and the human skin is adopted. Note that the Yagi antenna as well as the Yagi radiators of the $\text{Yagi}_{\text{SAMC}}$ and $\text{Yagi}_{\text{DAMC}}$ antennas are kept 11.5 mm apart from the skin in the following comparisons. To analyze the effects of the SAMC and DAMC structures on the antenna performance, three body locations, namely the chest, arm, and leg, are studied for antenna placements evaluating the extent of frequency detuning, SAR value, and antenna radiation efficiency.

5.6.1 Frequency Detuning

The return losses of the antennas under free-space and on chest conditions are shown in Figure 5.15. It is observed that the impedance match deteriorates significantly for the Yagi antenna without AMC when it is placed close to the body. Note that the Yagi antenna is placed at a separation distance of 11.5 mm apart from the skin. If the Yagi antenna is placed

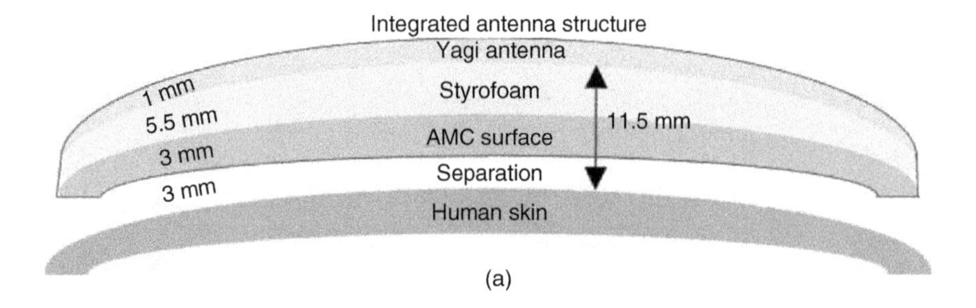

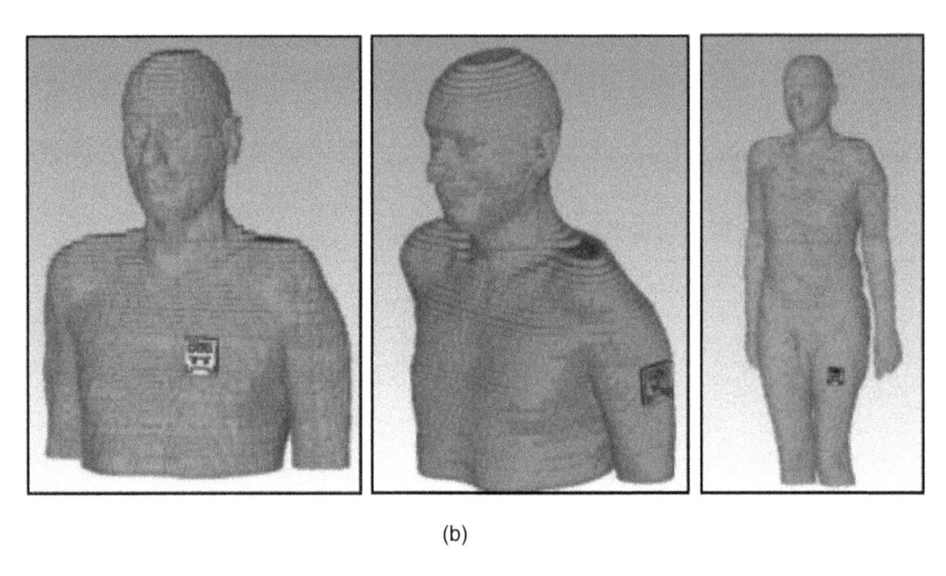

Figure 5.14 (a) Cross-sectional view of different layers with thicknesses for antenna placements on human body, (b) truncated voxel human model used for the on-body antenna performance analyses at locations of chest, upper arm, and thigh with a separation distance of 3 mm between the skin and the antenna.

at a closer separation of 3 mm from the skin, a very significant downward shift in the resonance frequency would be observed. In contrast, the AMC-backed antennas retain their impedance matching. A relatively small shift of resonant frequency to a lower frequency at 2.4 GHz is observed for $\text{Yagi}_{\text{SAMC}}$ when placed on chest as compared to its resonance in free space at 2.45 GHz. The $\text{Yagi}_{\text{DAMC}}$ shows almost no frequency shift on chest making it nearly 100% tolerant to the positioning of the human body. Hence, the effects of the DAMC in shielding the antenna from the human body while not affecting its impedance-matching characteristics has proven.

5.6.2 SAR and Antenna Efficiency

Since the wearable antenna needs to operate in close proximity to the human body, the SAR is a crucial factor to evaluate. Unless otherwise defined, the "SAR values" hereinafter

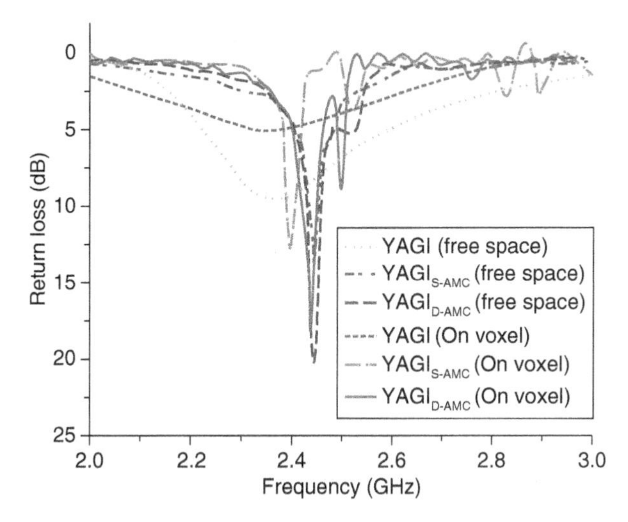

Figure 5.15 Simulated return losses of the antennas without and with AMC structures under the free-space and on-chest conditions.

in this chapter refer particularly to those averaged over 10 g of tissue, and the SAR values of an antenna are calculated under the condition that the antenna is fed by a peak power of 1 W. Referring to the Yagi antenna with no AMC structure, its antenna efficiency drops from 86% in free space to 29.4% when being placed on the human chest. And its peak SAR value is 4.2 W/kg. (The safety threshold of the SAR value is 2 W/kg.) In contrast, the antenna efficiency is improved to 74.8% with the peak SAR value being reduced to 1.24 W/kg for the $\text{Yagi}_{\text{SAMC}}$ antenna. The $\text{Yagi}_{\text{DAMC}}$ shows a further reduction in peak SAR value to 0.71 W/kg and a further enhancement in antenna efficiency to 79.0%. The peak SAR values for different placements on body are summarized in Table 5.3 for the three antennas. It is observed that the peak SAR values of the antennas with AMC structures on arm and leg also reduce significantly compared with those with no AMC structures, which is similar to the observations of the chest placement. Regarding a specific antenna, its peak SAR value varies with the antenna placement irrespective of the amount of tissue that the SAR values are averaged over. Note that if the Yagi antenna with no AMC structure is placed at a separation distance of 3 mm from the skin, its peak SAR values would become very high (included in Table 5.3). The AMC structures reduce the radiation from the antenna into the body, lower the peak

Table 5.3 Simulated peak SAR values of the three antennas at different placements on body.

	Chest (W/kg)		Arm (W/kg)		Leg (W/kg)	
Antenna configurations	**1-g tissue**	**10-g tissue**	**1-g tissue**	**10-g tissue**	**1-g tissue**	**10-g tissue**
YAGI	11.4	4.2	18.3	6.76	3.04	1.75
YAGI (3 mm)	23.6	11.1	34.3	14.7	24.7	9.47
$\text{YAGI}_{\text{S-AMC}}$	2.06	1.24	1.95	1.31	0.271	0.134
$\text{YAGI}_{\text{D-AMC}}$	1.47	0.714	1.81	1.16	0.211	0.111

SAR value, and improve the antenna radiation efficiency, and hence make antenna tolerant to being placed on human body. This gives a reliable indication of benefits of AMC being used as an antenna ground.

5.6.3 Radiation Patterns on A Human Body

The radiation characteristics of the $Yagi_{DAMC}$ antenna are discussed in this section. As shown in Figure 5.16, the antenna is placed on the right chest. As discussed earlier, almost

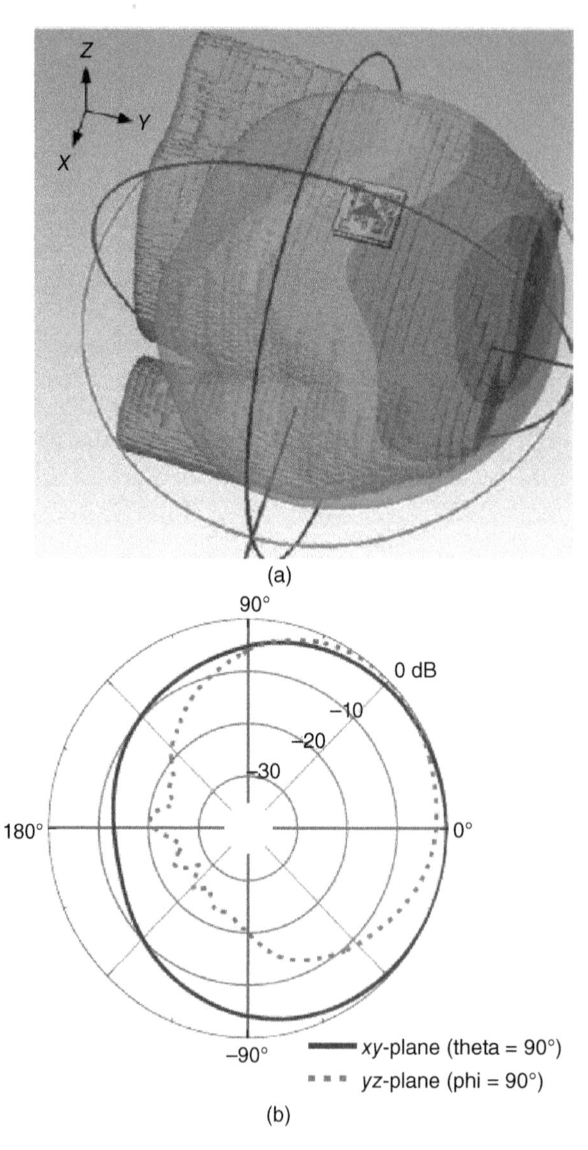

Figure 5.16 (a) Simulated 3D radiation pattern, (b) normalized radiation pattern in *xy* and *yz* planes, of the $Yagi_{DAMC}$ antenna at 2.45 GHz when it is placed on right chest of the truncated voxel human model.

no shift in resonant frequency is observed for the $\text{Yagi}_{\text{DAMC}}$ antenna when it is placed on chest compared to in free space. The normalized radiation patterns of $\text{Yagi}_{\text{DAMC}}$ antenna in *xy* and *yz* planes are shown in Figure 5.16(b). It is observed that, when the $\text{Yagi}_{\text{DAMC}}$ antenna is placed on the chest compared to in free space, the direction of the main beam shifts by 10° in *yx* plane toward the +*z*-axis and by 30° in *xz* plane toward the +*x*-axis, where the coordinate system is illustrated in Figure 5.16(a).

5.7 Antenna Performance Under Deformation

Since the on-body textile/latex antennas usually conform to the shape of the body, it is difficult to keep the antennas in flat state when they are worn on a human body. The performance of the $\text{Yagi}_{\text{DAMC}}$ antenna is hence studied under deformation. As adopted in [27], a bending radius of 40 mm in the *E* plane of the antenna is chosen to approximate an adult's arm for this bending study, which is shown in the insets of Figure 5.17. For a wearable antenna, a bending radius of 40 mm could be the most stringent situation of deformation. Latex material is highly flexible and can conform to any bending radii. Figure 5.18 demonstrates the flexibility of the $\text{Yagi}_{\text{DAMC}}$ antenna by placing it on the upper arm of an adult. The metal strips on the latex substrate are manufactured by screen-printing of silver ink. Since bending a Yagi antenna in H plane will fundamentally change its antenna characteristics, it should be avoided when wearing the proposed antenna on body. Meanwhile, for practical applications of a wearable antenna, bending in one plane is sufficient. Hence,

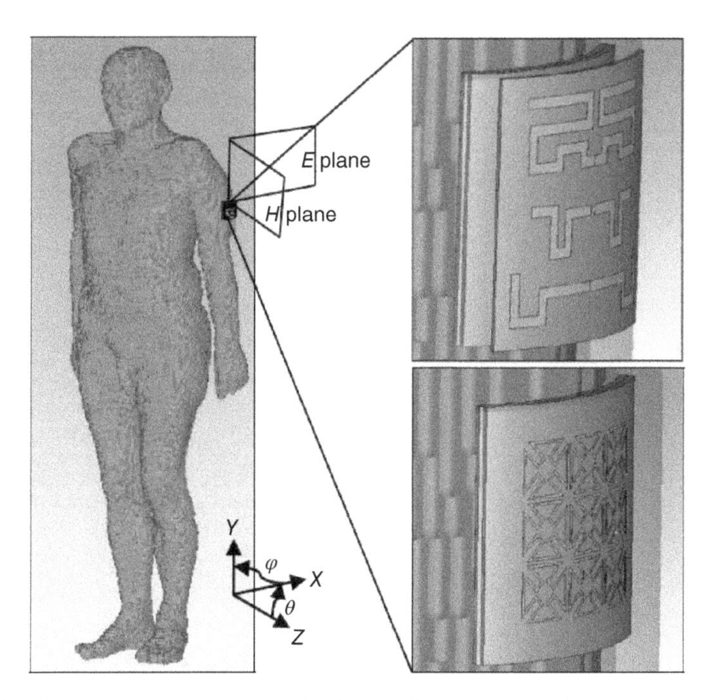

Figure 5.17 3D voxel human model used for the bending study of the $\text{Yagi}_{\text{DAMC}}$ antenna, of which the arm has a radius of 40 mm.

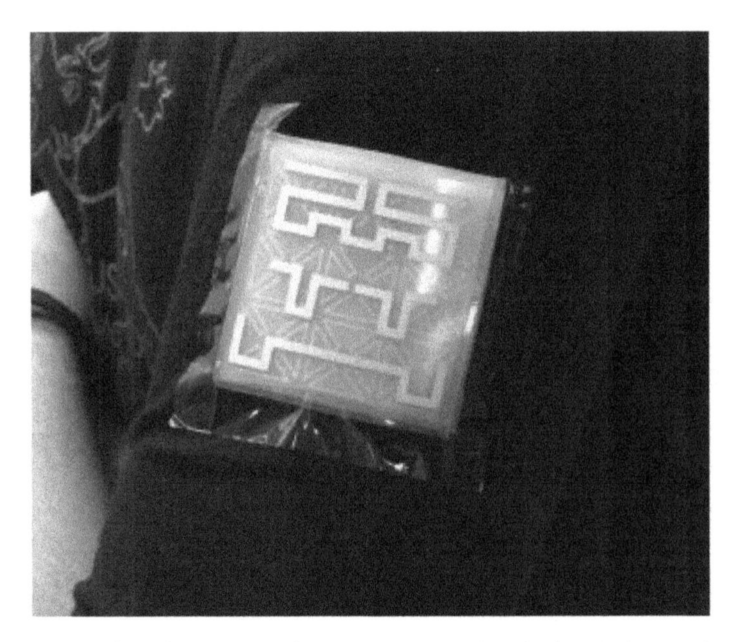

Figure 5.18 Photo of the fabricated $Yagi_{DAMC}$ antenna based on the latex substrate demonstrating the flexibility by bending on upper arm of an adult.

the return loss and radiation pattern under bending in *E* plane are analyzed for the $Yagi_{SAMC}$ and $Yagi_{DAMC}$ antennas.

Figure 5.19 shows the simulated return losses of the $Yagi_{SAMC}$ and $Yagi_{DAMC}$ antennas when they are flat and bent over the adult arm with a 40-mm radius of curvature. It is observed that the bending deformation has almost no effect on the impedance-matching

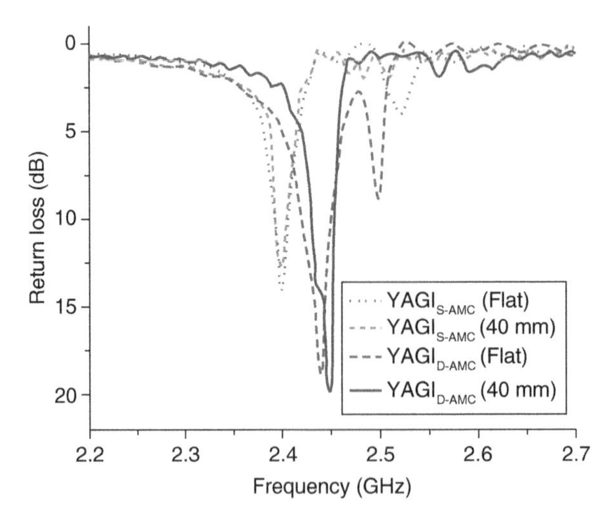

Figure 5.19 Simulated return losses of $Yagi_{SAMC}$ and $Yagi_{DAMC}$ antennas placed on adult arm in flat and bending (40-mm radius) states.

characteristics of the $\text{Yagi}_{\text{SAMC}}$ antenna, whereas a minor shift in resonance frequency is observed for the $\text{Yagi}_{\text{DAMC}}$ antenna. In addition, for the $\text{Yagi}_{\text{SAMC}}$ and $\text{Yagi}_{\text{DAMC}}$ antennas, their 10 dB-return loss bandwidths under bent state are slightly shrunk compared to those under flat states, respectively. Hence, it is demonstrated that impedance matching bandwidths of the Yagi antennas shrunk to some extent when being bent in *E* plane.

The normalized radiation patterns of the $\text{Yagi}_{\text{DAMC}}$ antenna in *yz* and *xy* planes are displayed in Figure 5.20(a) and (b), respectively. It is observed from Figure 5.20(a) that bending the antenna to a stringent condition of conforming to an adult upper arm of radius 40 mm

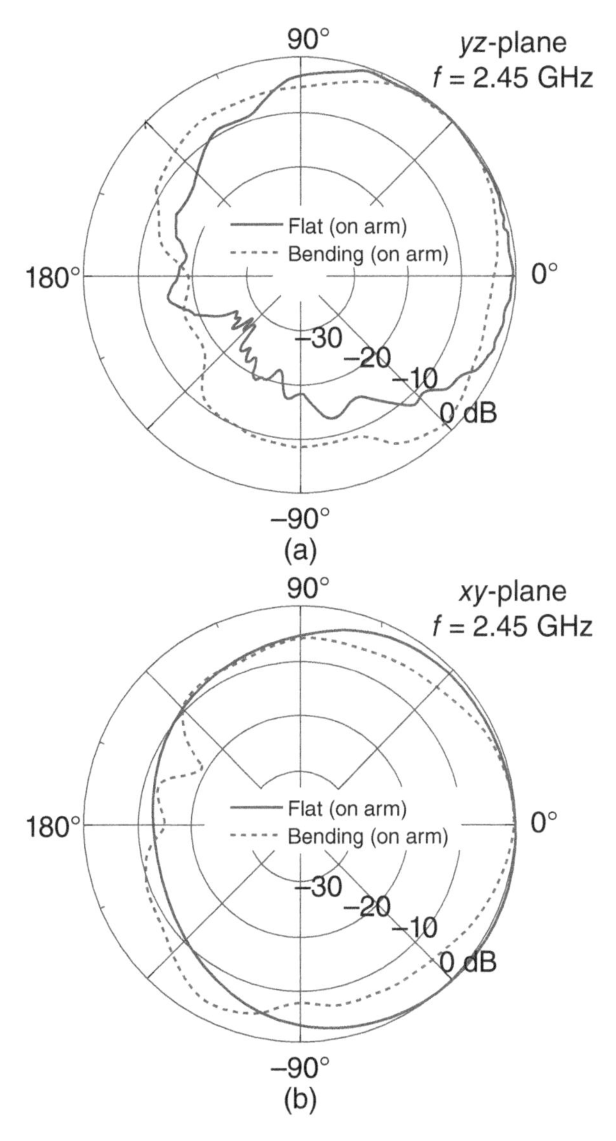

Figure 5.20 Normalized simulated radiation patterns for the $\text{Yagi}_{\text{DAMC}}$ antenna in flat and bent states in (a) *yz* ($\varphi = 90°$) plane, (b) *xy* ($\theta = 90°$) plane.

results in distortion in radiation pattern: the gain in end-fire direction drops several decibels. Nevertheless, as shown in Figure 5.18(b), the more important parameter – the front-to-back ratio of the $\text{Yagi}_{\text{DAMC}}$ antenna in *xy* plane preserves under the bent condition as compared to that of the flat condition. This can be attributed to the fact that the bending along the H plane for an end-fire radiation antenna will not affect the radiation pattern along the *xy* plane significantly.

5.8 Experiment

The fabricated antenna prototype including the Yagi radiator and the AMC structures are shown in Figure 5.21. They are fabricated by screen-printing silver ink on the flexible latex substrates. The prototypes are then sliced for the overall lateral dimensions of 50 mm × 50 mm as used in the simulations. The simulated characteristics such as return loss, radiation pattern, and gain have been discussed for the printed circuit board (PCB)-based and latex-based Yagi antennas previously. Commercially available flexible Styrofoam of 5.5 mm thickness is used to separate the Yagi radiator from the AMC structures.

5.8.1 Return Loss

The return losses of the antennas are measured by using the vector network analyzer (VNA) R&S ZVL13. The measured and simulated return losses of the $\text{Yagi}_{\text{SAMC}}$ as well as the $\text{Yagi}_{\text{DAMC}}$ antennas in free space are displayed in Figure 5.22. For both the $\text{Yagi}_{\text{SAMC}}$ and $\text{Yagi}_{\text{DAMC}}$ antennas, slight discrepancy is observed between the simulated and measured results. The measured 10 dB-return loss bandwidth is 40 MHz (2.43–2.47 GHz) for the $\text{Yagi}_{\text{SAMC}}$ antenna and 45 MHz (2.425–2.47 GHz) for $\text{Yagi}_{\text{DAMC}}$ antenna.

5.8.2 Radiation Pattern Measurement

Figure 5.23(a) and (b) display the normalized radiation patterns in *xy* ($\theta = 90°$) plane and *yz* ($\varphi = 90°$) plane, respectively, for the $\text{Yagi}_{\text{SAMC}}$ and $\text{Yagi}_{\text{DAMC}}$ antennas. The antennas (plane of the substrates) are oriented in *xy* plane and measured at 2.45 GHz in free space.

Figure 5.21 Fabricated antenna with AMC structures based on latex substrates: (left) Yagi radiator, (middle) single AMC layer with 3 × 3 units, and (right) double AMC layer with 3 × 3 units on the upper surface and 4 × 4 units on the lower surface.

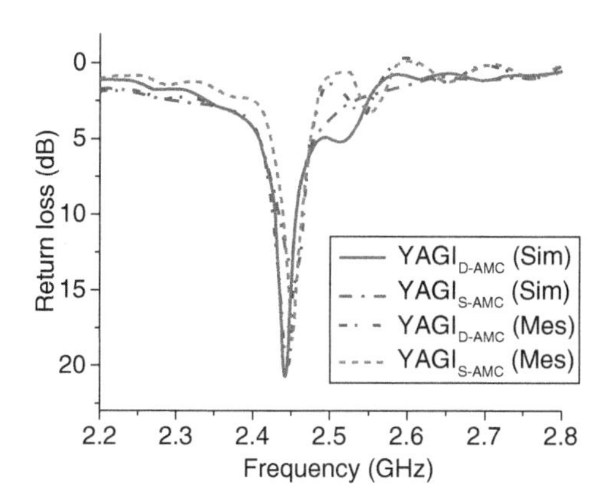

Figure 5.22 Measured and simulated return losses of the $Yagi_{SAMC}$ and $Yagi_{DAMC}$ antennas in free space.

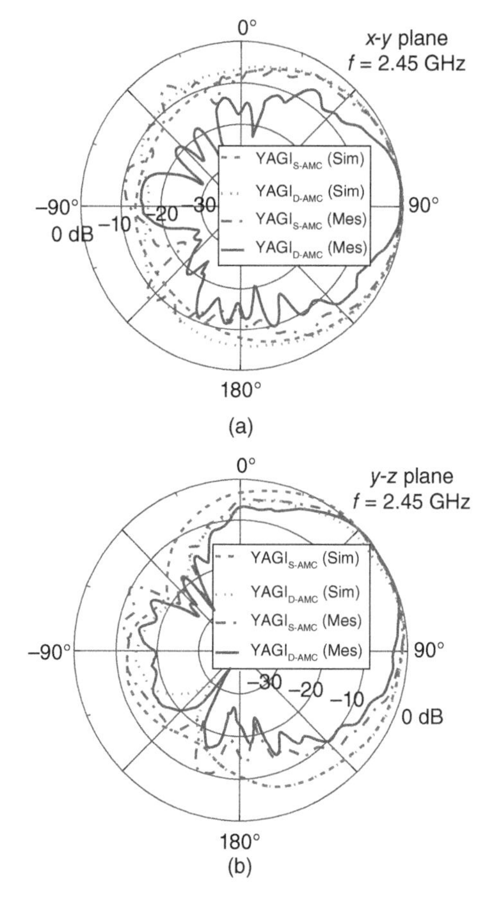

Figure 5.23 Measured and simulated radiation patterns of the $Yagi_{SAMC}$ and $Yagi_{DAMC}$ antennas based on the latex substrate at 2.45 GHz in (a) *xy* plane ($\theta = 90°$), (b) *yz* plane ($\varphi = 90°$). The antennas are oriented in *xy* plane.

Referring to Figure 5.23(a), the measured and simulated patterns of the $\text{Yagi}_{\text{SAMC}}$ agree well, while those of the $\text{Yagi}_{\text{DAMC}}$ do not. The measured results show that both the $\text{Yagi}_{\text{SAMC}}$ and $\text{Yagi}_{\text{DAMC}}$ antennas have end-fire radiation patterns with a front-to-back ratio higher than 12 dB.

As shown in Figure 5.23(b), an upward shift in the main beam direction from the AMC substrates is observed in the measured radiation pattern compared with the simulated one for both the $\text{Yagi}_{\text{SAMC}}$ and $\text{Yagi}_{\text{DAMC}}$ antennas. Since the end-fire direction of these AMC-backed Yagi antennas is very sensitive to the separation distance between the Yagi radiator and the AMC structures, these upward shifts can be attributed to the changes in the separation distance that occurred during the layer-by-layer manual integration of the flexible Yagi radiator, Styrofoam, and AMC structures.

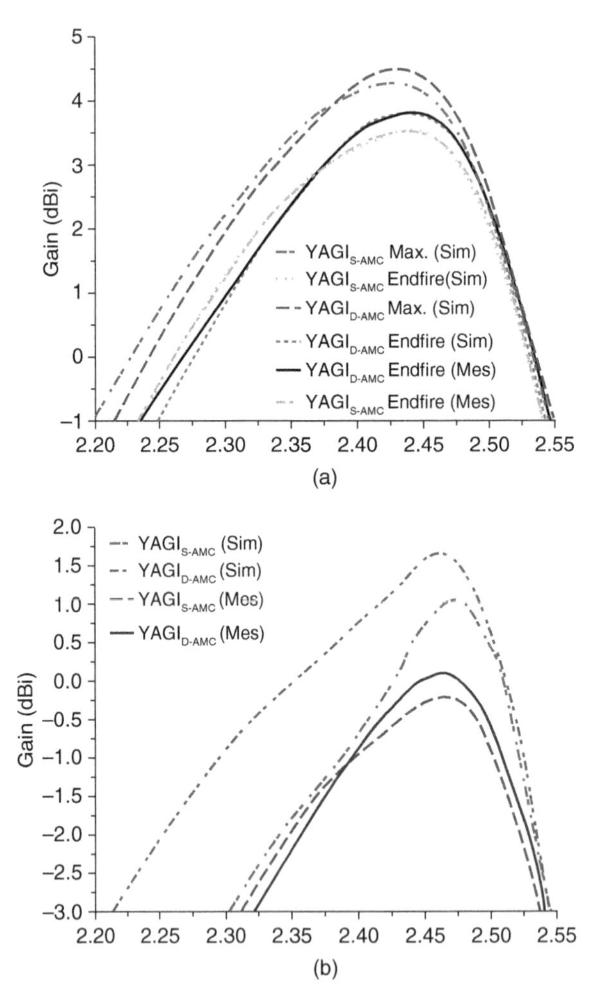

Figure 5.24 Measured and simulated endfire gains (the peak gain in the plane of the antenna) of the $\text{Yagi}_{\text{SAMC}}$ and $\text{Yagi}_{\text{DAMC}}$ antennas based on (a) the PCB substrates, (b) the latex substrate. In (a) for those PCB-based antennas, the simulated maximum gains (gain in the main beam direction) are also provided.

5.8.3 Gain Measurement

The gain of the $Yagi_{SAMC}$ and $Yagi_{DAMC}$ antennas was measured versus frequency when these antennas were placed in free-space. Figure 5.24(a) and (b) display the gains of the PCB-based antennas and the latex-based antennas, respectively. It is observed in Figure 5.24(a) that, for the end-fire gain (gain in the plane of the antenna), the measured results are in good agreement with those simulated ones. The measured end-fire gain at 2.45 GHz is 3.8 dBi for the $Yagi_{SAMC}$ antenna and 4.12 dBi for the $Yagi_{DAMC}$ antenna. Meanwhile, the simulated maximum gain (gain in the main beam direction) at 2.45 GHz is 4.44 dBi for the $Yagi_{SAMC}$ antenna and 4.78 dBi for the $Yagi_{DAMC}$ antenna.

As for the antennas based on the latex substrates, as displayed in Figure 5.24(b), relatively large disagreements are observed between the simulated and measured endfire gains. At the frequency of 2.45 GHz, the measured endfire gain for the $Yagi_{SAMC}$ antenna is −0.2 dBi; in contrast, the simulated endfire gain is 1.02 dBi. Also, at the frequency of 2.45 GHz, the $Yagi_{DAMC}$ antenna has a measured endfire gain of 0.12 dBi, whereas the simulated end-fire gain is 1.81 dBi. These disagreements in the simulated and measured results can be accounted for the shift in the direction of main beam, which is caused by the errors that occurred during the in-house fabrication of the latex-based antennas, as discussed in previous subsection. The measured results could be in good agreement with the simulated ones if the fabrication accuracy can be ensured.

5.9 Conclusion

In this chapter, we started with an overview of conventional wearable antennas. Following that, we explored the integration of a planar Yagi radiator with single-layer and double-layer AMC structures tailored for wearable applications. The purpose of implementing the AMC structures was to decrease the body's radiation exposure, subsequently improving the antenna's resilience to positioning on the human body. Using CST Microwave and RF Studio and a human voxel model, we analyzed various characteristics of the AMC-based Yagi antennas, such as return loss, gain, frequency shifting, radiation efficiency, and peak SAR value, in both free-space and on-body contexts.

The on-voxel analysis revealed that the double-layer AMC enhances antenna features such as the front-to-back ratio, frequency detuning, radiation efficiency, and peak SAR level, when compared to a Yagi antenna backed by a single-layer AMC or no AMC at all. Both the Yagi radiators and AMC structures were manufactured using PCB technology to validate the concept, and also using flexible latex for body-worn applications. We also evaluated the flexible antennas under a state of bending, the results of which confirmed that they function correctly when bent to a degree that aligns with an adult's upper arm. Therefore, the $Yagi_{DAMC}$ antenna built on flexible latex substrates offers an adequate bandwidth and radiation pattern, making it suitable for WBAN applications in the 2.4 GHz ISM band.

The future trajectory of wearable antennas leans toward the creation of designs that are not only flexible, resilient, and comfortable, but also multifunctional. This progression exploits the considerable strides made in the realm of materials science [34], adapting effectively to the distinctive challenges intrinsic to body-centric communications. Moreover, a crucial aspect warranting careful attention lies in the customization and optimization of these antennas, aligned with specific applications [35]. This ensures a more refined and efficient utilization of this technology. A variety of critical trends and requirements serve as an invaluable guide for researchers:

(1) *Flexible and Conformal Designs*: As wearable antennas need to adapt to various shapes and postures of the human body the trend is toward creating more flexible and conformal designs. These would maintain their performance characteristics despite being stretched, folded, or bent, providing a seamless blend with everyday clothing and accessories.
(2) *Advanced Materials*: The invention of novel materials such as conductive textiles, flexible electronics, and metamaterials, opens up possibilities for new antenna designs. These materials are lightweight, flexible, and can be directly incorporated into fabrics, offering advantages such as comfort, durability, and washability.
(3) *Body-centric Communications*: Given the proximity of wearable antennas to the human body, designs are being optimized for body-centric communications. This includes body area networks (BANs) and off-body communications where data can be transmitted to nearby devices or networks.
(4) *Multi-functionality*: There is a growing trend toward developing multifunctional wearable antennas that can operate across various frequency bands. This would enable wearers to connect with multiple wireless services or devices using a single wearable antenna.
(5) *Integration with Smart Clothing*: Wearable antennas are being integrated with smart clothing, enabling various applications such as health monitoring, location tracking, and communication. Smart clothing with embedded wearable antennas could monitor physiological parameters such as heart rate or body temperature, track users' location, or facilitate communication with other devices.
(6) *Personalized Designs*: Personalization of wearable antennas to suit individual requirements, body shapes, and preferences is also a future trend. This would ensure better comfort, esthetic appeal, and performance for each user.
(7) *Sustainable Practices*: There is an increased focus on sustainable practices in wearable antenna design. This includes the use of eco-friendly materials and manufacturing processes, as well as creating designs that are durable and require less energy to operate.

References

1 Chung, H.U., Rwei, A.Y., Hourlier-Fargette, A. et al. (2020). Skin-interfaced biosensors for advanced wireless physiological monitoring in neonatal and pediatric intensive-care units. *Nature Med.* 26 (3): 418–429.

2 Kennedy, T.F., Fink, P.W., Chu, A.W. et al. (2009). Body-worn E-textile antennas: the good, the low-mass, and the conformal. *IEEE Trans. Antennas Propag.* 57 (4): 910–918.

3 Klemm, M. and Troester, G. (2006). Textile UWB antennas for wireless body area networks. *IEEE Trans. Antennas Propag.* 54 (11): 3192–3197.

4 Morris, S.E., Bayram, Y., Lanlin, Z. et al. (2011). High-strength, metalized fibers for conformal load bearing antenna applications. *Antennas Propag. IEEE Trans.* 59 (9): 3458–3462.
5 Lanlin, Z., Zheyu, W., and Volakis, J.L. (2012). Textile antennas and sensors for body-worn applications. *Antennas Wireless Propag. Lett. IEEE* 11: 1690–1693.
6 Wang, Z., Lee, L.Z., Psychoudakis, D., and Volakis, J.L. (2014). Embroidered multiband body-worn antenna for GSM/PCS/WLAN communications. *Antennas Propag. IEEE Trans.* 62 (6): 3321–3329.
7 Wang, Z., Zhang, L., Bayram, Y., and Volakis, J.L. (2012). Embroidered conductive fibers on polymer composite for conformal antennas. *Antennas Propag. IEEE Trans.* 60 (9): 4141–4147.
8 Yuehui, O. and Chappell, W.J. (2008). High frequency properties of electro-textiles for wearable antenna applications. *IEEE Trans. Antennas Propag.* 56 (2): 381–389.
9 Locher, I., Klemm, M., Kirstein, T., and Troster, G. (2006). Design and characterization of purely textile patch antennas. *IEEE Trans. Adv. Pack.* 29 (4): 777–788.
10 Koski, K., Sydanheimo, L., Rahmat-Samii, Y., and Ukkonen, L. (2014). Fundamental characteristics of electro-textiles in wearable UHF RFID patch antennas for body-centric sensing systems. *Antennas Propag. IEEE Trans.* 62 (12): 6454–6462.
11 Kaivanto, E.K., Berg, M., Salonen, E., and de Maagt, P. (2011). Wearable circularly polarized antenna for personal satellite communication and navigation. *IEEE Trans. Antennas Propag.* 59 (12): 4490–4496.
12 Vallozzi, L., Van Torre, P., Hertleer, C. et al. (2010). Wireless communication for firefighters using dual-polarized textile antennas integrated in their garment. *IEEE Trans. Antennas Propag.* 58 (4): 1357–1368.
13 Serra, A.A., Nepa, P., and Manara, G. (2012). A wearable two-antenna system on a life jacket for cospas-sarsat personal locator beacons. *IEEE Trans. Antennas Propag.* 60 (2): 1035–1042.
14 Zhu, S. and Langley, R. (2009). Dual-band wearable textile antenna on an EBG substrate. *IEEE Trans. Antennas Propag.* 57 (4): 926–935.
15 Feresidis, A.P., Goussetis, G., Shenhong, W., and Vardaxoglou, J.C. (2005). Artificial magnetic conductor surfaces and their application to low-profile high-gain planar antennas. *IEEE Trans. Antennas Propag.* 53 (1): 209–215.
16 Goussetis, G., Feresidis, A.P., and Vardaxoglou, J.C. (2006). Tailoring the AMC and EBG characteristics of periodic metallic arrays printed on grounded dielectric substrate. *IEEE Trans. Antennas Propag.* 54 (1): 82–89.
17 Foroozesh, A. and Shafai, L. (2011). Investigation into the application of artificial magnetic conductors to bandwidth broadening, gain enhancement and beam shaping of low profile and conventional monopole antennas. *IEEE Trans. Antennas Propag.* 59 (1): 4–20.
18 Kim, S., Ren, Y.-J., Lee, H. et al. (2012). Monopole antenna with inkjet-printed EBG array on paper substrate for wearable applications. *IEEE Antennas Wireless Propag. Lett.* 11: 663–666.
19 Raad, H.R., Abbosh, A.I., Al-Rizzo, H.M., and Rucker, D.G. (2013). Flexible and compact AMC based antenna for telemedicine applications. *IEEE Trans. Antennas Propag.* 61 (2): 524–531.

20 Chahat, N., Zhadobov, M., Le Coq, L., and Sauleau, R. (2012). Wearable endfire textile antenna for on-body communications at 60 GHz. *IEEE Antennas Wireless Propag. Lett.* 11: 799–802.

21 Hussain, A.M., Ghaffar, F.A., Park, S.I. et al. (2015). Metal/polymer based stretchable antenna for constant frequency far-field communication in wearable electronics. *Adv. Funct. Mater.* 25 (42): 6565–6575.

22 Salman, S., Wang, Z., Colebeck, E. et al. (2014). Pulmonary edema monitoring sensor with integrated body-area network for remote medical sensing. *Antennas Propag. IEEE Trans.* 62 (5): 2787–2794.

23 Zhu, X.-Q., Guo, Y.-X., and Wu, W. (2016). A compact dual-band antenna for wireless body-area network applications. *IEEE Antennas Wireless Propag. Lett.* 15: 98–101.

24 Zhu, X.-Q., Guo, Y.-X., and Wu, W. (2016). Miniaturized dual-band and dual-polarized antenna for MBAN applications. *IEEE Trans. Antennas Propag.* 64 (7): 2805–2814.

25 Tong, X., Liu, C., Liu, X. et al. (2018). Switchable ON-/OFF-body antenna for 2.45 GHz WBAN applications. *IEEE Trans. Antennas Propag.* 66 (2): 967–971.

26 Hall, P.S., Yang, H., Nechayev, Y.I. et al. (2007). Antennas and propagation for on-body communication systems. *IEEE Antennas Propag. Mag.* 49 (3): 41–58.

27 Bai, Q. and Langley, R. (2009). Wearable EBG antenna bending and crumpling. In: *2009 Loughborough Antennas & Propagation Conference*, 201–204.

28 Agarwal, K., Nasimuddin, A., and A. (2013). Wideband circularly polarized AMC reflector backed aperture antenna. *IEEE Trans. Antennas Propag.* 61 (3): 1456–1461.

29 IEEE (2006). *IEEE Standard for Safety Levels With Respect to Human Exposure to Radio Frequency Electromagnetic Fields, 3 kHz to 300 GHz*, IEEE Std C95.1-2005 (Revision of IEEE Std C95.1-1991), 1–238.

30 IEEE (2012). IEEE Standard for Local and metropolitan area networks—Part 15.6: Wireless body area networks, IEEE Std 802.15.6-2012, pp. 1-271.

31 Agarwal, K., Guo, Y.-X., and Salam, B. (2016). Wearable AMC backed near-endfire antenna for on-body communications on latex substrate. *IEEE Trans. Comp. Pack. Manuf. Technol.* 6 (3): 346–358.

32 Latti, K., Kettunen, M., Strom, J., and Silventoinen, P. (2007). A review of microstrip T-resonator method in determining the dielectric properties of printed circuit board materials. *IEEE Trans. Instrum. Meas.* 56 (5): 1845–1850.

33 Amey, D.I. and Curilla, J.P. (1991). Microwave properties of ceramic materials. In: *1991 Proceedings 41st Electronic Components & Technology Conference*, 267–272.

34 Li, Y., Tian, X., Gao, S.P. et al. (2020). Reversible crumpling of 2D titanium carbide (MXene) nanocoatings for stretchable electromagnetic shielding and wearable wireless communication. *Adv. Funct. Mater.* 30 (5): 1907451.

35 Wang, H., Feng, Y., Hu, F., and Guo, Y.-X. (2022). A wideband dual-polarized ring-loaded cross bowtie antenna for wireless capsule endoscopes: design and link analysis. *IEEE Trans. Antennas Propag.* 70 (9): 7843–7852.

6

Investigation and Modeling of Capacitive Human Body Communication

6.1 Introduction

Recently, the wireless body area network (WBAN) has gained extensive attention due to its unprecedented capability and boundless potential to build adaptable platforms for continuous monitoring and recording of critical physiological parameters and signals [1]. In accordance with IEEE Std. 802.15.6, three distinct physical layers (PHYs) are encompassed within the purview of the WBAN: namely, the narrow band (NB) PHY, the ultra-wideband (UWB) PHY, and the human body communication (HBC) PHY [2]. The NB and UWB PHYs mainly operate at sub-GHz and GHz frequency bands and necessitate the use of on-body antennas for signal transmission and reception. Typically, air serves as the transmission medium, and the transmission range can span up to 10 m [3].

In contrast, HBC is a unique communication methodology that leverages the human body as a conduit for transmitting electrical signals. HBC technology capitalizes on the conductive properties of the human body to enable signal transmission between devices, such as sensors or medical implants, circumventing the need for conventional wireless communication techniques such as radio frequency (RF). Ordinarily, the HBC PHY operates at 21 MHz with a bandwidth of 5.25 MHz, and signal transmission is realized through electromagnetic (EM) coupling rather than radiation. Within HBC systems, the electrical fields are contained around a human body, resulting in diminished interference to sensor nodes and external devices and users, thereby augmenting security. Moreover, HBC PHY typically yields lower path losses and power consumption in body channels, particularly in long-range transmissions, making it a superior feature of HBC over the NB and UWB PHYs.

HBC shows promising applications in the medical sector for wireless monitoring of vital signs, and in the domain of human–machine interfaces, for controlling electronic devices via body movements. Its potential for secure communication applications is also being explored, as signal transmission is confined to the human body, making it challenging to intercept. In practical scenarios, the central hub of a healthcare-monitoring system should have the capacity to collect physiological parameters from sensors distributed across the body. These include electroencephalogram (EEG) sensors to monitor brain functionality, electromyogram (EMG) sensors to monitor muscle performance, electrocardiogram (ECG) sensors to monitor cardiac activity, and inertial sensors to monitor walking speed, among others. Certain sensor nodes may be located at a considerable distance from the central hub,

Antennas and Wireless Power Transfer Methods for Biomedical Applications, First Edition.
Yongxin Guo, Yuan Feng and Changrong Liu.

which could lead to high path losses if the NB or UWB PHYs are utilized. Additionally, since on-body antennas situated on different parts of the body need to communicate with the central hub, these antennas must be specifically designed to radiate in particular directions and to maintain unobstructed structures. Conversely, the HBC PHY can facilitate directionless coverage of the entire body. Furthermore, the electrodes in HBC systems are uniform, which simplifies the design, fabrication, and implementation processes, offering considerable flexibility. However, HBC usage is constrained by factors such as signal attenuation, interference, and the requirement for device-to-skin contact.

This chapter delves into a thorough investigation of HBC, introducing two of its types: galvanic and capacitive couplings. An in-depth literature review of capacitive HBC follows, encompassing an overview of various research methodologies and pertinent challenges. Subsequently, we present a novel measurement system, coupled with numerical simulation and a circuit model. The chapter culminates with a discussion shedding light on significant issues related to the practical application and system integration of capacitive HBC.

6.2 Galvanic and Capacitive Coupling HBC

The HBC, also known as body channel communication (BCC) or intra-body communication (IBC), was first proposed by Zimmerman [4]. Based on the coupling mechanism, HBC can be divided into two categories, i.e., galvanic coupling and capacitive coupling. In galvanic coupling, two electrode pairs are attached to a human body, as shown in Figure 6.1(a). The two electrodes at the transmitter (TX) are used to excite alternating currents, which propagate between them. Meanwhile, part of the alternating currents spreads through the human body. The human body acts as a part of the electrical circuit itself, making the signal propagation mostly independent of body movements or posture. This method can provide reliable communication links and it has a low power consumption. Owing to the absorption of lossy tissues, a largely attenuated signal is detected by the electrode pair at the receiver (RX). Therefore, the currents are confined within the human body in galvanic coupling, and no ground plane is required. Moreover, the operation frequency of this approach is usually below 1 MHz, and the transmission distance range is less than 15 cm [5].

The capacitive coupling uses the electric field that is naturally present in and around a human body. The transmitter applies a voltage that causes a change in this electric field, which the receiver can detect. In this case, the signal does not propagate through the body, but around it. Therefore, only one electrode of the TX (or RX) electrode pair is attached to the body, which is designated as the signal (SIG) electrode. On the other hand, the floating electrode is designated as the ground (GND) electrode, as shown in Figure 6.1(b). The transmission of signals is realized by two paths: the forward path between the two SIG electrodes through the body and the parasitic return path. The return path is formed by the coupling between the external ground plane and the GND electrodes. In addition, there also exists cross-coupling between the two GND electrodes. The ground plane is required as a reference in the capacitive HBC. As a result, the capacitive coupling can be affected by the surrounding environment and body movements, leading to potential signal instability. In addition, the operation frequency of the capacitive coupling ranges from 1 MHz to 100 MHz, and the entire human body can be covered by this approach.

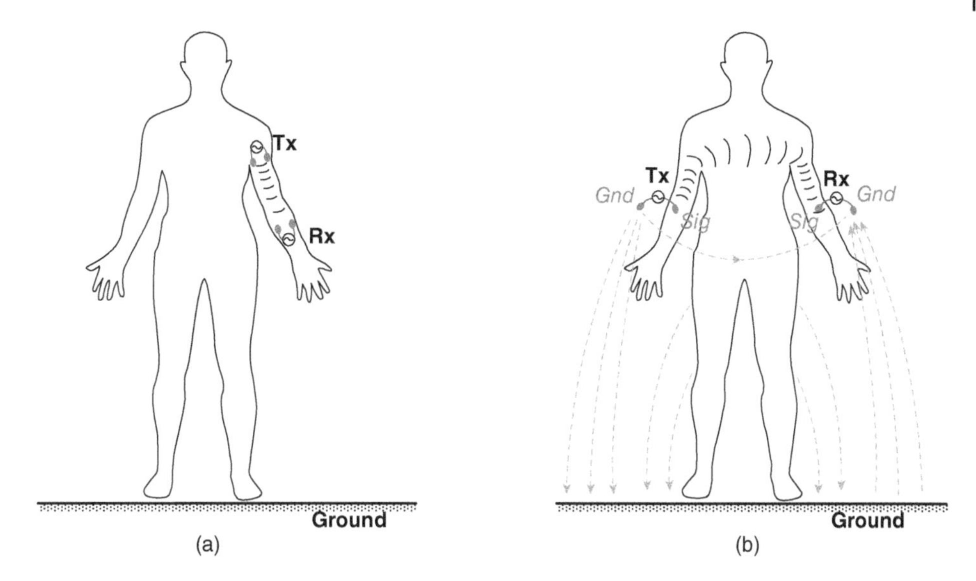

Figure 6.1 Two different coupling mechanisms of HBC: (a) Galvanic coupling, (b) capacitive coupling.

Both methods have their strengths and weaknesses, and the choice between them depends on the specific application and constraints. These techniques are being actively researched for applications in health monitoring, personal authentication, wearable device communication, and more. The development of effective HBC technologies holds great promise for enhancing user experience and promoting the advancement of personalized healthcare and secure, convenient communication systems. In this chapter, we will focus on the study of the capacitive HBC.

6.3 Capacitive HBC

Over the years, substantial investigations have been conducted on capacitive HBC, leading to a myriad of methodologies that effectively characterize transmission properties. These methodologies, encompassing experimental measurements, numerical models, circuit models, and theoretical analysis, are outlined herein.

6.3.1 Experimental Characterizations

Experimental characterizations involve direct empirical assessments of the capacitive HBC using various measurement tools and equipment. By conducting real-world tests with human subjects or models, researchers can obtain accurate and practical data about how capacitive HBC performs under different conditions. These empirical data are valuable for validating theoretical models and understanding practical challenges. As discussed above, the capacitive HBC system includes a parasitic return path, which should be preserved in order to characterize the HBC channels accurately. While in early works, the measurement was usually performed by connecting the TX and RX electrode pairs directly to a vector network analyzer (VNA) [6]. In this scenario, the two GND electrodes were connected to the internal ground of VNA, hence the capacitive return path was shorted, which could enhance the transmission performance significantly. In [7], a battery-powered signal generator (SG) with a programmable frequency synthesizer was used as TX to emulate the parasitic return path, whereas the RX electrode was connected to a spectrum analyzer (SA) or an oscilloscope (OSC), which enlarged the ground plane of the receiving device. This measurement setup was also adopted in [8]. To sum up, the grounding issue needs to be carefully addressed to obtain accurate HBC channel performance.

In 2009, Xu et al. proposed a new measurement setup by using a pair of baluns (FTB-1-1 A15 from Mini-Circuit) to isolate the TX and RX electrodes from the measurement instruments [9–11]. Such measurement setup has been widely accepted by researchers subsequently. In [12], the capacitive HBC within the frequency range from 100 kHz to 100 MHz was measured using FTB-1-6 baluns. The effects of electrode configuration and type, transmission distance, test person, and body movement and position were thoroughly investigated. In [13], the limb joint effects were studied, and it was found that the limb joint could increase the attenuation of the body channel. In 2013, Callejon et al. performed a comprehensive measurement of the capacitive HBC [14]. Their work demonstrated that a larger GND electrode could lead to higher transmission results, since the coupling between the GND electrodes and ground plane was enhanced. Similarly, the transmission performance would be improved when GND electrodes moved closer to the ground plane. In [15], Bea et al. proposed that the balun transformer can result in a coupling capacitance of approximately 2 pF to the ground plane. In view of this issue, a portable SG was used as the TX, and only one FTB-1-6B balun was used in this work (Figure 6.2).

From the above discussion, it can be seen that balun transformers have been widely utilized in the measurement of capacitive HBC channels. However, the effect of balun

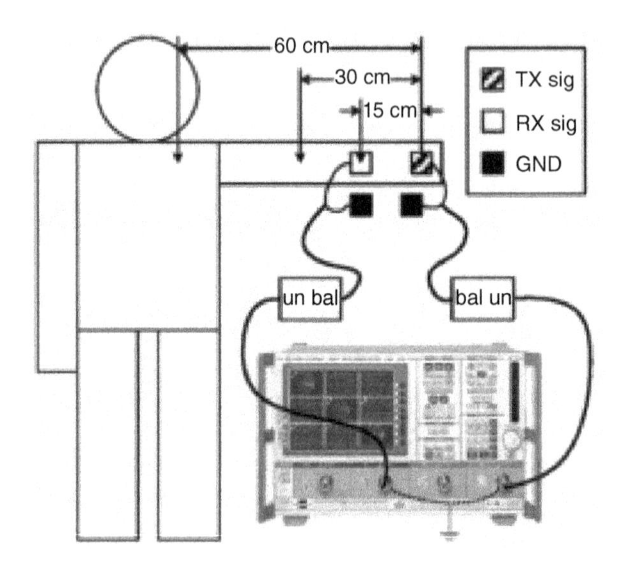

Figure 6.2 Measurement setup of capacitive HBC using a pair of baluns. Source: Xu et al. [9]/IEEE.

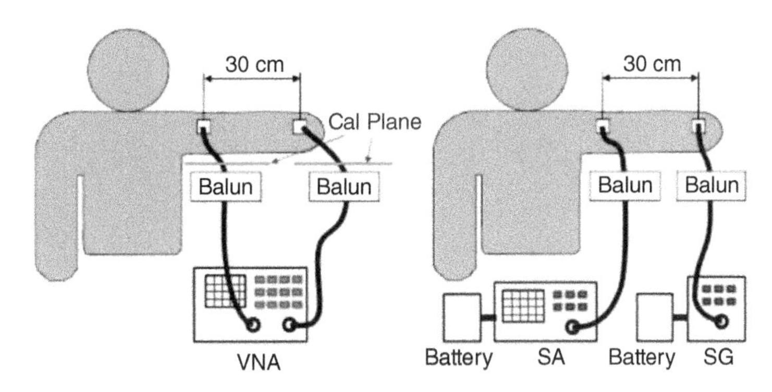

Figure 6.3 (a) Balun transformers used in [16], (b) measurement setup. Source: Sakai et al. [16]/from IEEE.

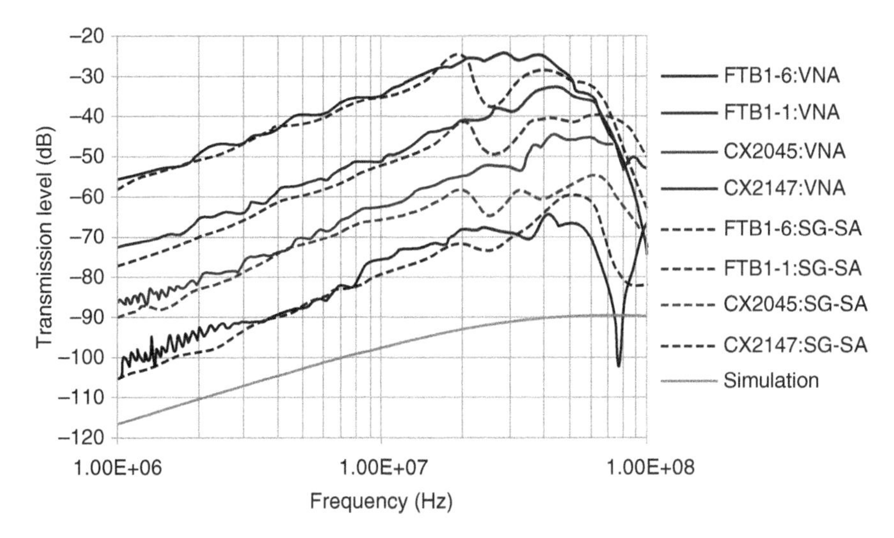

Figure 6.4 Measured transmission levels with different baluns. Source: Sakai et al. [16]/from IEEE.

is rarely investigated. In [16], the measurements were performed using different balun transformers, including FTB-1-6, FTB-1-1 from Mini-Circuits, and CX2045, CX2147 from Pulse Electronics. Besides, two measurement setups are for comparison, including one with a VNA, and the other with a signal generator and spectrum analyzer power by batteries, as shown in Figure 6.3. Besides, two measurement setups are for comparison, including one with a VNA, and the other with a SG and SA powered by batteries. As shown in Figure 6.4, different baluns could lead to a notable difference of 40 dB. It was proposed in [16] that such difference was caused by the inter-winding capacitances of baluns, which allowed the common-mode current to pass through. Therefore, the parasitic common-mode path needs to be eliminated in order to characterize the capacitive HBC channels accurately.

In [17], portable TX and RX devices powered by batteries were used in the experiments in order to emulate the real capacitive HBC environment, as shown in Figure 6.5. The TX device was a SG based on AD9953 using the direct digital synthesizer (DDS)

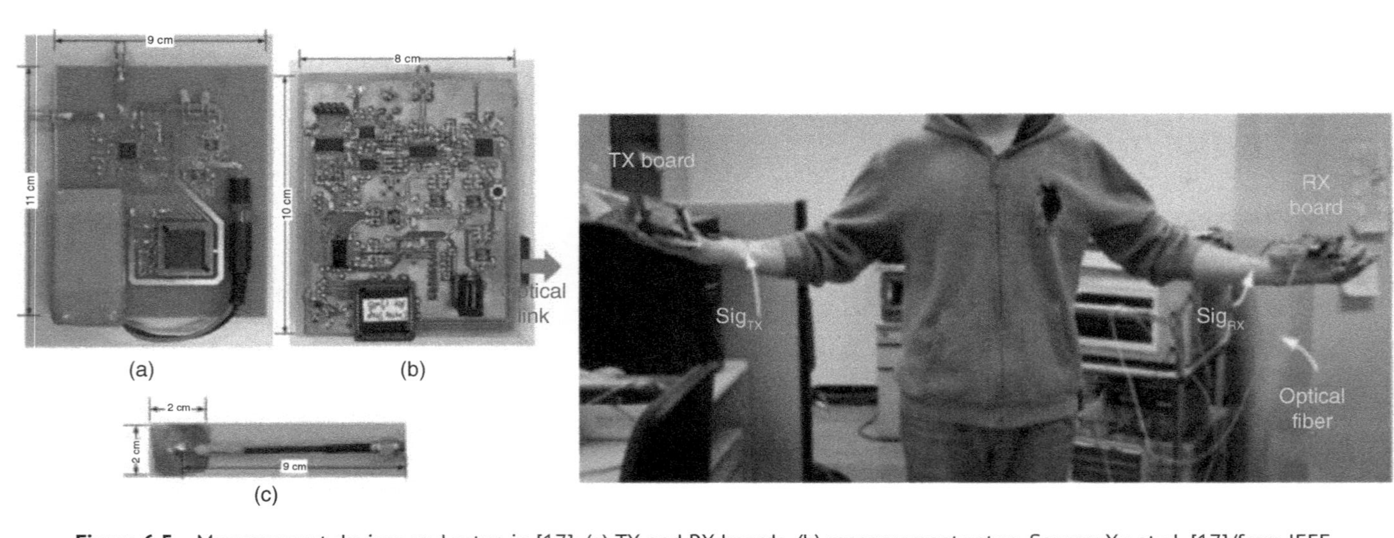

Figure 6.5 Measurement devices and setup in [17]: (a) TX and RX boards, (b) measurement setup. Source: Xu et al. [17]/from IEEE.

method. The RX device was developed based on analog-to-digital converters (ADC) with a sensitivity of −140 dB/Hz. The receiving signals measured by the ADC were transmitted by an optical fiber, which could float the GND electrode of the RX. Note that no balun was used in this work, hence the effect of balun can be eliminated completely. In addition, the measured results in [17] were validated by the numerical simulation, which will be discussed later.

6.3.2 Numerical Models

In addition to verifying the measurement results, the numerical simulation can visualize the electrical field distribution inside and around a human body, which helps to interpret the operation principles of the capacitive HBC. Various methods have been utilized to perform the simulation.

In [18], Fujii et al. utilized the finite-difference time-domain (FDTD) method to analyze the electric field distribution inside and around a human arm. It is demonstrated that a rectangular parallelepiped arm model with homogenous dielectric parameters is sufficient to analyze the capacitive HBC. In [19], a cylinder arm model was simulated using the FDTD method, and it was revealed that the electric field propagating along the arm exhibited the feature of surface wave, which almost decayed exponentially with distance. Moreover, the surface wave had a smaller attenuation constant than the radial component, indicating that the tissue–air interface is more favorable for the electric field transmission than air.

In [10], a multilayered arm model was simulated in ANSYS using the circuit-coupled finite element method (FEM), as shown in Figure 6.6. To represent the parasitic return path, a cross-coupling capacitor (*Cx*) was added between the GND_{TX} and GND_{RX} electrodes. This model can explain the capacitive HBC channels to some extent. However, the coupling between the external ground plane and the TX or RX GND electrodes is ignored. In [20],

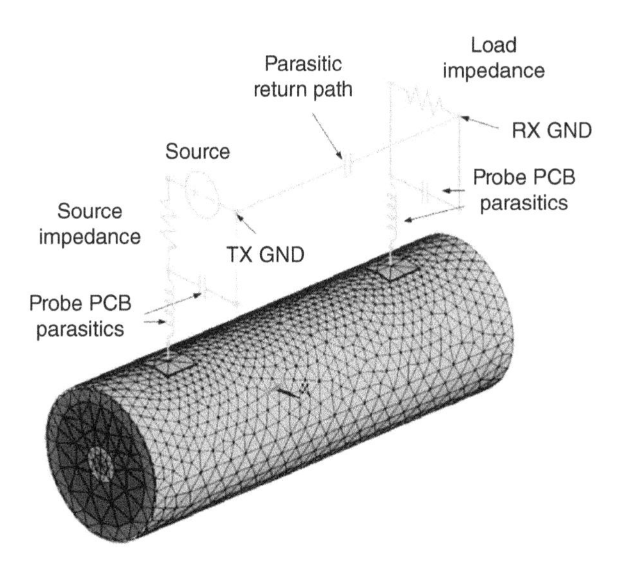

Figure 6.6 Circuit-coupled FEM model proposed in [10]. Source: Xu et al. [10]/from IEEE.

Xu et al. had improved the circuit-coupled FEM model by incorporating the TX board, human body, and ground plane. The coupling capacitance (C_{Gt_G}) from the GND_{TX} electrode to the ground plane was quantitatively determined. Besides, the simulation revealed that the capacitances Cx and C_{Gt_G} could be reduced by the shielding effect of the human body. While in the experiments, the SIG_{RX} electrode was directly connected to a grounded SA, which was contradictory to practical HBC environment.

As shown in Figure 6.7, an FEM model incorporating all measurements was established in [17], which coincided well with the measurement setup plotted in Figure 6.5(b). A good agreement between the simulated and measured results was achieved. Besides, the shielding effect of the body on the coupling capacitances C_{Gt_G} and C_{Gr_G} was determined by the simulation. In addition, the simulation also revealed that the HBC channel could be enhanced when the surrounding environment was closely coupled to the GND_{TX} and GND_{RX} electrodes. However, the cross-coupling capacitor (Cx) was not considered in this model.

6.3.3 Circuit Models of Capacitive HBC

In circuit models, the capacitive couplings within the HBC channels are represented by specific components, which can provide an intuitive explanation of the operation mechanism. Compared with the numerical models, circuit models consume much less time and thus are more efficient. Besides, the circuit models can facilitate the design of transceivers for practical applications (Figure 6.8).

In 2007, Cho et al. proposed a circuit model where the human body was divided into 10-cm-long arm and torso unit blocks, and each unit block was modeled as an *RC* parallel network [7]. The *RC* model of the body has been widely accepted subsequently. In [17], the coupling capacitances C_{Gt_G} and C_{Gr_G} were added to the circuit model to emulate the return path through the ground plane. Besides, the *RC* values of the unit block in contact with the skin were modified to explain the electrode–skin impedance (Z_{ES}). In [5], the effect of test fixtures was investigated by an equivalent circuit model, which included all the measurement devices, such as the coaxial cable, transitions, and baluns. While in this work, the effect of the human body was not considered when extracting the cross-coupling capacitor (Cx) was added between the GND_{TX} and GND_{RX} electrodes.

In [21], Callejon et al. proposed a distributed circuit model by studying the electrical properties of skin. Besides, the path loss was calculated using the transmission line theory. This model was improved in [22] by taking the coupling capacitances into account. In this model, the values of the coupling capacitances to the ground plane were chosen empirically.

In [23], an equivalent circuit composed of capacitors was proposed by Haga et al. The capacitances denoted the coupling between the human body and electrodes, which were derived by solving the capacitance matrix. An improved model incorporating the conductors was proposed in [24], and the values are determined using the method of moment (MoM). While in these two models, the human body was modeled as a node.

6.3.4 Theoretical Analysis

In [15], the authors proposed that the electric field on a human body is composed of the quasi-static field, induction field, and surface wave. Furthermore, the proportion of

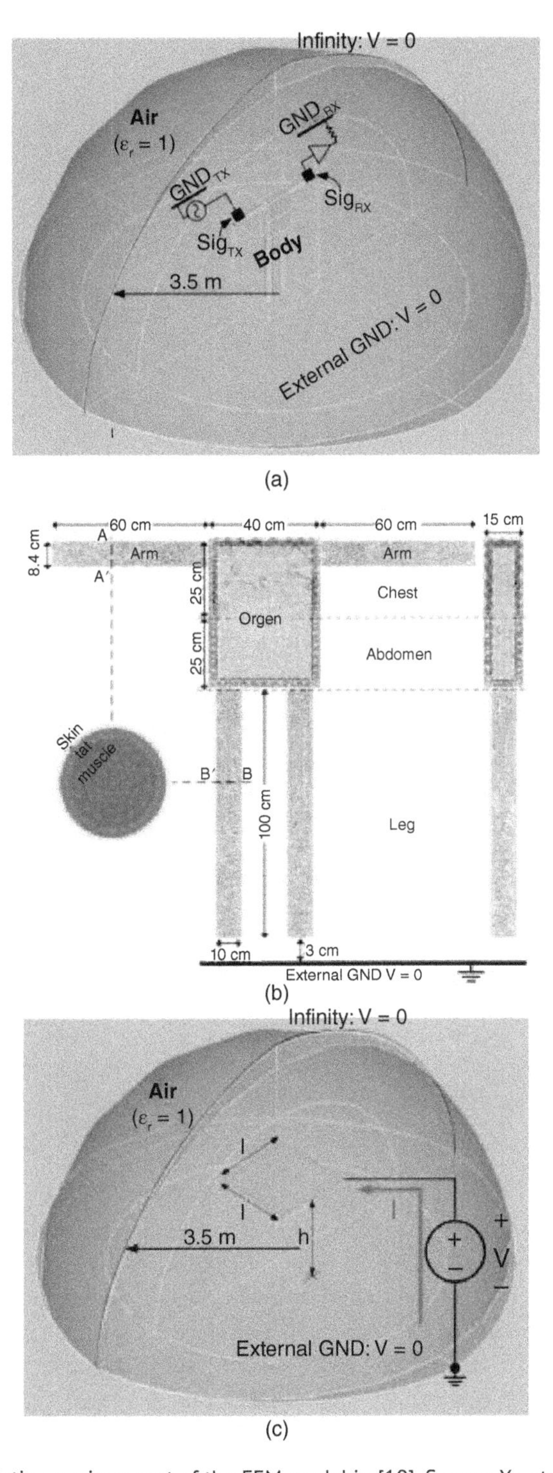

Figure 6.7 Simulation environment of the FEM model in [10]. Source: Xu et al. [10]/from IEEE.

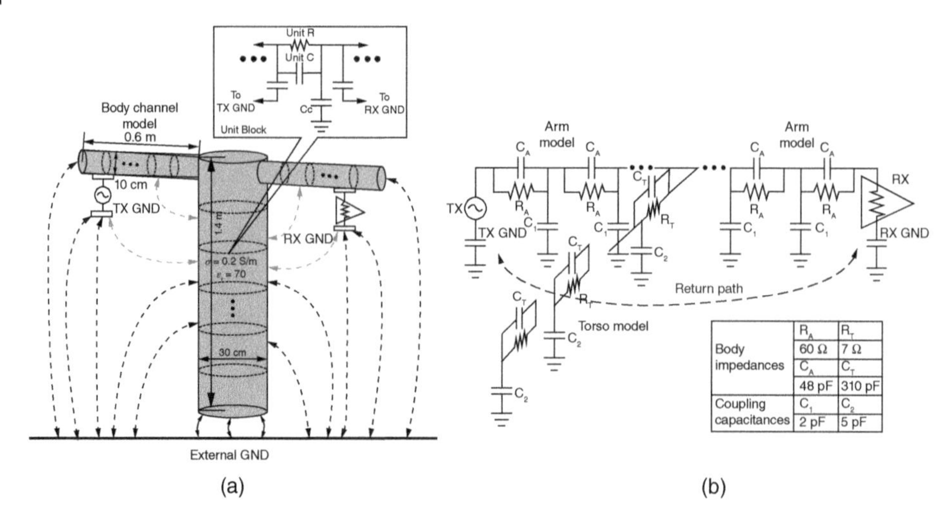

Figure 6.8 Equivalent circuit model proposed in [7]: (a) unit blocks of the human body, (b) equivalent circuit and parameter values. Source: Cho et al. [7]/from IEEE.

each component is determined by both signal frequency and separation distance. A path loss equation was also derived in this work and validated through measured results. Meanwhile, [25] theoretically analyzed the effects of electrode configuration on capacitive HBC channels. The study demonstrated that a vertical electrode pair at the transmitter (TX) and a horizontal electrode pair oriented longitudinally at the receiver (RX) constituted the optimal configuration.

6.4 Investigation and Modeling of Capacitive HBC

In this section, a systematic study of the capacitive HBC is presented. First, the measurement of HBC channels performed using novel devices is introduced. Then a numerical model of the capacitive HBC is established, which can include all measurement devices. Based on the analysis of the simulated electric field distribution, an equivalent circuit is proposed, and the parameter values are extracted using the FEM.

6.4.1 Measurement Setup and Results

As discussed above, the measurement results of the capacitive HBC channels are influenced by the measurement setup. To emulate the practical HBC environment, portable TX and RX devices powered by batteries are applied in this work. Figure 6.9 shows the photograph and block diagram of the portable TX device. The TX device is a SG developed based on AD9952, which can generate sine waves using the DDS method. The output signal frequency is below 120 MHz, and the output power is −2 dBm when terminated with a 50-Ω load. Besides, the signal frequency can be controlled by keys through the microcontrol unit (MCU).

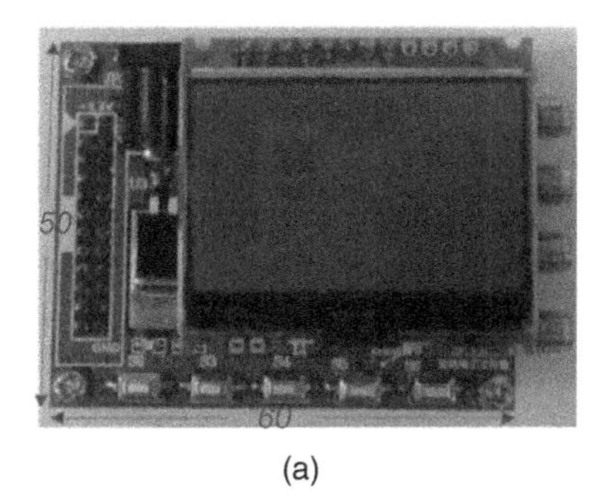

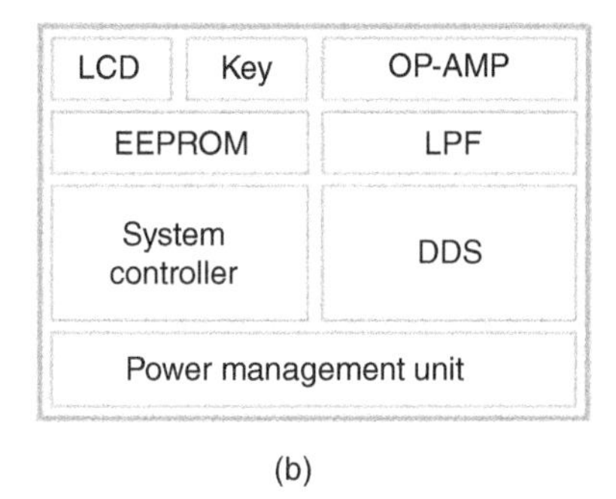

(a) (b)

Figure 6.9 Portable TX device: (a) photograph (Unit: mm), (b) block diagram of the TX.

Figure 6.10 Portable power detector.

Usually, a SA or an OSC is used as the RX device to detect the receiving signals. In this scenario, balun transformers are required to isolate the GND_{RX}. However, it is difficult to incorporate all the measurement instruments into EM models, especially the balun transformer. In view of this issue, a portable power detector powered by batteries is utilized in this work, as shown in Figure 6.10. The RX device is mainly composed of a logarithmic power detector and a voltage meter. The logarithmic power detector can detect and convert the receiving power (P_{in}) to a static output voltage (V_{out}), which can be easily measured by the simple voltage meter.

Figure 6.11 shows the measured curve between V_{out} and P_{in}. It can be seen that for P_{in} greater than −70 dBm, V_{out} is found to increase linearly with P_{in}. For easy reference, the measured results are fit by a linear curve using the least square method (LSM), which can be written as follows:

$$V_{out} = 0.0227 \times P_{in} + 1.8956. \tag{6.1}$$

Note that in (1), V_{out} is in unit of Volt and P_{in} is in unit of dBm. Besides, it is found that the curve between V_{out} and P_{in} is independent of the signal frequency within the frequency band of interest.

Figure 6.12 shows the schematic view of the proposed measurement system. As can be seen, the TX electrode pair is fed by the portable SG, and the SIG electrode is attached to

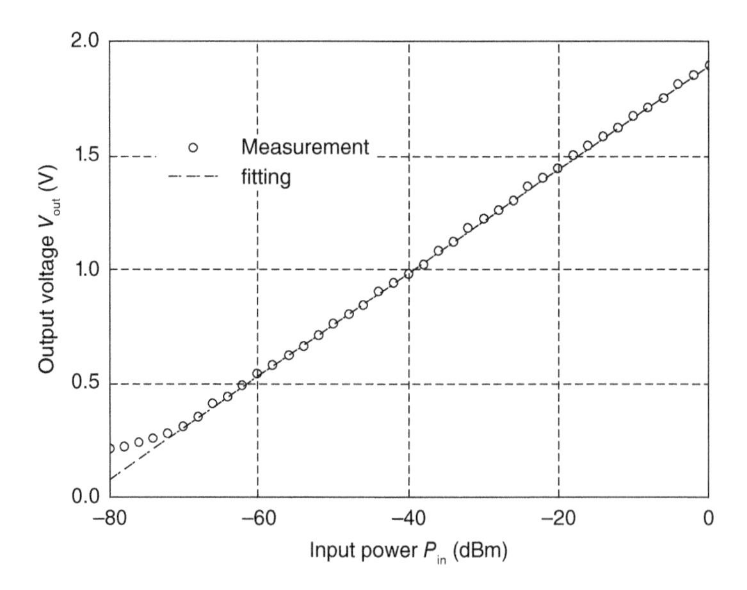

Figure 6.11 Measured curve between the input power (P_{in}) and output voltage (V_{out}).

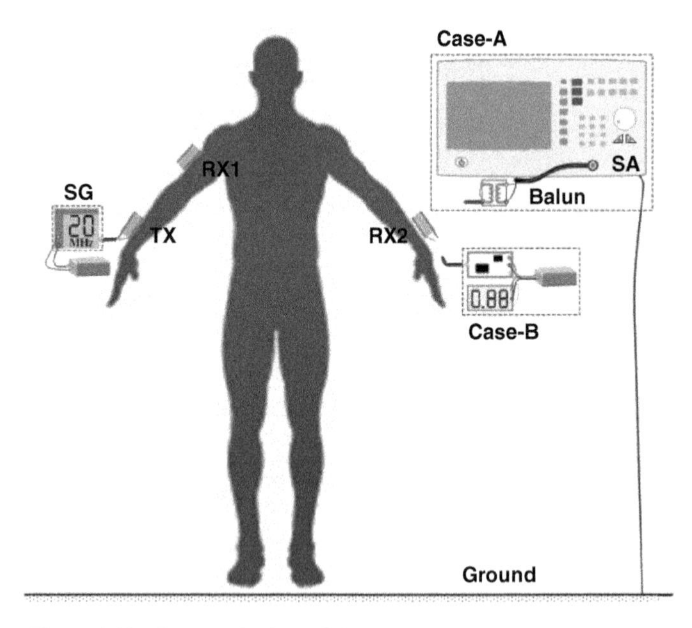

Figure 6.12 Schematic view of the proposed measurement system.

the lower arm. Besides, two receiving configurations are used in the experiments to verify the portable Rx device:

(1) Case-A: The receiving power is measured by a standard SA. In addition, an FTB-1-1balun transformer is used to isolate the internal ground of SA from the parasitic return path.
(2) Case-B: The portable RX device is used at the receiving terminal.

Note that in the Case-B setup, no balun transformer is required, hence the effect of balun can be completely eliminated. In addition, the parasitic effects of cables and power-lined devices coupled to the measurement system can be removed. Moreover, the entire measurement system of Case-B setup can be incorporated in EM models, which will be discussed in detail later.

The dimension of the electrode is 4 cm × 4 cm, and it is fabricated on a Rogers 5880 substrate ($\varepsilon_r = 2.2$, $\tan\delta = 0.0009$) with thickness of 0.51 mm. In the experiments, two receiving positions are selected to study the transmission performance along the body, namely, RX1 at the upper part on the same arm as TX, and RX2 at the lower part on the other arm. The distances from the TX electrode pair to RX1 and RX2 along the body surface are $d_{RX1} = 25$ cm and $d_{RX2} = 120$ cm, respectively.

To enhance the transmission performance of the HBC system, matching networks should be applied to the two electrode pairs [26, 27]. It is straightforward that when the SIG electrode is attached to the human body, the input impedance of the electrode pair will change. Previously, the input impedance was measured by directly connecting the electrode pair to a VNA. In this scenario, the GND electrode will be connected to the internal ground of VNA through the outer conductor of the coaxial cable. Such a measurement fixture is contradictory to practical HBC scenario where the GND electrode is floating. In view of this issue, a balun transformer (FTB-1-1) is used in this work to measure the input impedance, as shown by the inset of Figure 6.13. Note that the electrode pair is connected to the balanced port of balun since this port is electrically isolated from the metallic case.

First, the S-matrix of the balun is measured by the VNA, which is denoted as $[S]$. Then the electrode pair is connected to balun, and the total reflection coefficient Γ_{in} of capacitive HBC channel can be measured. According to [28], the reflection coefficient Γ_l of the load, i.e., the electrode pair herein, can be calculated as:

$$\Gamma_l = \frac{\Gamma_{in} - S_{11}}{(\Gamma_{in} - S_{11}) \times S_{22} + S_{12} \times S_{21}} \tag{6.2}$$

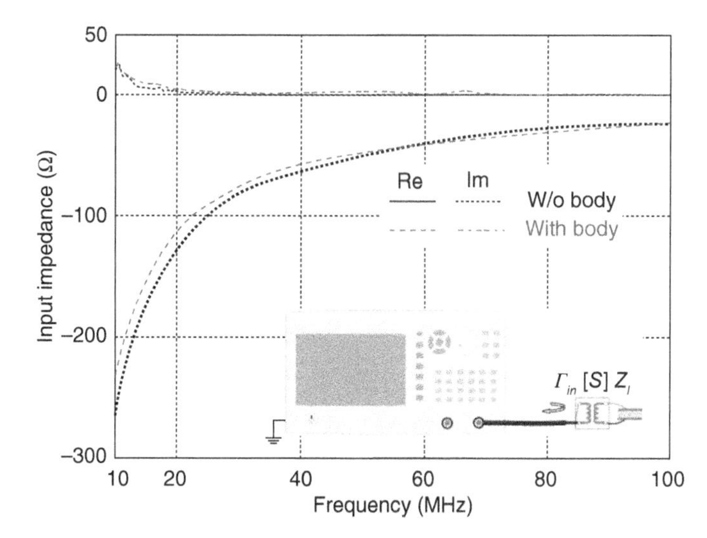

Figure 6.13 Schematic view of the proposed measurement system.

Then the input impedance of the electrode pair can be easily obtained based on the calculated Γ_1 as:

$$Z_1 = Z_0 \times \frac{1+\Gamma_1}{1-\Gamma_1} \tag{6.3}$$

Similarly, the input impedance can be obtained when the SIG electrode is attached to the body. Figure 6.13 shows the calculated input impedances of the electrode pairs with and without the human body. As can be seen, the input impedances are quite similar for the two scenarios. The measured results below 10 MHz are not plotted because these results are disrupted by a parasitic resonance at 3 MHz, which is caused by the magnetizing inductance of balun and the capacitance of the electrode pair. In Figure 6.13, the measured resistance of the input impedance decreases as frequency increases, and the resistance values are quite low at frequencies above 20 MHz. Besides, the measured negative reactance increases gradually as frequency increases, which exhibits the capacitive property of the electrode pair. It can be easily verified that the measured reactance curve without the human body corresponds to a capacitance of 62.5 pF, which agrees well with the value of the electrode pair. When the human body is attached to the SIG electrode, the measured reactance values are slightly higher. This phenomenon is mainly caused by the coupling between the GND electrode and the human body, which will be explained in more detail later. At 21 MHz, the measured input impedance is $2.34 - j \cdot 120\,\Omega$ for the case without body and the value is $4.06 - j \cdot 107.7\,\Omega$ for the case with body.

Based on the measured input impedance, matching networks of the electrode pair at 21 MHz can be designed, which consists of a lumped inductor ($L_p = 1\,\mu$H) and a series one ($L_s = 2.7\,\mu$H), as shown by the inset of Figure 6.14. Besides, the measured reflection coefficients of the electrodes pair with and without matching networks are also plotted, which are calculated using the method shown in Figure 6.13. As can be seen, good impedance

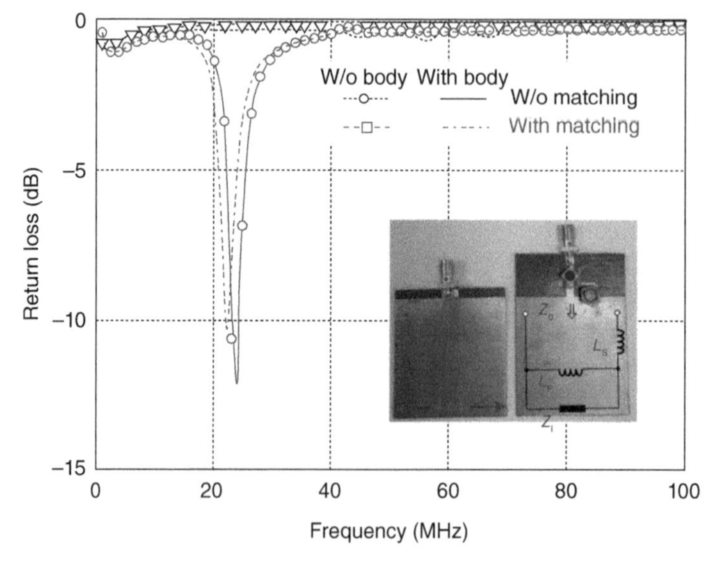

Figure 6.14 Measured impedance matching results of the electrode pairs with and without matching network. The inset shows the fabricated electrodes.

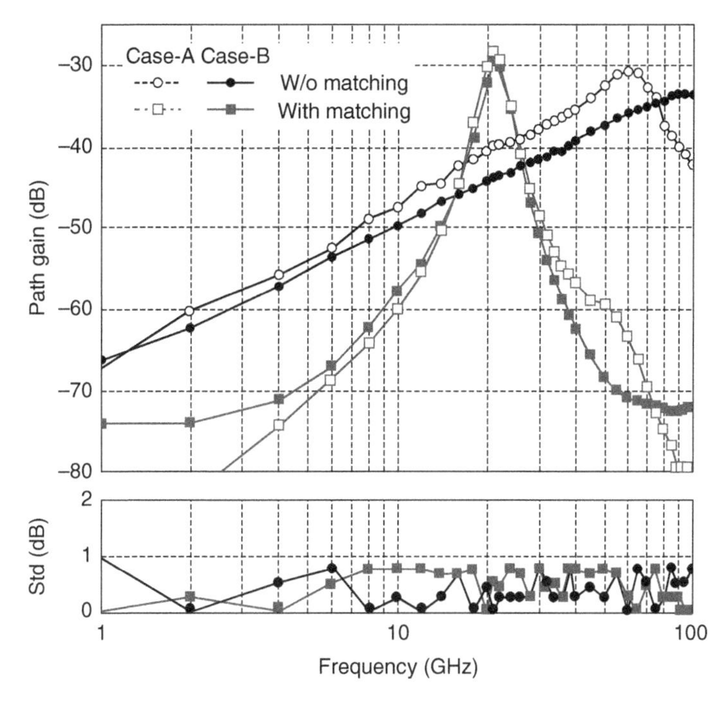

Figure 6.15 Measured path gains and standard deviations at $d_{Rx1} = 25$ cm.

matching can be obtained when matching networks are utilized. Besides, the resonant frequency slightly decreases when the SIG electrode is attached to the body.

To verify the validity of the proposed matching network, the measurements of path gains are performed on a human body, which has a height of 173 cm and a weight of 65 kg. The measured path gains at different receiving positions, i.e., Rx1 and Rx2, are plotted in Figures 6.15 and 6.16, respectively. The experiments are performed in an anechoic chamber, since the portable power detector is relatively susceptible to external interference due to its simple structure. Besides, the experiments are conducted over different days in order to verify the stability of the measurement devices.

As can be seen from Figure 6.15, a reasonable agreement is obtained between the measured results using Case-A and Case-B setups. For the case without matching network, the measured path gains using Case-A setup are about 3 dB higher than that using Case-B setup, which could be caused by the common mode current flowing through the balun transformer [16]. Generally, the measured path gain increases with frequency at a slope of about 20 dB/dec. For Case-A setup, the obtained path gain reaches its peak at 60 MHz, which is −30.8 dB. While for Case-B setup, the peak value is −33.5 dB at 90 MHz.

When matching networks are applied to the electrode pairs, the obtained path gains at 21 MHz can be improved significantly. The standard deviations of the measured path gains are less than 2 dB. Generally, a good agreement has been achieved between the two measurement setups. Owing to the limited measurement range of the proposed power detector shown in Figure 6.11, the measured minimum path gain is about −74 dB. With matching networks, the measured peak path gains occur at 21 MHz, and the values are

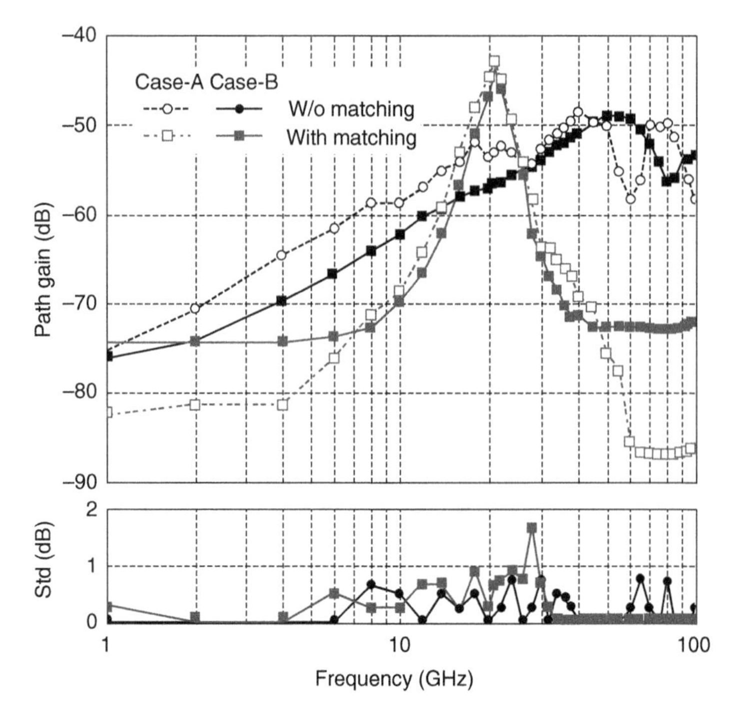

Figure 6.16 Measured path gains and standard deviations at $d_{Rx2} = 120$ cm.

−28.3 dB for Case-A and −29.2 dB for Case-B, respectively. Compared with the scenarios without matching network, the measured path gains at 21 MHz have increased by 11.7 dB and 14.6 dB for the two setups. Moreover, it should be mentioned that the impedance matching of the electrode pair will be affected when it is connected to other devices, such as the proposed SA, the power detector, and the balun. Hence, it can be inferred that if such effects are taken into consideration when designing the matching networks, the measured path gains can be further improved.

The path gain at a longer transmission distance can be similarly measured and analyzed. As shown in Figure 6.16, it is obvious that the path gains are lower at $d_{Rx2} = 120$ cm. For the case without matching network, the measured peak path gain occurs at 40 MHz for Case-A setup and at 55 MHz for Case-B setup. Besides, dips are observed on the measurement curves, which occur at 60 MHz for Case-A and 80 MHz for Case-B, respectively. When matching networks are utilized, the path gains at 21 MHz are improved greatly as predicted, and the measured peak values are −42.9 dB and −44.0 dB for the two measurement setups, respectively. Compared with the result at $d_{Rx1} = 25$ cm, the measured path gain at 21 MHz has decreased by 14.8 dB for the Case-B measurement setup. The decrease in the transmission performance is mainly caused by the reduced coupling between the GND electrodes of TX and RX, which will be explained in detail later.

6.4.2 Simulation Setup and Results

In order to verify the measurement results and analyze the mechanism of the signal transmission, numerical simulation of the capacitive HBC is performed. The simulation setup of the numerical model is described first, and the simulation results are analyzed subsequently. An equivalent circuit is then proposed and a good agreement between the numerical and circuit models is achieved.

To emulate the dispersive property of human tissues, the electrical properties of different tissues are calculated based on the multiple Cole–Cole equations presented in [29] as below:

$$\varepsilon_r^*(\omega) = \varepsilon_\infty + \sum_n \frac{\Delta\varepsilon_n}{1 + (j\omega\tau_n)^{(1-\alpha_n)}} + \frac{\sigma_i}{j\omega\varepsilon_0} \tag{6.4}$$

where $\varepsilon_r^*(\omega)$ is the complex dielectric constant, ε_∞ is the permittivity at infinite frequency, $\Delta\varepsilon_n$ and τ_n are the strength and relaxation time of the Debye dispersion model. α_n is the distribution parameter. σ_i is the static permittivity. The obtained relative permittivity and conductivity of each tissue in the frequency range of 1–100 MHz are plotted in Figure 6.17.

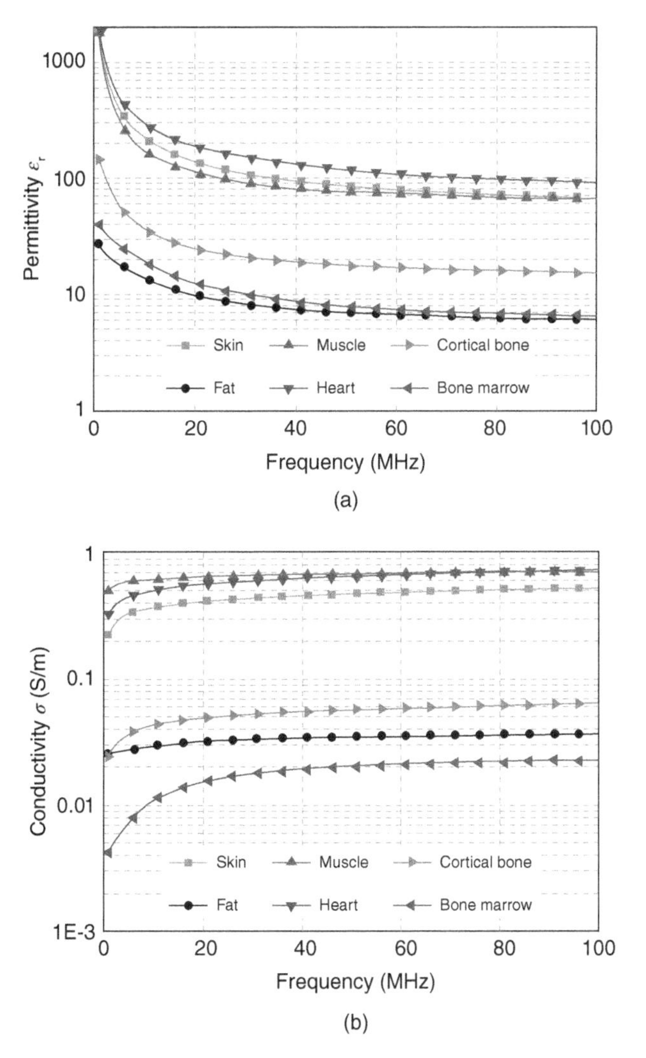

Figure 6.17 Calculated electrical properties of different human tissues: (a) relative permittivity, (b) conductivity.

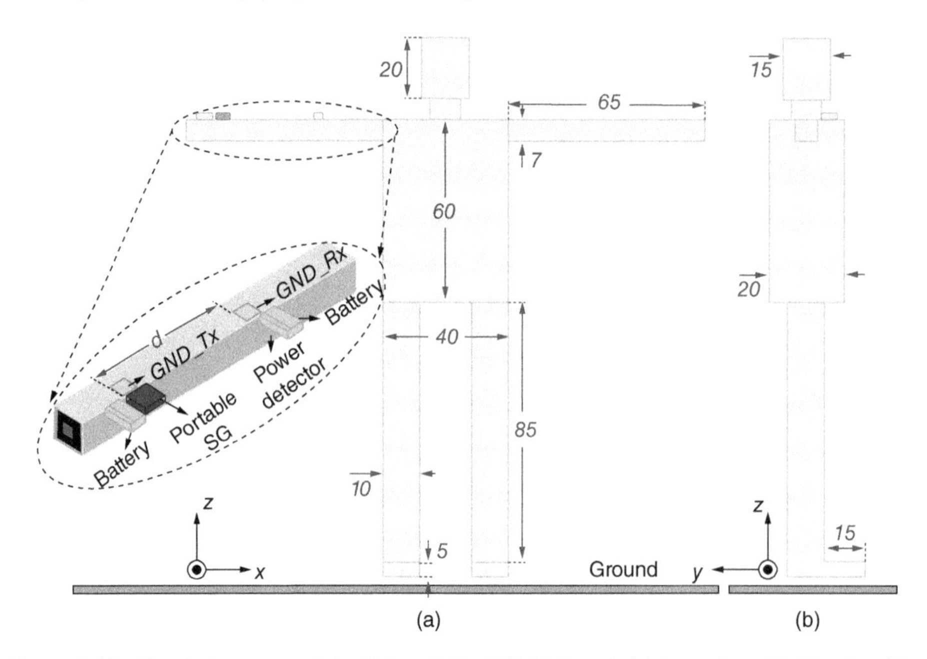

Figure 6.18 Simulation setup of the EM model in CST (Unit: cm): (a) front view, (b) side view. The inset shows the multilayer arm model and the measurement fixture of the electrodes, TX and RX devices, and batteries.

Table 6.1 The thicknesses of different tissue layers (unit: mm).

	Skin	Fat	Muscle	Cortical bone	Bone marrow	Heart
Arm	1.5	3.5	18	5	5	—
Head	1.5	3.5	10	5	5	50
Torso	1.5	5.5	13	5	5	70
Leg	1.5	8.5	27.5	6	6.5	—

The proposed human body model is composed of six kinds of tissues, i.e., skin, fat, muscle, cortical bone, bone marrow, and heart [20]. Besides, the torso and head of the body model are assumed to be filled with heart for simplicity. The EM model is established and simulated in CST Studio Suite [30]. The structure of the body model is plotted in Figure 6.18, as well as the test fixture of Case-B setup. Note that all the measurement devices are included in the numerical model. The thicknesses of the tissue layers in different parts of the body model are listed in Table 6.1. Besides, a ground plane with size of 2 m × 3 m is placed below the body model at a distance of 2 cm to emulate the parasitic return path.

Figure 6.19 shows the comparison between the simulated and measured path gains at transmission distances of 25 cm and 120 cm when matching networks are applied to the electrodes. As can be seen, the peak values of the simulated path gains occur at 21 MHz, and the simulation results coincide very well with the measurement. Whereas the simulated path gains at other frequencies are much lower than the measured results. For instance,

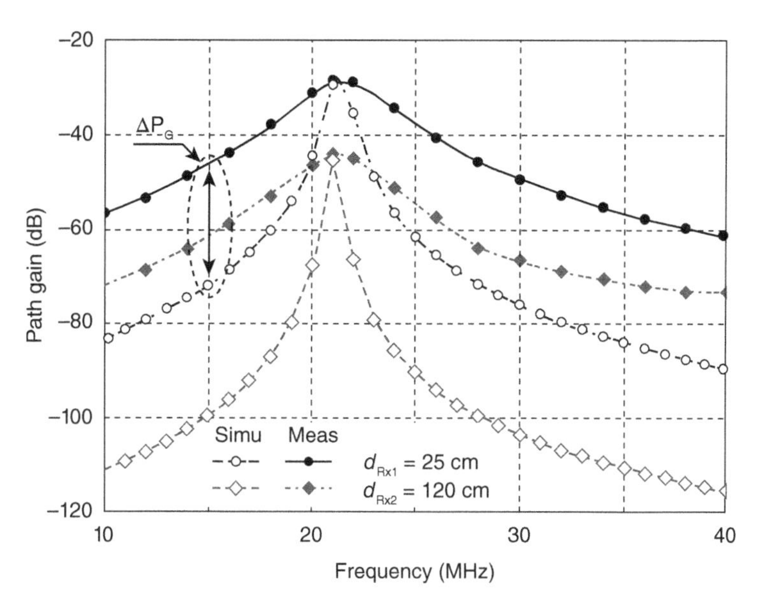

Figure 6.19 Simulated and measured path gains for transmission distances of 25 cm and 120 cm.

the measured path gain at 15 MHz is $P_{GM} = -47.3$ dB at the transmission distance of 25 cm, while the simulated result is $P_{GS} = -71.8$ dB. Thus, the deviation ΔP_G between the simulated and measured results is 24.5 dB. Such deviation is mainly caused by the impedance mismatch of the numerical model, which is explained as follows:

For the case without matching network, the simulated input impedance at 15 MHz is $Z_{ls} = 0.82 - j{\cdot}148.6\,\Omega$, while the measured result is $Z_{lm} = 9.61 - j{\cdot}152.8\,\Omega$. It is obvious that the simulated reactance coincides well with the measured result plotted in Figure 6.13, whereas the simulated resistance is much smaller than the measured value. The difference between the simulated and measured resistances could be caused by the simple numerical human model, where many tissue layers are omitted, such as the stratum corneum, epidermis, and so on. When matching networks are applied, the simulated reflection coefficient at 15 MHz is −0.00415 dB, while the measured result is −0.615 dB. Due to the poor impedance matching of the numerical model, an overwhelming majority of the input power will be reflected. Then, the matching network of the numerical model is slightly adjusted to tune the input impedance at 15 MHz to $140.8 - j{\cdot}420.7\,\Omega$, which is almost equal to the measured result with matching network. In this condition, the simulated path gain is increased to −44.2 dB, which is quite close to the measured result. Hence, it can be concluded that the deviation between the simulated and measured path gains is basically caused by the lower values of the simulated resistances, which results in lower quality factor and different impedance matching characteristics between the EM model and the experiment.

In addition to verifying the measurement results, the EM simulation can also help to interpret the operation principle of capacitive HBC. Figure 6.20 shows the simulated electric field distribution at 21 MHz around the body. Note that the Rx device and electrode pair are not included in the simulation. It can be seen that the simulated E-fields around the human

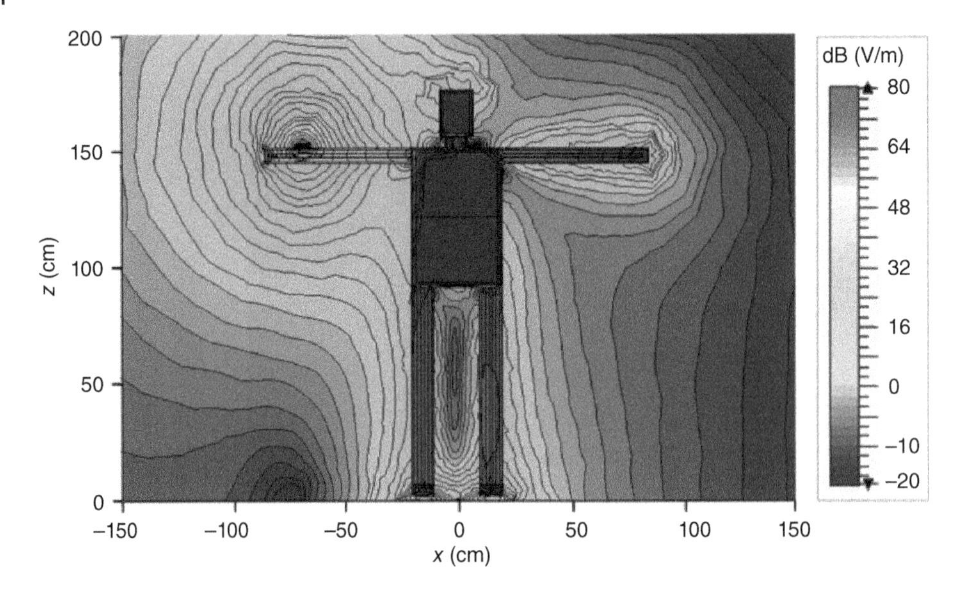

Figure 6.20 Simulated electric field distribution at 21 MHz around the human body without Rx devices.

body are much stronger than those in the air. Moreover, the E-fields at the extremities of the body, such as hands, feet, and head, are stronger than those at other parts. This property of capacitive HBC can enhance the transmission performance between different extremities of the body greatly, which is considered a prominent feature over the NB and UWB PHYs.

Figure 6.21 shows the simulated electric field at 21 MHz at the transmission distance of 25 cm. The ground plane and other parts of the body model are not shown. As can be seen, the electric field distribution can be divided into two parts, i.e., inside and around the arm model, which indicates the forward and parasitic return paths, respectively [4]. Besides, the simulated E-field outside the arm is much stronger than that inside the arm. In the proximity of TX and RX, part of the E-fields flows out of the arm and returns to the corresponding GND electrodes, which indicates the coupling between the human body and the GND electrodes. Figure 6.22 shows a detailed analysis of the coupling mechanism of capacitive HBC based on the simulated E-fields. The schematic diagram in Figure 6.22 can help to interpret the circuit model greatly and it will be explained in further detail later.

To study the propagation property of the E-field inside the arm model, the simulated E-fields are normalized and plotted in Figure 6.23. The extracted E-fields are distributed at the middle of the muscle layer, i.e., on line AA' in Figure 6.21, since the muscle layer occupies a major proportion of the arm model. The starting point of the x-axis is set at the left end of arm. It can be seen that the Ex component is dominant over Ey and Ez components. Moreover, the decibel values of Ex component in Figure 6.23 almost decrease linearly when the observation point is relatively far from TX electrode pair, which implies that the magnitude of Ex decays exponentially with distance. At the center of TX electrode, the phase of Ex component reverses. Besides, the phase of Ex almost decreases linearly as distance increases. While outside the arm model, the dominant field is Ez component, which coincides well with the finding in [19].

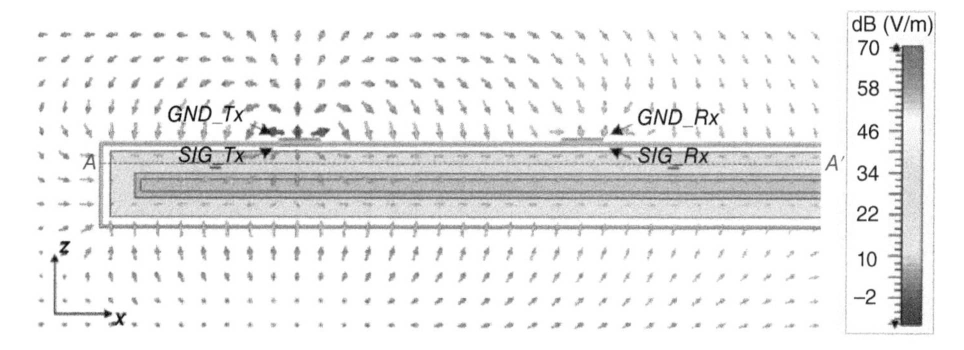

Figure 6.21 Simulation electric field at 21 MHz distributed around and inside the human arm at transmission distance of 25 cm.

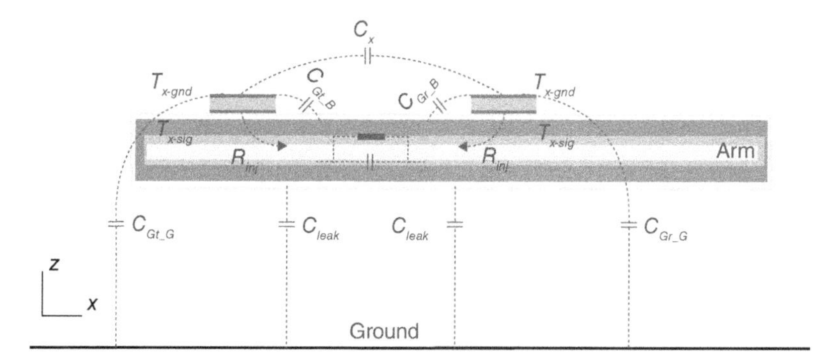

Figure 6.22 Analysis of the coupling mechanism of the capacitive HBC based on the simulated electric field distribution.

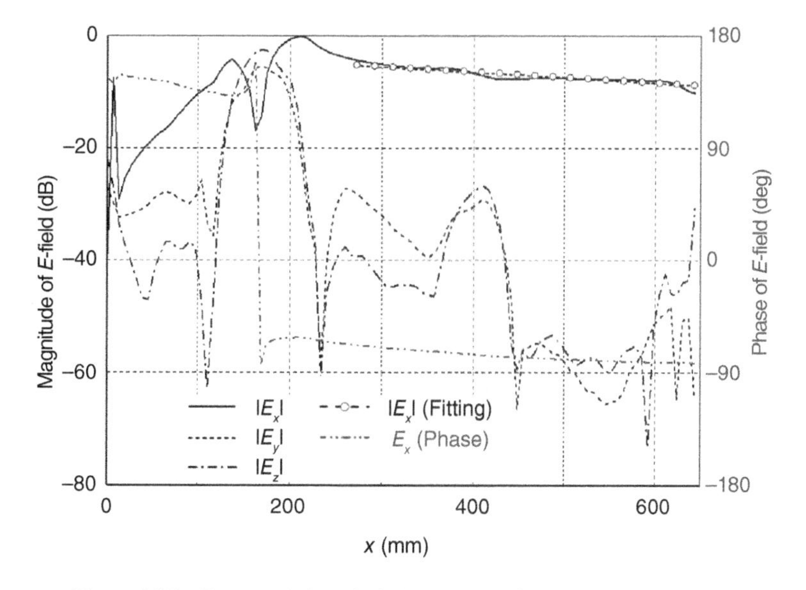

Figure 6.23 Extracted electric field on line *AA′* inside the arm model.

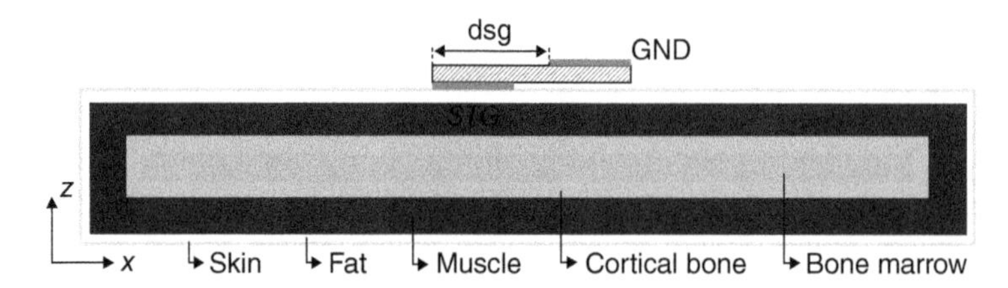

Figure 6.24 Simulation setup to investigate the effect of electrode configuration.

Figure 6.24 shows the simulation setup of the numerical model to investigate the effect of electrode configuration. In the simulation, the SIG electrode is attached to the skin layer, while the GND electrode is shifted along the x-axis, which creates a deviation of d_{sg} between them. Note that only the arm model is sufficient to study the input impedance of the electrode pair. The dimension of the arm model is the same as in Figure 6.18.

Figure 6.25 shows the simulated input impedances of the electrode pair for different values of d_{sg}. It can be seen that as d_{sg} increases, the real part of input impedance increases significantly, while the imaginary part is decreased. A rough explanation can be made to interpret such phenomenon. When the E-field flows from the SIG electrode to the GND electrode through the arm, it will be attenuated by the loss of the tissues. As d_{sg} increases, the loss of tissues will be enhanced, which causes an increase in the real part of input impedance. On the other hand, the capacitance of the electrode pair is reduced as d_{sg} increases, which leads to the decrease of the imaginary part of input impedance. To sum up, it can be concluded from the above analysis that the electrode configuration has a great effect on the impedance matching of the electrode pair, thus more investigation needs to be conducted to determine the optimal electrode structure.

6.4.3 Equivalent Circuit Model

The parasitic return path has been investigated in previous literatures, such as in [5, 17, 20]. In [5], the capacitance between the GND electrode and the ground plane and the cross capacitance between the GND electrodes are extracted using the FEM simulation. However, the influence of the human body is not considered. To determine the capacitances among the measurement devices, the human body, and the ground plane, numerical simulation is performed using the Ansys Maxwell software [31]. The human body model and the test fixture built up in Maxwell are the same as that in Figure 6.18. The body model is segmented into unit blocks with length of 10 cm. In addition, the tissue properties at 21 MHz are adopted here.

Based on the analysis of the simulated E-field distribution, an equivalent circuit of the capacitive HBC system is proposed, as shown in Figure 6.26. The electrode is simply modeled as a series of capacitor C_e and resistor R_e. Owing to the finite conductivity of human tissues, coupling capacitors exist between the human body and the GND and SIG electrodes of TX and RX, which are modeled as capacitors C_{Gt_B} and C_{St_B} for TX and C_{Gr_B} and C_{Sr_B} for RX, respectively. Moreover, the values of C_{St_B} and C_{Sr_B} are much larger than C_{Gt_B} and C_{Gr_B} since the SIG electrodes are directly attached to the body. In addition, injection current exists between the body and the SIG electrodes, as

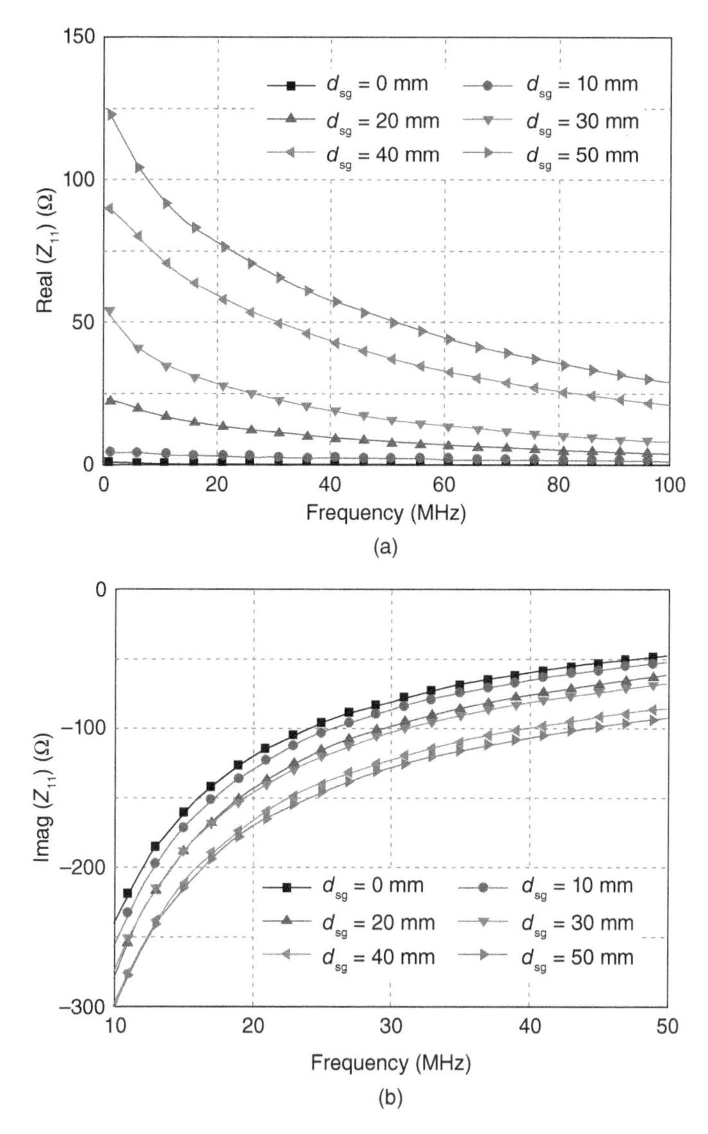

Figure 6.25 Simulated input impedances of the electrode pair for different values of d_{sg}: (a) real part, (b) imaginary part.

denoted by the resistor R_{inj}. The parameter R_{inj} is taken as the electrode–skin impedance in [5, 8]. The capacitors C_{Gt_G} and C_{Gr_G} represent the coupling between the floating GND electrodes to the external ground plane. Since the TX board is much larger than the RX devices, the value of C_{Gt_G} is larger than C_{Gr_G}. The cross-coupling between the GND electrodes is denoted by the capacitor C_x. The parameter values of the proposed equivalent circuit and the matching network are listed in Table 6.2.

It has been widely accepted that the circuit model for each unit block of the body can be an *RC* parallel network and a capacitor C_{leak} for the leakage to the ground plane. Besides, the

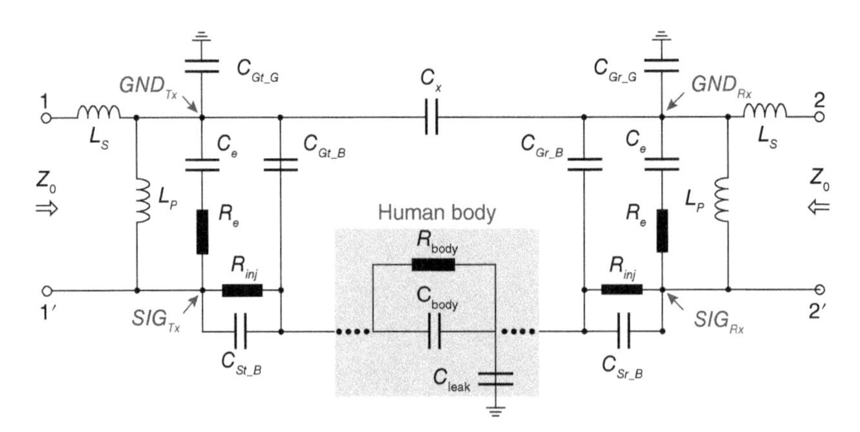

Figure 6.26 Structure of the proposed equivalent circuit.

Table 6.2 The parameters of the proposed equivalent circuit.

L_s	L_p	C_e	R_e	C_{Gt_G}
1 μH	2.7 μH	64.5 pF	0.5 Ω	1 pF
C_{Gt_B}	R_{inj}	C_{St_B}	C_{Gr_G}	C_{Gr_B}
5.7 pF	250 Ω	318 pF	0.4 pF	4.4 pF
C_{Sr_B}	C_x (d_{Rx1} = 25 cm)		C_x (d_{Rx1} = 120 cm)	
355 pF	0.078 pF		0.0075 pF	

Table 6.3 The parameters of the human body model.

	Arm	Head	Torso	Leg
R_{body} (Ω)	32	7.9	2.2	15.5
C_{body} (pF)	47	364	1295	95.5
C_{leak} (pF)	0.7	0.55	2	1.5

parameter C_{leak} of each unit block of the same body part should be different. Whereas, it is demonstrated by simulation that the human body model only has minor effect on the performance of the equivalent circuit. Hence, the parameter C_{leak} of each unit block of the same body part is set uniform. The parameters of the human body model are listed in Table 6.3.

Figure 6.27 shows the simulated path gains obtained using the numerical and equivalent circuit models at the transmission distances of 25 cm and 120 cm. The equivalent circuit model is simulated in ADS. As can be seen, the simulation results using the two models coincide very well, which demonstrates that the proposed circuit model can interpret the mechanism of the capacitive HBC very well.

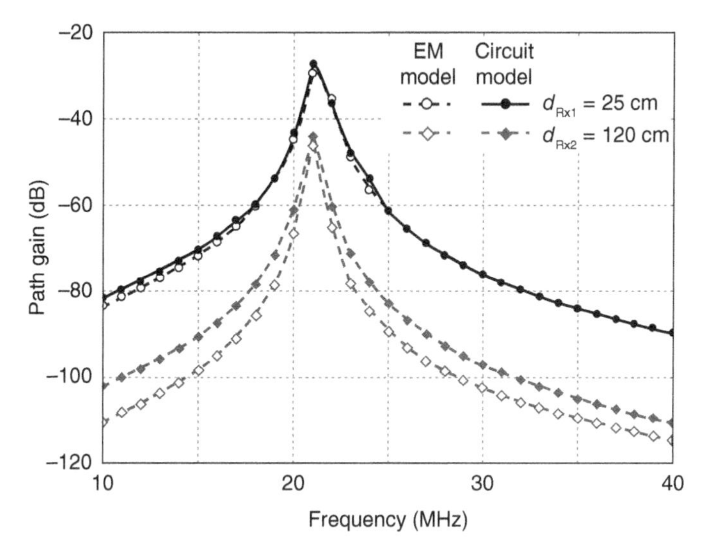

Figure 6.27 Simulated path gains obtained from the numerical model and the equivalent circuit model at different transmission distances.

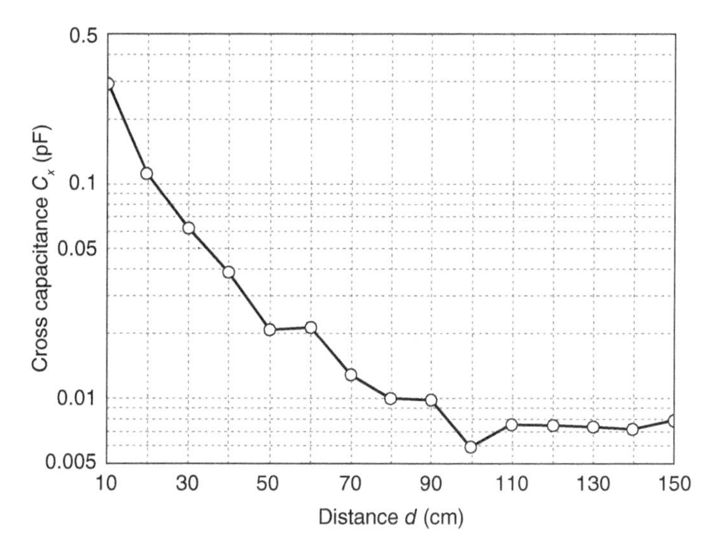

Figure 6.28 Simulated cross capacitances (Cx) at different separation distances between the GND electrodes of TX and RX.

The transmission capability of the capacitive HBC also needs to be studied. Figure 6.28 shows the simulated C_x results at different separation distances (d) between the GND electrodes. In the simulation, the TX device is fixed while RX device is shifted along the arms. It can be seen that for d less than 100 cm, C_x decreases sharply as d increases. At $d = 100$ cm, the RX electrode pair is placed on the shoulder and the minimum value of C_x occurs, which could be caused by the severe shielding effect of the head and torso. While for d larger than

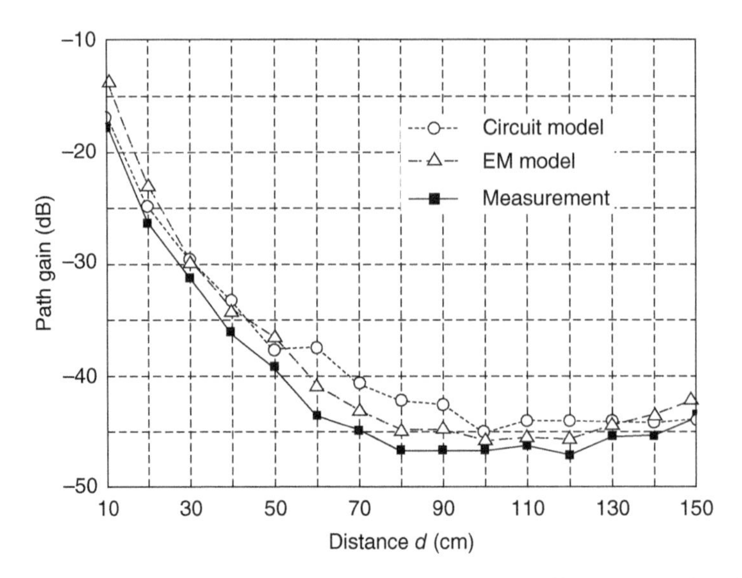

Figure 6.29 Simulated and measured path gains at 21 MHz at different separation distances between the GND electrodes of TX and RX.

100 cm, C_x is almost kept constant as d increases. At $d = 150$ cm, the shielding effect of the human body is further reduced, and the value of C_x at $d = 150$ cm is even slightly higher in spite of the longer separation distance.

Figure 6.29 shows the simulated path gains at 21 MHz using the circuit model based on the extracted C_x values along the arms. Besides, the simulated results are compared with the measured and EM model results. A reasonable agreement is obtained in Figure 6.29, which reveals that the transmission performance of HBC channels can be effectively modeled by extracting the C_x values at different parts of the body. The obtained path gains along the arms are higher than −50 dB, and the minimum value occurs at about $d = 100$ cm. Besides, the path gain between the extremities of the arms is relatively higher, which agrees well with the simulated E-field distribution in Figure 6.20.

6.5 Conclusions: Other Design Considerations of HBC Systems

In the previous sections, we introduced the design methodologies related to capacitive HBC, establishing a solid foundation for understanding this unique communication technique. Also, a systematic study of the capacitive HBC is proposed. In an endeavor to replicate a realistic HBC scenario, we employed portable devices for our experiments. Notably, our proposed measurement system eliminates the need for balun transformers, thereby effectively eradicating any impacts due to baluns. The veracity of this setup has been further confirmed through another set of measurements utilizing a standard SA and balun. Moreover, we introduced a novel method for measuring the input impedance of the entire HBC channel. Guided by the results obtained from these measurements, we designed matching networks to enhance the transmission properties of our HBC system. This innovative approach to system design and analysis underscores our commitment to refining HBC methodologies and promoting optimal communication performance. In a concerted effort to validate our measurement results and facilitate the analysis of the signal transmission mechanism, we conducted numerical simulations of the capacitive HBC. Notably, the maximum path gain at 21 MHz, as obtained from our measurements, agrees well with that in the simulation. Moreover, we found that deviations between simulated and measured path gains at other frequencies were attributable to lower resistances of the simulated input impedances. Our exploration was further extended to the effect of electrode configuration, where we discovered that offsetting the Signal and Ground electrodes led to an increase in simulated resistance. This interesting observation carries significant implications as it points toward the possibility of reducing the Q-factor and broadening the impedance and path gain bandwidths. Delving deeper into the mechanism, we scrutinized the simulated E-field distribution around and inside the body model and put forth an equivalent circuit model. Guided by the FEM, we extracted the circuit parameters and considered the effect on the human body. We also examined the transmission capability along the human body. The strong alignment between the models and the measurements corroborates the effectiveness of the proposed equivalent circuit in elucidating the operating principle of capacitive HBC. This analysis highlights the intricate interplay of various factors within the HBC system and sets the stage for future enhancements.

Nevertheless, to expedite the pragmatic utilization of capacitive HBC, various challenges need to be addressed. These include the universalization of channel characteristics, more suitable signal modulation techniques for communication, and the miniaturization of transceiver modules through integrated circuit design.

6.5.1 Channel Characteristics

As elucidated in the preceding sections, a considerable body of work has already been undertaken to detail the modelling and characterization of communication channels. However, there are still more factors that need to be modeled and taken into consideration. To fully realize capacitive HBC, a wide-range series of experiments should be conducted. These are essential for validating the reproducibility of the models and for adjusting the effects that individual variations may have on signal modeling. Tailored channel models will need to be developed for specific applications, such as communication for common head-mounted devices, chest-implanted devices, and wearable devices. Furthermore, it is noteworthy that most of the current channel research and modeling is predominantly centered on static channel attributes. As such, dynamic signal models, especially those influenced by human movements, require more comprehensive examination and should be given increased attention. In [32], researchers have embarked on modeling and studying the dynamic channel. They proposed that the burst features are the primary characteristics of dynamic changes of the fading channel in HBC, which can be described by the proposed three-state Fritchman model. In [33], based on measurements of the channel under both static and dynamic conditions, researchers summarized the time-invariant characteristics of the channel under static conditions and the time-varying characteristics under dynamic conditions. A probabilistic model based on a filter was proposed to describe the crucial features of the capacitive body channel. These studies have provided data and modeling references for the practical application of capacitive HBC. Research on dynamic channel characteristics should receive more attention in the future.

6.5.2 Modulation and Communication Performance

In the field of capacitive HBC, understanding and employing suitable modulation techniques are of paramount importance. The selection of an apt modulation technique significantly steers the overall performance of the system, touching on aspects such as data transmission rate, power efficiency, and system reliability.

Recent advancements in HBC systems have seen the adoption of digital signal transmission, including the IEEE Standard 802.15.6 for the WBAN. Several studies have probed the application of conventional modulation techniques, largely borrowed from the realm of RF communication, for HBC. For instance, reference [3] presents a comprehensive examination of these techniques. In another study, [34] proposed a frequency-selective digital transmission protocol conforming to the IEEE 802.15.6 WBAN standards.

On the other hand, reference [35] discusses the architecture of high-data-rate HBC systems, using multilevel coding applied to Walsh codes. A structurally simplified and spectrally efficient digital transmission methodology for narrowband digital transmission was introduced in [36]. Additionally, a scalable data rate HBC architecture using deep learning approaches was proposed in [37].

Addressing signal degradation in body channels represents a significant challenge in HBC systems. An effective method for mitigating this issue involves encoding the transmitted signal or mapping it to orthogonal sequences to achieve spreading gains. This methodology distinctly contrasts with conventional modulation techniques, which have been borrowed from RF communications for use in HBC systems. As detailed in a number of studies: orthogonal frequency-division multiplexing has been proposed in [33], frequency shift keying has been examined in [3], and ON–OFF keying has been explored in [38]. These digital transmission strategies facilitate the implementation of HBC systems with remarkably simplified structures, eliminating the need for digital-to-analog signal converters, RF modules, and antennas. This streamlined approach promotes a more efficient design, aiding in the advancement and applicability of HBC systems.

6.5.3 Systems and Application Examples

Implementing capacitive HBC in practical scenarios poses significant challenges, particularly in terms of system integration and the stability of communication links. In previous chapters, we have discussed various implanted antenna design solutions for capsule endoscopy applications, aiming to address the wireless transmission issues inherent in capsule endoscopy. In current capsule applications, the majority of designs adopt such traditional RF communication system approaches. However, in recent years, some researchers have proposed the application of capacitive HBC in the realm of capsule endoscopy. As illustrated in the Figure 6.30, it represents a system realized in one such study [39, 40]. Implementing capacitive HBC in capsule endoscopy could potentially offer a more efficient and body-friendly solution, eliminating the need for RF and possibly reducing power consumption, and size of the capsule, as well as the localization. This innovative approach

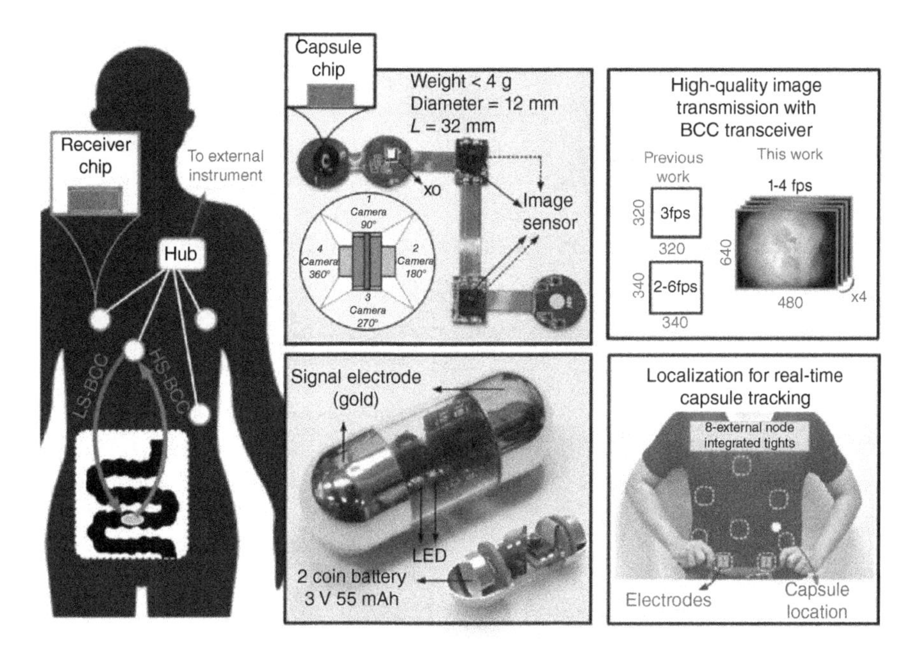

Figure 6.30 Capsule endoscope system: 80 Mb/s body-channel communication transceiver and sub-cm range capsule localization.

could enhance the functionality and comfort of capsule endoscopy, making it an intriguing area for future research. Likewise, this work [39, 40] compellingly demonstrates the feasibility of implementing capacitive HBC in real-world medical implants, thereby underscoring the importance of continued research in this field. In system-level design, considerations should extend beyond channel modeling and modulation scheme design. Crucially, advancements in chip technology should be harnessed to create miniaturized transceiver designs. In the future, HBC designs should be increasingly guided by system requirements, investigating system optimization designs under specific applications. The fusion of HBC and cutting-edge semiconductor technologies opens a realm of possibilities for creating small, efficient, and effective communication systems that can be integrated into a wide range of devices, from wearable technology to medical implants. Moreover, the development of such systems should always consider user needs, including factors such as energy consumption, signal reliability, and user comfort. This user-centric approach, coupled with rigorous technical design, promises to accelerate the practical application and impact of HBC in our daily lives.

References

1 Hall, P.S. and Hao, Y. (2006). *Antennas and Propagation for Body-Centric Wireless Communications*. Norwood, MA, USA: Artech House.

2 IEEE Std. *IEEE Standard for Local and Metropolitan Area Networks–Part 15.6: Wireless Body Area Networks*, IEEE Std. 802.15.6TM-2012 (2012).

3 Seyedi, M., Kibret, B., Lai, D.T.H., and Faulkner, M. (2013). A survey on intrabody communications for body area network applications. *IEEE Trans. Biomed. Eng.* 60 (8): 2067–2079.

4 Zimmerman, T.G. (1995). *Personal Area Networks (PAN): Near-Field Intra-Body Communication.* M.S. Thesis. Cambridge, MA, USA: School of Architecture Planning, Massachusetts Institute of Technology.

5 Pereira, M.D., Alvarez-Botero, F.R.d., and Sousa. (2015). Characterization and modeling of the capacitive HBC coupling. *IEEE Trans. Instrum. Meas.* 64 (10): 2626–2635.

6 Ruiz, J.A., Xu, J., and Shimamoto, S. (2006). Propagation characteristics of intra-body communications for body area network. In: *Consumer Communication and Networking Conference, 2006. CCNC 2006. 3rd IEEE*, vol. 1, 509–513.

7 Cho, N., Yoo, J., Song, S.-J. et al. (2007). The human body characteristics as a signal transmission medium for intrabody communication. *IEEE Trans. Microw. Theory Tech.* 55 (5): 1080–1086.

8 Cho, N., Roh, T., Bea, J., and Yoo, H.-J. (2009). A planar MICS band antenna combined with a body channel communication electrode for body sensor network. *IEEE Trans. Antennas Propag.* 57 (10): 2515–2522.

9 Xu, R., Zhu, H., and Yuan, J. (2009). Characterization and analysis of intra-body communication channel. In: *Proceedings of the IEEE Antennas and Propagation Society in International Symposium, Charleston, SC*, 1–4.

10 Xu, R., Zhu, H., and Yuan, J. (2009). Circuit-coupled FEM analysis of the electric-field type intra-body communication channel. In: *Proceedings of the IEEE Biomedical Circuits and Systems Conference*, 221–224.

11 Xu, R., Zhu, H., and Yuan, J. (2009). High-speed intra-body communication for personal health care. In: *Proceedings of the 2009 31st IEEE Annual International Conference on Engineering in Medicine and Biology Society, Minnesota, MN*, 709–712.

12 Lucev, Z., Krois, I., and Cifrek, M. (2012). A capacitive intrabody communication channel from 100 kHz to 100 MHz. *IEEE Trans. Instrum. Meas.* 61 (12): 3280–3289.

13 Seyedi, M.H., Kibret, B., Lai, D., and Faulkner, M. (2013). An empirical comparison of limb joint effects on capacitive and galvanic coupled intra-body communications. In: *8th IEEE International Conference on Intelligent Sensors, Sensor Networks and Information Processing (ISSNIP), Melbourne, Australia*, 213–218.

14 Callejon, M.A., Naranjo-Hernandez, D., Reina-Tosina, J., and Roa, L.M. (2013). A comprehensive study into intrabody communication measurements. *IEEE Trans. Instrum. Meas.* 62 (9): 2446–2455.

15 Bae, J., Cho, H., Song, K. et al. (2012). The signal transmission mechanism on the surface of human body for body channel communication. *IEEE Trans. Microw. Theory Techn.* 60 (3): 582–593.

16 Sakai, J., Wu, L.-S., Sun, H.-C., and Guo, Y.-X. (2013). Balun's effect on the measurement of transmission characteristics for intrabody communication channel. In: *Proceedings of the IEEE MTT-S International Microwave Workshop Series on RF and Wireless Technologies for Biomedical and Healthcare Applications (IMWS-BIO)*, 1–3.

17 Xu, R., Ng, W.C., Zhu, H. et al. (2012). Equation environment coupling and interference on the electric-field intrabody communication channel. *IEEE Trans. Biomed. Eng.* 59 (7): 2051–2059.

18 Fujii, K., Takahashi, M., and Ito, K. (2007). Electric field distributions of wearable devices using the human body as a transmission channel. *IEEE Trans. Antennas Propag.* 55 (7): 2080–2087.

19 Wang, J., Nishikawa, Y., and Shibata, T. (2009). Analysis of on-body transmission mechanism and characteristic based on an electromagnetic field approach. *IEEE Trans. Microw. Theory Tech.* 57 (10): 2464–2470.

20 Xu, R., Zhu, H., and Yuan, J. (2011). Electric-field intrabody communication channel modeling with finite-element method. *IEEE Trans. Biomed. Eng.* 58 (3): 705–712.

21 Callejon, M., Roa, L., Reina-Tosina, J. et al. (2012). Study of attenuation and dispersion through the skin in intra-body communications systems. *IEEE Trans. Inf. Technol. Biomed.* 16 (1): 159–165.

22 Callejon, M., Naranjo-Hernandez, D., Reina-Tosina, J. et al. (2012). Distributed circuit modeling of galvanic and capacitive coupling for intrabody communication. *IEEE Trans. Biomed. Eng.* 59 (11): 3263–3269.

23 Haga, N., Saito, K., Takahashi, M. et al. (2012). Proper derivation of equivalent-circuit expression of intra-body communication channels using quasi-static field. *IEEE Trans. Commun.* E95-B (1): 51–58.

24 Haga, N., Saito, K., Takahashi, M. et al. (2013). Equivalent circuit of intrabody communication channels inducing conduction currents inside the human body. *IEEE Trans. Antennas Propag.* 61 (5): 2807–2816.

25 Bae, J. and Yoo, H.-J. (2015). The effects of electrode configuration on body channel communication based on analysis of vertical and horizontal electrical dipole. *IEEE Trans. Microw. Theory Techn.* 63 (4).

26 Wu, L.-S., Sakai, J., Sun, H.-C., and Guo, Y.-X. (2013). Matching network to improve the transmission level of capacitive intra-body communication (IBC) channels. In: *Proceedings of the IEEE MTT-S International Microwave Workshop Series on RF and Wireless Technologies for Biomedical and Healthcare Applications (IMWS-BIO)*, 1–3.

27 Wang, H. et al. (2015). A 5.4-mW 180-cm transmission distance 2.5-Mb/s advanced techniques-based novel intrabody communication receiver analog front end. *IEEE Trans. Very Large Scale Integr. (VLSI) Syst.* 23 (12): 2829–2841.

28 Pozar, D.M. (2011). *Microwave Engineering*, 4ee, 164. New York: Wiley, ch. 4, sec. 5.

29 Gabriel, S., Lau, R., and Gabriel, C. (1996). The dielectric properties of biological tissues: III. Parametric models for the dielectric spectrum of tissues. *Phys. Med. Biol.* 41 (11): 2271–2293.

30 CST Microwave Suite Ver. 2013. Available: http://www.cst.com/2013.

31 Ansys Maxwell Ver. 2013. [Online]. Available: http://www.ansys.com/Products/Electronics/ANSYS-Maxwell.

32 Nie, Z.D., Ma, J.J., Li, Z.C. et al. (2012). Dynamic propagation channel characterization and modeling for human body communication. *Sensors* 12 (12): 17569–17587.

33 Sun, W., Zhao, J., Huang, Y. et al. (2019). Dynamic channel modeling and OFDM system analysis for capacitive coupling body channel communication. *IEEE Trans. Biomed. Circuits Syst.* 13 (4): 735–745.

34 IEEE Working Group for WPAN (2012). *IEEE Standard for Local and Metropolitan Area Networks—Part 15.6: Wireless Body Area Networks, Standard 802.15*. IEEE Working Group for WPAN.

35 Ho, C.K. et al. (2014). High bandwidth efficiency and low power consumption Walsh code implementation methods for body channel communication. *IEEE Trans. Microw. Theory Techn.* 62 (9): 1867–1878.

36 Kang, T.-W. et al. (2016). Highly simplified and bandwidth-efficient human body communications based on IEEE 802.15.6 WBAN standard. *ETRI J.* 38 (6): 1074–1084.

37 Ali, A., Inoue, K., Shalaby, A. et al. (2019). Efficient autoencoder-based human body communication transceiver for WBAN. *IEEE Access* 7: 117196–117205.

38 Zimmerman, T.G. (1996). Personal area networks: near-field intrabody communication. *IBM Syst. J.* 35 (3–4): 609–617.

39 Jang, J. et al. (2018). 4-Camera VGA-resolution capsule endoscope with 80 Mb/s body-channel communication transceiver and Sub-cm range capsule localization. In: *IEEE International Solid-State Circuits Conference-(ISSCC)*, 282–284.

40 Jang, J. et al. (2018). A four-camera VGA-resolution capsule endoscope system with 80-Mb/s body channel communication transceiver and sub-centimeter range capsule localization. *IEEE J. Solid-State Circ.* 54 (2): 538–549.

7

Near-field Wireless Power Transfer for Biomedical Applications

7.1 Introduction

In this era of technological innovation, the technology of wireless power transfer (WPT) has become deeply ingrained in our daily lives. Its application is readily apparent in commonplace consumer electronics such as mobile phones and smart watch chargers. However, the genesis of this transformative technology is often overlooked – it was originally deployed to address the unique challenges presented by implantable medical devices (IMDs), providing an efficient solution to transcutaneous power transfer without the need for intrusive wires.

IMDs, designed to monitor, assist, or therapeutically treat dysfunctional organs, have become increasingly prevalent in contemporary medical practice. The advent of wireless power delivery has allowed these devices to operate seamlessly and safely, circumventing the need for transdermal or percutaneous wires, which are often unwieldy and pose a high risk of infection, particularly in the realm of chronic animal model research.

Over the past two decades, the rapid growth and evolution of various IMDs, including but not limited to cardiac [1], cochlear [2], spinal [3], retinal [4], brain [5], and cortical [6] implants, has been remarkable. These devices have been meticulously refined to possess enhanced functionality, optimal form factor, and suitable packaging for prolonged implantation.

However, powering these devices continues to present significant challenges. Traditional packaged batteries offer a finite lifespan, contingent on the functionality and usage of the device [7]. Each battery replacement requires an invasive surgical intervention, posing additional risks and burdens to the patient. The development and deployment of sustainable, long-term wireless power solutions that supersede the limitations of traditional battery capacity have become of paramount importance [8]. This propels the drive toward the creation and refinement of innovative power solutions, with the potential to revolutionize the IMD landscape and profoundly enhance patient quality of life.

There are several WPT techniques that have been explored for use in IMDs, with the two primary categories being near-field and far-field techniques. Near-field WPT, the more commonly used of the two, involves techniques such as resonant inductive coupling [9–11] and capacitive coupling [12, 13]. These methods involve the transfer of power through magnetic fields generated between closely located coils or through electric fields between capacitive plates, respectively. On the other hand, far-field techniques [14, 15], such as microwave

Antennas and Wireless Power Transfer Methods for Biomedical Applications, First Edition.
Yongxin Guo, Yuan Feng and Changrong Liu.

power transfer, operate on the principle of electromagnetic (EM) radiation, where power is transferred over longer distances. Furthermore, an emerging approach in WPT for IMDs is ultrasonic power transfer [16–19], which utilizes acoustic waves for energy transmission. This technique can offer efficient power transfer with smaller device size, making it a promising direction for future research. Hybrid systems combining multiple WPT methods have also been proposed to leverage the strengths of individual techniques and address their limitations. These hybrid methods aim to achieve optimized power transfer efficiency (PTE), system miniaturization, and improved biocompatibility.

In this chapter, we predominantly delve into the exploration and understanding of near-field WPT techniques. We undertake a thorough analysis of the circuit and system models of resonant inductive WPT and capacitive WPT links. A comprehensive study of wireless power links utilizing inductive coupling is presented, accompanied by experimental validations that detail a design methodology for creating efficient inductive links. This contributes toward a guideline grounded in fundamental principles, serving to design efficient WPT links. Furthermore, we shed light on the flexible subcutaneous implant wireless power delivery that utilizes near-field capacitive coupling (NCC). Beginning with theoretical conceptualization and modeling, we navigate through simulations, design implementation, and actual experiments using animal models to provide an in-depth understanding of this approach.

7.2 Resonant Inductive Wireless Power Transfer (IWPT) and IWPT Topologies

The near-field resonant inductive coupling (NRIC) strategy is a well-vetted power transfer technique, receiving validation through U.S. FDA-approved implants [20, 21]. This system functions by using a transmitting coil (TX) positioned near the skin to produce a varying magnetic field. Subsequently, an electromotive force (EMF) is induced within the receiving coil (RX), which is located within the body, as illustrated in Figure 7.1 [22]. Both inductive wireless power transfer (IWPT) and resonant IWPT methods are grounded on near-field inductive coupling for the realization of WPT. The distinction lies in the fact that in resonant IWPT, the transmitting (TX) and receiving (RX) coils operate in a resonant state, while in IWPT, they do not. Currently, resonant IWPT has emerged as the preferred method for WPT that employs inductive coupling due to its higher PTE facilitated by magnetic resonance [23].

7.2.1 Resonances in IWPT

We start with the Resonant IWPT. The performances of the power transfer systems are usually bound by application requirements and constraints. With respect to the implantable applications, the inductive power links are restricted at least by poor coupling due to the large separation distance between the TX and RX coils (comparable with or larger than the RX dimensions), presence of the lossy tissues, strict form factor requirements for medical implants, and transmitter receiver misalignments. Nevertheless, in some other applications such as the wireless chargers for mobile devices, the devices are placed directly on

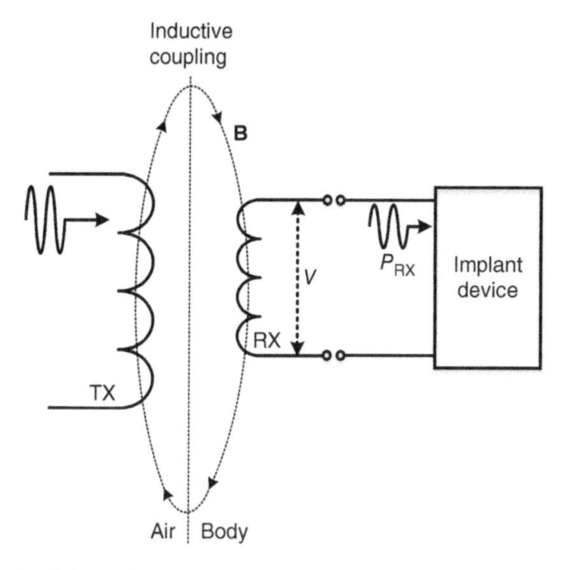

Figure 7.1 Schematic of the IWPT to wirelessly charge an implant device. The power is wirelessly transferred from the TX coil outside of body to the RX coil inside of the body in the form of inductive coupling. Source: Agarwal et al. [22]/IEEE.

a charging pad/console, which houses the transmitter; hence, the separation distance and the misalignment would not be a problem. The coupling between the TX and RX coils is much stronger and the size constraints are loosen as compared with those in implantable applications, respectively. Therefore, the inductive power transfer is designed for a variety of applications each with a different set of requirements. The design of such inductive power links requires diligent analysis and optimization that caters specific application needs. In the subsequent sections, the two typical topologies of the IWPT are introduced and analyzed [11]. A design methodology for inductive power links is then presented with an emphasis on coupling optimization.

Referring to Figure 7.2, a TX coil TX is driven by a current I. An RX coil RX is placed in the vicinity of the TX coil. The separation distance between the TX and RX coils is denoted as d. The magnetic field generated by the TX coil is partially intercepted by the RX coil, through which a linkage of magnetic flux between the TX and RX coils is created.

The EMF induced in the RX coil is given by [24, 25]:

$$\text{Induced EMF} = \oint \mathbf{Edl} = -\frac{d}{dt}\iint \mathbf{BdS} = -\frac{d\Phi}{dt} \tag{7.1}$$

where **E** is the electric field at the element **dl** of the RX coil, **S** is the area enclosed by the RX coil, on which the magnetic field generated by the TX coil is **B**. Φ is the magnetic flux linked by the RX coil. The induced EMF around the RX coil is proportional to the time rate of change of the magnetic flux Φ. The sign is specified by Lenz's law, which states that the induced current (and accompanying magnetic flux) is in such a direction as to oppose the change of flux through the coil. The TX coil is usually excited by a sinusoidal current in practical applications.

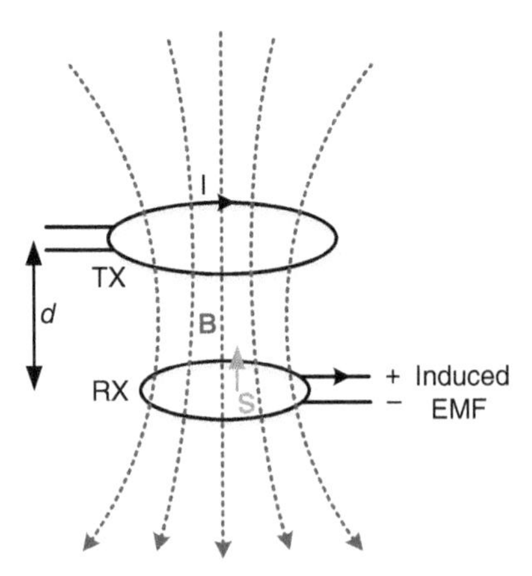

Figure 7.2 Illustration of the Faraday's law of induction. Source: Agarwal et al. [22]/IEEE.

The following challenges can be inferred for applications where the inductive power transfer is required.

(1) The changing rate of Φ is proportional to the amplitude and frequency of the excitation current. The amplitude is generally limited by safety restrictions imposed by the regulatory bodies such as FCC and CE. The frequency should also be within the range licensed by FCC for WPT.
(2) A larger RX coil yields a higher magnetic flux linked by the coil and hence higher EMF can be obtained. However, size of the RX is typically limited by the end applications/product requirements, for example, the size of the coil used in an implant for wireless charging is usually in the range of millimeter. In addition, based on the magnetic field pattern generated by the TX coil, there could be diminishing returns on the induced EMF when the RX size exceeds a certain limit.
(3) Magnetic flux linkage between TX and RX coils is maximized when the direction of $\mathbf{B}$ is normal to the area of the RX coil (aligned with the area vector $\mathbf{S}$). However, the orientations of the TX and RX coils are again limited by the end applications/product requirements. For example, the orientation of the RX coil with respect to the TX coil in wirelessly powered retinal implants keeps changing as the RX coil is in a rotating eyeball.
(4) A smaller separation distance between the TX and RX coils lead to a higher magnetic flux linked by the RX coil and hence a higher induced EMF can be achieved. Nevertheless, the TX–RX separation distance is also limited by the practical applications. For example, the TX and RX coils can be very close to each other (less than 5 mm) in wireless charging of mobile phones. In contrast, the separation distance between the TX and RX coils depends on the position of the implant and is typically larger than 20 mm in implantable applications.

The transfer efficiency of an inductive power link can be improved by using resonance at the receiver [23, 26]. Meanwhile, the resonant circuit is desired to provide impedance matching to the load. Two types of resonant circuits are illustrated in Figure 7.3. The shunt

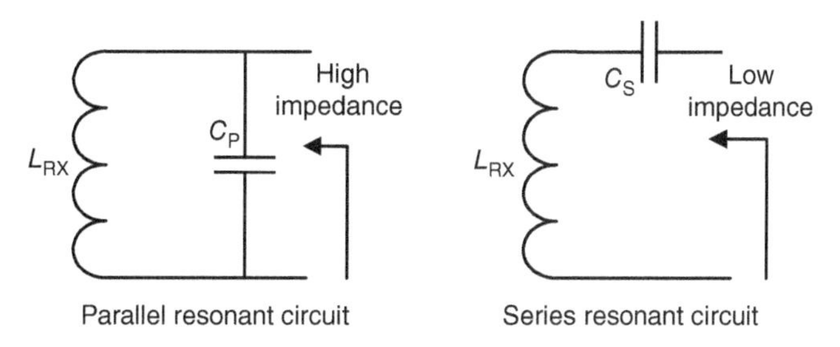

Figure 7.3 The series and parallel resonant circuits in the inductive power links.

resonant circuit presents itself as a high-impedance source to the load; on the contrary, the series resonant circuit presents itself as a low-impedance source to the load. In addition, the shunt resonant circuit provides a relatively large voltage swing that aids the rectifier circuitry; in contrast, the series resonance circuit delivers power using larger current at a relatively low voltage, which makes it unsuitable for low-power applications.

For a given pair TX and RX coils, it can be shown that there is a frequency boundary between the two topologies for a specific load. This frequency is defined as the *cross-over frequency* beyond which the series resonance topology is more efficient than the parallel resonance topology and vice versa. The cross over frequency is given by [11]:

$$f_c = \frac{R_L}{2\pi L_{RX}}\left(\sqrt{1 - k^2\frac{Q_{TX}}{Q_{RX}}}\right) \tag{7.2}$$

where R_L is the load resistance, Q_{TX}/Q_{RX} is the ratio of unloaded quality factors of the TX and RX coils, respectively, and L_{RX} is the inductance of the RX coil. Please note that the quality factors of the coils vary with frequency and the cross-over frequency is obtained by feeding values of the quality factor at a few frequencies. The predominant usage of the low frequency (at most in a few MHz range) in the IWPT favors the shunt resonant circuit, which is corroborated by its widespread usage. Nevertheless, it is suggested in [27] that, compared with the frequency ranges below several MHz, the sub-GHz or GHz ranges be more favorable to wirelessly charge a cm-sized or mm-sized implant, respectively. The WPT operating in the sub-GHz and GHz frequency ranges show its potential for deep-implanted miniaturized receivers.

At frequencies higher than f_c, the series resonance topology is desirable as it provides a better PTE than its shunt counterpart. In fact, the series resonant method has some additional inherent advantages like the better response to fluctuations in coupling coefficient and the ability to operate with a smaller load (i.e., suitable for matching to the circuits of 50 Ω characteristic impedance), which will be shown in the following context.

At this juncture, there are two resonant topologies which can be used for IWPT, each with own advantages and disadvantages. There is a clear need to identify which topology to resort to give an application requirement. This should be seen in the light of the fact that the operating frequency, load, coils, and the coupling coefficient are different for various applications.

7.2.2 Resonant IWPT Topologies

By analyzing the topologies of the IWPT, we are actually trying to answer the fundamental question – "*given a pair of coils and their orientations in space, is there a maximum efficiency of power transfer from one coil to the other across various loads, operating frequencies, and resonance types. If so, how can it be achieved?*" In this section, we answer this fundamental question which paves a methodological way for designing an optimal IWPT link for a given application. The various IWPT links are analyzed using equivalent circuit models of inductors making the results analogical to the inductor realization. Using the results from the analysis, the IWPT link can be optimized from the viewpoint of the topology, load, and operating frequency.

The inductive coupled links use the mutual inductance between the primary and secondary coils to transfer power. The PTE and the magnitude of the power transferred to the load can be improved using resonant tuning [9, 18, 28–30] and matching techniques [28] at the primary and secondary sides, respectively. Based on the type of resonances (series/parallel) used at the secondary side as well as the type of compensations (series/parallel) used at the primary side, four different topologies can be formulated. For convenience, the four topologies can be represented as SS/SP/PS/PP, where the first alphabet denotes the type of compensation at the primary side and the second alphabet denotes the type of resonance used at the secondary side. It should be noted that the type of the resonance at the secondary side influences the PTE irrespective of the type of the primary compensation, which has been demonstrated in [31]. That is to say the SS and SP topologies are independent, whereas the characteristics of the PS and PP topologies can be derived from the former ones. Hence, we will only analyze the topologies of SS and SP, which are shown in Figure 7.4 and the results remain established for the topologies of PS and PP, respectively.

7.2.3 Power Transfer Efficiency

PTE is a key metric to evaluate the performance of an IWPT link. It is one of the main design objectives in many applications. The PTE is defined as the ratio of power delivered to the load to the total power fed into the TX coil. The power fed into the TX coil, besides the part dissipated in surrounding environment, incurs ohmic losses in the TX and RX coils. The PTE can be computed by modeling the receiver side as an equivalent load for the transmit side. Specifically, the receiver can be modeled as reflected impedance at the transmit side in our analyses. The receiver impedance as reflected to the transmitter side for the SS and SP topologies, i.e., $Z_{SS}{}^{\mathrm{r}}$ and $Z_{SP}{}^{\mathrm{r}}$, have be demonstrated to be equal to [11]:

$$Z_{\mathrm{SS}}{}^{\mathrm{r}} = \frac{\omega^4 k^2 L_{\mathrm{TX}} L_{\mathrm{RX}} C_S^2 (R_{\mathrm{RX}} + R_{\mathrm{L}})}{\omega^2 C_S^2 (R_{\mathrm{RX}} + R_{\mathrm{L}})^2 + (1 - \omega^2 L_{\mathrm{RX}} C_S)^2} + j\frac{(1 - \omega^2 L_{\mathrm{RX}} C_S)\omega^3 k^2 L_{\mathrm{TX}} L_{\mathrm{RX}} C_S}{\omega^2 C_S^2 (R_{\mathrm{RX}} + R_{\mathrm{L}})^2 + (1 - \omega^2 L_{\mathrm{RX}} C_S)^2} \tag{7.3}$$

$$\begin{aligned} Z_{\mathrm{SP}}{}^{\mathrm{r}} = {} & \frac{\omega^2 k^2 L_{\mathrm{TX}} L_{\mathrm{RX}} (R_{\mathrm{RX}} + R_{\mathrm{L}} + \omega^2 C_{\mathrm{P}}^2 R_{\mathrm{RX}} R_{\mathrm{L}}^2)}{[R_{\mathrm{RX}} + R_{\mathrm{L}}(1 - \omega^2 L_{\mathrm{RX}} C_P)]^2 + [\omega L_{\mathrm{RX}} + \omega C_P R_{\mathrm{RX}} R_{\mathrm{L}}]^2} \\ & + j\frac{\omega^3 k^2 L_{\mathrm{TX}} L_{\mathrm{RX}} [C_{\mathrm{P}} R_{\mathrm{L}}^2 (1 - \omega^2 L_{\mathrm{RX}} C_{\mathrm{P}})) - L_{\mathrm{RX}}]}{[R_{\mathrm{RX}} + R_{\mathrm{L}}(1 - \omega^2 L_{\mathrm{RX}} C_{\mathrm{P}})]^2 + [\omega L_{\mathrm{RX}} + \omega C_{\mathrm{P}} R_{\mathrm{RX}} R_{\mathrm{L}}]^2} \end{aligned} \tag{7.4}$$

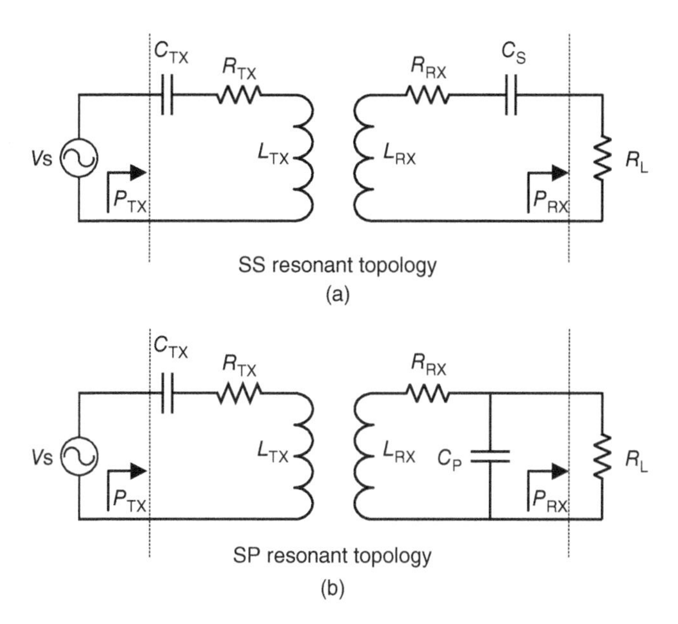

Figure 7.4 (a) The SS, (b) the SP resonant topologies used in the IWPT. Source: Jegadeesan et.al. [11]/IEEE.

where ω is the operating angular frequency, k is the coupling coefficient. The definitions of the other symbols in (7.3) and (7.4) can be found in Figure 7.4. Based on the principle of power sharing, the total PTE of the two topologies can be computed as follows:

$$\eta_{SS} = \left\{1 + \frac{R_{RX}}{R_L} + \frac{R_{TX}}{k^2 L_{TX} R_L}\left\{L_{RX} + \frac{1}{\omega^2}\left[\frac{(R_{RX}+R_L)^2}{L_{RX}} - \frac{2}{C_S}\right] + \frac{1}{\omega^4 L_{RX} C_S^2}\right\}\right\}^{-1} \quad (7.5)$$

$$\eta_{SP} = \left\{\begin{array}{c} 1 + \dfrac{R_{RX}}{R_L} + \omega^2 C_P^2 R_{RX} R_L \\ + \dfrac{R_{TX}}{k^2 L_{TX} L_{RX} R_L}\left\{\left[\dfrac{(1-\omega^2 L_{RX} C_P)R_L + R_{RX}}{\omega}\right]^2 + [L_{RX} + C_P R_L R_{RX}]^2\right\}\end{array}\right\}^{-1} \quad (7.6)$$

For now, we will assume the fact that resonance maximizes efficiency of inductive power links as followed in [9, 29, 30, 32, 33] and derive the efficiency expression under resonance conditions. We will validate this assumption and in fact identify cases where this assumption fails in the next section. We can show that under resonant tuning conditions, the efficiency expressions (7.5) and (7.6) reduce to:

$$\eta_{SS} = \left[1 + \frac{R_{RX}}{R_L} + \frac{R_{TX}(R_{RX}+R_L)^2}{k^2\omega^2 L_{TX} L_{RX} R_L}\right]^{-1} \quad (7.7)$$

$$\eta_{SP} = \left\{1 + \frac{R_{RX}}{R_L} + \frac{R_{RX} R_L}{\omega^2 L_{RX}^2} + \frac{R_{TX}}{k^2 L_{TX} L_{RX} R_L}\left[\left(\frac{R_{RX}}{\omega}\right)^2 + \left(L_{RX} + \frac{R_L R_{RX}}{\omega^2 L_{RX}}\right)^2\right]\right\}^{-1} \quad (7.8)$$

7.2.4 Experimental Verification

The efficiency expressions under resonant conditions are verified experimentally using a pair of coils printed on FR-4 substrate, which are shown in Figure 7.5(a). The parameters of the two coils at 3 MHz are presented in Table 7.1. The effective inductance and the series resistance of the two coils were extracted from the measured one-port S parameters using the vector network analyzer (VNA) HP8753D.

The two coils were then stacked together and separated by a distance of 10 mm using spacers as shown in the Figure 7.5(b). The coupling coefficient was measured using the two-port S parameters obtained from the aligned coil system. The values of the parameters extracted from measurement at 3 MHz are also provided in Table 7.1.

The efficiency expressions were computed using MATLAB for the measured parameters for frequencies from 1 to 7 MHz and two different loads (50 Ω, 100 Ω) using (7.7) and (7.8). The efficiency values in (7.7) and (7.8) were actually evaluated by substituting the measured parameters (self-inductance, resistance, and mutual inductance) of the coil, which were obtained using a network analyzer (HP8753D) as shown in Table 7.1.

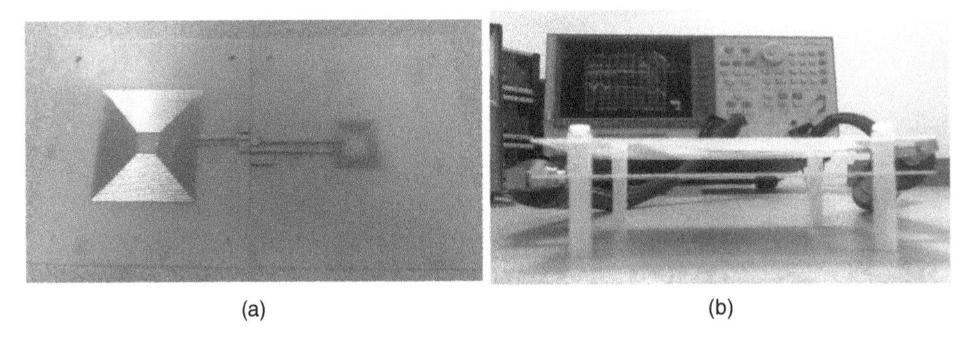

(a) (b)

Figure 7.5 (a) Square planar coils printed on FR-4 substrates for the experimental validation, (b) the measurement setup of the two-coil IWPT system. Source: Jegadeesan et.al. [11]/IEEE.

Table 7.1 Parameters of the Tx and Rx coils operating at 3 MHz.

	TX coil	RX coil
Number of turns	20	11
Internal diameter	10 mm	10 mm
External diameter	49 mm	20.4 mm
Width of trace	0.5 mm	0.2 mm
Pitch of spiral	1 mm	0.5 mm
L effective	12.8 μH	2.84 μH
R effective	4.47 Ω	2.80 Ω
Q factor	53.9	19.11
Coupling (k)	0.173	

Source: Jegadeesan et.al. [11]/IEEE.

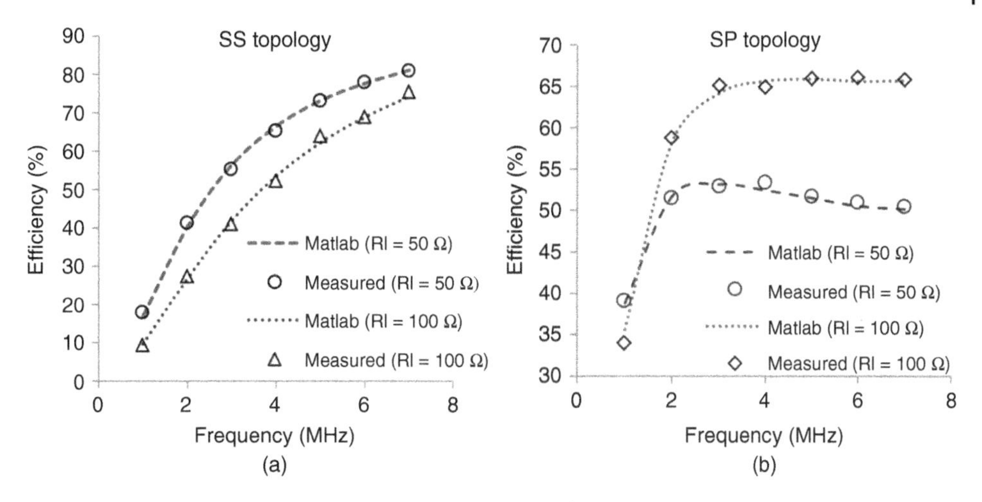

Figure 7.6 Comparison of the calculated and measured PTEs under the resonant condition for (a) the SS topology, (b) the SP topology. Source: Jegadeesan et.al. [11]/IEEE.

To verify the calculated efficiencies experimentally, the same set up shown in Figure 7.5(b) was used, and the primary side was powered using an analog signal generator (Agilent E8257D) and the secondary side was connected to an oscilloscope (HP 54616C). The power drawn from the source and power delivered to the load were measured at tuned resonant frequencies from 1 to 7 MHz after accounting for the input reflection at the primary side. The efficiency was then computed as a ratio of received power to the transmitted power. The experiment was repeated for two different loads (50 Ω and 100 Ω). The computed efficiency (specified as MATLAB) is compared with the measured efficiency values as shown in Figure 7.6. From the graphs, it can be seen that the results agree well with each other. The very accurate prediction of the efficiency is due to the usage of measured values of the coils in evaluating (7.7) and (7.8), which were in turn derived without making approximations.

7.2.5 Limitations of the Resonance Tuning

Resonant tuning is used to maximize the efficiency in the previous section. The use of resonant tuning has been misunderstood in the context of inductive power links. Many prior works [9, 29, 30, 32, 33] take for granted that resonance is an absolute necessity to maximize power efficiency. In fact, this assumption is valid for almost all wireless power links and might never fail for most applications. Though it can be considered as a parasitic case in the context of wireless power links, it is nevertheless required to point out the possibility of a non-resonant link operating more efficiently than the resonant one when the coupling is very strong, and the parallel resonant topology is used.

This anomaly can be explained using the reflected impedance of the secondary side at the primary side, which are shown in (7.3) and (7.4). It is found from the equations that, for the SS topology, the imaginary part of the reflected impedance is zero under the resonance irrespective of the coupling coefficient. This is not true for the SP topology as the secondary side presents a capacitive load to the primary side at resonance and it vanishes only under weak coupling conditions (typically $k < 0.5$). Therefore, the SP topology under strong coupling condition requires a coupling-dependent tuning method and resonance does not clearly maximize the PTE.

The maximum efficiency under strong coupling can be obtained by maximizing the efficiency expressions with respect to the secondary capacitor C_s and C_p. The capacitances that maximize the efficiency for the SS and SP topologies are obtained as follows:

$$C_S = \frac{1}{\omega^2 L_{RX}} \tag{7.9}$$

$$C_p = \frac{1}{\omega R_{RX}(k^2 Q_{TX} + Q_{RX} + Q_{RX}^{-1})} \tag{7.10}$$

where Q_{TX} and Q_{RX} are the quality factors of the primary and secondary coils at the operating frequency, respectively. For a coil of inductance L and self-resistance R, it is given by $Q = \omega L/R$.

It is clearly evident from (7.9) and (7.10) that SS topology has maximum efficiency at resonance irrespective of the nature of coupling and the SP topology has maximum efficiency when the secondary side is non-resonating. When the coupling becomes weak, the expression in (7.10) reduces to the resonance mode of operation.

We simulated two planar spiral inductors as shown in Figure 7.7 using HFSS. We extracted the inductance and resistance of the coils at different frequencies and fed them into ADS (circuit model) which carried out the efficiency calculations for the two choices of tuning capacitors as given in (7.9) and (7.10) to obtain the graph shown in Figure 7.8. It can be seen from Figure 7.8 that non-resonant tuning method as proposed by (7.10) has a better efficiency than the shunt resonant tuning method. The improvement in efficiency from shunt resonant to shunt non-resonant mode is significant only when

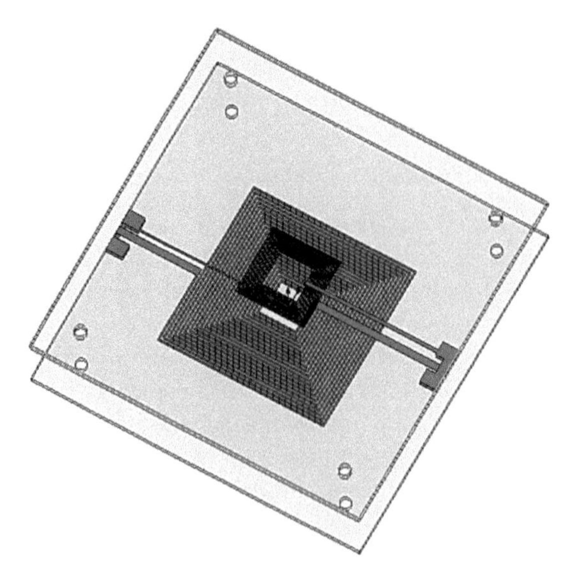

Figure 7.7 Limitations of the resonant tuning method in tightly coupled inductive power links of SP topology: simulation setup in HFSS. Source: Jegadeesan et.al. [11]/IEEE.

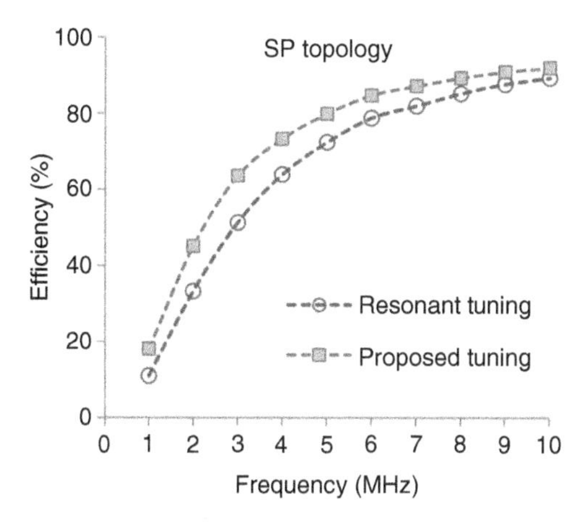

Figure 7.8 Comparison of efficiencies obtained using resonant tuning and the proposed tuning method simulated using HFSS and ADS. Source: Jegadeesan et.al. [11]/IEEE.

the coupling is strong ($k > 0.5$). Most of the inductive power transfer link. have coupling coefficients typically less than 0.5 and hence for all practical purposes we can assume that the resonant tuning will maximize efficiency, which justifies the ubiquitous use of resonant tuning in today's power links. Thus, it is required to keep in mind for future applications that the non-resonant tuning will be able to provide better efficiency than the resonant tuning method for tightly coupled circuits.

7.3 IWPT Topology Selection Strategies

7.3.1 For Applications With a Fixed Load

For a given application, there is a need to identify which topology operates more efficiently. We compare the efficiencies of the series and parallel resonant power transfer methods in this section to identify under what criterion does one method outperforms the other. We infer the following from the analysis. The efficiency increases sharply with frequency for the SS topology. This is justified from (7.7) as the dominant term in the denominator reduces as the square of frequency. For the SP topology, the efficiency has a local maximum with respect to frequency and hence the efficiency dies down at higher frequencies. From these two inferences, we conjuncture the existence of frequency boundary between the topologies, that specifies which topology outperforms the other at a particular frequency of interest.

By comparing the peak efficiencies of the SS and SP topologies for the same choice of coils, it is found that the frequency at which both the topologies have the same efficiency is:

$$f_c = \frac{R_L}{2\pi L_{RX}}\left(\sqrt{1 - k^2 \frac{L_{TX}R_{RX}}{L_{RX}R_{TX}}}\right) = \frac{R_L}{2\pi L_{RX}}\left(\sqrt{1 - k^2 \frac{Q_{TX}}{Q_{RX}}}\right) \tag{7.11}$$

The f_c is defined as the cross-over frequency, which has been mentioned in (7.2), and the definitions of other parameters can be found in Figure 7.4. When the operating frequency is larger than the cross-over frequency f_c, the SS topology provides a higher efficiency than the SP topology and vice versa. It is to be noted that the Q_{TX} and Q_{RX} that are used in (7.11) are frequency-dependent quantities. However, the ratio of Q_{TX} and Q_{RX} does not change much with frequency (as long as we are operating at frequencies much lower than the self-resonant frequency of either coils), and hence the ratio can be theoretically evaluated at any standard frequency to find the cross-over frequency. Thus, we can infer that series resonant method is more efficient for smaller loads and high frequency of operation, whereas parallel resonant method works well for larger loads and low frequency of operation.

When the inductive link is loosely coupled ($k \ll 1$), the cross-over frequency reduces to:

$$f_c = \frac{R_L}{2\pi L_{RX}} \tag{7.12}$$

The cross-over frequency in (7.11) applies only to a special case of loosely coupled coils and is not independent of the coupling coefficient k as claimed in that literature.

In order to verify (7.11), the two fabricated planar square spiral coils shown in Figure 7.5(b) were characterized to obtain their inductance, series resistance and coupling at a fixed separation of 10 mm. We then made measurements on efficiency after using proper tuning circuits at the secondary side by following the same procedure as has been mentioned in Section 7.2.4. The frequency was swept from 1 to 8 MHz for two different loads (50 and 100 Ω) to obtain the cross-over frequency. The cross-over frequency was then obtained by modeling the inductive link in HFSS and the abstracted parameters are fed into its equivalent circuit model in ADS to obtain the simulated efficiencies. The cross-over frequency was then calculated using the theoretical model [34–37] for inductors on PCB using MATLAB. Figure 7.9 shows the comparison of cross-over frequencies for two different loads. It can be seen the results correlate well with each other.

Thus, given a pair of inductors, the most efficient topology can be chosen by computing the cross-over frequency and comparing it with the frequency of operation. The predominant use of low frequency of operation in implants had favored the SP topology and hence the SS topology has not been considered so far as a viable method to transfer power in earlier

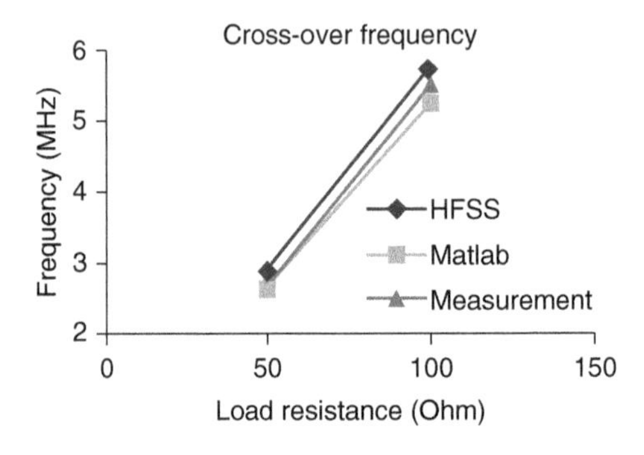

Figure 7.9 Verification of the cross-over frequency for the SS and SP resonant topologies used in inductive links. Source: Rangarajan Jegadeesan et.al. [11]/IEEE.

works. With the push toward high-frequency operation in wireless power links, it is imperative that the SS topology can provide better efficient links in the foreseeable future. Also, from (7.7), we can see that for the SS topology with smaller loads the dominant term in the denominator is the second term and hence the efficiency fluctuations with respect to changes in coupling coefficient are smaller than its parallel counterpart.

7.3.2 For Applications With a Variable Load

The concept of cross-over frequency was presented in the previous section, and it is noticeable that the choice of topology depends on the load impedance used. In applications where the load impedance is fixed and matching networks cannot be implemented (due to cost/space constraints), the cross-over frequency presents a very simple method to identify the dominant topology. However, in applications where there is freedom to choose a load, the topology selection problem becomes difficult as we can choose a load in favor of either topologies, and hence left with the question of which topology to use under any loading conditions. We address this scenario as follows: The efficiency is a function of load impedance as is evident from (7.7) and (7.8). This raises a question, whether the efficiency can be further improved by proper choice (if allowed to be chosen) of an optimal load. In applications where the load can be chosen or matched to maximize the PTE, we strive to find out the optimal load for both the topologies and we compare the efficiency between the two topologies under optimal loading conditions. The optimal load for the series and shunt resonant topologies can be computed as:

$$R_{\mathrm{L}}^{\mathrm{S}} = R_{\mathrm{TX}}\sqrt{1 + k^2 Q_{\mathrm{TX}} Q_{\mathrm{RX}}} \tag{7.13}$$

$$R_{\mathrm{L}}^{\mathrm{P}} = R_{\mathrm{TX}} Q_{\mathrm{RX}}^2 \sqrt{\frac{1 + k^2 Q_{\mathrm{TX}}/Q_{\mathrm{RX}}}{1 + k^2 Q_{\mathrm{TX}} Q_{\mathrm{RX}}}} \tag{7.14}$$

The efficiency expressions for the optimized load for both the topologies take the same form as shown below:

$$\eta = \left(1 + \frac{1}{r} + \frac{1}{k^2 Q_{\mathrm{TX}} Q_{\mathrm{RX}}}(2 + r + r^{-1})\right)^{-1} \tag{7.15}$$

The values of r for the two topologies are given in (7.16) and (7.17), respectively.

$$r_{\mathrm{ss}} = \sqrt{1 + k^2 Q_{\mathrm{TX}} Q_{\mathrm{RX}}} \tag{7.16}$$

$$r_{\mathrm{sp}} = \sqrt{\frac{1 + k^2 Q_{\mathrm{TX}} Q_{\mathrm{RX}}}{1 + k^2 Q_{\mathrm{TX}}/Q_{\mathrm{RX}}}} \tag{7.17}$$

The expressions for optimal load and efficiency under optimal loads are verified using simulations that were done using Ansys HFSS (Coil extraction) and Keysight ADS (Circuit simulation). We used the HFSS simulation setup similar to the one in Figure 7.7 (with a separation of 10 mm between the coils) and extracted the inductor parameters and fed them

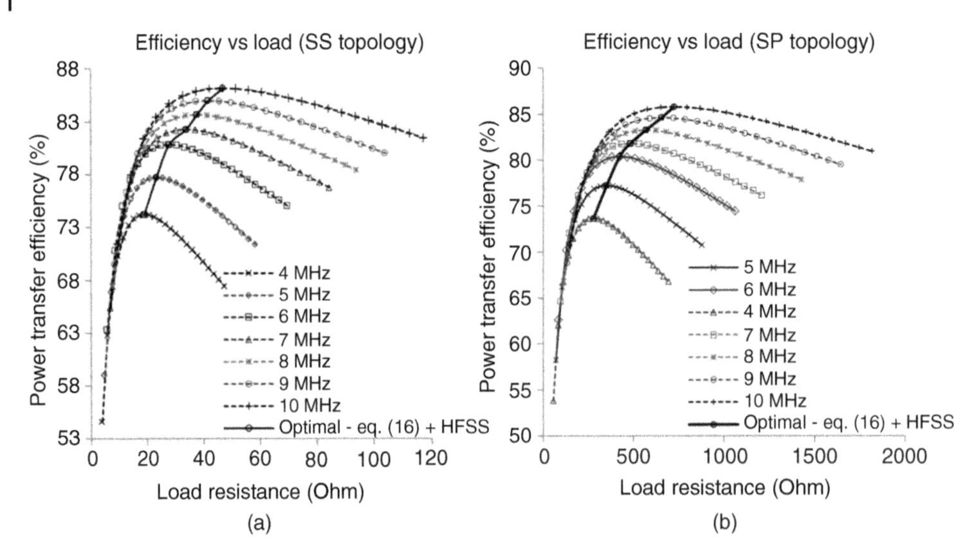

Figure 7.10 The PTEs versus load resistance for various operating frequency and the maximum efficiency operating points for (a) the SS topology, (b) the SP topology. Source: Jegadeesan et.al. [11]/IEEE.

into a circuit simulation setup in ADS for loads varying from 1/5th to 2.5 times of calculated optimal load and for frequencies ranging from 4 to 10 MHz. The maximum efficiency operating points for the above curves are then theoretically computed with the parameter values extracted from HFSS (shown as solid black line with circles) and are plotted as shown in Figure 7.10. The maximum efficiency points corresponding to optimized loads for each frequency match well with the corresponding computed values from (7.15) as can be seen from Figure 7.10.

It can be seen from Figure 7.10 that the optimal load increases with frequency for a given pair of coils for both the topologies. However, if we keep increasing the frequency of operation, the efficiency will drop beyond a certain point where the quality factors of the TX and RX coils, namely Q_{TX} and Q_{RX}, start to reduce, as they get closer to their respective self-resonant frequencies. It should also be noted that the efficiency is relatively stable with changes in the load when the operating frequency is high; hence, we infer that the wireless links operating at a high-frequency are relatively more robust to load variations than its low-frequency counterparts.

Since the efficiency for the inductive links with both series and parallel topologies are known under optimal loading conditions, it can be shown that, *under optimal loading conditions, the series topology always outperforms the parallel topology; and both the topologies have almost equal performance under loose coupling conditions.*

To verify this important result, the efficiencies for the same pair of coils as shown in Figure 7.7 under different coupling conditions are simulated using the HFSS and the ADS, and the results are shown in Figure 7.11. For weakly coupled links with optimized loads, both the series and shunt resonant topologies have nearly equal efficiencies. As the coupling coefficient increases, the dominance of the series resonant topology becomes prominent as evident from Figure 7.11.

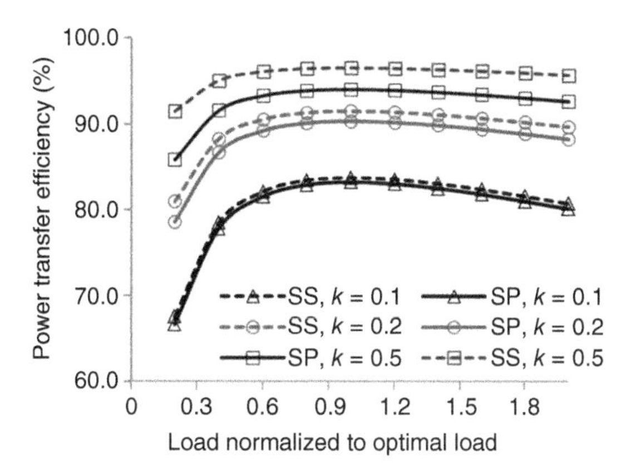

Figure 7.11 The PTEs versus load resistance (normalized to the optimal load) under various coupling conditions. Source: Jegadeesan et.al. [11]/IEEE.

Therefore, given a link with the freedom to choose a load, the series topology needs to be selected for strongly coupled coils. For weakly coupled links, both the topologies operate at the similar efficiencies, and the topology needs to be chosen based on the specific advantages suitable for a particular application.

7.3.3 Optimal Operating Frequency

With all the analysis in place, we try to answer whether there is an optimal frequency of power transfer between two coils with an optimal load and dominant topology. The optimal frequency of operation, however, depends on the type of fabrication used to realize the inductor, and hence no generic closed form expression for optimal frequency of operation can be provided. We proceed as follows: First, we know from Section 7.3.2 that the SS topology is more efficient than the SP topology. Thus, the optimal frequency of operation is that frequency which maximizes the PTE of the SS topology. From (7.15), it is shown that maximizing (7.15) equals maximizing the objective function $g(f)$ shown below. If $g(f)$ is maximized, the efficiency in (7.15) is also maximized for a given coupling between the coils:

$$g(f) = Q_{\mathrm{TX}} Q_{\mathrm{RX}} \tag{7.18}$$

Thus, the optimal frequency of operation reduces to that frequency which maximizes the product of quality factors of the coils. The quality factors of the coils can be obtained by evaluating the inductance and resistance of the coils at different frequencies using the appropriate theoretical model for the inductors. For square planar inductors, the closed-form expressions can be used from [34, (2)] [35, (5)], [36, (6–8)], and [37, ch. 4.10.1].

All the aforementioned work makes use of the inductor parameter values (inductance and self-resistance) and mutual inductances, which were either measured using a network analyzer or simulated using HFSS. However, theoretical modeling of the square planar coils realized on PCB can also be used to obtain the inductor parameters. We refer to [34, (2)] [35, (5)], [36, (6–8)], and [37, ch. 4.10.1] for a completely theoretical approach. The models are accurate for operating frequencies away from the self-resonant frequency of the coil. Since the wireless power links typically operate away from the self-resonant frequency of

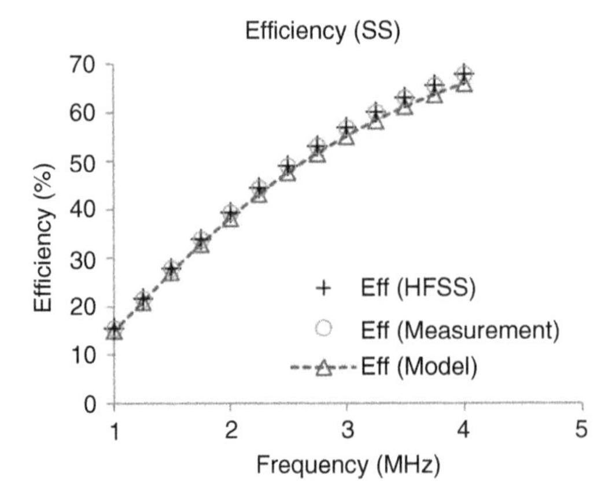

Figure 7.12 Comparison of efficiency values by evaluating (7.7) using inductor parameters from models and measurements. Source: Jegadeesan et.al. [11]/IEEE.

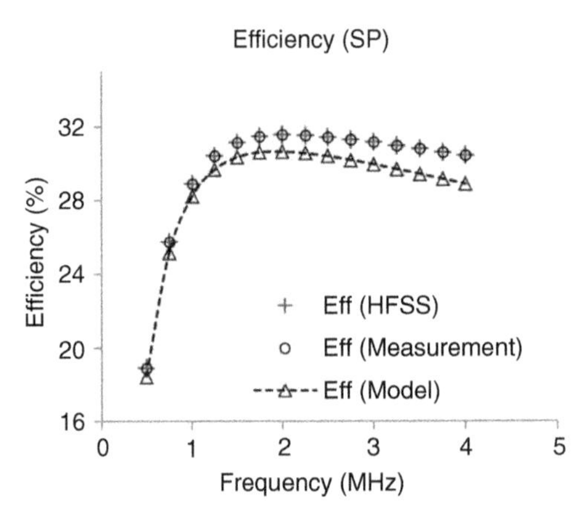

Figure 7.13 Comparison of efficiency values by evaluating (7.7) using inductor parameters from models and measurement. Source: Jegadeesan et.al. [11]/IEEE.

the coil (so that Q factor is large), the models are applicable. The accuracy of those models of inductors resulted in an error <5% for the wireless PTE, which is acceptable for engineering applications. The comparison between the results obtained using parameter values from models and measurement has been done by evaluating (7.7) and (7.8) as shown in Figures 7.12 and 7.13.

The coils used for this comparison study were 16 turn square planar inductors fabricated on an FR-4 substrate, with an internal diameter of 7 mm and an external diameter of 25.4 mm. The pitch of the inductor spiral is 0.6 mm, and the trace width is 0.2 mm. The TX and RX coils were assumed to be identical and separated by a distance of 10 mm.

7.3.4 Upper Limit on Power Transfer Efficiency

Given a pair of coils and their orientation in space, it is now possible using (7.13)–(7.18) to calculate the maximum PTE of operation from one coil to another across all loads, topology,

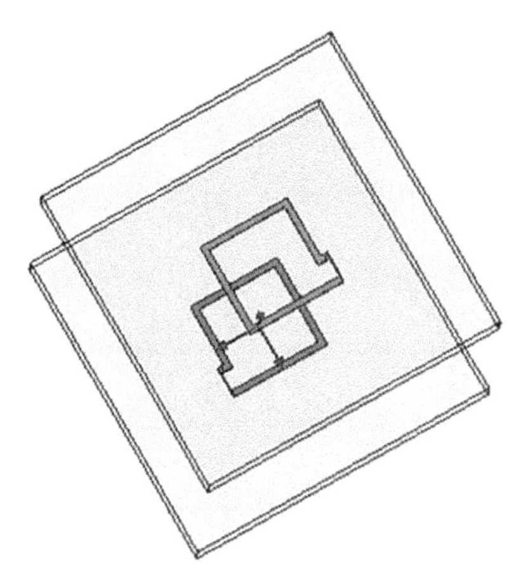

Figure 7.14 The simulation model of the pair of coils that are chosen to study the maximum PTE. Source: Jegadeesan et.al. [11]/IEEE.

and frequency of operation. To demonstrate this, as shown in Figure 7.14, we consider a pair of identical single-turn square spiral inductors built on an FR4 substrate. The coils have an external size of 3 mm, and the width of the copper trace is 0.2 mm. The two coils are separated by a distance of 7 mm. To arrive at the upper bound on PTE, we follow the four steps shown below.

Step 1: Identify the Optimal Frequency of Operation: The models for the inductors in Figure 7.14 are extracted from our earlier work [38]. The function $g(f)$ in (7.18) is maximized to obtain the optimal operation frequency at 3.01 GHz.

Step 2: Topology Identification: Since the small single-turn inductors have a very low inductance and their separation is larger than their dimensions, the coupling is very weak and based on our inference in Section 7.3.2, the SS and SP topologies would operate at the same efficiency. Hence, we can choose either of the topologies, and we will proceed with the use of the SS topology.

Step 3: Optimal Load: The optimal load at 3.01 GHz for the SS topology is then computed using (7.16) as 2.1 Ω.

Step 4: Maximum Efficiency: The maximum efficiency is then simply computed from (7.15) as 35.67%. The resonating capacitance for the SS topology is computed for 3.01 GHz as 0.37 pF.

The maximum PTE versus frequency was then simulated using HFSS and ADS as mentioned in Section 7.2.4 for this link and the result is shown in Figure 7.15. It is observed that the optimal frequency of operation obtained by maximizing the function (7.18) is in the vicinity of 3 GHz, which coincides with the maximal efficiency obtained using simulations, thereby verifying the result (7.18).

We now summarize that, for the structure given in Figure 7.14, the most efficient (35.67%) way to transfer power from one coil structure to another is to connect a series resonating capacitor of 0.37 pF to either coils and terminating or matching the receiving coil to 2.1 Ω and operating the link at a frequency of 3.01 GHz.

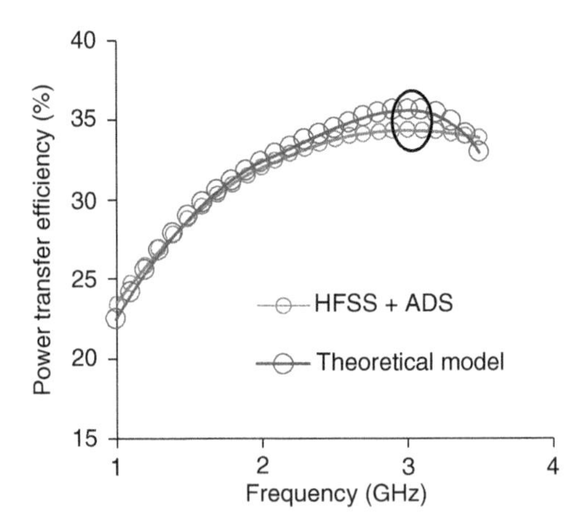

Figure 7.15 PTEs versus frequency calculated by and HFSS + ADS and theoretical model.

7.4 Capacitive Wireless Power Transfer (CWPT)

Besides the resonant IWPT, the CWPT utilizing the NCC provides another distinct option to wirelessly transfer power. It has unique characteristics compared with the resonant IWPT, which enable it to find applications in wirelessly charging medical implants [13]. For instance, as will be explained in the following context, compared with that of the resonate IWPT, the PTE of the CWPT is less sensitive to the deformations of the TX or RX plates. Therefore, these plates can be made flexible and conformal. The NCC was initially proposed for industrial applications [39, 40] and used later in biomedical applications [12, 41]. The NCC and the NIC are compared for biomedical implants from the point view of biosafety in [42], where it is found that the NCC links have better performance than the resonant NIC links on EM interference.

As illustrated in Figure 7.16, the alternating current (AC) source is situated outside the body, with the implant device placed inside. Two parallel-plate capacitors are utilized to establish a current loop between the AC source and the implant device, eliminating the need for wires to penetrate the skin. For each capacitor, the conductive plates are attached to the top and bottom surfaces of the skin respectively. Power is then wirelessly transmitted in the form of displacement currents IDisp across the skin. The capacitive wireless power transfer (CWPT), serving as the capacitive counterpart of the IWPT, operates based on this principle of displacement currents. This power transfer relies on the mutual capacitance between the conductors located on each side of the skin (thickness D). The completion of a current loop, through which power is delivered to the load, requires two such capacitors. Given the small value of the capacitor (<1 pF), formed by two conductors separated by a few millimeters (approximately 5 mm), the power transfer capability is somewhat constrained. However, by choosing the appropriate operating frequency and input matching, the power transfer capability can be optimized with minimal tissue losses.

Consider a pair of conductors of area A and placed on each side of the skin with a very small separation ($D < 5$ mm) for the two TX–RX pairs, as shown in Figure 7.16. When the plates are excited by a time-varying voltage $V(t)$, currents at the conductor discontinuity

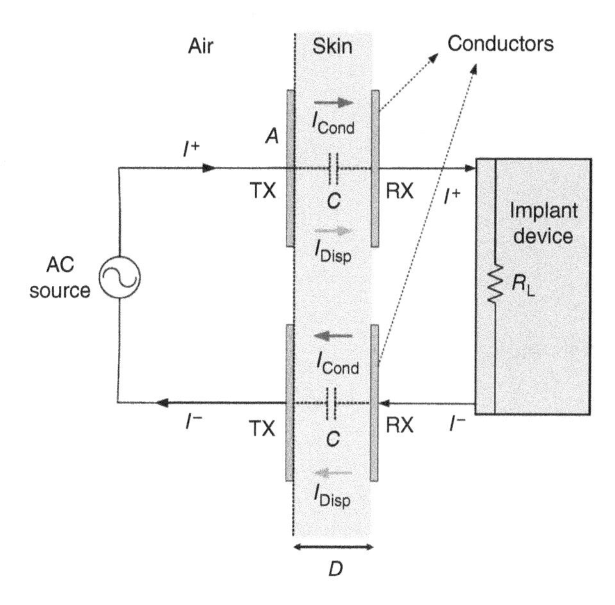

Figure 7.16 Schematic of the CWPT to wireless charge an implant device. The displacement currents generated between the TX and RX metal plates, namely the TX and RX respectively, enable wireless transfer of power to the implanted RX. Source: Jegadeesan et.al. [13]/IEEE.

are supported by the displacement currents between the conductors. Due to the electric fields between the conductors, conduction currents are also induced in the skin and the surrounding tissues. The magnitudes of the displacement currents and conduction currents are given by:

$$I_{\text{disp}} = \varepsilon_0 \varepsilon_r(\omega) A \frac{\partial \mathbf{E}}{\partial t} \tag{7.19}$$

$$I_{\text{disp}} = \frac{V(t)\sigma(\omega)A}{D} \tag{7.20}$$

where ε_0 is the permittivity of the free space, $\varepsilon_r(\omega)$ and $\sigma(\omega)$ are the frequency-dependent relative permittivity and conductivity of the skin, respectively. It is critical to minimize conduction currents that contribute to undesirable losses, while optimizing displacement currents for efficient power transfer. On one hand, from Eq. (7.19), it is apparent that we can enhance the displacement currents – responsible for WPT – by expanding the effective conductor area, by accelerating the change rate of the electric fields, by elevating the relative permittivity of the medium between the conductors, or by employing a combination of these strategies. On the other hand, Eq. (7.20) reveals that curtailing the conduction current requires reducing the conductor area, minimizing the medium's conductivity, and diminishing the excitation voltage. Therefore, a balance must be struck to satisfy these contrasting demands, with the goal of achieving a high-PTE. It is important to acknowledge that the conductor area is subject to limitations, dependent on the implant location and the specific application at hand.

As illustrated in Figure 7.16, the power source fuels the NCC link. Due to the substantial mismatch at the TX terminals, reflected power is a concern, and thus incurred losses necessitate compensation. Introducing a matching network can enhance power transfer through the link at the desired operational frequency. The power delivered to the load correlates directly with the square of the input current. This means that monitoring the transmitted current provides immediate insight into the power delivered to the implant, thus eliminating the need for backchannel telemetry, which is typically needed in NRIC links for closed-loop control of the power delivered to the implant. This section primarily concentrates on the modeling, optimization, and validation of NCC links in an animal model; therefore, the implementation of closed-loop power control using input current sense falls outside of its scope and would not be discussed further.

7.4.1 NCC Link Modeling

The NCC enables efficient link designs by finding the optimal trade-off between different loss sources (tissue loss, conductor loss, and return loss). However, since tissue properties directly relate to the performance of the link and its characteristics (skin thickness, fat content, and so on) vary from person-to-person and from one location on the body to another, it is not possible to arrive at a single accurate theoretical model for the NCC link. Nevertheless, a simplified model allows us to predict the performance of the NCC link given typical transmitter–receiver separations and identify the frequency bands in which the link operates efficiently. Based on the simplified model and the predicted frequency band, we can then build an *optimized* NCC link for any specific body locations and tissue thicknesses with the help from full-wave EM simulation tools such as the CST Microwaves & RF Studio.

To construct an equivalent circuit model of the NCC link, we initiate with the assembly of metallic patches. These patches create two capacitors, one on the implant side and the other on the external side. Power transmission via these capacitances incurs losses in the tissue due to the conduction current, compounded by the ohmic losses within the patches. These tissue and conductor losses can be represented by equivalent resistances incorporated into the circuit in series with the patches' capacitance. The patches and lead wires together contribute to the system's self-inductance, which can be represented as series inductors, as depicted in Figure 7.17. It is important to note that the validity of the equivalent circuit model shown in Figure 7.17 is contingent on the wavelength at the operational frequency being considerably larger than the physical dimensions of the patches. Upon computation

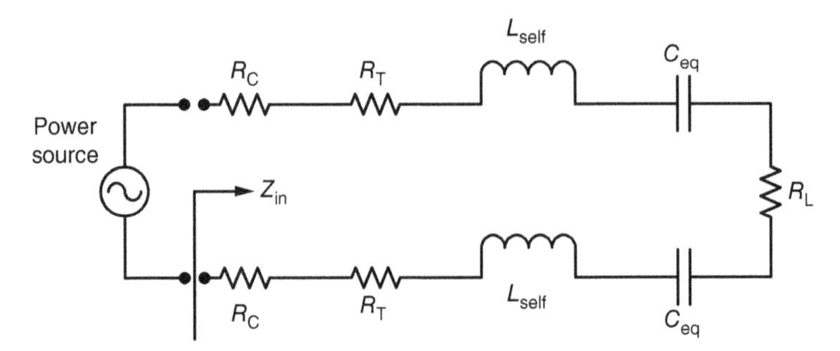

Figure 7.17 Equivalent circuit model of the NCC link, where R_C denotes the conductor losses, R_T denotes the tissue losses, L_{Self} denotes the self-inductance of the patches and lead wires, C_{eq} denotes the equivalent capacitance between the patches, and R_L is the load resistance. Source: Jegadeesan et.al. [13]/IEEE.

of the individual component values in the equivalent circuit model, we can then assess the performance of the NCC link. As illustrated in the forthcoming sections, the values of these components can be determined accordingly.

7.4.1.1 Tissue Model

To estimate the tissue losses and the equivalent capacitance derived from the transmitter and receiver patches in the NCC link, we need tissue models. Based on the strength, frequency, and exposure time, electric fields cause varied effects in tissues. These effects are further influenced by the distinct electrical properties (ε_r, σ, μr) of different tissue types. Delving into the interactions between an incident electric field and the atomic or molecular level of tissues offers lucid insights into diverse loss mechanisms. An alternating electric field impacting the skin incites three main phenomena: atomic polarization, molecular dipole distortion, and conduction [43].

Polarization reversal caused by alternating fields comes with loss due to molecular collision, dependent on the relaxation resistance presented by the tissue. The relaxation losses are mitigated at lower frequencies, due to fewer polarization reversals per unit time. Conversely, at extremely high frequencies, molecules lack ample time for complete polarization, resulting in lower relaxation losses. Nonetheless, when the frequency of the alternating electric field aligns with the tissue's relaxation resonance, relaxation losses reach their peak [44]. Therefore, it is advantageous to operate the power link at frequencies diverging from the tissue's natural relaxation resonance (typically above tens of gigahertz).

Conduction losses, which typically escalate with frequency, owe their variability to the dispersive nature of the tissue. Although precise predictions of the tissue loss mechanisms provide valuable insights, their derivation can be tedious and complex. This complexity is further amplified when we consider the temperature dependence of the dielectric properties. Consequently, we strive to model losses collectively at a macroscopic level, as opposed to investigating individual mechanisms. Existing dielectric models, from simple approximations (Debye [45]) to precise predictions (Cole–Cole [46]), can estimate tissue properties. In this instance, we employ the Cole–Cole dispersion model to calculate the dielectric properties of tissues, which will subsequently be used to compute the equivalent circuit parameters. While developing the model, we make the following assumptions, which are deemed valid.

(1) The skin comprises of three layers: stratum corneum, epidermis, and dermis. Nevertheless, it can be considered a homogeneous layer, as the dielectric properties of these three layers are similar.
(2) The electric fields are well confined between the conductors (skin is directly sandwiched between the conductors) and the fringing fields are minimal. Hence, we can neglect the effects on surrounding tissues.

From the Cole–Cole relaxation model, the dielectric tissue dispersion can be represented as:

$$\varepsilon_r(\omega) = \varepsilon_\infty + \sum_{n=1}^{4} \frac{\Delta\varepsilon_n}{1 + (j\omega\tau_n)^{(1-\alpha_n)}} + \frac{\sigma_i}{j\omega''_0}. \tag{7.21}$$

In (7.21), $\Delta\varepsilon_n$ is the dielectric dispersion range for the nth relaxation region, σ_i is the static ionic conductivity, τ_n is the relaxation time constant of the tissue corresponding to the nth

Table 7.2 Human skin (wet) relaxation model parameters.

ε_∞	$\Delta\varepsilon_1$	τ_1 (ps)	α_1	$\Delta\varepsilon_2$	τ_2 (ns)	α_2
4	39	7.96	0.10	280	79.58	0.00
$\Delta\varepsilon_3$	τ_3 (μs)	α_3	$\Delta\varepsilon_4$	τ_4 (ms)	α_4	σ
3E4	1.59	0.16	3E4	1.592	2E-1	4E-4

Source: Adapted from Gabriel et al. [45].

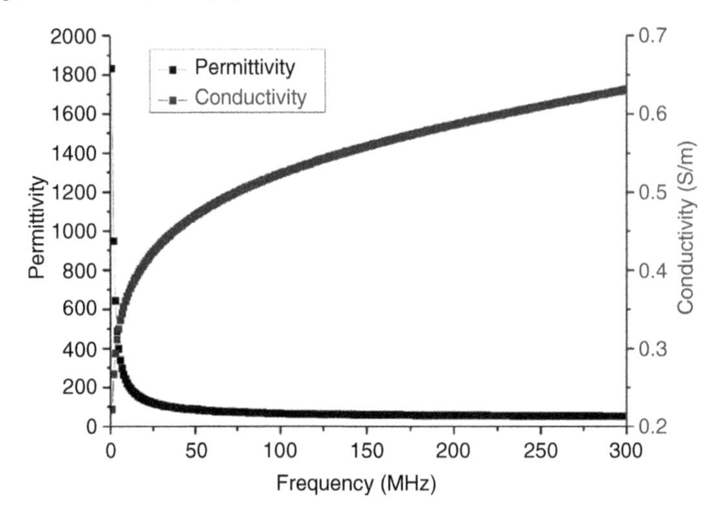

Figure 7.18 Conductivity and permittivity profile of wet skin in the operating range of a capacitive power transfer link. Source: Jegadeesan et.al. [13]/IEEE.

relaxation region over the frequency spectrum (1 MHz to 100 GHz), α_n is the distribution parameter for the nth relaxation region of the tissue, and ε_∞ is the relative permittivity of the tissue at a frequency much higher than the first relaxation resonance of the tissue.

Using (7.21) in the relaxation region, as listed in Table 7.2, we computed the permittivity and conductivity of skin as shown in Figure 7.18, with the parameters as described in [45] (tabulated in Table 7.2). As can be seen, the conductivity increases with frequency, and the permittivity decreases with frequency. It is noted that the conductivity computed above includes both the relaxation and ionic conduction losses. From (7.21), we see that at frequencies much lower than the first relaxation resonance of the tissue, the permittivity is almost independent of frequency and the losses are contributed by the ionic drift and low frequency polarization mechanisms.

We use the Cole–Cole model with all the relaxation mechanisms as shown in (7.21) for all the analysis that follows.

7.4.1.2 Tissue Loss

The tissue losses are represented as an equivalent loss resistance R_T, as illustrated in Figure 7.17. In estimating this, we postulate that the losses primarily occur within the volume between the TX and RX patches, the region where the electric fields are concentrated. Meanwhile, losses outside this volume are assumed to be minimal. Drawing from fundamental principles, the dielectric loss resistance can be expressed as follows:

$$R_{\mathrm{T}} = \mathrm{Real}\left(\frac{-jD}{\omega''_{0}\varepsilon_{r}(\omega)A}\right) \tag{7.22}$$

where A is the area of the metallic patches, and D is the separation between the patches.

By using (7.21) in (7.22), we computed the equivalent tissue loss resistances for three different NCC links (Table 7.3) as shown in Figure 7.19. The tissue thickness generally varies between 1 and 3 mm and is different for different people and different areas in the body. A tissue thickness of $D = 3$ mm comparable with the experimental model (NHP) was used in our study. The performance variation of the NCC link with tissue thickness can be studied by varying the values of D in (7.22).

As can be seen, the losses tend to decrease with frequency. However, when operating at higher frequencies, more power is reflected at the transmitter input (as shown in Figure 7.20), thus making the power transfer extremely inefficient. Hence, we inferred the existence of an optimal frequency at which a balance is struck between reflection loss and tissue losses.

Table 7.3 Dimensions of the NCC link pairs used in the study.

NCC link pairs ↓	TX patch sizes	RX patch sizes
1	40 mm × 40 mm	20 mm × 20 mm
2		15 mm × 20 mm
3		10 mm × 20 mm

Source: Jegadeesan et.al. [13]/IEEE.

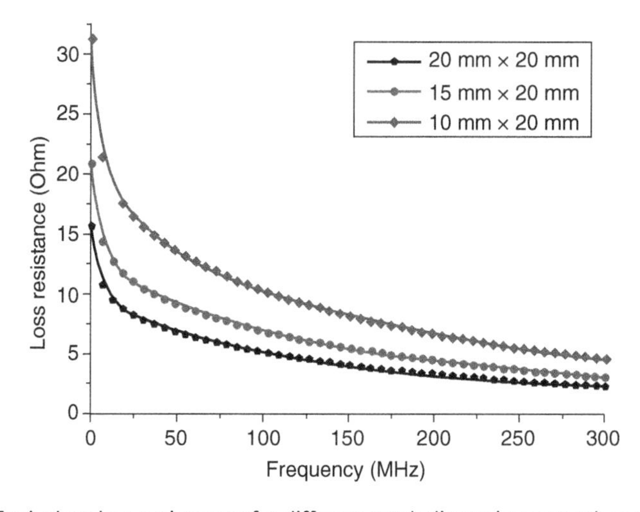

Figure 7.19 Equivalent loss resistances for different patch dimensions over the usable frequency range. Source: Jegadeesan et al. [13]/IEEE.

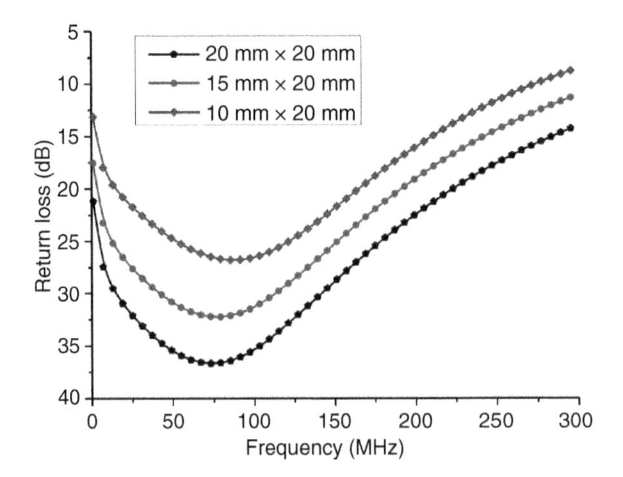

Figure 7.20 Return loss for the three different capacitive patches with dimensions of 10 mm × 20 mm, 15 mm × 20 mm, and 20 mm × 20 mm. Source: Jegadeesan et.al. [13]/IEEE.

7.4.1.3 Conductor Loss (R_C)

In the context of NCC links, conductor losses arise due to the inherent resistance of the leads and the surface resistance of the metallic patches. Notably, the latter typically outweighs the former. At operating frequencies in the MHz to sub-GHz range, these conductor losses are generally insignificant. For NCC links, such losses can be represented as an equivalent series resistance. Owing to their substantially lower magnitude compared to tissue losses, conductor losses are typically disregarded. The equivalent series resistance can be expressed as follows:

$$R_C = \frac{\sigma}{\delta_s} \tag{7.23}$$

where σ is the conductivity and δ_s is the loss tangent of the metallic patches.

7.4.1.4 Self-inductance

The self-resonance of the capacitor in an NCC link is governed by the inductance of the conductors (i.e., metallic patches). The quality factor of the capacitor drops significantly at this self-resonant frequency, prompting the operation to take place at frequencies lower than the capacitor's self-resonant frequency. The series inductance stands as one of the principal elements that restrict the operating frequency of the link. As outlined in [47], the self-inductance can be computed using the following expression:

$$L_{\text{Self}} = 2l\left(\ln\left(\frac{2l}{w+t}\right) + 0.5 + 0.2235\left(\frac{w+t}{l}\right)\right) nH \tag{7.24}$$

where l is the length, w is the width, and t is the thickness of the metal patch sheet.

7.4.1.5 Equivalent Capacitance

The equivalent capacitance between the transmitting and receiving patches plays a pivotal role, as it directly influences the power transfer capability of the NCC link. This equivalent capacitance represents the lossless capacitance formed between the transmitting and receiving patches. It can be illustrated that the equivalent capacitance equals:

$$C_{\text{eq}} = \text{Real}\left(\frac{\varepsilon_0 \varepsilon_r(\omega) l w}{D}\right). \tag{7.25}$$

7.4.1.6 Return Loss

Return loss due to the reflection at the patches can be modeled by computing the input impedance of the NCC link. Since the operating frequency of the NCC link is lower than the resonance frequency of the cavity formed by the two patches on each side of the skin, we can assume that the energy transmission primarily occurs through the capacitive coupling between the patches. Thus, the return loss can be modeled by computing the input impedance of the power transfer link, as shown in Figure 7.17. For a matched output load of 50 Ω for the NCC link, the return loss can be simply computed by the mismatch between this input impedance and the standard reference impedance (50 Ω). The input impedance is a function of the capacitance between the two patches, the self-inductance of the metallic patches and lead wires, the tissue and conductor loss, and the implant load.

The patches have self-inductance, due to the magnetic fields created by the currents flowing through the surface. Ideally, the capacitive impedance monotonically reduces with frequency. However, the self-inductance of the patches breaks this trend by introducing a self-resonance, and has to be taken into account while computing the input impedance. The self-inductance of a sheet of metal with width "w," length "l," and thickness "t" is known from the established formulas as given by:

$$Z_{\text{C}} = \frac{1}{j\omega C_{\text{eq}}}, Z_L = j\omega L_{\text{self}} \tag{7.26}$$

The input impedance and the return loss of the NCC link are hence computed as:

$$Z_{\text{in}} = 2Z_c + 4Z_L + 4R_C + 2R_T + R_L \tag{7.27}$$

$$\text{Return loss} = 20\log\left(\frac{Z_{\text{in}} - Z_0}{Z_{\text{in}} + Z_0}\right) dB \tag{7.28}$$

We computed the return loss of NCC links made of patches with three different dimensions, as mentioned earlier using the procedure described above. The plot is shown in Figure 7.20. As can be seen, the return loss is low at a specific frequency determined by the patch dimensions and the separation between the implant and external patches. For a given patch, we chose the band of frequencies in which the return loss was lower so that maximum power can be transferred to the implant.

7.4.1.7 Power Transfer Efficiency

Taking all the losses into account, the PTE of an NCC link can be calculated as the ratio of power received at the implant to the power transmitted from the source. It is worth mentioning that the R_{L} is not considered in the efficiency expression as we only consider the efficiency degradation due to losses in the link excluding at the input port. It will, however, be considered when computing the maximum power deliverable at the load based on the specific absorption rate (SAR) limitations. The PTE of the NCC link is calculated from (7.29).

$$\text{PTE} = \frac{P_{\text{Load}}}{P_{\text{Load}} + P_{\text{Tissue}} + P_{\text{Cond}}}(1 - |\Gamma|^2) = \frac{R_{\text{L}}}{R_L + R_{\text{T}}}(1 - |\Gamma|^2) \tag{7.29}$$

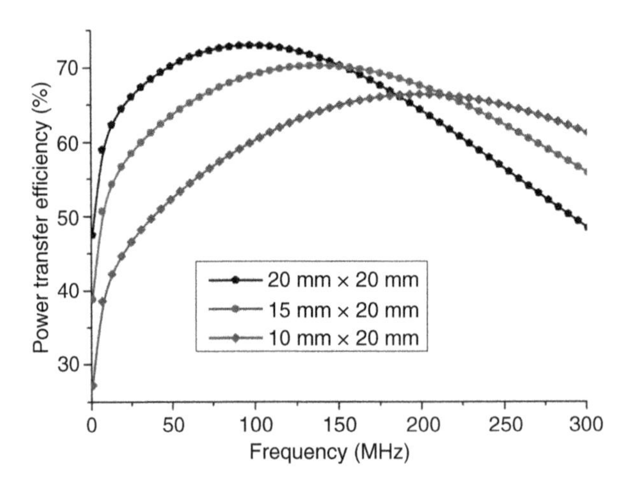

Figure 7.21 PTEs of the NCC links with capacitive patches of different dimensions. Source: Jegadeesan et.al. [13]/IEEE.

where R_L is the load resistor, R_T is the tissue loss, and Γ is the reflection coefficient. This PTE also does not include source losses, which are generally low. Class E amplifiers [48] with operating efficiency of over 90% have been reported for use in the sub-GHz range and can be used as a powering source. The PTEs of the NCC link with different capacitive patches were computed using (7.29) and shown in Figure 7.21. As can be seen, the PTE of the NCC link varies with the patch dimensions, where a larger patch achieves a higher PTE compared with a smaller patch.

7.4.1.8 Power Transfer Limit

For a specific PTE, increasing the transmitted power (P_{tx}) amplifies the power delivered to the implant. However, the transmit power is constrained by tissue losses, measured by a metric known as the SAR. SAR represents the amount of EM energy absorbed per unit mass of tissue, expressed in W/kg, and it is typically averaged over a 10 g tissue sample to monitor localized heating [49]. Safety guidelines mandate an EM wave exposure limit to a SAR value of 2 W/kg, a standard to which all powering schemes must adhere [50]. To calculate the power transfer limit for the NCC scheme, we examined a tissue volume corresponding to a mass of 10 g. For a heterogeneous layer consisting of skin and fat, this equates to the volume illustrated in Figure 7.22, based on skin mass density data [51].

The volume under consideration incorporates both the transmitter (TX) and receiver (RX) patches, along with the fields generated by them, and it is typically localized around the patch area. Thus, it is reasonable to assume that the total power loss (PL) in the tissue for the NCC link is confined to this volume. Therefore, the average SAR for this volume (equivalent to 10 g) directly equals half of the tissue loss (accounting for two sets of patches) for the NCC configuration. The SAR can be estimated as follows:

$$\text{PL (in Watt)} = (1 - \eta) \times P_{tx} \tag{7.30}$$

$$\text{SAR} = \frac{(1 - \eta) \times P_{tx}}{2 \times 0.010} \tag{7.31}$$

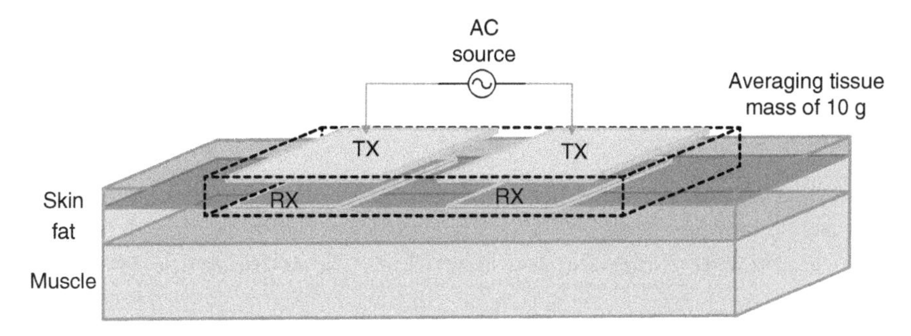

Figure 7.22 The three-layer tissue model illustrating the tissues in the vicinity of the TX–RX capacitive patches. Source: Jegadeesan et.al. [13]/IEEE.

This is a conservative estimation as there might be minimal losses outside of this volume. Using the SAR limit of 2 W/kg, the limit on power transmitted through the NCC link can be computed as:

$$\frac{(1-\eta) \times P_{\mathrm{tx}}^{\mathrm{limit}}}{2 \times 0.010} = 2 \ \mathrm{W/kg} \tag{7.32}$$

$$P_{\mathrm{tx}}^{\mathrm{limit}}(\text{in Watt}) = \frac{0.04}{1-\eta} \tag{7.33}$$

Table 7.4 enumerates the power transmission limit and maximum power accessible at the implant for three different patch sizes (20 mm × 20 mm, 20 mm × 15 mm, and 20 mm × 10 mm). The table reveals that using the NCC method, over 100 mW of power can be wirelessly delivered to the implant given the transmitted powers (in mW), all while complying with SAR safety guidelines. Hence, as long as the power required at the implant is under 100 mW, the NCC links can serve as a viable alternative to inductively coupled NRIC links, presenting the added benefit of more flexible implementations.

To further enhance the PTE and select an appropriate operating frequency, the NCC links can be in resonance tuned by compensating the link's equivalent capacitance with an inductor. Our derived model reveals that the PTE is significantly high even without any resonance tuning at a frequency (sub-GHz for the typical patch dimensions required for implantable devices) that is large enough to bypass the capacitive impedance, yet small enough to steer clear of the self-resonant frequency of the capacitance (between the patches). The PTE bandwidth of this frequency range tends to be broad, indicating minimal detuning effects caused by the patches' flexion. Therefore, we suggest employing the NCC link in nonresonant mode within a specified frequency range (sub-GHz range) to

Table 7.4 Power transmission limits of NCC power links.

Patch dimensions (mm × mm) →	20 × 20	20 × 15	20 × 10
Maximum transmit power (mW)	188.8	178.1	163.2
Maximum available power at implant (mW)	137.8	125.2	108.4
Optimal frequency (MHz)	98	138	190

Source: Jegadeesan et.al. [13]/IEEE.

attain a good PTE. Concurrently, there would be minimal performance degradation when the patches undergo bending deformation in the tissue environment.

7.4.2 Full-wave Simulation

The performance of the NCC link was scrutinized using a three-layer tissue model in CST MWS with respective tissue thicknesses of the skin layer (2.5 mm), fat layer (6 mm), and muscle layer (10 mm), as illustrated in Figure 7.22. The tissue properties were directly imported as defined in CST's heterogeneous voxel-based anatomical models. The TX patches, placed on the skin's outer side, and the implanted RX patches, situated at the skin–fat interface, were encapsulated within a biocompatible polydimethylsiloxane (PDMS) substrate layer. The electrical properties of PDMS (dielectric constant and loss tangent) were evaluated using the ring resonator method and found to be 2.8 and 0.012, respectively, as detailed in [52].

Figure 7.23(a) displays the reflection coefficient for three different implant RX patch dimensions (10 mm × 20 mm, 15 mm × 20 mm, and 20 mm × 20 mm). The size of the TX patches remains constant at 40 mm × 40 mm for all the TX–RX NCC link configurations, agreed with the theoretical model used earlier. We observe that the reflection coefficient corresponds to the patch sizes and separation (skin thickness) between the TX–RX link pairs. As the implant patch dimensions increase, the optimal frequency decreases. The TX patches, given their large size, fully overlap the implant patches, thus we have not conducted a TX–RX misalignment study in this context, and will reserve it for future work. The reflection coefficient graph indicates the usable operational frequency band, helping to minimize reflection losses at the TX input. Figure 7.23(b) plots the forward gain $|S_{21}|$, indicating that the amount of power transferred is dependent on the size of the implant patches. However, the actual PTE will be lower than the simulated values, as the simulations cannot account for all losses encountered in WPT through the tissue environment.

SAR simulations, in accordance with IEEE Std. C95.1 [49], were conducted for all three NCC link pairs. The simulations considered a TX power of 1 W, 500 mW, and the maximum

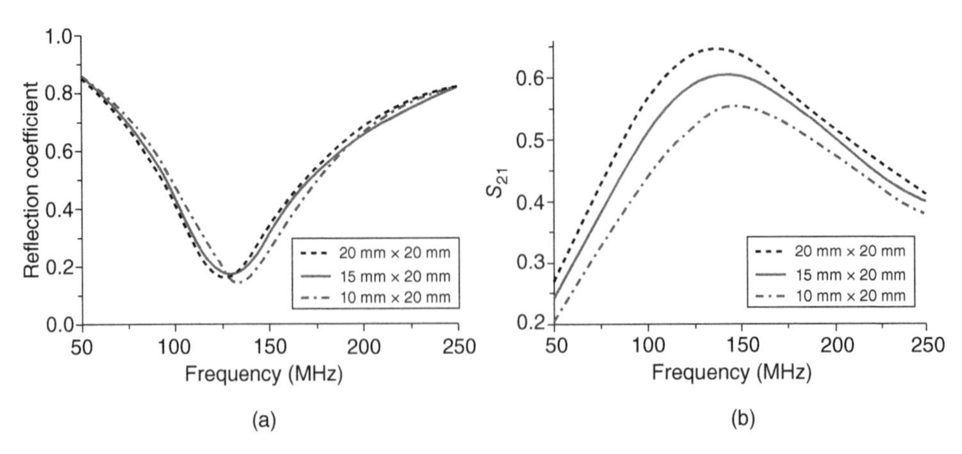

Figure 7.23 Simulated CST MWS results for three implants RX patch dimensions: (a) reflection coefficient graph, (b) $|S_{21}|$ graph. Source: Jegadeesan et.al. [13]/IEEE.

Table 7.5 Localized peak SAR values (average over 10 g tissue).

Input power ↓	20 × 20 mm^2	15 × 20 mm^2	10 × 20 mm^2
1 W	8.02	9.28	10.4
500 mW	4.01	4.64	5.19
SAR ≤ 2 W/kg (TX power)	2.01 (250 mW)	2.09 (225 mW)	2.07 (200 mW)

Source: Jegadeesan et.al. [13]/IEEE.

power limit to maintain peak localized SAR values ≤2 W/kg (averaged over 10 g tissue). The results are summarized in Table 7.5. Larger RX patches, with a higher $|S_{21}|$ value as seen in Figure 7.23(b), dissipate less transmitted power in the tissues for the same transmit power, leading to lower SAR values. Thus, a 20 mm × 20 mm implant RX patch can transmit more power than a 15 mm × 15 mm patch while adhering to the SAR limits as listed in Table 7.5. The simulated SAR plot (averaged over 10 g tissue) for a 20 mm × 20 mm implant RX patch is depicted in Figure 7.24(a) for skin, fat, and muscle layers, with a TX power of 250 mW. The results indicate that we can safely transmit approximately 250 mW of power while complying with international safety standards.

To evaluate the temperature variation induced by the EM fields in the tissue, we performed a co-simulation using the thermal stationary solver in CST and CST MWS. The initial body temperature is assumed to be 37 °C (310.15 K), and the ambient temperature is set at 293.1 K. Figure 7.24(b) displays the temperature variation plot obtained for the transmit power of about 250 mW. The temperature of the tissue surrounding the TX–RX patches increased from the initial temperature of 310.15 K to 310.814 K, an increase of 0.664 °C. As it is crucial that the temperature of the surrounding tissue does not rise by more than

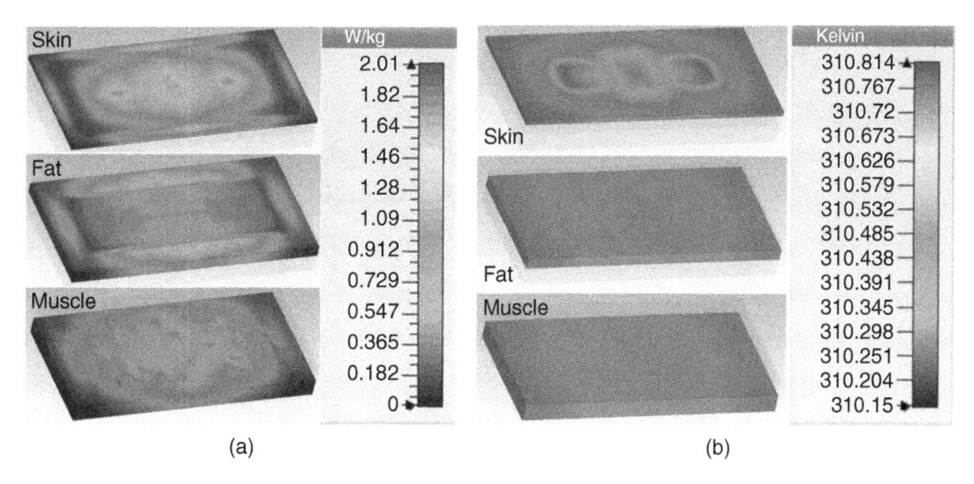

Figure 7.24 Simulation results obtained from the three-layer tissue model in CST for skin layer (2.5 mm), fat layer (6 mm), and muscle layer (10 mm) with 250 mW of transmit power to the 20 mm × 20 mm RX implant patches: (a) SAR plot (average over 10 g tissue), (b) temperature variation plot. Source: Jegadeesan et.al. [13]/IEEE.

1–2 °C due to the NCC power transfer mechanism, this temperature variation plot gives us a measure of thermal safety associated with this powering scheme.

7.4.3 Optimal Link Design

Optimal NCC links can be tailored based on the expected separation and power handling capacity. The link design process involves determining the optimal operating frequency for a given separation to minimize losses. We utilized CST Microwave Studio to simulate the NCC power link's performance and identify the ideal operational range. The optimal operating frequency is one that minimizes tissue losses. However, a closed-form expression for the optimal frequency cannot be derived, given that the dielectric properties of the tissues are frequency-dependent and require experimental characterization. From our theoretical modeling insights and the CST MWS simulation results, the optimal operating frequency was identified to lie between 100 and 160 MHz, with minimal losses around 130 MHz (refer to Figures 7.20, 7.21 and 7.23).

The TX and RX patches are made of thinly cut copper tape, each with a thickness of 0.035 mm, and tailored to the required dimensions. To ensure chronic in vivo operation, the metallic conductors need to be encapsulated in a biocompatible material [53]. In our design, we encapsulated the copper patches within PDMS, as previously described. The PDMS was prepared by mixing silicone elastomer (SYLGARD 184) with its curing agent in a 10 : 1 mass ratio, stirring well, and then degassing in a vacuum chamber for 20 min to prevent the formation of air bubbles and leaks. This semi-cured PDMS was then poured over the mold to encapsulate the thin copper patches, degassed again, and then cured in an oven at 70 °C for 2 h. The total thickness of the implant sheet encapsulated in PDMS prepared using this process was less than 2 mm. We prepared three different sizes of RX patches to create three different NCC links, each with the same TX size. Figure 7.25 depicts the fabricated TX and RX patches. To prevent direct contact with any bodily fluids during experiments, the TX patches were also encapsulated in Kapton tape.

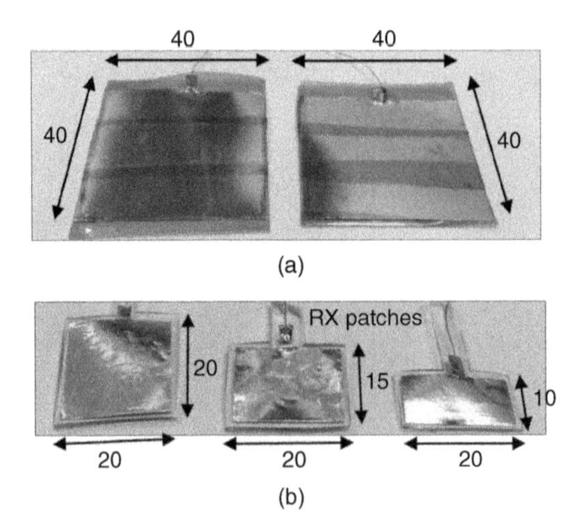

Figure 7.25 Fabricated copper tape patches: (a) TX pair of 40×40, (b) PDMS encapsulated RX pairs of dimensions 20 × 20, 15 × 20, and 10 × 20 (mm^2). Source: Jegadeesan et.al. [13/IEEE.

7.5 CWPT: Experiments in Nonhuman Primate Cadaver

The designed NCC links were evaluated in a non-human primate (NHP) cadaver, as depicted in Figure 7.26. The acute implantations and NHP experiments were approved by the Institutional Animal Care and Use Committee of the National University of Singapore. The experiments were conducted in the forearm, a location that provided sufficient space for acute testing of the NCC link. For these experiments, we utilized three NCC links, each comprising implant receivers of varying dimensions (200, 300, and 400 mm^2) as shown in Figure 7.25. We conducted two sets of experiments. In the first, we characterized the PTE of the NCC link for different implant patch dimensions. In the second, we examined the performance of the NCC link under varying bending deformations.

7.5.1 Study on Power Transfer Efficiency

The receiver (RX) patches for the NCC link, located on the implant side, were inserted under the skin in the forearm of the cadaver through a 25 mm incision. The external transmitter (TX) patches were then positioned on the skin directly above the implanted patches to establish the NCC link. An RF signal generator (Rhode and Schwartz SGS100A), with an input power of 25 dBm ($\approx$316 mW), powered the input of the NCC link. We opted for larger external patches (40 mm $\times$ 40 mm) to circumvent any performance issues arising from misalignment between the implant and external patches. The implanted patches were wired externally through the incision and connected to a 50 Ω load.

Accurately measuring the received power at the implant can be challenging due to ground loop coupling between the transmitting power source and the measuring oscilloscope, as they share the same ground through the power sockets. This loop can be eliminated using a differential probe. To attain accurate power measurements and eradicate ground looping, we utilized a rectifier (HSMS828) and matching network to convert the received power into a DC voltage, which was then measured using a low-frequency differential probe. The rectifier with matching network was independently characterized for rectification efficiency over a frequency band of 90–160 MHz. This conversion efficiency was used to recalculate the AC power received at the implant. The PTE of the link was subsequently computed as the ratio of power input from the signal generator to the AC power received at the implant terminals.

We measured the PTE of three different capacitive power transfer links in the above-mentioned manner, and the results are depicted in Figure 7.27(a)–(c). As illustrated in Figure 7.27(c), the PTE is highest in the sub-GHz range from the frequency range of 100–150 MHz for centimeter-sized patches, aligning with our earlier theoretical model. It is important to note that the efficiencies do not necessarily peak at the points of minimum reflection coefficient (comparison between the graphs in Figure 7.27[a] and [c]). This is because the PTE does not account for the reflection losses (please refer to [(7.29)], where reflection loss is omitted). The reflection losses occur at the TX source, and the TX–RX efficiency generally only accounts for the losses from the transmitter to the receiver. Most WPT research, regardless of the powering scheme, disregards the return losses at TX when calculating the PTE of the link. Another reason for the discrepancy is that tissue and conductor losses do not necessarily minimize at the same frequency where the reflection

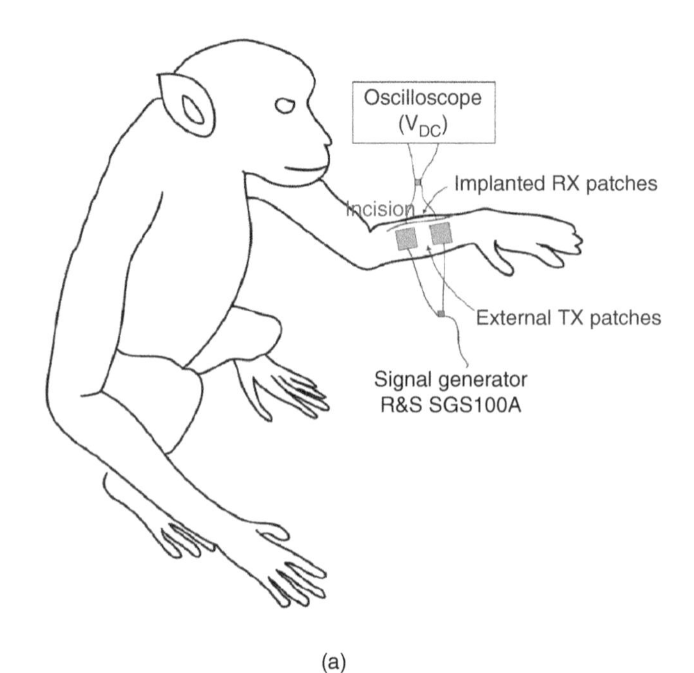

(a)

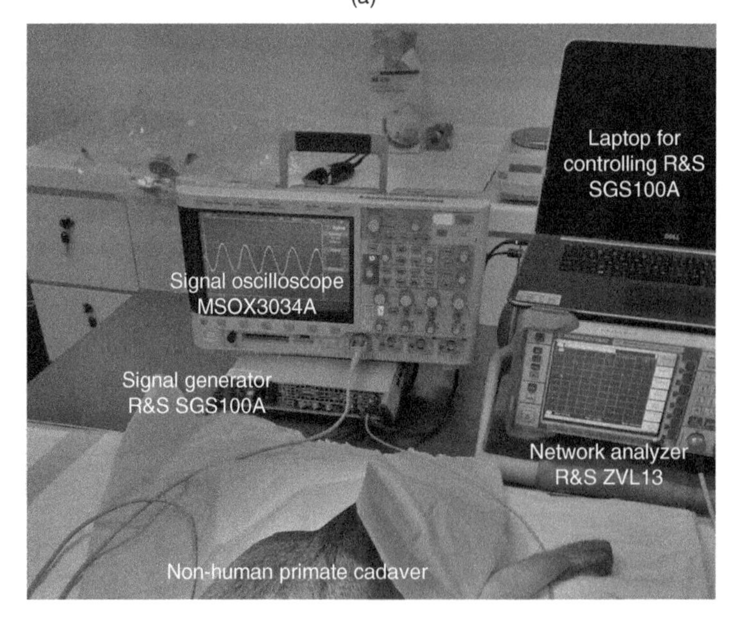

(b)

Figure 7.26 Measurement setup in an NHP cadaver experiment used to validate the PTE of the capacitive power transfer link: (a) diagram of the setup, (b) photo from the actual NHP experiment. Source: Jegadeesan et.al. [13]/IEEE.

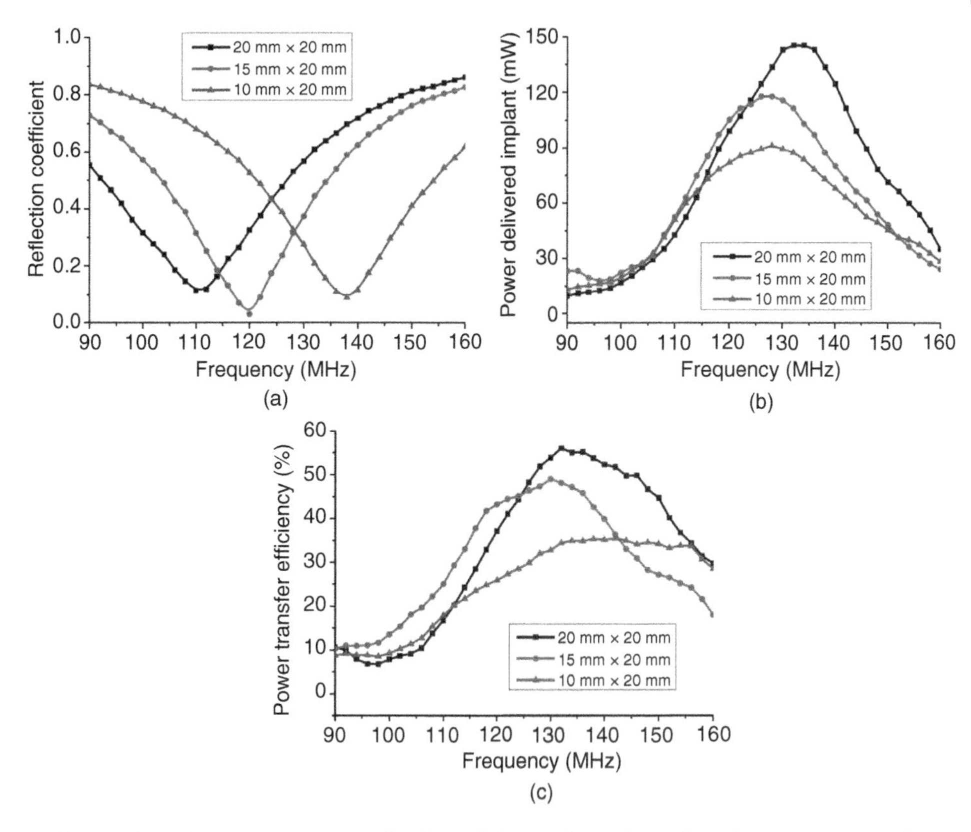

Figure 7.27 Measurement results for three different dimensions of the implanted copper RX patches: (a) reflection coefficient graph, (b) power delivered to the implant graph (mW), (c) PTE graph (without considering the reflection losses at the TX input). The TX–RX metallic patches were separated by ~3 mm thick skin layer of the NHP cadaver. Source: Jegadeesan et.al. [13]/IEEE.

losses are low. Consequently, the overall efficiency does not peak at the minimum reflection coefficient point. The peak PTE of the three NCC links was proportional to the implant dimensions, with the largest patch (20 × 20) achieving 56% efficiency and the smallest (10 × 20) achieving 35% efficiency. The power delivered to the implant graph has also been plotted in Figure 7.27(b).

7.5.2 Flexion Study

One of the primary benefits of the NCC link is that the bending of the patches exerts minimal effect on the PTE. Using a 20 mm × 20 mm implant patch, we conducted PTE measurements. The flexion study was undertaken at three distinct bending radii (20, 30, and 40 mm) for the implant sheets. Custom 3D printed acrylic sheets with bending radii of 20, 30, and 40 mm were utilized in the study, as depicted in Figure 7.28. The implant sheets were affixed to these acrylic sheets using polyamide, and the bent sheets were subsequently inserted under the skin through a 25 mm incision. The external patches were then conformally positioned over the skin and secured with tapes (refer to Figure 7.29 for a flexion demonstration; TX patches were later affixed to a cadaver's skin for the experiment). We measured the PTE of the NCC link using the same previous method for the three different

Figure 7.28 3D printed acrylic sheets for realizing the bending radii of 20, 30, and 40 mm inside the animal tissue during NHP cadaver experiments. Source: Jegadeesan et.al. [13]/IEEE.

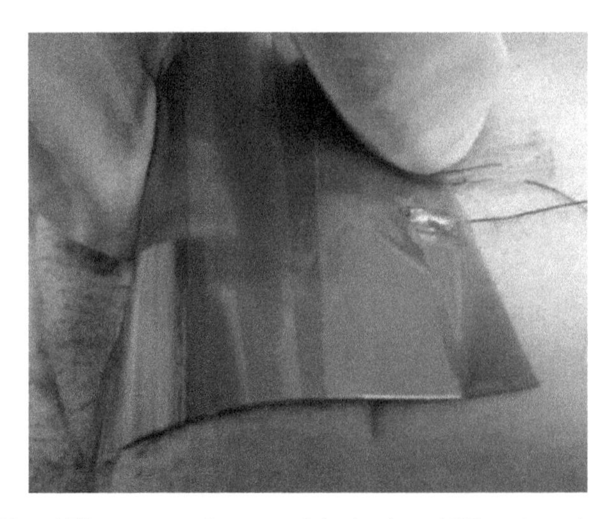

Figure 7.29 Flexed TX copper patch on top of the implanted RX patch on the arm of an NHP cadaver for demonstrating the level of flexion provided by the NCC powering scheme [13].

bending conditions. Figure 7.30 showcases the reflection coefficient, power delivered, and PTE graphs for different bending radii.

Examining the impact of bending deformation on the implanted patches, we concluded that the optimal operating frequency shifts slightly to a lower frequency upon bending, compared to the flat patch configurations. Additionally, we noticed that the NCC link managed to achieve an efficiency exceeding 30%, even under a severe bending radius such as 20 mm, a condition highly unlikely to occur in the postimplantation tissue environment. The principal benefit of utilizing the capacitive power link lies in the fact that the capacitance between the patches remains fairly consistent with flexion, thereby avoiding dramatic detuning as observed in NRIC links. Given that the electric fields are localized and confined between the patches, this approach promises superior EMI performance.

It is noteworthy to mention that the PTE without bending is nearly identical to that with bending at 120 MHz, suggesting that operating a link at this region minimizes power fluctuations at the implant for a 20 mm × 20 mm patch.

7.6 Summary

In this section, we dive deep into the exploration of near-field WPT techniques, shedding light on the principles and practical application of resonant IWPT and CWPT. These

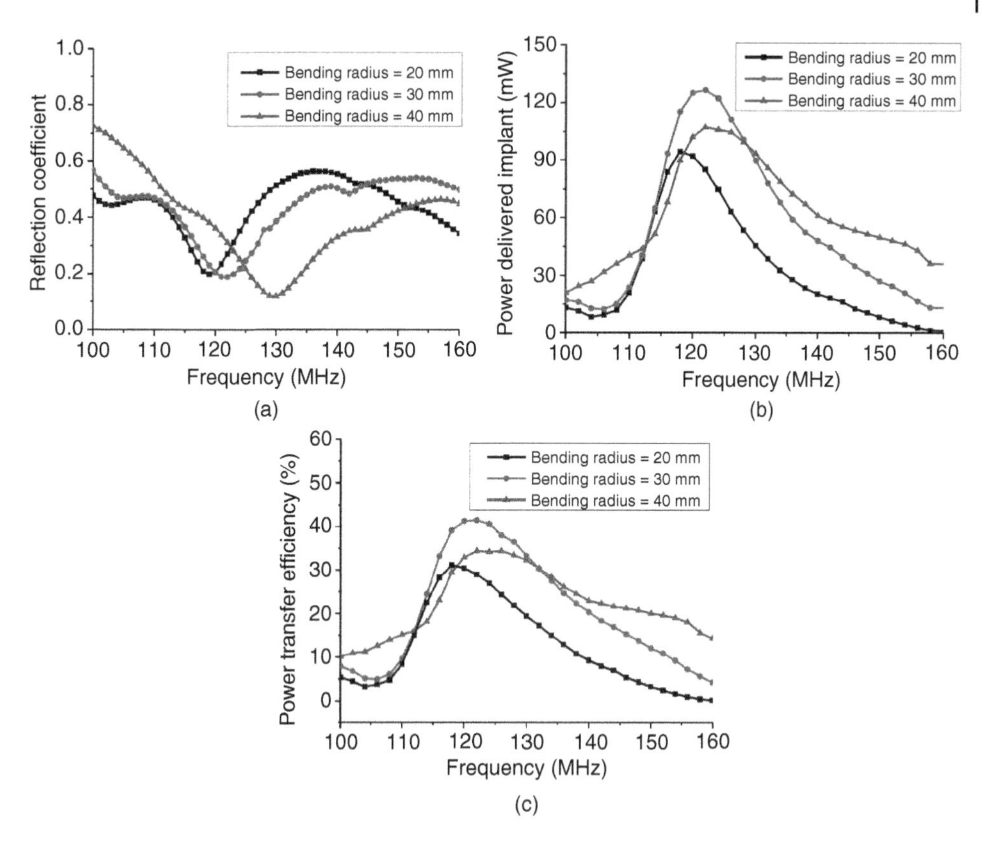

Figure 7.30 Measurement results for three different flexion levels (bending radius = 20, 30, and 40 mm) of the implanted copper RX patches: (a) reflection coefficient graph, (b) power delivered to the implant graph (mW), and (c) PTE graph (without considering the reflection losses at the TX input). The reflection coefficient variation is minimal for various degrees of flexion. The TX–RX metallic patches were separated by ~3 mm thick skin layer of the NHP cadaver. Source: Jegadeesan et.al. [13]/IEEE.

strategies are analyzed in the context of their viability and effectiveness in the field of biomedical applications.

In addition to the near-field techniques, mid-field, and far-field WPT methodologies, as well as ultrasonic power transfer, are attracting growing attention from academia and industry. However, their applicability is generally more restricted to low-power settings. When dealing with devices with significant power requirements, the NRIC and NCC strategies stand out as the most effective solutions. The exploration will continue in Chapter 8, where we delve into the potential of far-field WPT techniques in the biomedical sphere. In Chapter 9, we will highlight the integration of the NRIC technique with neurostimulator systems, complemented by a comparative animal-based study with NCC links.

Beyond technical effectiveness, the development of WPT systems for biomedical devices must adhere to stringent safety standards, taking into account electrical, biological, physical, and EM interference safety issues. These standards, regulated by the FDA, are crucial checkpoints that a device must pass, necessitating successful preclinical and clinical trials before approval for human usage. With the increasing implementation of wireless power technologies in IMDs, the path toward safer usage, standardization, and regulatory approval is expected to expand, ushering in wider adoption.

Lastly, in the era of miniaturization where electronic implants are becoming increasingly compact, the design of efficient near-field WPT to cater to high-power applications is an evergreen research topic, warranting continuous investigation and innovation.

References

1 DiMarco, J.P. (2003). Implantable cardioverter-defibrillators. *New England J. Med.* 349 (19): 1836–1847.

2 Zeng, F.-G., Rebscher, S., Harrison, W. et al. (2008). Cochlear implants: system design, integration, and evaluation. *IEEE Rev. Biomed. Eng.* 1: 115–142.

3 North, R. (2008). Neural interface devices: Spinal cord stimulation technology. *Proc. IEEE* 96 (7): 1108–1119.

4 Weiland, J. and Humayun, M. (2008). Visual prosthesis. *Proc. IEEE* 96 (7): 1076–1084.

5 Coffey, R.J. (2009). Deep brain stimulation devices: a brief technical history and review. *Artif. Org.* 33 (3): 208–220.

6 Lebedev, M.A. and Nicolelis, M.A. (2006). Brain–machine interfaces: past, present and future. *Trends Neurosci.* 29 (9): 536–546.

7 Kindermann, M., Schwaab, B., Berg, M., and Frohlig, G. (2001). Longevity of dual chamber pacemakers: device and patient related determinants. *Pacing Clin. Electrophysiol.* 24 (5): 810–815.

8 Hauser, R.G. (2005). The growing mismatch between patient longevity and the service life of implantable cardioverter-defibrillators. *J. Amer. College Cardiol.* 45 (12): 2022–2025.

9 Jow, U.-M. and Ghovanloo, M. (2009). Modeling and optimization of printed spiral coils in air, saline, and muscle tissue environments. *IEEE Trans. Biomed. Circuits Syst.* 3 (5): 339–347.

10 RamRakhyani, A.K., Mirabbasi, S., and Chiao, M. (2011). Design and optimization of resonance-based efficient wireless power delivery systems for biomedical implants. *IEEE Trans. Biomed. Circuits Syst.* 5 (1): 48–63.

11 Jegadeesan, R. and Guo, Y.-X. (2012). Topology selection and efficiency improvement of inductive power links. *IEEE Trans. Antennas Propag.* 60 (10): 4846–4854.

12 Sodagar, A.M. and Amiri, P. (2009). Capacitive coupling for power and data telemetry to implantable biomedical microsystems. In: *Proceedings of the 4th International IEEE EMBS Conference on Neural Engineering*, 411–414.

13 Jegadeesan, R., Agarwal, K., Guo, Y.X. et al. (2017). Wireless power delivery to flexible subcutaneous implants using capacitive coupling. *IEEE Trans. Microw. Theory Tech.* 65 (1): 280–292.

14 Liu, C., Guo, Y.-X., Sun, H., and Xiao, S. (2014). Design and safety considerations of an implantable rectenna for far-field wireless power transfer. *IEEE Trans. Antennas Propag.* 62 (11): 5798–5806.

15 Falkenstein, E., Roberg, M., and Popovic, Z. (2012). Low-power wireless power delivery. *IEEE Trans. Microw. Theory Tech.* 60 (7): 2277–2286.

16 Ozeri, S. and Shmilovitz, D. (2010). Ultrasonic transcutaneous energy transfer for powering implanted devices. *Ultrasonics* 50 (6): 556–566.

17 Suzuki, S.-N., Kimura, S., Katane, T. et al. (2002). Power and interactive information transmission to implanted medical device using ultrasonic. *Jpn. J. Appl. Phys.* 41 (5S): 3865–3866.

18 Maleki, T., Cao, N., Song, S.H. et al. (2011). An ultrasonically powered implantable micro-oxygen generator (IMOG). *IEEE Trans. Biomed. Eng.* 58 (11): 3104–3111.

19 Seo, D. et al. (2016). Wireless recording in the peripheral nervous system with ultrasonic neural dust. *Neuron* 91 (3): 529–539.

20 Hochmair, I. et al. (2006). Med-el cochlear implants: state of the art and a glimpse into the future. *Trends Amplif.* 10 (4): 201–219.

21 Patrick, J.F., Busby, P.A., and Gibson, P.J. (2006). The development of the nucleus freedom cochlear implant system. *Trends Amplif.* 10 (4): 175–200.

22 Agarwal, K., Jegadeesan, R., Guo, Y., and Thakor, N.V. (2017). Wireless power transfer strategies for implantable bioelectronics. *IEEE Rev. Biomed. Eng.* 10: 136–161.

23 Kurs, A., Karalis, A., Moffatt, R. et al. (2007). Wireless power transfer via strongly coupled magnetic resonances. *Science* 317 (5834): 83–86.

24 Hayt, W.H. and Buck, J.A. (2001). Time-varying fields and Maxwell's equations. In: *Engineering Electromagnetics*, 322–347. McGraw-Hill.

25 Jackson, J.D. (1999). Magnetostatics, Faraday's law, quasi-static fields. In: *Classical Electrodynamics*, 3ee, 174–236. John Wiley & Sons, Inc.

26 Tesla, N. (1907). *Apparatus for Transmitting Electrical Energy*. US 1119732.

27 Poon, A.S.Y., O'Driscoll, S., and Meng, T.H. (2010). Optimal frequency for wireless power transmission into dispersive tissue. *IEEE Trans. Antennas Propag.* 58 (5): 1739–1750.

28 Wang, C.S., Covic, G.A., and Stielau, O.H. (2004). Power transfer capability and bifurcation phenomena of loosely coupled inductive power transfer systems. *IEEE Trans. Ind. Electron.* 51 (1): 148–157.

29 Uei-Ming, J. and Ghovanloo, M. (2007). Design and optimization of printed spiral coils for efficient transcutaneous inductive power transmission. *IEEE Trans. Biomed. Circuits Syst.* 1 (3): 193–202.

30 Cannon, B.L., Hoburg, J.F., Stancil, D.D., and Goldstein, S.C. (2009). Magnetic resonant coupling as a potential means for wireless power transfer to multiple small receivers. *IEEE Trans. Power Electron.* 24 (7): 1819–1825.

31 Jegadeesan, R. and Guo, Y.X. (2010). A study on the inductive power links for implantable biomedical devices. In: *2010 IEEE Antennas and Propagation Society International Symposium*, 1–4.

32 Silay, K.M., Dehollain, C., and Declercq, M. (2008). Improvement of power efficiency of inductive links for implantable devices. In: *Proceedings of the Research in Microelectronics and Electronics, PRIME*, 229–232.

33 Fu, W.Z., Zhang, B., and Qiu, D.Y. (2009). Study on frequency-tracking wireless power transfer system by resonant coupling. In: *Proceedings of the IEEE 6th International Power Electronics and Motion Control Conference (IPEMC'09)*, 2658–2663.

34 Mohan, S.S., Hershenson, M.D., Boyd, S.P., and Lee, T.H. (1999). Simple accurate expressions for planar spiral inductances. *IEEE J. Solid-State Circ.* 34 (10): 1419–1424.

35 Zolog, M., Pitica, D., and Pop, O. (2007). Characterization of spiral planar inductors built on printed circuit boards. In: *2007 30th International Spring Seminar on Electronics Technology (ISSE)*, 308–313.
36 Ban-Leong, O., Dao-Xian, X., Pang-Shyan, K., and Fu-jiang, L. (2002). An improved prediction of series resistance in spiral inductor modeling with eddy-current effect. *IEEE Trans. Microwave Theory Tech.* 50 (9): 2202–2206.
37 Paul, C.R. (2009). *Inductance: Loop and Partial*, 220. John Wiley & Sons.
38 Jegadeesan, R. and Guo, Y. (2011). Evaluation and optimization of high frequency wireless power links. In: *2011 IEEE International Symposium on Antennas and Propagation (APSURSI)*, 400–403.
39 Theodoridis, M.P. (2012). Effective capacitive power transfer. *IEEE Trans. Power Electron.* 27 (12): 4906–4913.
40 Kline, M., Izyumin, I., Boser, B., and Sanders, S. (2011). Capacitive power transfer for contactless charging. In: *2011 Twenty-Sixth Annual IEEE Applied Power Electronics Conference and Exposition (APEC)*, 1398–1404.
41 Jegadeesan, R., Guo, Y.X., and Je, M. (2013). Electric near-field coupling for wireless power transfer in biomedical applications. *IEEE MTT-S Int. Microwave Workshop Ser. RF Wireless Technol. Biomed. Healthcare Appl. (IMWS-BIO)* 2013: 1–3.
42 Al-Kalbani, A.I., Yuce, M.R., and Redoute, J.-M. (2014). A biosafety comparison between capacitive and inductive coupling in biomedical implants. *IEEE Antennas Wireless Propag. Lett.* 13: 1168–1171.
43 Feldman, Y., Puzenko, A., and Ryabov, Y. (2006). *Dielectric Relaxation Phenomena in Complex Materials. Fractals Diffusion and Relaxation in Disordered Complex Systems: Advances in Chemical Physics*. Hoboken, NJ, USA: Wiley.
44 Foster, K.R. and Schwan, H.P. (1989). Dielectric properties of tissues and biological materials: a critical review. *Critical Rev. Biomed. Eng.* 17 (1): 25–104.
45 Gabriel, S., Lau, R.W., and Gabriel, C. (1996). The dielectric properties of biological tissues: III. Parametric models for the dielectric spectrum of tissues. *Phys. Med. Biol.* 41 (11): 2271–2293.
46 Gabriel, S., Lau, R.W., and Gabriel, C. (1996). The dielectric properties of biological tissues: II. Measurements in the frequency range 10 Hz to 20 GHz. *Phys. Med. Biol.* 41 (11): 2251–2269.
47 Terman, F.E. (1945). *Radio Engineers Handbook*. New York, NY, USA: McGraw-Hill.
48 Sokal, N.O. and Sokal, A.D. (1975). Class E-A new class of high-efficiency tuned single-ended switching power amplifiers. *IEEE J. Solid-State Circuits* SSC-10 (3): 168–176.
49 IEEE (2006). IEEE standard for safety levels with respect to human exposure to radio frequency electromagnetic fields, 3 kHz to 300 GHz. *IEEE Standard* C95 (1-2005): 1–238.
50 Agarwal, K. and Guo, Y.-X. (2015). Interaction of electromagnetic waves with humans in wearable and biomedical implant antennas. *Proc. Asia–Pacific Symp. Electromagn. Compat.* 154–157.
51 ICRP (2009). Adult reference computational phantoms. *Ann. ICRP* 39 (2).
52 Jegadeesan, R., Nag, S., Agarwal, K. et al. (2015). Enabling wireless powering and telemetry for peripheral nerve implants. *IEEE J. Biomed. Health Inf.* 19 (3): 958–970.
53 Hassler, C., Boretius, T., and Stieglitz, T. (2011). Polymers for neural implants. *J. Polymer Sci. B Polymer Phys* 49 (1): 18–33.

8

Far-field Wireless Power Transmission for Biomedical Application

8.1 Introduction

In the preceding chapter, we delved into the complexities and considerations surrounding near-field wireless power transmission (WPT). Despite its many advantages, near-field WPT often necessitates the use of wearable devices to forge a power transmission link bridging the internal and external realms of the body. This reliance on external apparatuses can impose limitations and inconveniences. For example, patients equipped with cochlear implants may confront challenges when required to disengage their external devices for activities such as bathing, thereby impeding the effective utilization of their cochlear implants. In view of these hurdles, far-field WPT emerges as a potentially powerful resolution.

Far-field WPT, with its capability to surmount the constraints associated with near-field WPT, brings forth immense promise. It facilitates power transmission over greater distances, thereby obviating the necessity for intimate proximity between the power source and the device being charged. This paradigm shift ushers in a new era for applications where the use of external devices is either impractical or inconvenient. Take, for instance, medical implants demanding uninterrupted power supplies, they stand to gain significantly from far-field WPT, freeing patients from the shackles of external devices and enabling them to participate in daily activities without any hindrance.

This chapter extensively covers the fundamental aspects, efficiency considerations, link design, and challenges of far-field wireless charging for implantable medical devices (IMDs) [1]. We explore methods to enhance charging efficiency while adhering to safety guidelines. One notable approach is the use of parasitic patches on the human body to amplify the WPT link, improving power reception [2]. We also address efficiency issues caused by antenna polarization and radiation direction misalignments. To address these concerns safely, we propose an innovative antenna alignment method using intermodulation in WPT antennas [3].

8.2 Far-Field EM Coupling

The far-field electromagnetic coupling (FEC) strategy operates on the principle of electromagnetic radiation, wherein a receiver (RX) antenna is situated at a considerable

Antennas and Wireless Power Transfer Methods for Biomedical Applications, First Edition.
Yongxin Guo, Yuan Feng and Changrong Liu.

distance ($R \geq 2D^2/\lambda$, where D is the largest dimension of the antenna and λ represents its corresponding wavelength at the operating frequency) from the transmitter (TX) antenna. Within the far-field zone of an antenna, the radiated fields constitute a plane wave, comprised solely of E_θ and H_ϕ components of the electric and magnetic fields. An external TX antenna broadcasts electromagnetic waves, with the radiated power characterized by the following equation:

$$P_T = \frac{1}{2} Re \oint_S (\vec{E} \times \vec{H}^*).ds \tag{8.1}$$

which in the far-field zone (only θ and ϕ components of the E and H fields present) reduces (8.1) to (8.2).

$$P_T \cong \frac{1}{2\eta} \oint_{4\pi} (|E_\theta|^2 + |E_\phi|^2) r^2 d\Omega \quad \text{and} \quad \eta = \frac{E_\theta}{H_\phi} \tag{8.2}$$

Here, $d\Omega = \sin\theta d\theta d\phi$ represents the element of solid angle ($0 \leq \theta \leq \pi$, $0 \leq \phi \leq 2\pi$), and η signifies the intrinsic impedance of the medium through which the wave travels [4]. When these radiated fields strike a tuned RX antenna situated inside the body, they generate a current across the antenna's terminals. This induced current is subsequently rectified and employed by the implanted device, as depicted in Figure 8.1. The power received (P_R) can be computed using the Friis transmission formula, expressed as follows:

$$P_R = \frac{G_T G_R \lambda_0^2}{(4\pi d)^2} (1 - |S_{11}|^2)(1 - |S_{22}|^2) e_p \times P_T \tag{8.3}$$

In the formula above, P_T represents the output power level of the TX, G_T is the gain of the TX antenna, G_R is the gain of the RX antenna, d denotes the distance between TX and RX, and e_p signifies the polarization mismatch of antennas. In practical situations, the value calculated by the Friis formula can be seen as the maximum possible received power. However, it is important to note that in a real-world radio communication system, numerous factors

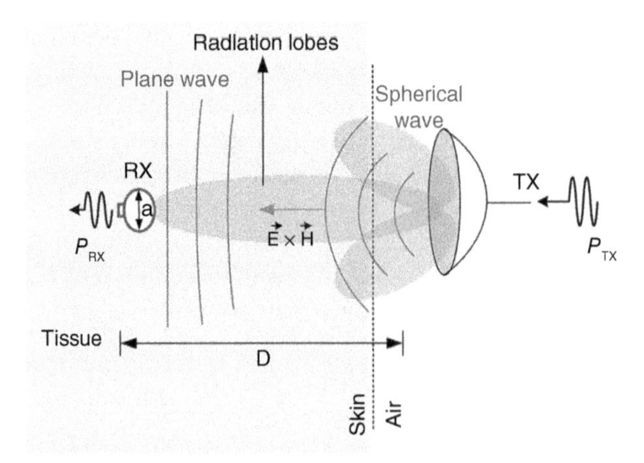

Figure 8.1 Schematic of the far-field electromagnetic coupling-based wireless power transfer method. The TX generates focused EM waves, which when incident on the implanted RX induces current used for driving the IMD. Source: Agarwal et al. [1]/IEEE.

can diminish the received power P_R [5]. In such cases, the received power can be estimated by using the transmission coefficient $|S_{21}|$, as indicated by the following equation [2, 6]:

$$P_R = |S_{21}|^2 \times P_T. \tag{8.4}$$

Over the past decade, the FEC wireless power strategy has been extensively explored for long-range power transmission in free space [7–9]. However, its application in powering biomedical implants has remained relatively under-researched. The design of a far-field-based wireless power transfer scheme for biomedical implants has been presented in [2], along with related safety concerns and studies in [10, 11]. A system-level design for a cardiovascular pressure measurement implant was introduced in [12]. In [13], separate on-chip transmitter and receiver antennas were proposed for wireless far-field power and telemetry of an implantable intraocular pressure monitoring system. A triple-band stacked implantable antenna was designed in [14] for FEC-based data telemetry, WPT, and wake-up control signals at operating frequencies of 402 MHz (Medical Implant Communication Service band), 433 MHz, and 2.4 GHz. The far-field method can support long-range power (a few tens of millimeters), enabling the powering of deep-seated implants using antennas much smaller than those used in near-field implants. However, given the operation at GHz frequencies, the power rectification losses are higher. Combined with the inherently low power transfer efficiency (PTE) of this scheme, it becomes less attractive to meet the power requirements of most IMDs designed for various applications.

8.2.1 Power Transfer Efficiency

The overall PTE of an FEC-based TX–RX system is contingent upon various factors such as the directivity and efficiency of the transmitter and receiver antennas, impedance mismatch losses, and the efficiencies of the RF-to-DC conversion circuits. Let us delve into each of these factors step by step.

(1) The miniaturization of the receiver antenna in an FEC system poses a significant challenge in achieving high antenna efficiency. Electrically small antennas (ESAs) ($a \leq \lambda/2\pi$, where 'a' is the radius of the sphere completely enclosing the antenna and λ is its corresponding wavelength) have a small radiation resistance, which hampers radiation efficiency and complicates matching it to a high-resistance feeding system [15]. Several antenna miniaturization techniques such as using a high-permittivity dielectric substrate/superstrate [16], lengthening the radiator's current path (e.g., spiral [14] or meandering [17]), inductive [18] or capacitive [19] loading for impedance matching, planar inverted-F antennas (PIFA) [20], and higher operating frequencies are employed to strike a balance between antenna efficiency and size reduction, as shown in Figure 8.2.
TX antennas, positioned outside the human body, can be electrically larger than the implanted receiver antennas. For efficient FEC links, the design preference lies in high-gain TX antennas (featuring high directivity and efficiency), and large bandwidth.
(2) According to the Chu–Harrington limit, the lower limit for the Q-factor is given by the following equation [21]:

$$Q_{min} = \frac{1}{(ka)^3} + \frac{1}{(ka)} \cong \frac{1}{BW} = \frac{f}{\Delta f}\left(k = \frac{2\pi}{\lambda}\right). \tag{8.5}$$

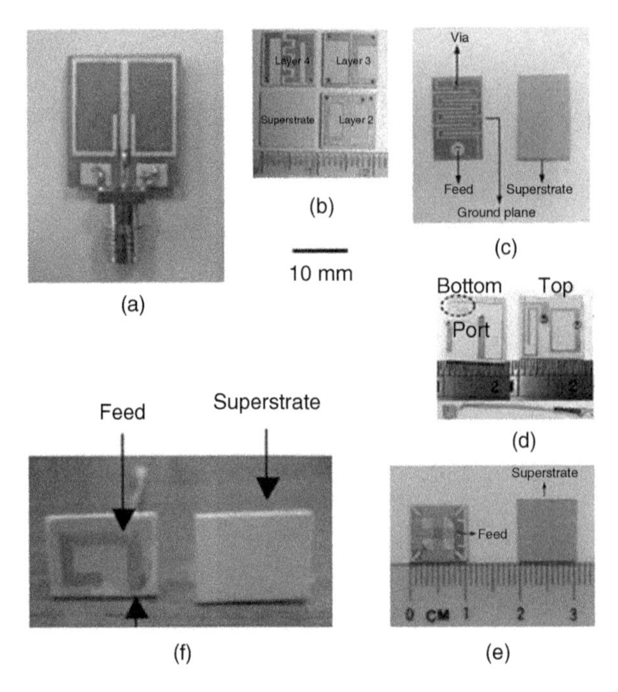

Figure 8.2 Commonly used antenna miniaturization techniques in FEC: (a) high dielectric antenna on ANMg (permittivity = 28) in Chien et al. [16], (b) spiral multiband antenna in Huang et al. [14], (c) meandering used in Liu et al. [17], (d) inductive loading used in Xu et al. [18], (e) capacitive loading used in Liu et al. [19], and (f) PIFA proposed in Kim et al. [20].

The Q-factor implies that there is a limit to the impedance bandwidth for a given radiation efficiency. Therefore, when the implanted receiver antenna is miniaturized, its bandwidth decreases and the radiation resistance becomes smaller relative to the loss resistances, leading to a reduction in radiation efficiency. ESAs typically exhibit low gains due to their diminished directivity and efficiency, making an impedance tuning/matching network crucial for minimizing impedance mismatch losses.

(3) The efficiencies of RF-to-DC conversion circuits, as indicated in Eq. (8.6), are lower at GHz frequencies, thereby contributing to power rectification losses. These losses further undermine the overall system efficiency of the FEC-based power transfer scheme.

$$\eta(\%) = \frac{P_{\mathrm{DC}}}{P_{\mathrm{R}}} \times 100\% = \frac{V_{\mathrm{DC}}^2}{R_{\mathrm{L}} P_{\mathrm{R}}} \times 100\% \tag{8.6}$$

where P_{DC} is the output DC power, P_{R} is the power received by the rectifier, V_{DC} is the output DC voltage, and R_{L} is the load resistance.

8.2.2 Link Design

The core of an FEC-based power link primarily comprises two far-field antennas [22] (TX, and RX; see Figure 8.1), which are optimized to deliver electromagnetic radiation at a specific operating frequency. The TX antenna is typically optimized in free space, while the RX antenna is optimized within a human voxel model. Certain electromagnetic simulation software, such as the finite integration technique-based CST [23], contain human anatomical datasets for analyzing the interaction of electromagnetic waves with the human body [10]. These are utilized for designing and conducting safety analysis of implantable antennas.

According to the FCC regulations for the maximum TX output power [24] in industrial, scientific, and medical (ISM) bands, the transmitted power cannot exceed 30 dBm (1 W). Also, the maximum isotropic radiated power (effective isotropic radiated power (EIRP)) should be ≤36 dBm (4 W). Given that the EIRP for a proposed wireless power link is described by Eq. (8.7), the transmitted power P_T is adjusted to balance the TX gain G_T and impedance mismatch losses L_{imp} (as in Eq. 8.8), ensuring the EIRP remains within acceptable limits.

$$[\text{EIRP}]_{\text{dBm}} = [P_T]_{\text{dBm}} + [G_T]_{\text{dB}} - [L_{\text{imp}}]\text{dB} \tag{8.7}$$

$$L_{\text{imp}}(\text{dB}) = -\,10\,\log(1 - |\tau|^2) \tag{8.8}$$

where τ is the appropriate reflection coefficient.

The maximum permissible exposure limit for uncontrolled exposure to an intentional radiator operating at a frequency [11] is 10 W/m^2. The power flux density at the distance d from the TX is given by the following equation:

$$[W_f]_{\text{W/m}^2} = \frac{\text{EIRP}}{4 \times \pi \times d^2} \leq 10\ \text{W/m}^2. \tag{8.9}$$

Given that the EIRP is restricted by FCC standards to 36 dBm, the minimum separation between the transmitter (TX) and receiver (RX) for FEC power transfer is 0.178 m, assuming a power flux density equal to 10 W/m^2. To ensure the safety limits are upheld, an analysis of peak specific absorption rate (SAR) and temperature arising from body-absorbed radiation is necessary. This should be conducted in line with the IEEE C95.1-1999 standards [25], serving as a guide to limit the TX and RX power levels in an FEC-based system.

8.2.3 Challenges and Solutions

The inherent low overall PTE in FEC systems necessitates addressing certain design challenges to improve system efficiency while adhering to safety standards. The challenges and potential solutions are elaborated below:

(1) Due to the safety considerations established by the FDA and FCC for far-field-based power transfer systems, the TX power, and consequently the P_r at the implant, is quite limited. This, coupled with the low RF-to-DC conversion circuit efficiency, significantly reduces the final DC received power level $(P_{dc})_{RX}$ that drives the IMDs. One possible approach to enhancing the directivity of the TX antenna toward the RX implant antenna involves using parasitic patches on the human body [2]. This would increase the received RF power levels at the implanted rectenna, while ensuring TX and RX power levels remain within safety limits. The rectifier circuit can also be optimized to maximize the RF-to-DC conversion circuit efficiency.
(2) The low overall PTE in FEC-based systems makes the far-field powering scheme challenging for IMDs. Technological advancements in ultra-low power electronics are decreasing the overall power requirements of implanted devices. Techniques such as

TX beam focusing using a planar immersion lens (similar to the metalens proposed for mid-field powering scheme [26]), or antenna array methods, can focus the TX beam toward the miniaturized RX antenna. This not only reduces unnecessary radiation and tissue losses but also enhances the overall PTE of the system.

8.3 Enhanced Far-field WPT Link for Implants

As previously stated, one of the significant obstacles with far-field wireless power transfer for implantable devices resides in the limited power reception, primarily due to safety constraints. In this section, we put forth an innovative solution to this challenge by proposing the design of an implantable rectenna, specifically purposed for far-field wireless power transfer. This rectenna is conscientiously designed with crucial safety standards as its cornerstone, thereby ensuring it adheres to the stringent regulatory benchmarks set for implantable devices.

8.3.1 Safety Considerations for Far-field Wireless Power Transmission

This section commences with a comprehensive discussion regarding safety considerations for far-field WPT. In Section 8.2.2, we have previously provided an in-depth exploration of the FCC regulations regarding maximum TX output power, and the maximum permissible exposure limit. Complementing these, other factors such as SAR and the limit of focalized temperature, warrant intricate considerations as well.

SAR is a measure of the rate at which energy is absorbed by the human body when exposed to a radio frequency (RF) electromagnetic field. SAR for electromagnetic energy can be calculated from the electric field within the tissue as:

$$\mathrm{SAR} = \frac{1}{V}\int_{\mathrm{tissue}} \frac{\delta(r)|E(r)^2|}{\rho(r)} dr \tag{8.10}$$

where σ is the tissue electrical conductivity, E is the root mean square (RMS) electric field, ρ is the tissue density, and V is the volume of the tissue. Maximum SAR values are limited to preserve patient safety. Two standards are referenced at this point. The IEEE C95.1-1999 standard restricts the SAR averaged over any 1 g of tissue in the shape of a cube (1-g average SAR) to less than 1.6 W/kg [25]. The IEEE C95.1-2005 standard restricts the SAR averaged over any 10 g of tissue in the shape of a cube (10-g average SAR) to less than 2 W/kg [27].

A temperature increase in body tissues can be caused by the absorbed power from an electromagnetic field. It is very important that the temperature of the tissue surrounding the implanted device does not increase more than 1–2 °C. The temperature of the body tissues can be modeled using the following bioheat equation in (8.11) [28]:

$$Cp\frac{\delta T}{\delta t} = \nabla \cdot (k\nabla T) + \rho \cdot \mathrm{SAR} + A - B(T - T_b) \tag{8.11}$$

where K is the thermal conductivity, Cp denotes the specific heat, A is the basal metabolic rate, B is the term associated with blood perfusion, ρ is the tissue density in kg/m^3, and $\nabla \cdot (k \nabla T)$ represents the thermal spatial diffusion term for heat transfer through conduction at temperature in degrees Celsius. Temperature studies inside the human body model can be done through co-simulation with CST Microwave Suite and a thermal solver [29, 30].

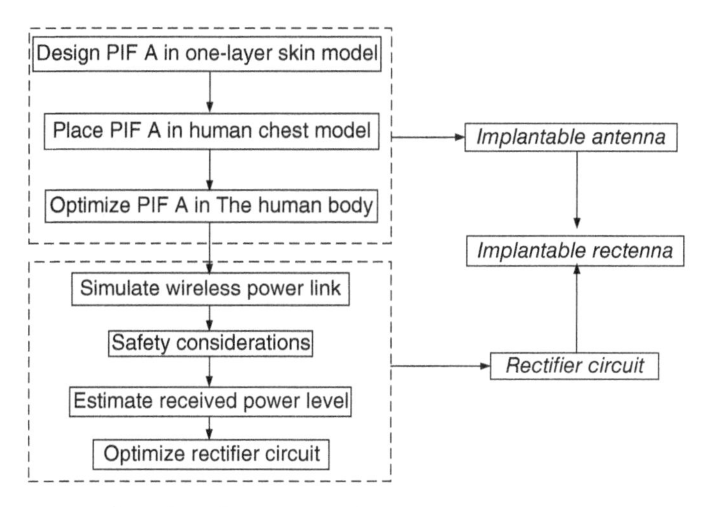

Figure 8.3 Design steps of the implantable rectenna.

8.3.2 Implantable Rectenna Design

An implantable rectenna, a device designed to receive and store energy from RF power, comprises an implantable antenna and a rectifier circuit. For effective wireless power delivery, a high-efficiency rectenna is essential. Figure 8.3 describes the detailed design steps for the implantable rectenna operating at 2.45 GHz. The process begins with the design of a miniaturized implantable PIFA. Upon optimizing the resonant frequency of the antenna within a human chest model, safety considerations are simulated for the wireless power link. Utilizing a parasitic patch on the human body optimally enhances the gain value of the implantable antenna. Based on these safety considerations, the power level received by the implantable PIFA can then be estimated. Subsequently, a rectifier circuit is designed, and an integrated rectenna solution is presented. Finally, a measurement setup for far-field WPT is proposed and comprehensively analyzed.

8.3.2.1 Implantable Antenna Configuration

One-layer skin model: An implantable antenna shown in Figure 8.4 is designed on a two-layer Rogers 3010 substrate with dielectric constant (ε_r) of 10.2, a thickness of 25 mil for each layer, and loss tangent ($\tan\delta$) of 0.0022. The patch and the ground are 3 mm × 8 mm and 4 mm × 8 mm in size, respectively. The folded ground is introduced to increase the length of the ground plane thus to reduce the antenna size. A one-layer skin simulation model is used to design the antenna for easy optimization, where the distance from the side of the structure to the edge of the skin is 4 cm on both sides, the distance from the top of the superstrate to the edge of the skin is 4 mm and the distance from the bottom of the substrate to the edge of the skin is 20 mm. The skin's electrical properties ($\varepsilon_r = 38$, $\sigma = 1.44$ S/m) at 2.4 GHz were used in the one-layer skin simulation model. Ansoft High Frequency Structure Simulator (HFSS) is used for the design and analysis. Simulated reflection coefficient of the implantable antenna in the one-layer skin model is shown in Figure 8.5. It can be seen that the impedance bandwidth of the antenna covers from 2.37 to 2.47 GHz for $|S_{11}|$ less than −10 dB. The peak realized gain in the one-layer skin model is ~−19 dBi at 2.45 GHz.

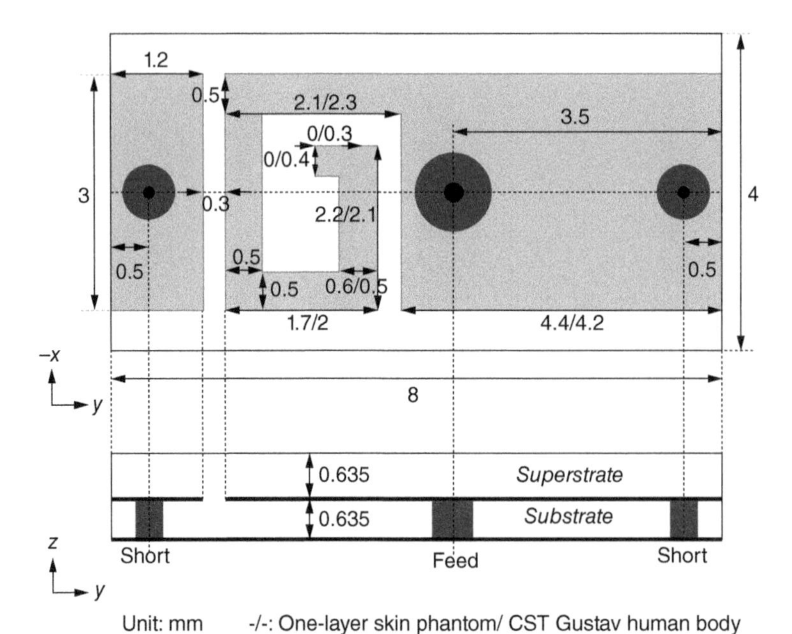

Figure 8.4 Geometry of the implantable antenna with a folded ground. Source: Liu et al. [2]/IEEE.

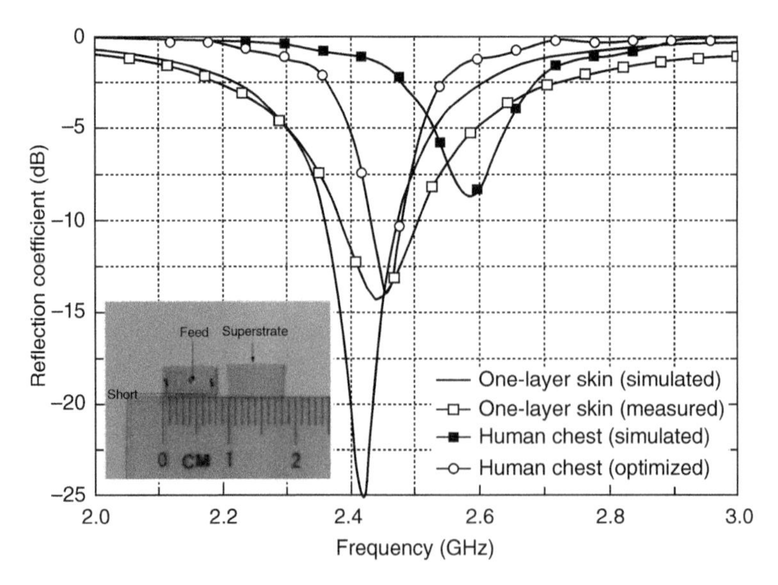

Figure 8.5 Simulated and measured reflection coefficient of the antenna in different models. Source: Liu et al. [2]/IEEE.

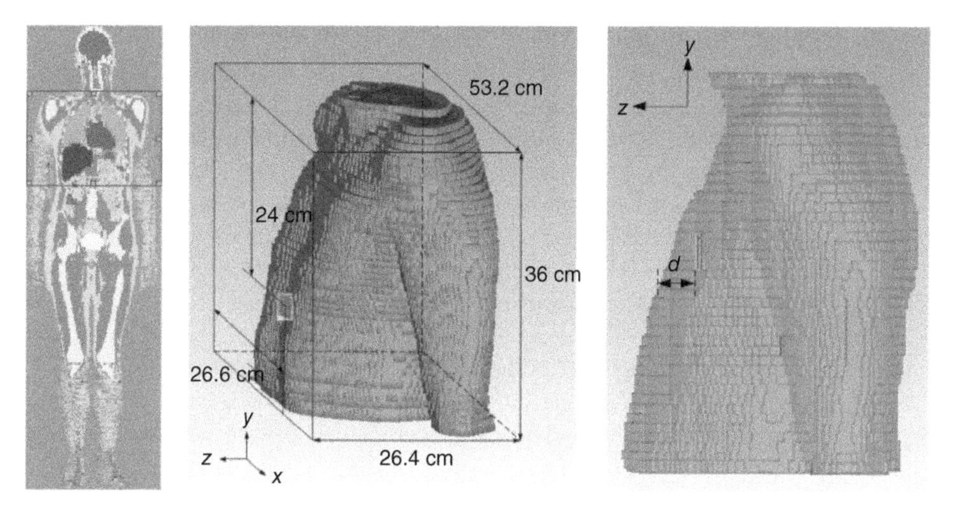

Figure 8.6 Three-dimensional Gustav Voxel human body used for the implantable antenna design in a human chest. Source: Liu et al. [2]/IEEE.

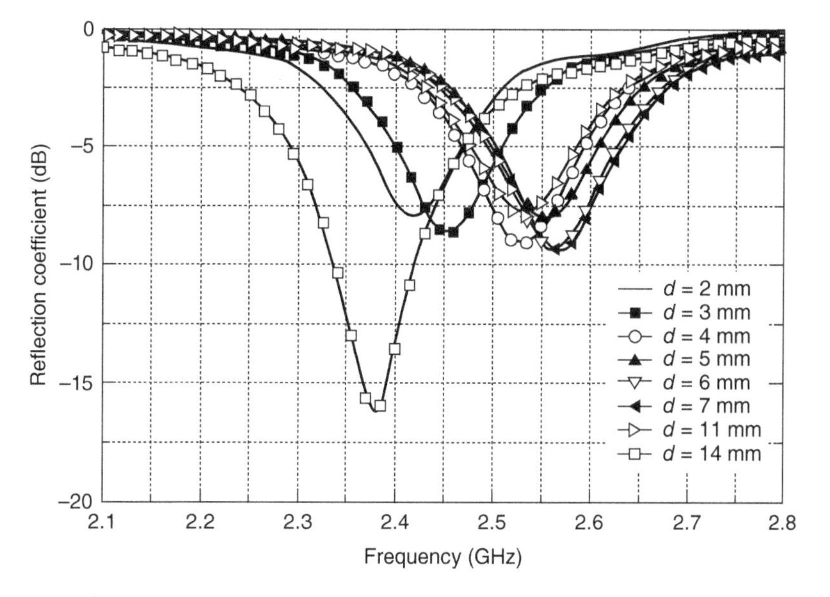

Figure 8.7 Simulated reflection coefficients in the Gustav human body with different implant depths. Source: Liu et al. [2]/IEEE.

Human body model: In order to study the design in a realistic environment and optimize the operating frequency at 2.45 GHz, the implantable antenna is evaluated within the voxel human body. Figure 8.6 shows the three-dimensional voxel Gustav human body used for the implantable design in a human chest. Numerical analyses are performed using the CST Microwave Suite [30] (Figure 8.7).

The resonant frequency of the antenna experiences a shift from 2.45 GHz to 2.59 GHz when the simulation environment is altered, also resulting in poorer matching. However, by adjusting the resonant length and the feeding point, the implantable antenna can operate at 2.45 GHz with satisfactory matching. Figure 8.2 showcases the geometric parameters in different simulation models with identical implant depths. The corresponding results are exhibited in Figure 8.3. The implantable antenna, designed for placement within the human chest, achieves an impedance bandwidth of 4.1% (ranging from 2.433 to 2.479 GHz).

To understand the frequency shift within the human body, the antenna is further investigated in the CST Gustav human body model, considering different implant depths. Figure 8.5 displays the simulation results, which indicate that the implant depth significantly influences the resonant frequency. This is due to changes in the surrounding environment with varied implant depths. In comparison to the antenna in the skin layer, the resonant frequency shifts upward in the fat layer and downward in the muscle layer.

Measurement: The antenna was evaluated using a homogeneous mixture of liquid skin phantom as proposed in [31]. The mixture consists of 58.2% deionized water, 36.7% Triton X-100 (polyethylene glycol mono phenyl ether), and 5.1% diethylene glycol butyl ether (DGBE). During the measurement process, a volume of skin material similar to the design was selected to illustrate the design concept. Figure 8.5 displays the measured reflection coefficient of the implantable antenna. As indicated by Figure 8.5, the measured impedance bandwidth of the implantable antenna spans from 2.395 to 2.498 GHz (4.2%) when $|S_{11}|$ is less than −10 dB.

8.3.2.2 Wireless Power Link Study

Figure 8.8 illustrates the simulation environment for the wireless power link, where a straightforward microstrip patch antenna is employed as a Tx antenna. The implantable antenna, on the other hand, serves as an Rx antenna, drawing energy from the ISM-band RF signals for far-field wireless power transfer. The CST Microwave Suite software is utilized to determine the received power level and ascertain safe power transfer conditions.

The transmitter output power (P_t) is set as 1 W. $|S_{21}|$ values with different distances between Tx and Rx at 2.45 GHz are shown in Figure 8.9. Also, simulated maximum 1-g/10-g average SAR values at 2.45 GHz are shown in Figure 8.9 when the output power of the transmitter equals 1 W.

In this study, the wireless power link was simulated using CST Microwave Suite and the simulation environment can contain most of the factors that affect the received power. In this case, the simulated $|S_{21}|$ values can be used to calculate the received power as

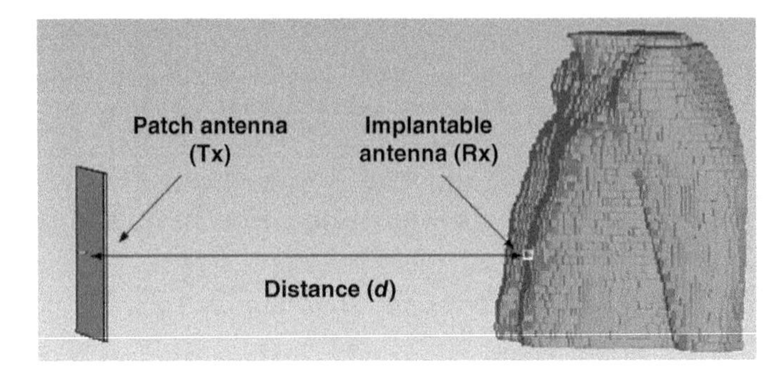

Figure 8.8 Simulated environment of the wireless power link. Source: Liu et al. [2]/IEEE.

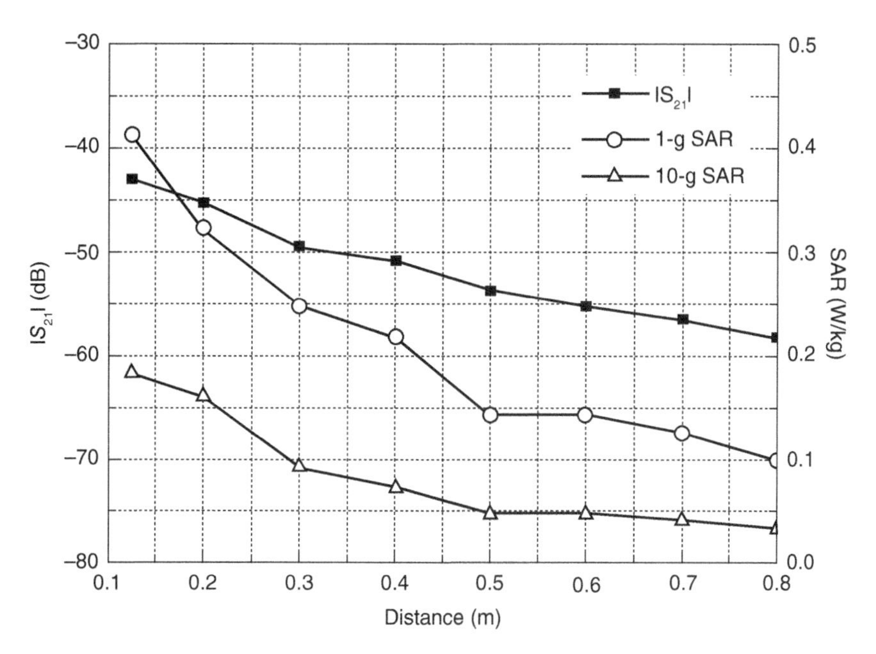

Figure 8.9 Simulated $|S_{21}|$ and maximum SAR values of the wireless power link. Source: Liu et al. [2]/IEEE.

$|S_{21}|^2 = P_r/P_t$. The detailed received power with different distances between Tx and Rx is shown in Figure 8.10. Note that the input power of the Tx (P_t) antenna is 1 W. In fact, P_t is restricted by the safety considerations.

8.3.2.3 Safety Concerns

A. FCC Rules for Maximum Transmit Output Power at ISM-bands

For the wireless power link, the EIRP can be calculated using the formula (8.7): [EIRP]_dBm = 30 dBm + 7 dB − 0.582 dB = 36.418 dBm. Note that the gain value at main radiation direction could change a little with different distances between the Tx and Rx, and we assume that the gain is 7 dBi at any distance. In order to satisfy the FCC standard in 2.4 GHz ISM-band, the output power of the transmitter should drop to 29.582 dBm (0.908 W).

B. Maximum Permissible Exposure (MPE) Limit

As for the wireless power link, the EIRP is limited by the FCC standard. The minimum distance between Tx and Rx is 0.178 m when the power flux density equals 10 W/m^2.

C. Specific Absorption Rate (SAR) Limit

The detailed 1-g/10 g-avg SAR distributions of the Gustav human body are shown in Figure 8.9 when the distance between Tx and Rx is 500 mm. In this study, the input power is set as 1 W. The maximum 1-g/10-g average SAR values are only 0.413 W/kg and 0.183 W/kg, respectively. Considering the FCC rules for maximum transmit output power in the ISM bands, the *EIRP* is limited to 36 dBm, where the transmitter input power is about 0.908 W.

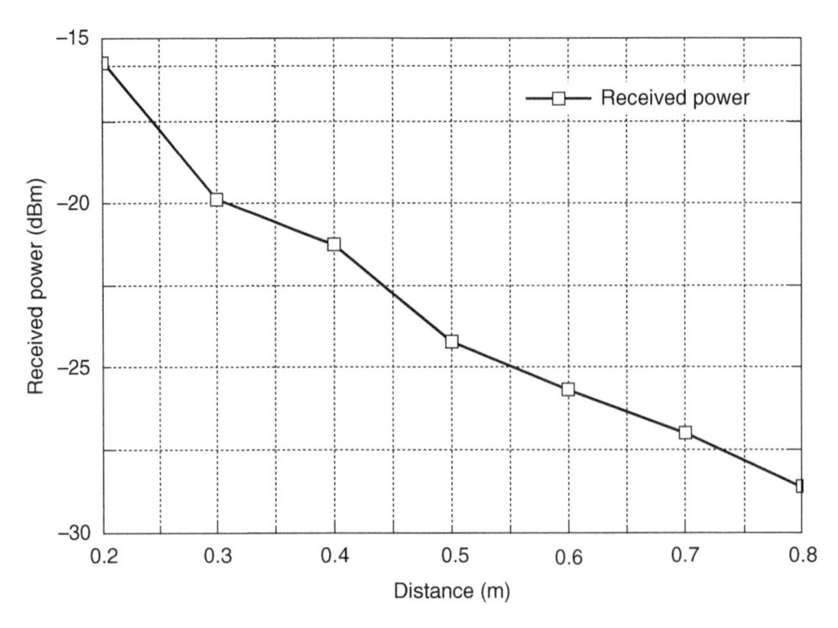

Figure 8.10 Calculated received power of the implantable antenna when input power equals 1 W. Source: Liu et al. [2]/IEEE.

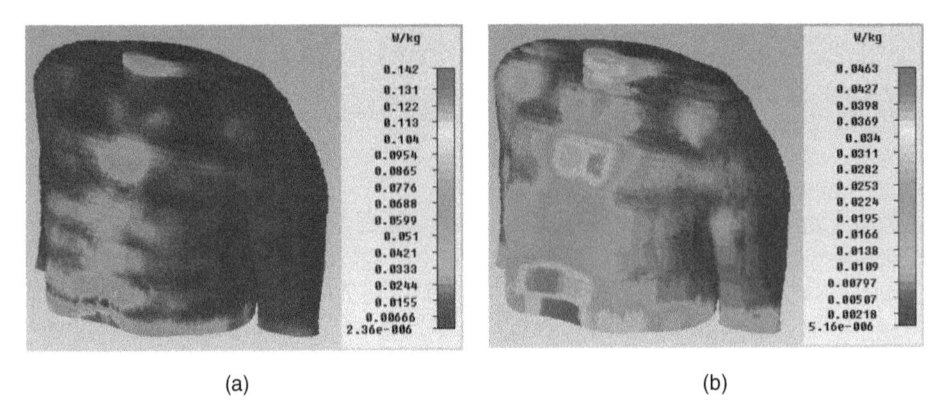

Figure 8.11 1-g/10 g-avg SAR distributions of the human body when the distance between Tx and Rx is 500 mm: (a) 1 g-avg SAR, (b) 10-g avg SAR. Source: Liu et al. [2]/IEEE.

In this condition, the maximum 1-g/10-g average SAR values should be less than the values shown in Figures 8.9 and 8.11.

D. Focalized Temperature Limit

The initial body temperature is considered to be 37 °C. Background temperature is set as 293.1 K. The input power of the Tx antenna is set as 1 W. Figure 8.12 shows the obtained temperature variation. It can be seen from Figure 8.12 that the temperature of the human body is increased from the initial temperature of 37 °C to 37.15 °C (310.3 K). Note that there

Figure 8.12 Simulated temperature variation. Source: Liu et al. [2]/IEEE.

is no significant temperature difference in the tissues. The reasons could be (1) the input power of the Tx antenna is not enough to cause a significant temperature increase due to the power absorption; (2) the whole temperature increase up to 37.15 °C may be due to the metabolic activities in the tissues.

8.3.2.4 Method to Enhance the Received Power

For the wireless power link to operate within the far-field region and meet safety considerations, the received power has to be limited, as shown in Figure 8.10. However, the amount of P_r by the implantable antenna is significantly influenced by the RF-to-DC conversion circuit efficiency. Achieving a rectifier circuit with acceptable conversion efficiency at low-input power poses a challenge for biomedical applications since conversion efficiency tends to increase with higher input RF power.

A method of enhancing the received power level involves increasing the directivity of implantable antennas. As these antennas must be small for biomedical applications, we can employ a parasitic patch over the human body to boost the wireless power link. Initially, we optimize the antenna with a parasitic patch in a single-layer skin model to maximize the implantable antenna's directivity. We then examine the misalignment effect of the parasitic patch, considering that its position on the human body may shift due to human activities. Finally, the optimized results are simulated in CST with the human body to evaluate the methodology.

Optimization of a parasitic patch: Based on the Yagi antenna theory, we could consider the implantable antenna as a driven element, with the parasitic patch acting as a director element. The size of the parasitic patch and its distance from the implantable antenna can be optimized to achieve the optimal amplitude phase of the induced current distribution on the director element. This results in the maximum forward gain and an optimal radiation pattern. Note that the feasibility of the parasitic patch also needs to be considered. The parasitic patch should be compact and wearable.

The simulation environment for the implantable antenna with a parasitic patch is depicted in Figure 8.13. Here, the parasitic patch is placed h_2 above the implantable

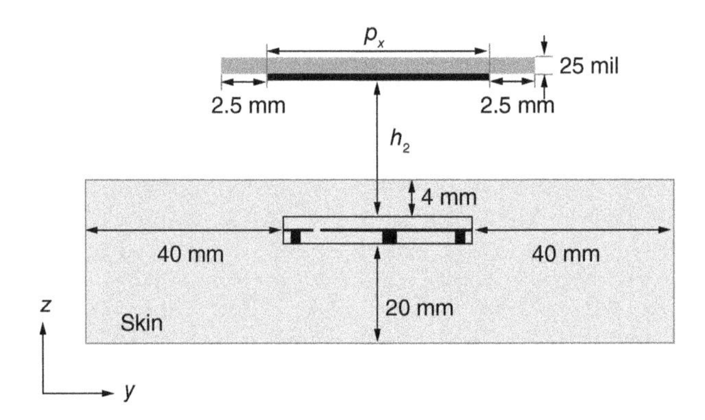

Figure 8.13 Geometry of the implantable antenna with the parasitic patch. Source: Liu et al. [2]/IEEE.

antenna. The sizes of the parasitic patch and its substrate are $P_y \times P_y$ and $(P_y + 5\,\text{mm}) \times (P_y + 5\,\text{mm})$, respectively. The same substrate material (Rogers RO3010) is used in this design. In future iterations, the parasitic patch and the substrate can be made from flexible materials commonly used in wearable antenna designs.

Two steps are made to get the maximum level of forward gain: Step 1: optimize h_2 with the fixed parasitic size $p_y = 30$ mm; Step 2: optimize p_y with the fixed h_2 (h_2 value is from step 1). The optimized results of both steps are shown in Figure 8.14. Figure 8.14(a) indicates that the increased speed of gain would be slow when h_2 is up to 14 mm and the enhanced gain would drop when $h_2 > 23$ mm. The maximum enhanced gain can be achieved when $h_2 = 19$ mm. After h_2 is fixed to 19 mm, we can get a maximum enhanced gain of nearly 9 dB when $d_y = 38$ mm, as shown in Figure 8.14(b).

Misalignment of the parasitic patch: For practical applications, the parasitic patch's misalignment is a realistic possibility. Based on the optimized results, two misalignment scenarios were modeled and studied as shown in Figure 8.15. Case 1: The parasitic patch is rotated by an angle α. Case 2: The parasitic patch is moved d_y away from the centerline.

The resonant frequencies and bandwidths are nearly unchanged, demonstrating that the misalignment of the parasitic patch has a minimal effect on the antenna's resonant behavior. However, the impact of the misalignment on the antenna's radiation cannot be ignored. As shown in Figure 8.16(a), the peak realized gain decreases as the rotation angle α increases. Figure 8.16(b) reveals that as the offset distance d_y increases, the parasitic patch causes beam steering. Considering the effects of the misalignment of the parasitic patch, as shown in Table 8.1, the realized gain can still increase to an acceptable level.

8.3.2.5 Wireless Power Link With the Parasitic Patch

After optimizing the size of the parasitic patch, the wireless power link can be studied, as shown in Figure 8.17. Figure 8.18 presents the simulated S-parameters of the wireless power link, both with and without the parasitic patch. The reflection coefficients of the implantable antenna show no significant difference. However, when adding the parasitic patch with optimized size (d_y) and distance (h_2), $|S_{21}|$ is enhanced from −54.2 dB to −41.3 dB at 2.45 GHz.

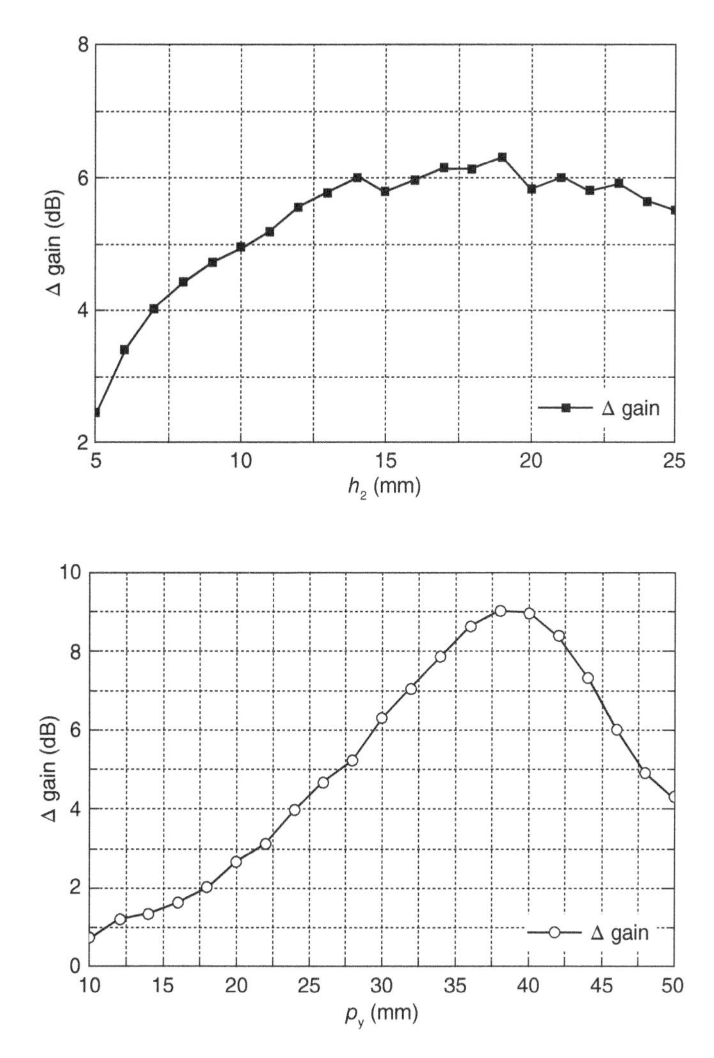

Figure 8.14 Enhanced gain with different heights (h_2) and patch sizes (p_x): (a) enhanced gain with different heights (h_2) when $p_y = 30$ mm, (b) enhanced gain with different patch sizes (p_y) when $h_2 = 19$ mm. Source: Liu et al. [2]/IEEE.

Figure 8.19 depicts the simulated 1-g/10 g-average SAR distributions within the human body with the parasitic patch when the distance between Tx and Rx is 500 mm at 2.45 GHz. When compared to Figure 8.11, we observe that while SAR values increase around the parasitic patch, they still remain within safe limits. Figure 8.20 displays the simulated temperature variation in the wireless power link with the parasitic patch. Table 8.2 provides a detailed comparison of the performance of the wireless power link with and without the parasitic patch. The received power level can be improved by approximately 12.9 dB without altering safety conditions when the parasitic patch is applied over the human body. This enhancement is slightly greater than the result optimized in the one-layer skin phantom, which is reasonable given the large effect on the human body.

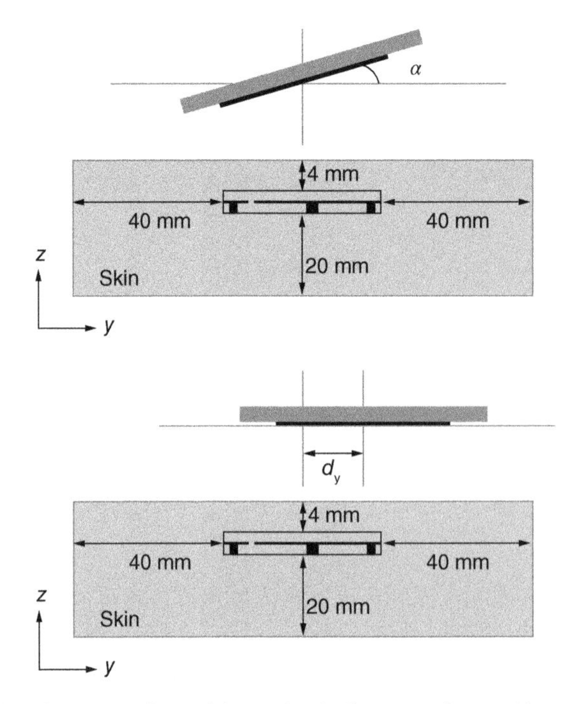

Figure 8.15 Geometry of parasitic patch misalignment. Source: Liu et al. [2]/IEEE.

8.3.3 Measurement and Discussion

After the design and testing of the implantable antenna, as well as the discussion of safety considerations, we must design a rectifier circuit to be integrated with the implantable antenna. This will allow for the testing of the transmission efficiency of the far-field wireless power link.

Efficient rectennas are crucial for wireless power delivery [7, 11, 13, 14, 32–38]. To enhance the conversion efficiency, optimal diode RF and DC impedances for efficient rectification were determined as a function of input power in [7], which was helpful for the optimization of the antenna design. Circularly polarized application rectennas were studied and designed in [32, 33]. A high-efficiency 2.45-GHz rectenna for low-input RF power was presented in [34]. Another work [35] designed a 2.45 GHz rectenna for an input power level of 0 dBm, and in [36], an RF-to-DC conversion efficiency of 15.7% at −20 dBm was reported.

However, few research groups have delved into the study of far-field wireless power links for biomedical applications. In [13], separate transmit and receive on-chip antennas were designed for implantable intraocular pressure monitoring applications. A triple-band implantable antenna was designed in [14] to accommodate data telemetry (402 MHz), WPT (433 MHz), and a wake-up controller (2.45 GHz). The optimal frequency range for WPT into dispersive tissue was studied in [38], with the findings suggesting the optimal frequency is above 1 GHz for a small receive coil and typical transmit-receive separations. The safety considerations of far-field powering were explored in [11] using a simplified theoretical analysis and FCC limits for radiating antennas. In [29], SAR, specific absorption

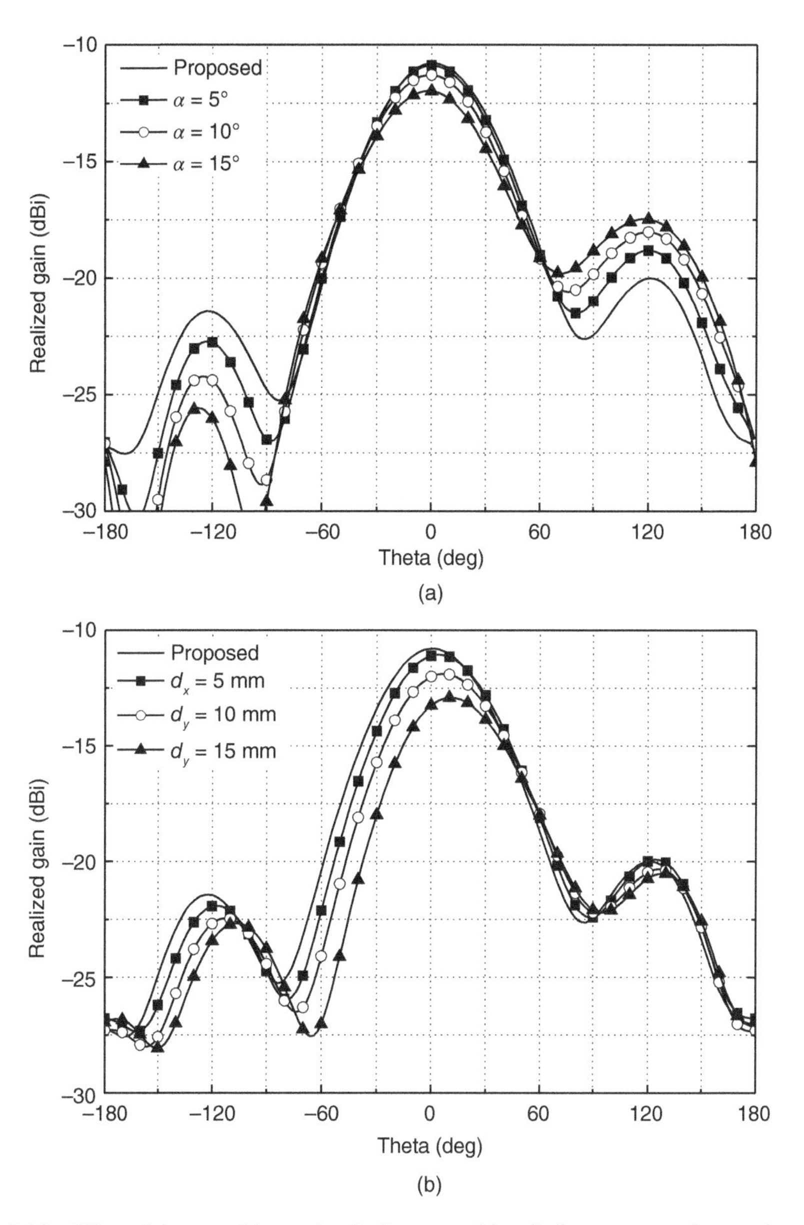

Figure 8.16 Effect of the parasitic patch misalignment: (a) radiation patterns of rotated parasitic patch, (b) radiation patterns of offset parasitic patch. Source: Liu et al. [2]/IEEE.

(SA), and temperature increase were analyzed to compare the compliance of the impulse radio ultra-wideband (IR-UWB) transmitting device with international safety regulations.

8.3.3.1 Rectifier Circuit Design

To implement WPT, we require a rectifier circuit that can convert the received RF power into DC power. Based on safety considerations, a received RF power level of approximately

Table 8.1 Performance comparisons for the parasitic patch misalignment at 2.4 GHz.

Rotate degree α (deg)	Realized gain (dBi)	ΔGain (dB)	Offset distance d_y (mm)	Realized gain (dBi)	ΔGain (dB)
0	−10.7	8.6	0	−10.7	8.6
5	−10.9	8.4	5	−11.1	8.2
10	−11.3	8.0	10	−12.0	7.3
15	−12.0	7.3	15	−13.2	6.1

Source: Liu et al. [2]/IEEE.

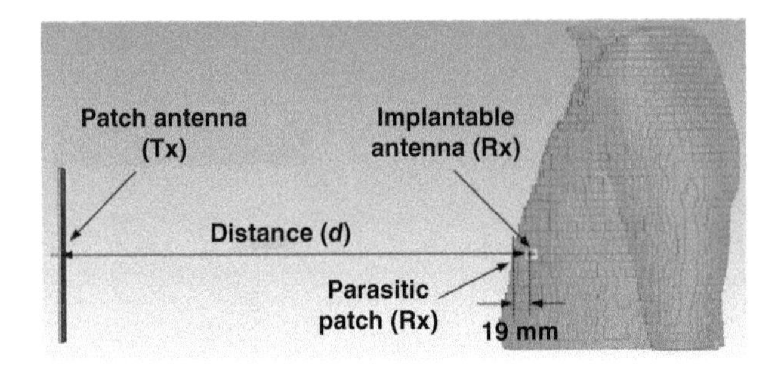

Figure 8.17 Simulated environment of wireless power link with parasitic patch. Source: Liu et al. [2]/IEEE.

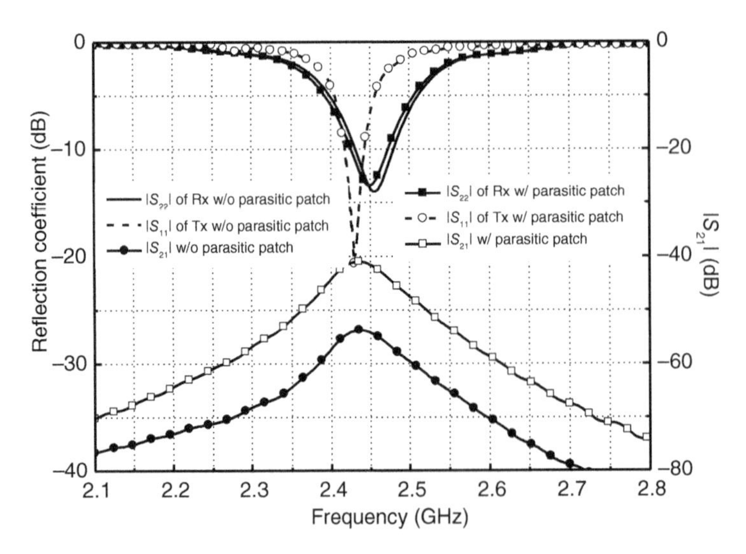

Figure 8.18 Simulated S-parameters of wireless power link with/without parasitic patch. Source: Liu et al. [2]/IEEE.

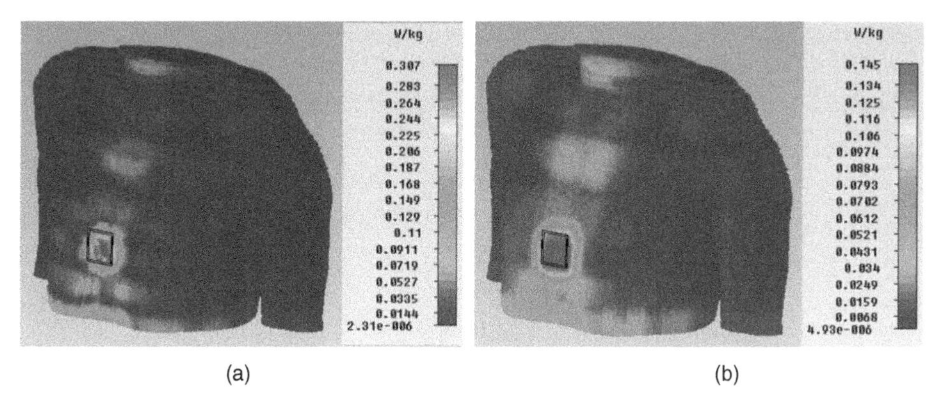

Figure 8.19 Simulated 1-g/10 g-avg SAR distributions of the human body with parasitic patch when the distance between the Tx and Rx is 500 mm: (a) 1 g-avg SAR, (b) 10-g avg SAR.

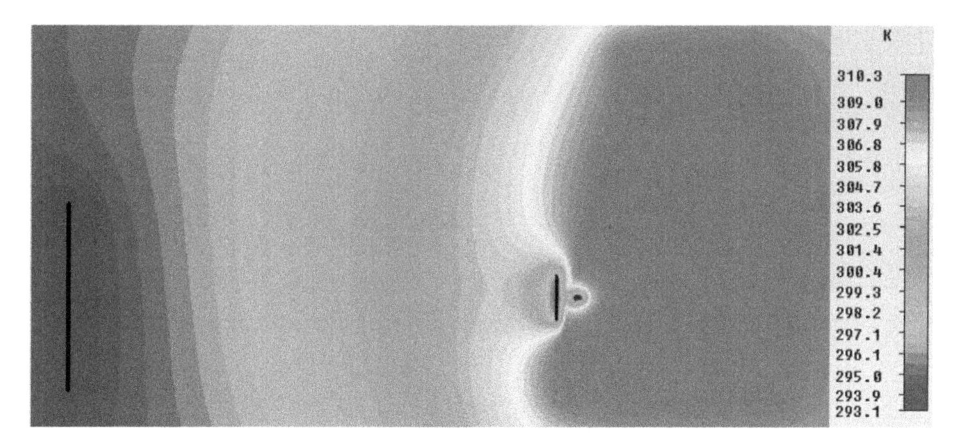

Figure 8.20 Simulated temperature variation of the wireless power link with parasitic patch. Source: Adapted from Wheeler [15].

Table 8.2 Performance comparisons of the wireless power link with/without the parasitic patch at 2.45 GHz.

Input power = 1 W	Without parasitic patch	With parasitic patch
1 g-avg SAR (W/kg)	0.142	0.307
10 g-avg SAR (W/kg)	0.0463	0.145
$\lvert S_{21} \rvert$ (dB)	−54.2	−41.3
Temperature increase (°C)	0.15	0.178

Source: Liu et al. [2]/IEEE.

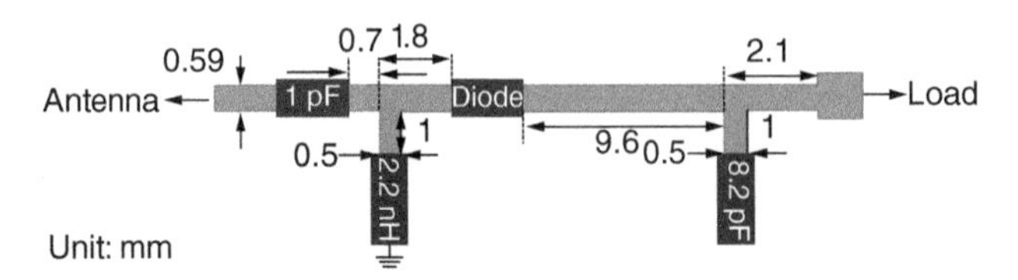

Figure 8.21 Schematic of a rectifier. Source: Liu et al. [2]/IEEE.

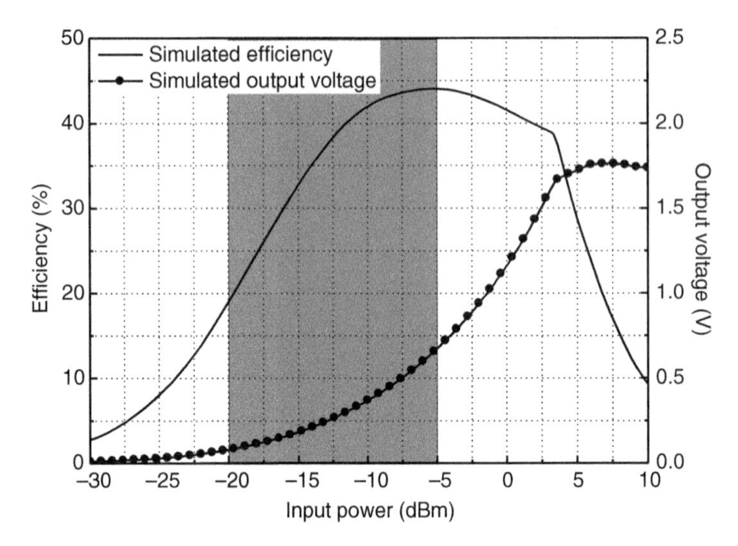

Figure 8.22 Simulated results of the proposed rectifier. Source: Liu et al. [2]/IEEE.

−11 dBm at a central operating frequency of 2.45 GHz can be established when the distance is 0.5 m with the parasitic patch.

The design of the rectifier circuit needs to be optimized to achieve acceptable conversion efficiency when the input power falls within the range of −20 to −10 dBm. We chose the Schottky HSMS-2852 diode due to its low built-in voltage, fast switching response, and high cutoff frequency. The Advanced Design System (ADS) software was utilized to simulate the rectifier. Figure 8.21 showcases the schematic of the detailed design circuit. The source impedance of the rectifier is set at 50 Ω, allowing easy integration with the proposed implantable antenna without impedance mismatch. Figure 8.22 presents the simulated conversion efficiency and output voltage values of the rectifier as functions of input power level with a 3250-Ω resistor. The simulation results demonstrated that the rectifier circuit can achieve a conversion efficiency of approximately 19% at −20 dBm and about 42% at −10 dBm.

8.3.3.2 Integration Solution of the Implantable Rectenna

Upon separately designing the antenna and the rectifier circuit, these components must be integrated to form a rectenna.

In this study, we utilized an interconnecting via to link the implantable antenna with the rectifier, with both components sharing the same ground plane. The configuration for

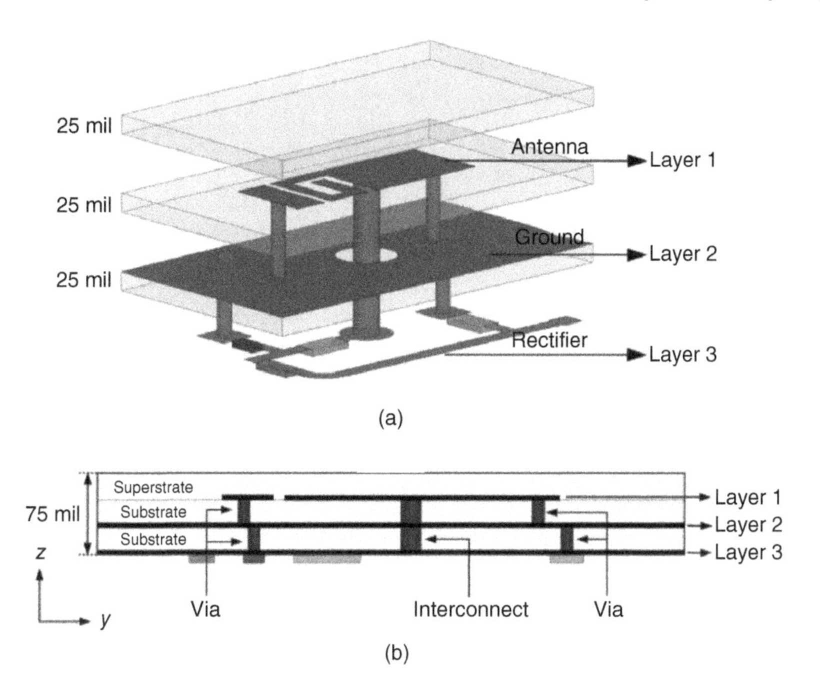

Figure 8.23 Configurations for integrating the rectifier as a rectenna: (a) geometry of the rectenna, (b) side view of the rectenna.

integrating the rectifier into a rectenna is illustrated in Figure 8.23. Note that at the 2.45 GHz ISM-Band, the use of the interconnecting via results in a negligible transition loss.

8.3.3.3 Measurement Setup

The photograph of the constructed rectenna is provided in Figure 8.24. The wireless power transfer measurement setup is depicted in Figure 8.25, where the proposed rectenna is embedded in minced pork and a horn antenna, with a peak gain of 7.6 dBi at 2.5 GHz, is positioned 0.3 m from the edge of the pork. We then measure the DC output voltage (V_{dc}) of the rectenna and subsequently calculate the total PTE.

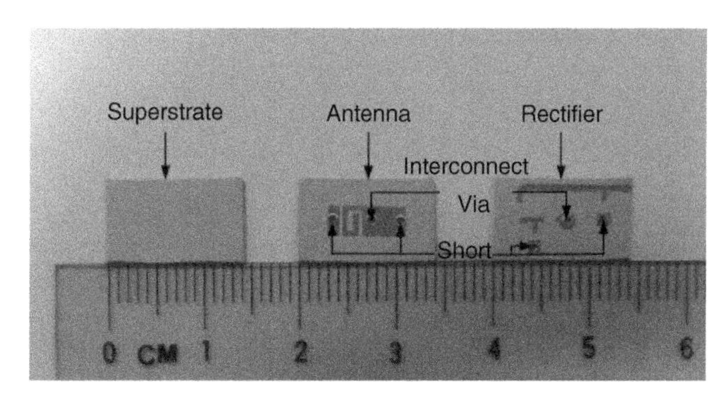

Figure 8.24 Photograph of fabricated rectenna. Source: Liu et al. [2]/IEEE.

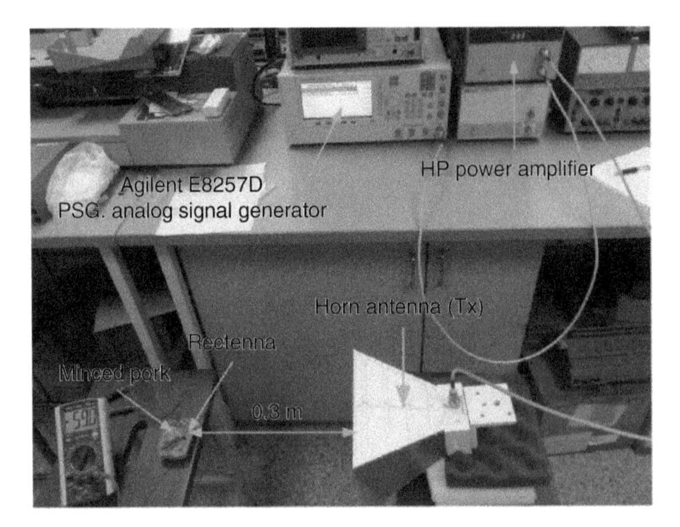

Figure 8.25 Measurement setup for wireless power transfer. Source: Liu et al. [2]/IEEE.

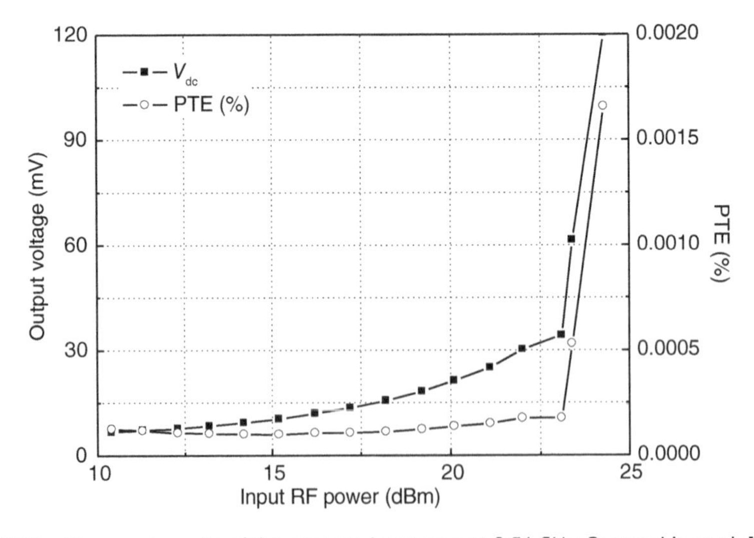

Figure 8.26 Measured results of the proposed rectenna at 2.51 GHz. Source: Liu et al. [2]/IEEE.

Detailed measurements at 2.51 GHz are recorded in Figure 8.26. It is important to note that in these measurements, the maximum input power of the horn antenna was limited to 24.3 dBm by our power amplifier. Comparison with simulated data (Figures 8.10 and 8.22, $|S_{21}| = \sim-50$ dB when $d = 0.3$ m, $V_{dc} = 9.4$ mV and 89 mV when input power is −30 dBm and −20 dBm, respectively) reveals only minor discrepancies between the simulated and measured output voltages when the input power level ranges between 10 and 25 dBm. Other factors such as variations in lumped elements such as capacitors and inductors, as well as fabrication errors, may lead to frequency shifts and reduced efficiency of the rectenna. To address this issue, distributed elements can be applied to design a rectifier with more precision, but the dimensions of the rectifier to be implanted alongside the antenna in the human body must be considered. The total power transmission efficiency is quite low due to substantial free space loss and biological tissue loss.

Note that the parasitic patch was not used in this measurement because there was no large human body to serve as a reflector. However, based on the simulated results in a human body model in CST (Figure 8.18), we anticipate a significant improvement in the received power when the proposed rectenna is implanted in the human body and the parasitic patch is placed over it.

8.4 WPT Antenna Misalignment: An Antenna Alignment Method Using Intermodulation

The efficiency of WPT for far-field antennas in IMDs is often undermined by misalignments in both antenna polarization and radiation direction. This necessitates the creation of a universal method to effectively rectify these misalignment issues specifically tailored for IMD applications.

In this section, we introduce an innovative method for antenna alignment that leverages intermodulation to address the misalignment concerns in WPT antennas in a manner that prioritizes safety [3]. Traditionally, the common solution to meet the power requirements of IMDs involves increasing the incident power density, which may raise safety issues for surrounding human tissues. Instead, our proposed method uses two-tone (2T) waveform excitation to augment rectification while simultaneously generating intermodulation, as shown in Figure 8.27. This intermodulation power is then fed back via a magnetic resonant coupling link.

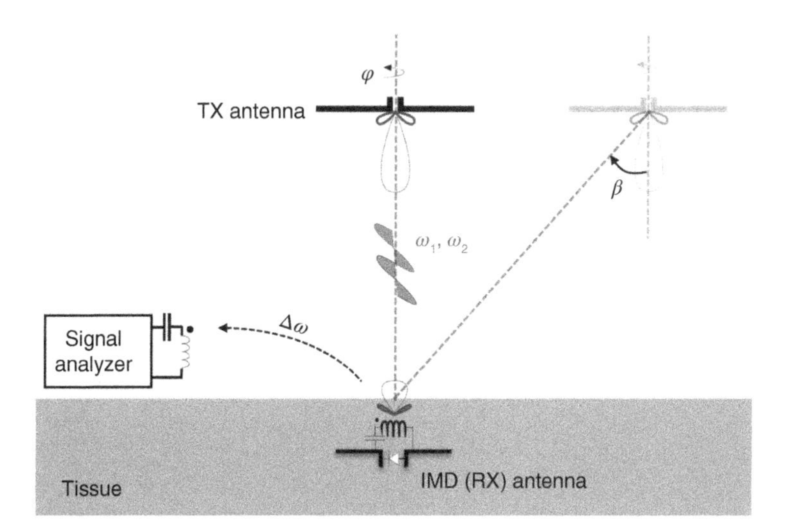

Figure 8.27 Proposed intermodulation-based WPT system for addressing the antenna polarization and radiation direction misalignments. Source: Zhang et al. [3]/IEEE.

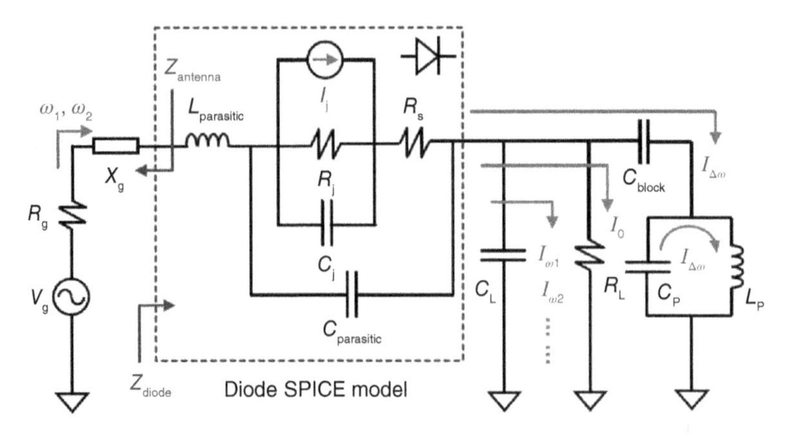

Figure 8.28 Small-signal equivalent circuit for single series-mounted rectifier under a 2T waveform excitation. Source: Zhang et al. [3]/IEEE.

8.4.1 Operation Mechanism

In this section, we will provide a theoretical analysis of RF-to-dc power conversion efficiency (PCE) enhancement and intermodulation generation under a 2T waveform excitation. Subsequently, we will explore the relationship between the generation of intermodulation and antenna misalignments.

8.4.1.1 PCE Enhancement and Intermodulation Generation

Figure 8.28 showcases a small-signal equivalent circuit for a singular series-mounted rectifier under the influence of a two-tone waveform excitation. The RX antenna and the rectifier's matching circuit can be modeled as a voltage source with an internal resistance R_g and a reactance X_g. The Schottky diode is depicted via its SPICE circuit model, which includes a parasitic inductance, Lparasitic, a parasitic capacitance, Cparasitic, a series resistance R_s, a small-signal junction capacitance C_j, a small-signal junction resistance R_j, and a current source I_j. A low-pass filter (LPF), composed of a load capacitance C_L and a load resistance R_L, is employed to stabilize the DC output voltage derived from the diode's nonlinear junction. It is important to note that C_L must be meticulously selected to short-circuit all harmonics [39], while keeping the DC and intermodulation generations unaffected.

Assuming that P_{rx} represents the received power of the 2T waveform excitation at frequencies ω_1 and ω_2, the corresponding generated voltage, V_g, can be expressed as follows, based on references [40, 41]:

$$V_g = \sqrt{4R_g P_{rx}}(\cos \omega_1 t + \cos \omega_2 t) \tag{8.12}$$

According to the Thévenin equivalent circuit theory, V_j has the following relationship with V_g:

$$V_j = \frac{Z_j}{R_s + Z_j} \cdot \frac{Z_{diode}}{Z_{antenna} + Z_{diode}} V_g = RV_g \tag{8.13}$$

where:

$$Z_{\text{diode}} = j\omega L_{\text{parasitic}} + \frac{1}{j\omega C_{\text{parasitic}} + \frac{1}{R_s + Z_j}} \tag{8.14a}$$

$$Z_{\text{antenna}} = R_g + jX_g \tag{8.14b}$$

$$Z_j = \frac{1}{\frac{1}{R_j} + j\omega C_j} \tag{8.14c}$$

and R is a constant for small signal [40, 41].

This junction voltage V_j generates I_j comprising DC I_0, intermodulation current $I_{\Delta\omega}$, and all harmonics. The relation between I_j with V_j satisfies:

$$I_j(V_j) = I_s(e^{\frac{qV_j}{nkT}} - 1) \tag{8.15}$$

where I_s is the saturation current, q is the elementary charge, n is an ideality factor, k is Boltzmann's constant, and T is the temperature. Typically, the current source I_j can be expanded using the Taylor expansion [42] at $Vj = 0$ for small signal. A fourth-order Taylor expansion [40, 41] is considered as it is enough to demonstrate the nonlinear rectification:

$$I_j \approx \alpha I_s V_j + \frac{\alpha^2 I_s}{2} V_j^2 + \frac{\alpha^3 I_s}{6} V_j^3 + \frac{\alpha^4 I_s}{24} V_j^4 \tag{8.16}$$

where $\alpha = q/nkT$. Since odd-order terms do not contribute to DC, only even-order terms are considered. By substituting (8.12) and (8.13) into (8.16), the DC I_0 can be calculated with:

$$I_0 = 2\alpha^2 I_s \mathrm{R}^2 R_g P_{\text{rx}} + \frac{3}{2}\alpha^4 I_s \mathrm{R}^4 R_g^2 P_{\text{rx}}^2 \tag{8.17}$$

To evaluate the PCE enhancement of a 2T waveform excited rectifier, the same P_{rx} continuous waveform (CW) excitation is thus considered, whose corresponding voltage is $V'g$:

$$V_g' = \sqrt{8R_g P_{\text{rx}}} \cos \omega_0 t \tag{8.18}$$

where $\omega_0 = (\omega 1 + \omega 2)/2$. By substituting (8.13) and (8.18) into (8.16), another DC I'_0 under CW excitations is:

$$I_0' = 2\alpha^2 I_s \mathrm{R}^2 R_g P_{\text{rx}} + \alpha^4 I_s \mathrm{R}^4 R_g^2 P_{\text{rx}}^2 \tag{8.19}$$

By comparing (8.17) and (8.19), it can tell that the DC generation is enhanced under the 2T waveform excitation. Since the load resistance R_L is fixed, the PCE of $I_0{}^2 R_L / P_{\text{rx}}$ enhancement is thus proved.

As illustrated in Figure 8.28, all harmonics are shorted by the load capacitance C_L. The intermodulation is chosen by a DC-block capacitance C_{block} and then drives a primary resonator (L_p and C_p). Similar to the DC generation, only even-order terms from (8.16) generate the intermodulation:

$$I_{\Delta\omega} = \left(4\alpha^2 I_s \mathrm{R}^2 R_g P_{\text{rx}} + \frac{8}{3}\alpha^4 I_s \mathrm{R}^4 R_g^2 P_{\text{rx}}^2\right) \cos \Delta\omega t \tag{8.20}$$

which shows a monotonic relation between the intermodulation current $I_{\Delta\omega}$ and P_{rx}.

8.4.1.2 Relation Between Intermodulation and Misalignments

For the alignment of an IMD antenna with the transmitting antenna, we need to establish a relationship between the intermodulation generation and the degree of antenna misalignment. Both radiation direction and polarization misalignment are considered and denoted by angles β and φ, respectively (refer to Figure 8.27). The vertical distance between the transmitting and IMD antennas remains constant. The antenna radiation pattern is a function of the deviation angle β, and for a directive antenna, it typically decreases as β increases. Also, a tilt angle φ between the transmitting and IMD antennas can result in polarization mismatch, reducing the received power by a ratio of $\cos^2\varphi$ [43, 44]. Thus, we can establish the far-field wireless power link between two misaligned antennas.

$$P_{\text{rx}}(\beta, \varphi) = P_{\text{tx}} \frac{\lambda^2}{(4\pi h)^2} G_{\text{tx}}(\beta) G_{\text{rx}}(\beta) \cos^2\beta \cos^2\varphi \tag{8.21}$$

where λ is the wavelength, h is the fixed vertical distance, Ptx is the TX power, Gtx(β) and Grx(β) are the antenna gains, and the subscripts "tx" and "rx" represent the TX and RX (IMD) antennas, respectively.

8.4.2 Miniaturized IMD Rectenna Design With NRIC Link

In this section, we design an IMD rectenna system alongside an NRIC link to demonstrate the proposed antenna alignment method. The system is divided into three components: (i) a miniaturized rectifier with an intermodulation readout; (ii) a co-designed IMD antenna and rectifier circuit; and (iii) the establishment of an NRIC link.

8.4.2.1 Miniaturized Rectifier With Intermodulation Readout

Figure 8.29(a) depicts a single series-mounted rectifier, which is composed of a low-threshold SMS7630 Schottky diode, a matching network (L_{m} and C_{m}), and an LPF (C_{L} and R_{L}). The values of the lumped elements, L_{m} and C_{m}, are optimized using the ADS to ensure a minimized circuit size and maximized incident power to the diode.

The load capacitance, C_{L}, is utilized to short-circuit all the harmonics without impacting the DC and intermodulation power. To segregate the DC and intermodulation currents from the harmonics, a DC-block capacitance (C_{block}) is introduced after the load resistance (R_{L}). The intermodulation power can then be measured from port 2 (P2). The rectifier is designed and fabricated on a high-dielectric 25-mil RO3010 substrate ($\varepsilon_{\text{r}} = 11.2$ and $\tan\delta = 0.0022$). Figure 8.29(b) illustrates the final circuit layout, with an inset showing the printed circuit board (PCB) realization.

The rectifier's performance is initially evaluated under CW excitation at 2.45 GHz. A signal generator at P1 produces an incident power ranging from −20 to −11 dBm, representing the received power P_{rx} from the IMD antenna. The output DC voltage (V_{o}) across the load resistance (R_{L}) is both simulated and measured. The RF-to-dc PCEs of the rectifier are computed by the following:

$$\text{PCE}\ (\%) = \frac{V_{\text{o}}^2}{R_{\text{L}}} \times \frac{1}{P_{\text{rx}}} \times 100\% \tag{8.22}$$

Figure 8.30(a) presents a comparison of the simulated and measured PCEs and V_{o}, where a significant consistency can be observed. A minimum measured efficiency of 10% at an

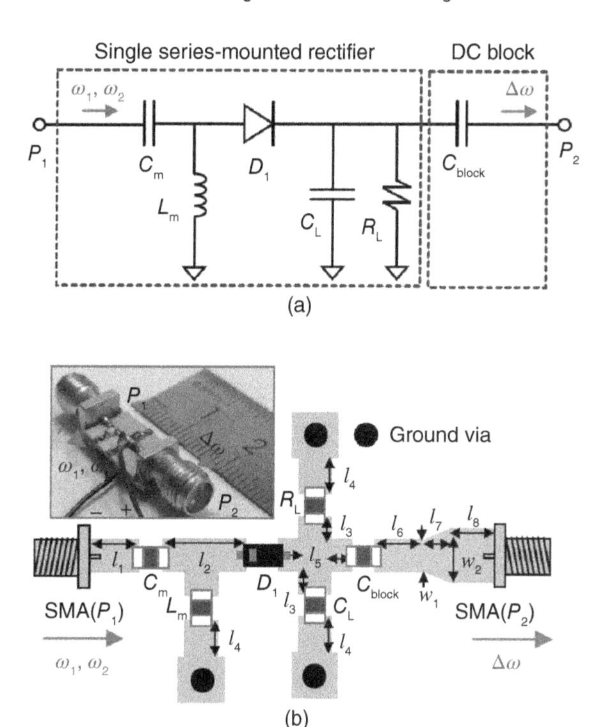

Figure 8.29 (a) Schematic of the single series-mounted rectifier with port 1 (P1) for 2T waveform excitation and port 2 (P2) for intermodulation readout, (b) layout of the rectifier. The design parameters are: w1 = 0.89 mm, w2 = 1.5 mm, l1 = 1.615 mm, l2 = 2.125 mm, l3 = l5 = 0.615 mm, l4 = 1 mm, l6 = 1.837 mm, l7 = 0.68 mm, l8 = 1.5 mm, Cm = 0.5 pF, Lm = 2.2 nH, CL = 100 pF, RL = 2 kΩ, and C_{block} = 4.7 μF. Inset: PCB realization. Source: Zhang et al. [3]/IEEE.

incident power of −20 dBm is achieved, verifying that the designed rectifier can operate at very low incident power levels. The discrepancy (~5%) between simulated and measured PCEs may arise from cable losses and fabrication variations. To assess the frequency bandwidth of the rectifier, DC output voltages at 2.44 and 2.46 GHz are also measured, as illustrated in Figure 8.30(a). The overlap of output DC voltages (V_o) indicates a bandwidth of over 20 MHz for the rectifier. Thus, the designed rectifier can properly function under a 2T waveform excitation with a 4-MHz frequency spacing.

A 2T waveform excitation at 2.448 and 2.452 GHz serves as the incident power, its power level identical to that of the CW excitation at 2.45 GHz. The intermodulation power is both simulated and measured at P2, as shown in Figure 8.30(b). For comparison purposes, the simulated and measured V_o under the CW excitation at 2.45 GHz are also provided in Figure 8.30(b). A good alignment between the simulated and measured intermodulation can be observed. The minor discrepancy may arise from cable losses, process variation, as well as the possible inaccuracy of the diode model in ADS. It is demonstrated that the intermodulation power increases steadily with the incident power (P_{rx}), which is consistent with the theoretical analysis in Eq. (8.20). Furthermore, both simulation and measurement reveal an enhancement of V_o due to 2T waveform excitation, as predicted by Eqs. (8.17) and (8.19).

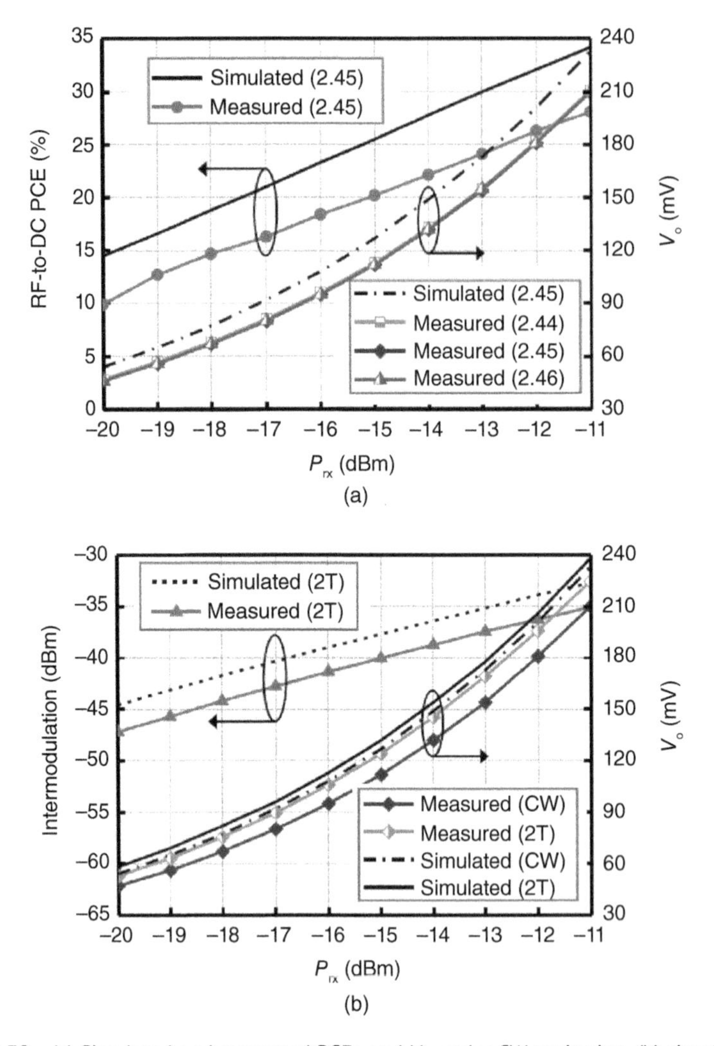

Figure 8.30 (a) Simulated and measured PCEs and V_o under CW excitation, (b) simulated and measured intermodulation power and V_o under CW and 2T excitations. "CW" represents the continuous waveform, while "2T" represents the 2T waveform [3].

8.4.2.2 IMD Antenna Codesigned With Rectifier Circuit

The IMD antenna investigated in this study is a patch antenna, as illustrated in Figure 8.31. A homogeneous cubic phantom is used to represent human tissue. Figure 8.31(a) depicts the simulation domain, where the IMD antenna is implanted at a depth of 2 mm. All simulations are carried out in CST, using human tissue electrical properties at 2.45 GHz of $\varepsilon_r = 52.729$ and $\sigma = 1.7388$ S/m.

The IMD antenna is designed on the high-dielectric 25-mil RO3010 for miniaturization purposes. The rectifier circuit is directly connected to the IMD antenna via a transmission line (wf = 0.89 mm and lf = 2 mm) with a matching network. The patch antenna is co-optimized with the rectifier in CST. Figure 8.31(b) presents the top and side views of

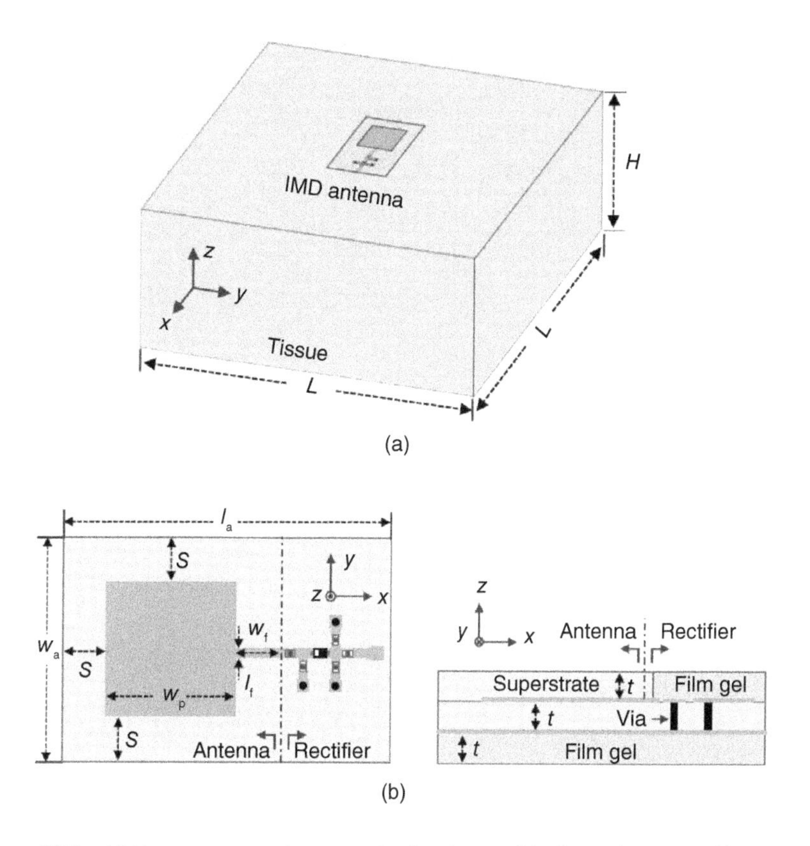

Figure 8.31 (a) Homogeneous phantom of cube shape with dimensions: $L = 100$ mm and $H = 27$ mm, (b) top and side views of the IMD patch antenna layout with its structure characteristics: lf = 2 mm, wf = 0.89 mm, $s = 2$ mm, wp = 15 mm, wa = 17 mm, la = 23 mm, and $t = 0.635$ mm. Source: Zhang et al. [3]/IEEE.

the IMD antenna layout. Given the high conductivity of human tissue, the patch antenna needs to be encapsulated with a biocompatible insulation layer. As shown in Figure 8.31(b), a superstrate of RO3010 is placed on top of the antenna. Meanwhile, polydimethylsiloxane is utilized as the biocompatible thin-film gel to insulate the IMD rectenna system from human tissue.

Figure 8.32 depicts the reflection coefficients and realized gains. There is a good agreement between the simulated and measured reflection coefficients. Moreover, a bandwidth spanning from 2.4 to 2.5 GHz is achieved, which should suffice for receiving the two-tone waveform excitation. The maximum antenna gain, approximately −7.5 dBi, is obtained at 2.45 GHz. Notably, the curve of the realized gain within the frequency range of interest from 2.44 to 2.46 GHz is quite flat, enabling identical interception of the two-tone waveform excitation.

Figure 8.33 illustrates the simulated radiation patterns. The minor cross-polarization, as shown in Figure 8.33(b), signifies the excellent linearity of the IMD antenna, making it an appropriate choice for demonstrating the proposed antenna alignment method. It

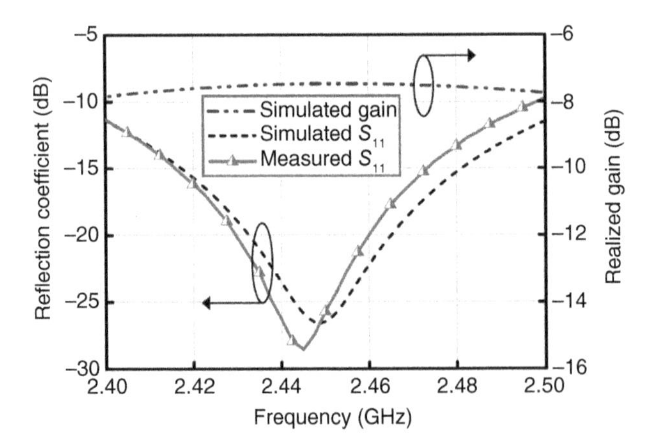

Figure 8.32 Reflection coefficients and simulated realized gain of IMD antenna. Source: Zhang et al. [3]/IEEE.

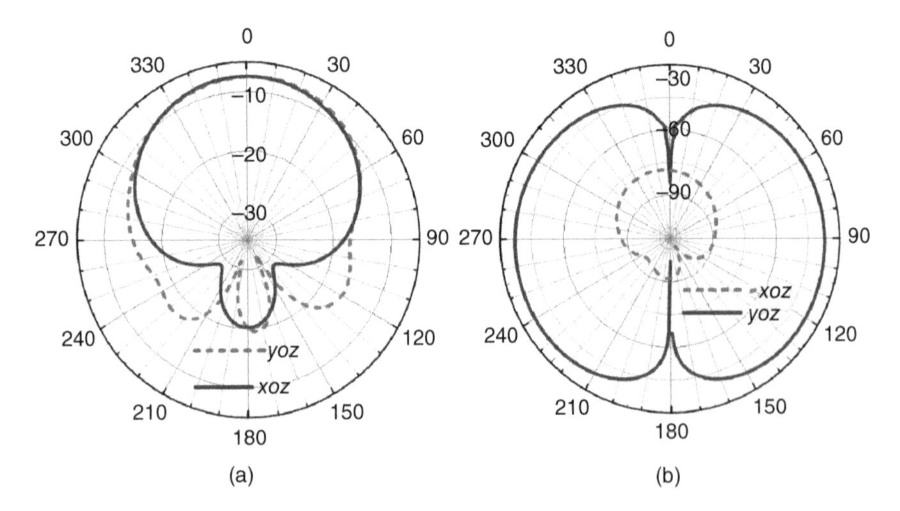

Figure 8.33 Simulated radiation patterns: (a) co-polarization, (b) cross-polarization. Source: Zhang et al. [3]/IEEE.

should be noted that the proposed method is also applicable to antennas with higher cross-polarization, such as PIFAs, because it aligns antennas by searching for the maximum intermodulation that corresponds to the direction of the peak co-polarization for L_P antennas.

8.4.2.3 NRIC Link Establishment

To enable the feedback of intermodulation, we employ a non-radiative inductive coupling (NRIC) link, the equivalent circuit of which is presented in Figure 8.34(a). The efficiency of the NRIC link between the primary and secondary coils is constrained by factors such as the operational intermodulation frequency ($\Delta\omega$), the mutual coupling coefficient (k), the self-inductances (L_p and L_s), and the parasitic resistances (r_p and r_s) [45, 46]. These can be

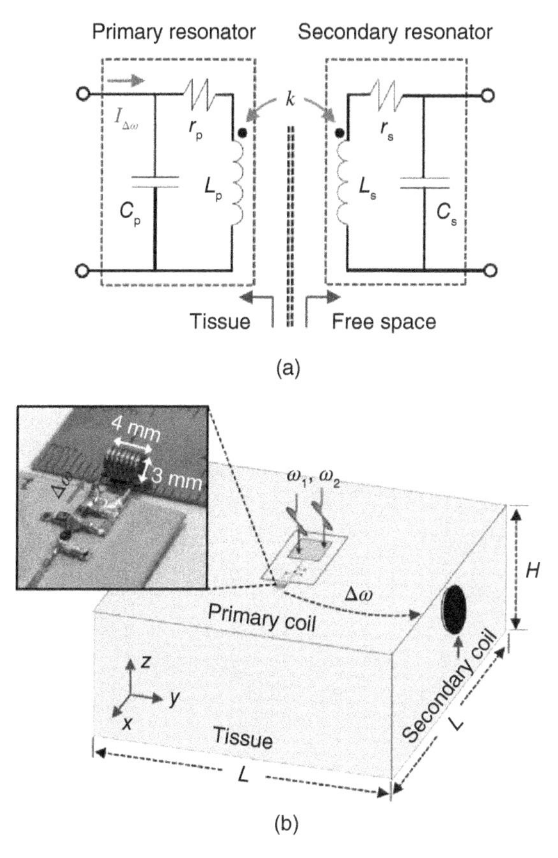

Figure 8.34 (a) Equivalent circuit of the NRIC link, (b) arrangement of the NRIC link. Source: Zhang et al. [3]/IEEE.

calculated using the following equation:

$$\eta = \frac{(\Delta\omega)^2 k^2 L_p L_s}{r_p r_s} = k^2 Q_p Q_s \tag{8.23}$$

where subscripts "*p*" and "*s*" denote the primary and the secondary coils, respectively, Q_p and Q_s are the quality factors.

Given the substantial distance separating the primary (internal) and secondary (external) resonators in realistic IMD applications, the mutual coupling coefficient (k) is notably weak. As a result, it is critical to carefully design and select primary and secondary coils to counterbalance the decrease in link efficiency brought on by this weak mutual coupling. As depicted in the inset of Figure 8.34(b), we fabricated a seven-turn solenoidal coil using a 0.5-mm-diameter copper wire. To enhance the quality factor (Q_p) of the size-constrained primary coil, we inserted a cylindrical ferromagnetic core. Consequently, the primary coil was measured to have an augmented inductance of around 0.31 μH ($r_p = 6.5\,\text{m}\Omega$). It was soldered onto the IMD rectenna, as displayed in Figure 8.34(b). We utilized a commercial coil from Wurth Electronics [47] as the secondary coil, which features two layers of 14-turn planar coils (with an outer diameter of 20 mm, and an inner diameter of 10 mm) along with a ferromagnetic disk (diameter is 20.5 mm). The secondary coil exhibits an inductance of 5.86 μH ($r_s = 0.12\,\Omega$).

Since the mutual coupling coefficient (k) is very weak, it has little effect on the resonant frequency. Both coils are tuned by using capacitances (C_p and C_s) at the intermodulation frequency ($\Delta\omega$) of 4 MHz

$$\Delta\omega = \frac{1}{2\pi\sqrt{L_p C_p}} = \frac{1}{2\pi\sqrt{L_s C_s}}. \tag{8.24}$$

When $I\Delta\omega$ drives the primary coil, the intermodulation power can be broadcasted through the human tissue and captured by the external secondary coil. It is worth mentioning that a high intermodulation frequency may enhance the link efficiency according to (8.13); however, a higher frequency also leads to higher loss when penetrating the human tissue. Hence, there is a trade-off between the NRIC link efficiency and the penetration loss.

8.4.3 Experimental Validation

8.4.3.1 Experimental Setup

A Rohde & Schwarz SMW200A signal generator and a ZVL spectrum analyzer are employed to produce a 2T waveform excitation at 2.448 and 2.452 GHz ($\Delta\omega = 4$ MHz), and to detect the intermodulation power, respectively. A 2×2 patch antenna array, designed on a 60-mil RO4003C substrate, serves as the TX antenna. Positioned at a height of $h = 400$ mm above the IMD rectenna, it provides a bandwidth from 2.43 to 2.46 GHz and achieves a gain exceeding 10 dBi within this bandwidth. Figure 8.35 presents the TX antenna's simulated radiation patterns at 2.448 and 2.452 GHz. It should be noted that the proposed system is also compatible with broadband or less-directive TX antennas, provided the antenna radiation patterns at 2Ts do not significantly differ.

The IMD rectenna system is implanted 2-mm deep in a cubic container filled with minced pork, with its maximum radiation direction aligned with the z-axis. Using the HP 85070A dielectric probe kit, the relative permittivity and conductivity of the minced pork are measured and found to be $\varepsilon_r = 50.2$ and $\sigma = 1.67$ S/m at 2.45 GHz. These values closely match the $\varepsilon_r = 52.729$ and $\sigma = 1.7388$ S/m used in CST simulations. The secondary coil is positioned at the external surface of the container.

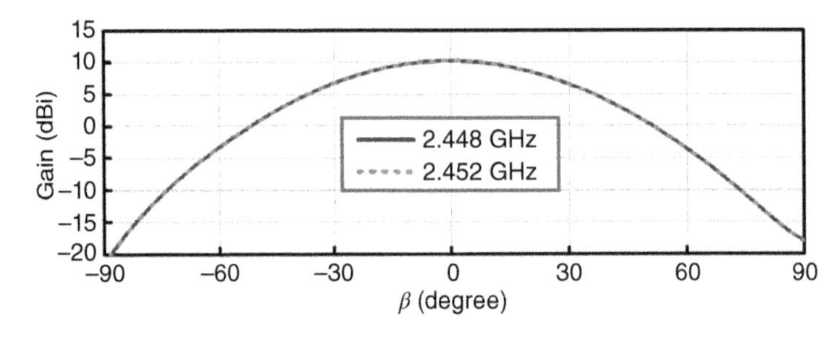

Figure 8.35 Simulated radiation patterns at 2.448 and 2.452 GHz of the TX antenna. Source: Zhang et al. [3]/IEEE.

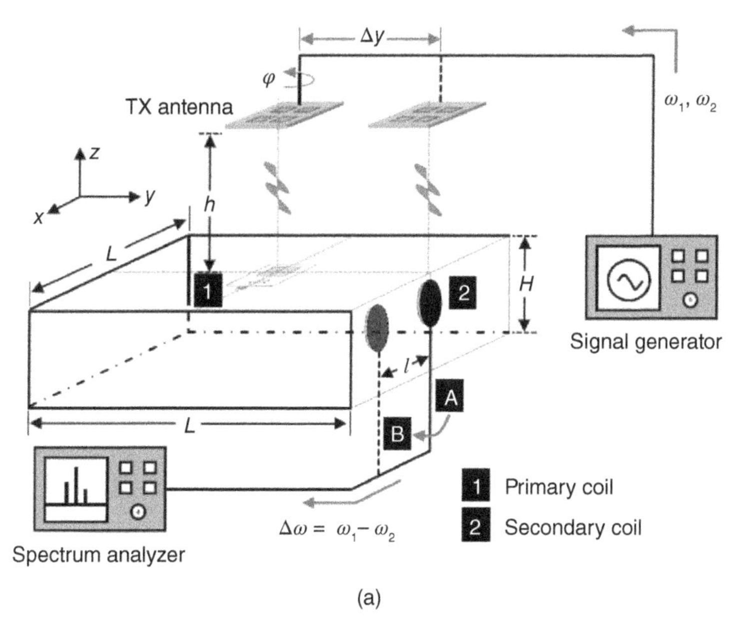

(a)

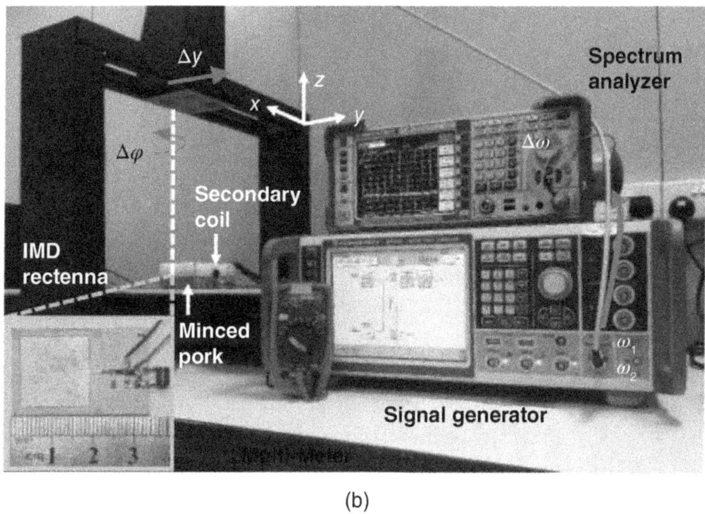

(b)

Figure 8.36 (a) Schematic, (b) realization of the experimental setup. Source: Zhang et al. [3]/IEEE.

Figure 8.36 illustrates the detailed experimental setup, and Table 8.3 outlines the experimental parameters. The free-space path loss (FSPL) is calculated as follows:

$$\text{FSPL(dB)} = -20\log\left(\frac{\lambda}{4\pi h}\right) \tag{8.25}$$

where λ is the wavelength and h is the height between the TX and IMD antennas.

When the TX antenna transmits at a power level of 15 dBm, the IMD rectenna is capable of receiving a power level of −14.8 dBm. This corresponds to a power density of 0.157 W/m^2, significantly below the maximum permissible exposure limit of 10 W/m^2 set

Table 8.3 Parameters in experiment.

Quantity	Specification
Phantom dimensions	$L = 100$ mm, $H = 27$ mm
Transmitted power	$P_{ts} = 15$ dBm
TX antenna gain	$G_{rs} = 10$ dBi
Free space path loss	FSPL - 32 dB
IMD antenna gain	$G_{cs} = -7.5$ dBi
Received power by rectifier	$P_{rs} \sim -14.8$ dBm
Primary resonator	$L_p = 0.31$ uH, $C_p = 5100$ pF
Secondary resonator	$L_s = 5.86$ uH, $C_s = 270$ pF

Source: Zhang et al. [3]/IEEE.

by the FCC [25, 27]. Given that the measured intermodulation power required to drive the primary coil is −40 dBm at $P_{rx} = -15$ dBm [see Figure 8.30(b)] when the TX transmits approximately 15 dBm of power, the corresponding average SAR near the primary coil is about 0.28 W/kg. This is notably lower than both the 1-g average SAR limit of 1.6 W/kg [25] and the 10-g average SAR limit of 2 W/kg [27].

8.4.3.2 Results and Discussion

In order to validate the proposed system, we conducted experiments to analyze the relationship between the intermodulation power and the degree of both antenna misalignments, as depicted in Figure 8.36. For the antenna polarization misalignment experiment, we varied the tilt angle φ from 0° to 90° in increments of 15°, while maintaining alignment of the radiation directions. The radiation direction misalignment experiment was carried out by shifting the TX antenna's position along the *y*-axis, with polarizations kept aligned. The deviation distance Δy ranged from 0 to 25 cm in 5 cm increments, sufficient for practical applications.

Intermodulation power was first sensed by the secondary coil located at position A, where the two coils are face-to-face. Output voltages (V_0) were also measured to display the rectified DC power level of the IMD rectenna. As shown in Figure 8.37(a), both V_0 and the intermodulation component decrease as the tilt angle φ increases; intermodulation power drops from −80 to −95 dBm as φ increases from 0° to 90°, following a function of $\cos 2\varphi$, consistent with (10). As depicted in Figure 8.37(b), both the intermodulation power and V_0 also decrease monotonically as the deviation distance Δy increases. Given that both the antenna gain and the deviation angle β impact the received power P_{rx} of the IMD rectenna in (10), the intermodulation and V_0 decrease more rapidly than the prior cases in Figure 8.37(a). These findings affirm that both antenna polarization and radiation direction misalignments can be addressed by seeking maximum intermodulation.

To further showcase the robustness of the proposed system, the secondary coil was moved from positions A to B, a distance of $l = 40$ mm along the *x*-axis, while other parameters remained unchanged. Observations from Figure 8.37 suggest that when the tilt angle exceeds 60° or the deviation distance surpasses 15 cm, distinguishing the intermodulation power from the noise floor becomes challenging. To improve the resolution, employing a higher Q and a larger secondary coil could offset the link efficiency decline resulting from

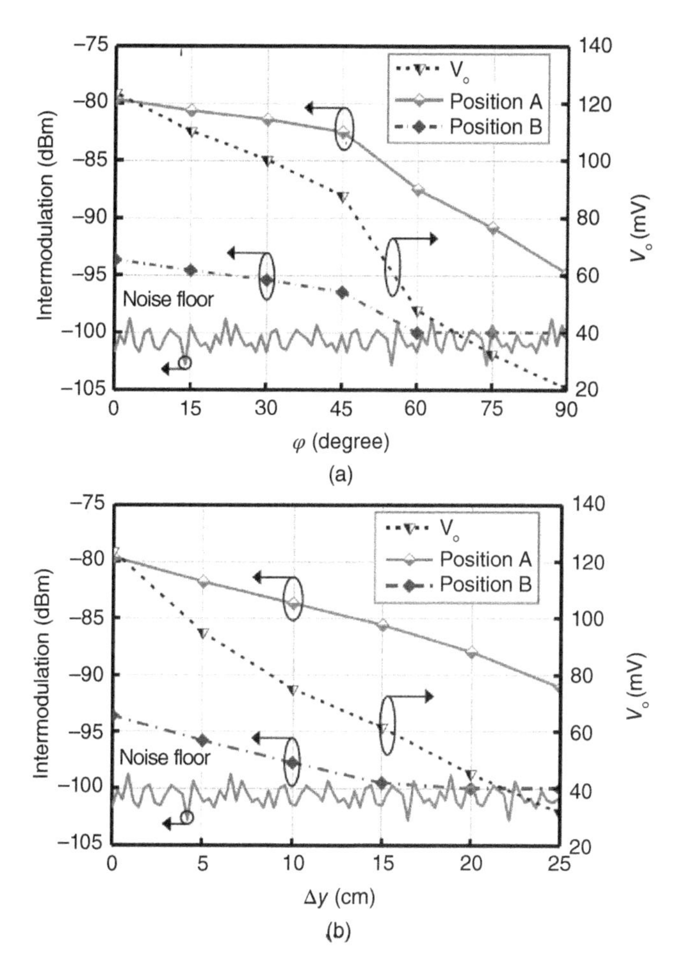

Figure 8.37 Intermodulation and output voltages measurements: (a) Polarization, (b) radiation direction misalignments. Source: Zhang et al. [3]/IEEE.

weak mutual coupling. In fact, when intermodulation is sufficiently strong, the secondary coil can be positioned anywhere near the phantom, provided it does not interfere with the TX and IMD antennas.

8.5 Summary

In summary, this chapter has shed light on the key aspects of far-field wireless charging for IMDs. We have discussed methods to improve charging efficiency, including the use of parasitic patches on the human body and the antenna alignment method using intermodulation. These techniques aim to enhance the power reception level and address efficiency issues caused by misalignments in antenna polarization and radiation direction. The proposed approaches provide safe and effective solutions to optimize wireless power transfer for IMDs, opening up possibilities for their practical implementation in healthcare applications.

References

1 Agarwal, K., Jegadeesan, R., Guo, Y.X., and Thakor, N.V. (2017). Wireless power transfer strategies for implantable bioelectronics. *IEEE Rev. Biomed. Eng.* 10: 136–161.
2 Liu, C., Guo, Y.-X., Sun, H., and Xiao, S. (2014). Design and safety considerations of an implantable rectenna for far-field wireless power transfer. *IEEE Trans. Antennas Propag.* 62 (11): 5798–5806.
3 Zhang, H., Gao, S.P., Ngo, T. et al. (2019). Wireless power transfer antenna alignment using intermodulation for two-tone powered implantable medical devices. *IEEE Trans. Microw. Theory Tech.* 67 (5): 1708–1716.
4 Balanis, C.A. (2005). *Antenna Theory: Analysis and Design*, 1. Wiley.
5 Pozar, D.M. (2009). *Microwave Engineering*. Wiley.
6 Warty, R., Tofighi, M.-R., Kawoos, U., and Rosen, A. (2008). Characterization of implantable antennas for intracranial pressure monitoring: reflection by and transmission through a scalp phantom. *IEEE Trans. Microw. Theory Tech.* 56 (10): 2366–2376.
7 Falkenstein, E., Roberg, M., and Popovic, Z. (2012). Low-power wireless power delivery. *IEEE Trans. Microw. Theory Tech.* 60 (7): 2277–2286.
8 Agarwal, K. et al. (2013). Highly efficient wireless energy harvesting system using metamaterial based compact CP antenna. *Proc. 2013 IEEE MTT-S Int. Microw. Symp. Digest* 1–4.
9 Shinohara, N. and Matsumoto, H. (1998). Experimental study of large rectenna array for microwave energy transmission. *IEEE Trans. Microw. Theory Tech.* 46 (3): 261–268.
10 Agarwal, K. and Guo, Y.-X. (2015). Interaction of electromagnetic waves with humans in wearable and biomedical implant antennas. *Proc. Asia-Pacific Symp. Electromagn. Compat.* 154–157.
11 Bercich, R. et al. (2013). Far-field rf powering of implantable devices: safety considerations. *IEEE Trans. Biomed. Eng.* 60 (8): 2107–2112.
12 Chow, E.Y., Chlebowski, A.L., Chakraborty, S. et al. (2010). Fully wireless implantable cardiovascular pressure monitor integrated with a medical stent. *IEEE Trans. Biomed. Eng.* 57 (6): 1487–1496.
13 Marnat, L., Ouda, M., Arsalan, K. et al. (2012). On-chip implantable antennas for wireless power and data transfer in a glaucoma-monitoring SoC. *IEEE Antennas Wireless Propag. Lett.* 11: 1671–1674.
14 Huang, F.-J., Lee, C.-M., Chang, C.-L. et al. (2011). Rectenna application of miniaturized implantable antenna design for triple-band biotelemetry communication. *IEEE Trans. Antennas Propag.* 59 (7): 2646–2653.
15 Wheeler, H.A. (1975). Small antennas. *IEEE Trans. Antennas Propag.* AP-23 (4): 462–469.
16 Chien, T.-F., Cheng, C.-M., Yang, H.-C. et al. (2010). Development of nonsuperstrate implantable low-profile CPW-fed ceramic antennas. *IEEE Antennas Wireless Propag. Lett.* 9: 599–602.
17 Liu, C., Guo, Y.-X., and Xiao, S. (2012). A hybrid patch/slot implantable antenna for biotelemetry devices. *IEEE Antennas Wireless Propag. Lett.* 11: 1646–1649.

18 Xu, L.-J., Guo, Y.-X., and Wu, W. (2014). Miniaturized dual-band antenna for implantable wireless communications. *IEEE Antennas Wireless Propag. Lett.* 13: 1160–1163.

19 Liu, C., Guo, Y.-X., and Xiao, S. (2014). Capacitively loaded circularly polarized implantable patch antenna for ISM band biomedical applications. *IEEE Trans. Antennas Propag.* 62 (5): 2407–2417.

20 Kim, J. and Rahmat-Samii, Y. (2004). Implanted antennas inside a human body: simulations designs and characterizations. *IEEE Trans. Microwave Theory Tech.* 52 (8): 1934–1943.

21 Chu, L.J. (1948). Physical limitations of omni-directional antennas. *J. Appl. Phys.* 19 (12): 1163–1175.

22 Hall, P.S. and Hao, Y. (2006). Antennas and propagation for body-centric wireless communications. *Artech House* 2.

23 Computer Simulation Technology. (2009).

24 Federal Communications Commission (1996). Understanding the FCC regulations for low-power non-licensed transmitters. *Office Eng. Technol.* .

25 IEEE (2006). IEEE standard for safety levels with respect to human exposure to radio frequency electromagnetic fields, 3 kHz to 300 GHz. *IEEE Std* C951-2005: 1–238.

26 Ho, J.S., Qiu, B., Tanabe, Y. et al. (2015). Planar immersion lens with metasurfaces. *Phys. Rev. B* 91 (12).

27 IEEE. IEEE standard for safety levels with respect to human exposure to radiofrequency electromagnetic fields, 3kHz to 300 GHz, IEEE Standard C95.1-2005. (2005).

28 Pennes, H.H. (1948). Analysis of tissue and arterial blood temperatures in resting forearm. *J. Appl. Physiol.* 1: 93–122.

29 Thotahewa, K.M.S., Redouté, J.-M., and Yuce, M.R. (2013). SAR, SA, and temperature variation in the human head caused by IR-UWB implants operating at 4 GHz. *IEEE Trans. Microw. Theory Tech.* 61 (5): 2161–2169.

30 CST Microwave Suite Version 2014. Available: https://www.3ds.com/products-services/simulia/products/cst-studio-suite/.

31 Yilmaz, T., Karacolak, T., and Topsakal, E. (2008). Characterization and testing of a skin mimicking material for implantable antennas operation at ISM band (2.4 GHz-2.48 GHz). *IEEE Antennas Wireless Propag. Lett.* 7: 418–420.

32 Harouni, Z., Cirio, L., Osman, L. et al. (2011). A dual circularly polarized 2.45-GHz rectenna for wireless power transmission. *IEEE Antennas Wireless Propag. Lett.* 10: 306–309.

33 Ali, M., Yang, G., and Dougal, R. (2005). A new circularly polarized rectenna for wireless power transmission and data communication. *IEEE Antennas Wireless Propag. Lett.* 4: 205–208.

34 Ren, Y.-J. and Chang, K. (2006). New 5.8-GHz circularly polarized retrodirective rectenna arrays for wireless power transmission. *IEEE Trans. Microw. Theory Tech.* 54 (7): 2970–2976.

35 Sun, H., Guo, Y.-X., He, M., and Zhong, Z. (2012). Design of a high-efficiency 2.45-GHz rectenna for low-input-power energy harvesting. *IEEE Antennas Wireless Propag. Lett.* 11: 929–932.

36 Akkermans, J.A.G., Beurden, M.C.V., Doodeman, G.J.N., and Visser, H.J. (2005). Analytical models for low-power rectenna desigin. *IEEE Antennas Wireless Propag. Lett.* 4: 187–190.

37 Vera, G., Georgiadis, A., Collado, A., and Via, S. (2010). Design of a 2.45 GHz rectenna for electromagnetic (EM) energy scavenging. *IEEE Radio Wireless Symp.* 61–64.

38 Poon, A.S., O'Driscoll, S., and Meng, T.H. (2010). Optimal frequency for wireless power transmission into dispersive tissue. *IEEE Trans. Antennas Propag.* 58 (5): 1739–1750.

39 Yu, C., Liu, C., Zhang, B. et al. (2011). An intermodulation recycling rectifier for microwave power transmission at 2.45GHz. *Prog. Electromagn. Res.* 119: 435–447.

40 Viikari, V., Seppa, H., and Kim, D.-W. (2011). Intermodulation read-out principle for passive wireless sensors. *IEEE Trans. Microw. Theory Techn.* 59 (4): 1025–1031.

41 Hannula, J.-M., Rasilainen, K., and Viikari, V. (2015). Characterization of transponder antennas using intermodulation response. *IEEE Trans. Antennas Propag.* 63 (6): 2412–2420.

42 Pozar, D.M. (2011). *Microwave Engineering*, vol. 11, 525–529. Wiley.

43 Sun, H. and Geyi, W. (2016). A new rectenna with all-polarization-receiving capability for wireless power transmission. *IEEE Antennas Wireless Propag. Lett.* 15: 814–817.

44 Zhang, H., Gao, S., Wu, W., and Guo, Y.-X. (2018). Uneven-to-even power distribution for maintaining high efficiency of dual-linearly polarized rectenna. *IEEE Microw. Wireless Compon. Lett.* 28 (12): 1119–1121.

45 RamRakhyani, A.K., Mirabbasi, S., and Chiao, M. (2011). Design and optimization of resonance-based efficient wireless power delivery systems for biomedical implants. *IEEE Trans. Biomed. Circuits Syst.* 5 (1): 48–63.

46 Jegadeesan, R., Nag, S., Agarwal, K. et al. (2015). Enabling wireless powering and telemetry for peripheral nerve implants. *IEEE J. Biomed. Health Inform.* 19 (3): 958–970.

47 WE-WPCC Wireless Power Transfer Transmitter Coil. (2018). Available: https://katalog.we-online.de/pbs/datasheet/760308101104.pdf.

9

System Design Examples: Peripheral Nerve Implants and Neurostimulators

9.1 Introduction

The dawn of the 21st century has witnessed remarkable advancements at the intersection of neuroscience [1], bioengineering [2], and digital technology [3]. A prominent fruit of this intersection is the development of neurostimulators and peripheral nerve implants, transformative technologies that have ushered in new therapeutic avenues for various debilitating neurological conditions.

Neurostimulators and peripheral nerve implants represent groundbreaking interventions that leverage electrical stimulation to modulate neuronal activity [4]. Through precise electrical stimulation, these devices can alleviate the symptoms of diverse neurological disorders, including Parkinson's disease [5], epilepsy [6], and chronic pain [7], thereby enhancing the quality of life for patients worldwide.

However, despite their therapeutic potential, these devices face significant challenges concerning power management and data communication. Traditional methods, predominantly wired connections, have been fraught with limitations, ranging from the risk of infections and device failures to the burden of frequent surgical replacements and patient discomfort.

As we venture into an era marked by the confluence of medical science and technology, the development of wireless capabilities has been a game-changer, particularly in the domain of neurostimulators and peripheral nerve implants. Wireless power transfer techniques enable these implants to function without the need for an internal battery, thereby minimizing the need for invasive surgical procedures and increasing device longevity [8]. Concurrently, wireless communication systems enable real-time data transmission, thus facilitating personalized treatment strategies, proactive health monitoring, and prompt medical intervention [9].

In the preceding chapters, we have undertaken a systematic review of antenna technologies and wireless power transfer methodologies, all tailored specifically to meet the exigencies of biomedical scenarios. As we progress into this chapter, we will bring together the practicality of peripheral nerve implants and actual neurostimulator products. Our discussion will encompass system design oriented toward practical applications, providing a thorough view of the theoretical concepts put to test in the real world. Furthermore, we will also delve into the challenges and complexities manufacturers

Antennas and Wireless Power Transfer Methods for Biomedical Applications, First Edition.
Yongxin Guo, Yuan Feng and Changrong Liu.

grapple with when these technologies are applied in the field. To that end, we will provide an overview of the solutions and mitigation strategies they have successfully employed to overcome these obstacles, ensuring the successful and efficient functioning of these implantable devices. This chapter aims to bridge the gap between theoretical understanding and its practical application, providing insights into the complex but intriguing world of biomedical wireless technology.

9.2 Wireless Powering and Telemetry for Peripheral Nerve Implants

9.2.1 Peripheral Nerve Prostheses

Peripheral nerve injuries often result in significant impairment of motor and sensory functions in the affected limbs. More often than not, traditional surgical repair techniques proposed for these nerve injuries fall short of restoring complete functional recovery. The most effective strategy lies in directly transmitting the nerve signals to the denervated muscles, thereby reestablishing functionality promptly, rather than awaiting nerve regeneration. Implementing such an approach necessitates a neural recording implant capable of recording and classifying nerve signals, which are then communicated to a stimulator implant that delivers patterned stimulation to execute functional actions such as hand grasp, among others. The system-level diagram of the nerve prosthesis is illustrated in Figure 9.1. In this study, we introduce a wireless platform designed for the recording and stimulator implants, a critical step toward building a comprehensive peripheral nerve prosthesis.

9.2.1.1 Stimulator Implant

Functional muscle stimulation is realized by administering current into the corresponding muscles via electrodes, following a specified sequence and strength [11]. The stimulator implant possesses the capacity for controllable multichannel accurate current drive, which facilitates the injection of charge to stimulate the targeted muscles. The external control unit energizes the implanted stimulator and dispatches control data, which defines the strength, duration, and type of stimulation, along with the stimulation pattern, to the implanted stimulator as portrayed in Figure 9.2. The stimulator unit confirms the receipt of the stimulator control data through backchannel impedance modulation.

9.2.1.2 Neural Recording

The neural recording implant operates by collecting neural signals directly from the nerve via an array of multichannel electrodes. The retrieved signals are then subjected to an amplification and digitization process before being wirelessly transmitted to an external control unit, as shown in Figure 9.3. This external control unit carries out dual functions. It wirelessly powers the implant unit and concurrently gathers nerve signals relayed from the implant. Upon receipt of the signals, it carries out a classification process, following which it transmits the corresponding control signals to the stimulator unit. This chain of actions ultimately results in the creation of a structured and patterned stimulation.

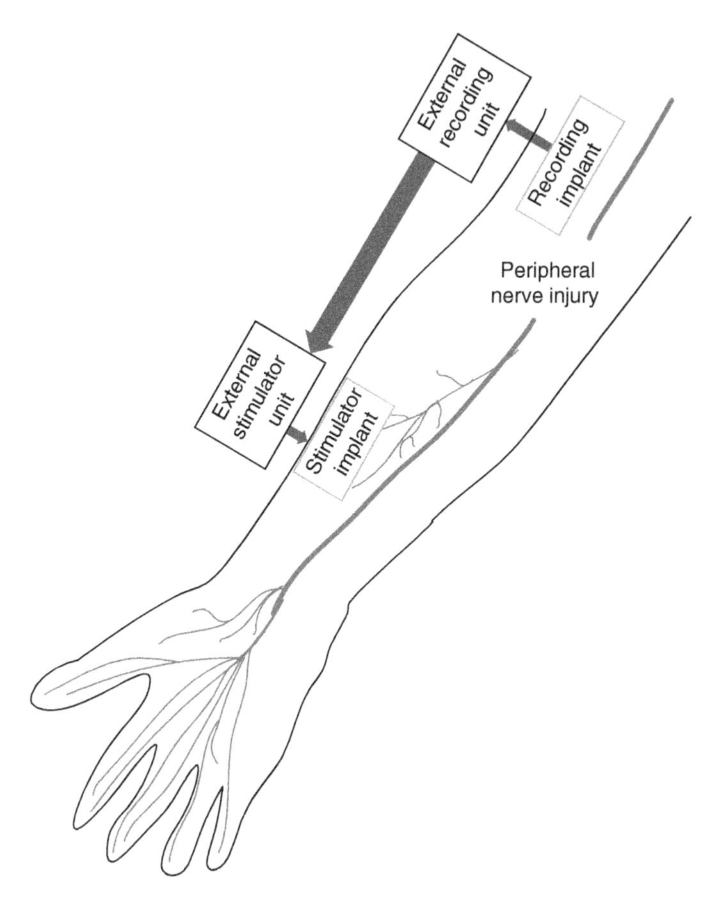

Figure 9.1 Proposed peripheral nerve prosthesis in hand. Source: Jegadeesan et al. [10]/IEEE.

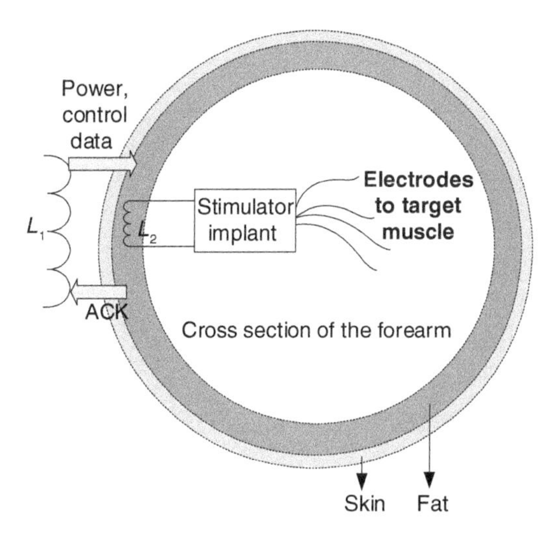

Figure 9.2 Wireless scheme for the muscle stimulator implant shown using a cross-sectional view of the forearm. The power delivery and bidirectional data telemetry are done using the inductors L_1 and L_2. Source: Jegadeesan et al. [10]/IEEE.

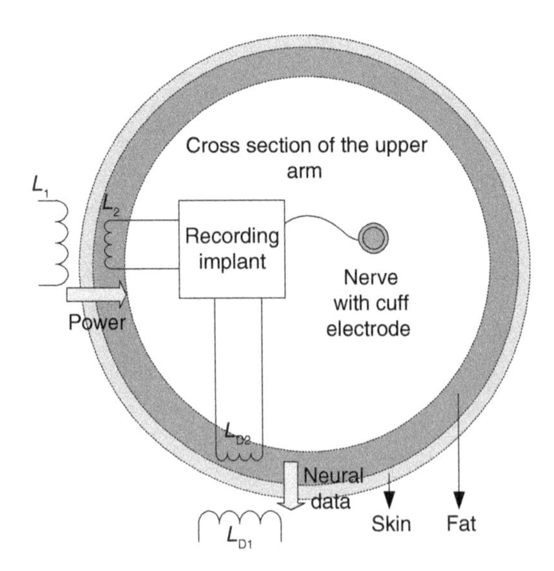

Figure 9.3 Wireless scheme for the neural recording implant shown using a cross-sectional view of the upper arm. The power is delivered using coils L_1 and L_2, where as data telemetry is performed using coils L_{D1} and L_{D2}. Source: Jegadeesan et al. [10]/IEEE.

9.2.1.3 Wireless Power Delivery and Telemetry Requirements

The generation of functional muscle stimulation demands a current of up to 10 mA, considering that electrode impedances can vary between 500 Ω and 1 kΩ. As a result, the stimulator implant necessitates a peak power delivery of 100 mW to ensure its sustained operation. Besides, the stimulator implant requires a bidirectional data telemetry of 4.8 kb/s, which translates to sending 16 bytes of data within a 50 ms listening window of the stimulator. This data encompasses details about pulse duration, pulse strength, phase, and packet frame bytes, all crucial for control signals receipt and provision of acknowledgment.

In contrast, the neural recording implant demands 35 mW of power to cater to amplification, analog-to-digital conversion, and digital logic functionality. A data link with a capacity of 1.3 Mb/s is essential to relay the neural signals (derived from eight channels, with 20 kHz data) from the implant to the external control unit. Table 9.1 below presents an all-encompassing view of the requirements for the wireless platform.

Table 9.1 Summary of requirements for the wireless platform used in peripheral nerve prostheses.

Implant type	Power requirement (mW)	Data to implant	Data from implant
Stimulation	100	4.8 Kbps	4.8 Kbps
Recording	35	NA	1.3 Mbps

Source: Jegadeesan et al. [10]/IEEE.

9.2.2 Wireless Platform for Peripheral Nerve Implants

9.2.2.1 Wireless Platform for Stimulator Implant

The selection of an appropriate wireless platform for the implant hinges upon the implant's power requirements. Both the stimulation and recording implants have substantial power needs (exceeding 10 mW), and the only viable and proven method to consistently transfer power to these implants is the near-field inductive-coupling scheme. Hence, this is the preferred choice for our implants. On the other hand, data requirements are significantly different, with the stimulator implant demanding bidirectional telemetry at a low bandwidth (4.8 kb/s) and the recording unit requiring high bandwidth telemetry (1.3 Mb/s for four-channel neural signal acquisition at 20 kHz) from the implant. For the stimulator, bidirectional telemetry can be easily accomplished by modulating the data onto the power link. However, for the recording implant, a separate data link is necessary to fulfill the bandwidth requirements. The following sections delve into the specifics of the wireless platform for powering peripheral nerve implants and carrying out telemetry.

Power for the implanted stimulator is supplied via an inductive power link comprising an external transmitting coil L_1 placed on the forearm and a receiving coil L_2 implanted just beneath the skin in the forearm, as depicted in Figure 9.2. The parallel resonant topology is employed to achieve the 10 V rectified voltage requirement for the stimulator. The power received at the implanted coil L_2 is utilized by the stimulator implant at two voltage levels. The current stimulator circuitry directly taps part of the power from the unregulated voltage (which should maintain at least 10 V under loading conditions for proper operation), while the digital logic circuitry derives its power from the 3.3 V regulated supply.

Control data are transmitted from the external stimulator unit to the implant using an on–off keying technique (refer to M_2 in Figure 9.5). This keying disturbance appears at the implant side as intermittent absences of power and is recovered as control data using a diode demodulator. The capacitor C_o (see Figure 9.4) intermittently powers the digital circuitry during power fluctuations caused by modulation. The received data are decoded by the digital logic, and the decoded command is used to set the stimulation parameters. The data are looped back to the transmitter as an acknowledgement (ACK) using impedance modulation influenced by the detuning capacitor C_m and switch M_4. This impedance modulation is reflected at the transmitting coil L_1 as voltage fluctuations. Given the substantial transmit power, the voltage fluctuations cannot be directly demodulated without extra circuitry for high-voltage protection of the demodulator. The class E amplifier is also detuned by the demodulator input, resulting in increased power losses. Consequently, a small reader coil L_{D1} is used to pick up the modulated data from the transmit coil, which is then demodulated and recovered using a logic detector as an acknowledgment. It is crucial to note that the coupling between the transmitting coil L_1 and the data reader coil L_{D1} should be kept to a minimum (coupling coefficient <0.05), just enough for data detection. A higher coupling would detune the class E amplifier from its optimal operating point.

The class E amplifier generates the power signal for the transmitting coil L_1 and is designed to deliver 100 mW at the implant at a separation of 5 mm. The load network, comprised C_{m1}, C_{m2}, C_{m3}, and L_m, is chosen based on a specific R_{load} (the equivalent load of the implant as reflected to the primary side) and a power requirement of 100 mW across this load. The timing requirements for class E operation are met based on the analysis provided in the classical work [12].

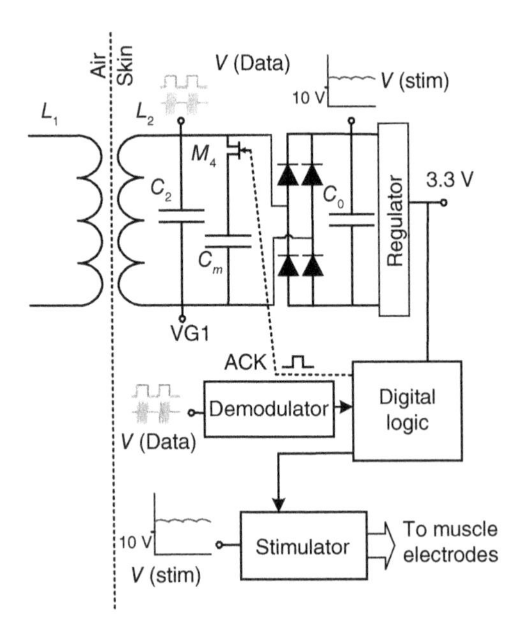

Figure 9.4 Schematic of the implant section of wireless platform for muscle stimulator. Source: Jegadeesan et al. [10]/IEEE.

The optimal frequency of operation was found to be 1 MHz based on the power level and the coil geometry. It is to be noted that the ACK data can be optionally used to request different transmit power levels from the transmitter during coil misalignments to keep the received power at the implant sufficient. The transmit power level can be adjusted using the switch M_3 (Figure 9.5), which modifies the class E amplifier gain.

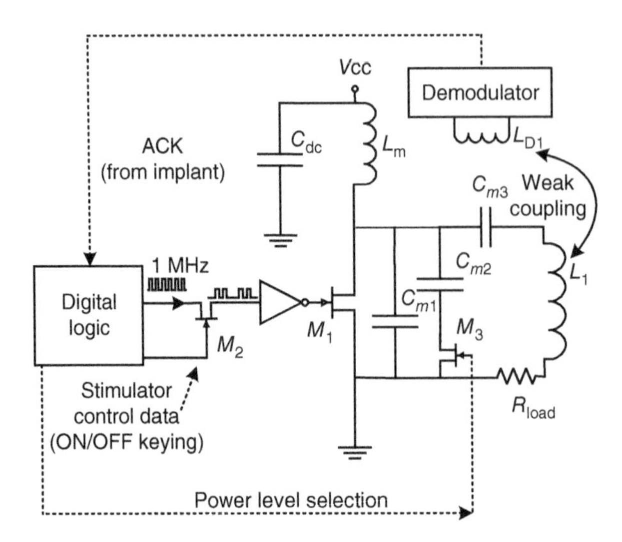

Figure 9.5 Schematic of the external section of wireless platform for muscle stimulator. Source: Jegadeesan et al. [10]/IEEE.

9.2.2.2 Wireless Platform for Recording Implant

The neural recording implant is powered using a near-field inductive power link, similar to the one employed by the stimulator implant. The power received at the implant coil L_2 is rectified and regulated to 3.3 V, which is used to power the entire recording implant.

The maximum data rate, constrained by the carrier-to-signal separation requirement of the decoder, is 125 kb/s for a carrier frequency of 1 MHz. Increasing the power carrier frequency to over 10 MHz could provide a sufficient data rate to transmit the neural recording data. However, high rectifier losses and a low power supply rejection ratio (PSRR) of recording amplifiers at high frequencies would result in an inefficient power transfer scheme and poor noise performance. Consequently, the recording implant employs a separate link to transfer neural signal data. The neural data from the amplifier is digitized by the analog-to-digital converter (ADC) and sequentially sent using the high-speed data link formed by the inductances L_{D1} and L_{D2}.

A digital signal synthesizer is used to generate a 43-MHz square-wave signal at 3.3 V_{PK-PK}, which is fed into the data reader coil L_{D1}. The data from the neural implant is load modulated at the implant side (see M_4, in Figure 9.6). This load modulation is recovered at the external unit as neural signal data at L_{d1}, as depicted in Figure 9.7. The design of the class E amplifier was conducted in a manner similar to the stimulator implant, but with a lower target power of 35 mW.

9.2.3 Design and Experiments

In an exploratory investigation, we examined the efficacy of near-field inductive-coupling links using rat subjects to assess performance discrepancies between air and tissue

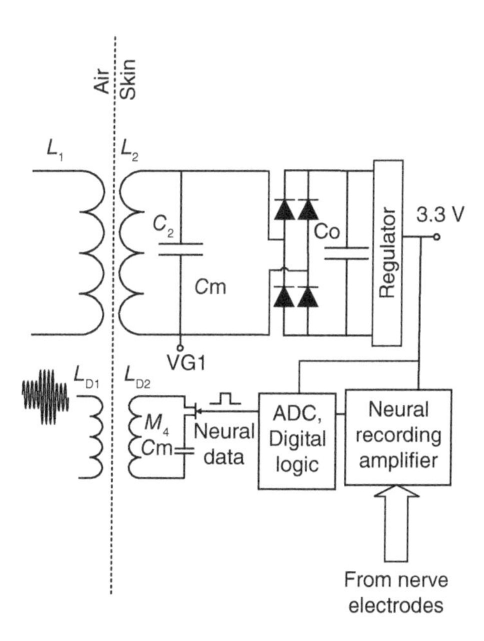

Figure 9.6 Schematic of the implant section of wireless platform for neural recording. Source: Jegadeesan et al. [10]/IEEE.

environments. The insights derived from this study have been instrumental in the design of the wireless link specifically tailored for peripheral nerve implants.

9.2.3.1 Power Transfer Characteristics in Tissue Environments

In order to understand the impact of tissues on power transfer efficiency (PTE) across varying transmitter-receiver separations and operating frequencies, an inductive power link, with specifications delineated in Table 9.2 and coils as depicted in Figure 9.8, was employed. The intent behind this investigation was to pinpoint the optimal frequency of operation that would curtail tissue-related losses and facilitate efficient power management. To prevent short circuits due to potential leakage of tissue fluid, the implant coil was safeguarded with polyimide tape.

The study was conducted on a single rat, following all the requisite guidelines outlined by the institutional animal care and use protocol. The rat was first anesthetized using a standard dose of ketamine and xylazine. Subsequently, a small incision was made in the rat's abdominal region, creating a passage for the implant-side coil. Following the precise positioning of the implant coil within the rat's body, the external coil was situated on the skin's surface, and an optimal position that ensured the coil alignment was established.

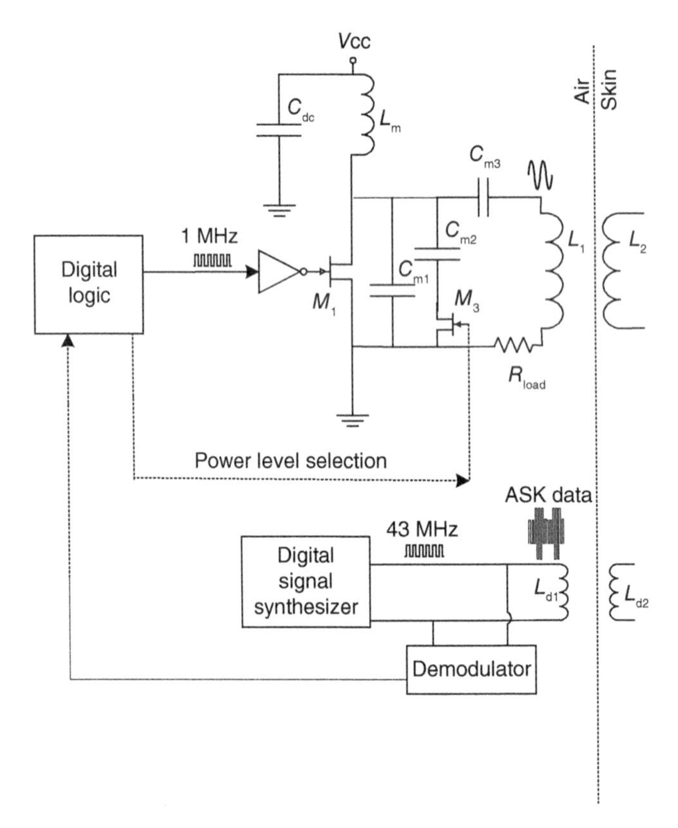

Figure 9.7 Schematic of the external section of wireless platform for neural recording. Source: Jegadeesan et al. [10]/IEEE.

Table 9.2 Wireless link components and measurement data.

	Transmitter coil	Receiver coil
Geometry	45 mm × 45 mm	8 mm × 8 mm
	0.5 mm trace width	0.1 mm trace width
	1 mm pitch	0.2 mm pitch
	20 turns	15 turns
Characteristics	10.1 μH at 5 MHz	1.2 μH
	13.75 μH at 13.56 MHz	(5 MHz to 17 MHz)
	16.7 μH at 17 MHz	
Optimal load for	800 ohm at 5 MHz	
the power link	3.3 k ohm at 13.56 MHz	
	4.8 k ohm at 17 MHz	

Source: Jegadeesan et al. [10]/IEEE.

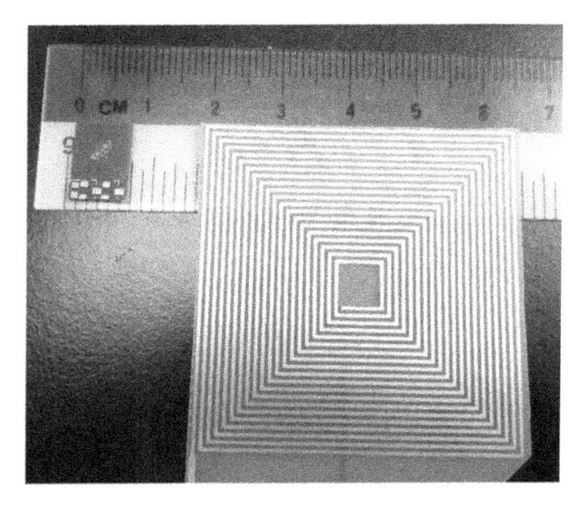

Figure 9.8 Planar PCB coils used in the inductive link study: implant coil (8 mm × 8 mm) on the left and the transmit coil (45 mm × 45 mm) on the right. Source: Jegadeesan et al. [10]/IEEE.

At this point, the incision was sutured, enclosing the implant coil within the rat's body, with two wires extruding for power readout. It is noteworthy that the thickness of the rat's abdominal skin, roughly 2 mm, closely parallels that of human skin in the arm, which fluctuates between 2 and 3 mm.

Power was supplied to the external coil using a signal generator. The power at the implant coil was then measured at the readout terminals with an optimal load (refer to Table 9.2). The PTE was computed as the quotient of the power delivered to the load to the power drawn from the source (note: The power drawn from the source was adjusted for input reflection, measured using a vector network analyzer (VNA), the R&S ZVL13).

Throughout the study, PTE was measured at three different depths by surgically creating pockets beneath the skin, beneath the fat, and beneath the muscle, as illustrated

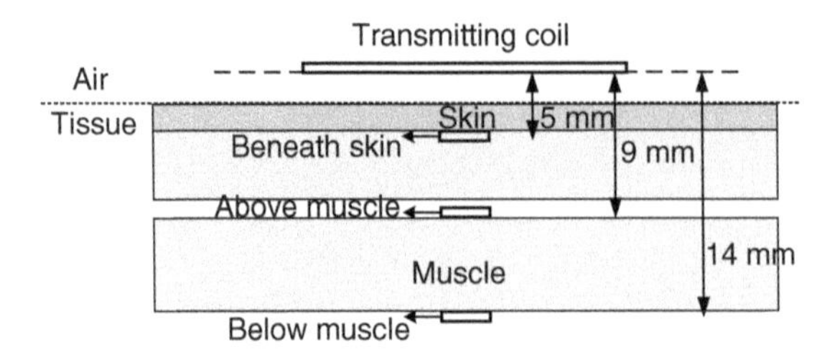

Figure 9.9 Three different implant depths used for measuring the power transfer efficiency between the transmitter and receiver coils in rat. Source: Jegadeesan et al. [10]/IEEE.

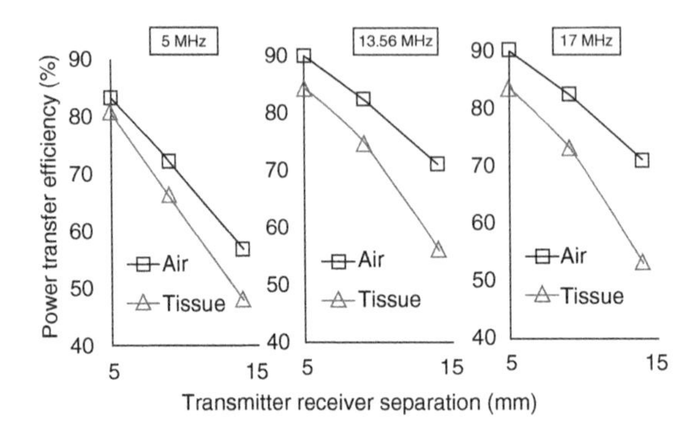

Figure 9.10 Graph showing measured power transfer efficiency versus separation at three different operating frequencies. The transmitter power is 250 mW. Source: Jegadeesan et al. [10]/IEEE.

in Figure 9.9. The separations corresponding to these three positions were approximately 5 mm, 9 mm, and 14 mm, respectively. These measurements were replicated at different frequencies (5, 13.56, and 17 MHz), and the results were juxtaposed with efficient measurements in air at similar separations, as can be seen in Figure 9.10.

From the experimental results, we drew several key insights that will guide the configuration of the power transfer link for the peripheral nerve implants:

(1) As expected from electromagnetic field theory, the PTE significantly declines with distance. Therefore, implanting the coil at greater depths is not advisable, making a subcutaneous link the most appropriate design for the power transfer link.
(2) The disparity in PTE between tissue and air environments is more pronounced at higher frequencies, indicating that tissue losses are more severe at these frequencies. This observation is also consistent with the Debye relaxation model of tissues, which demonstrates frequency dependence. Consequently, we deduce that a lower frequency of operation is desirable when managing large power requirements. Additionally, neural signal amplifiers exhibit good PSRR at lower frequencies, enabling better suppression of the carrier frequency. However, it should be noted that while choosing a lower frequency of operation decreases the overall efficiency of the power transfer link (due to increased coil losses in both the transmitter and the receiver), it significantly reduces tissue losses. This trade-off may need to be considered when transferring high power to the implant.

(3) Square planar coils, due to their sharp edges, are not suitable for long-term implantation. Hence, a circular coil with smooth edges is recommended for the design of the power transfer link.

9.2.3.2 Power Transfer Link for Peripheral Nerve Implants

Based on the preliminary findings from our rat model study, a low-frequency subcutaneous power transfer link is suggested, as it would permit the handling of larger power requirements, ideally suiting the needs of our peripheral nerve implant. Nevertheless, managing this significant power using planar coils on a printed circuit board (PCB) is a challenge due to the thin traces and low-quality factor of the coils. Therefore, we opted for coils made of laminated AWG32 copper wire. This decision, however, introduces the challenge of fabricating the coils in a way that is repeatable and resistant to deformation postimplantation. Our solution involved creating wire-wound coils using plastic molds with a release mechanism (to minimize coil variations). These were then encapsulated with polydimethylsiloxane (PDMS) as depicted in Figure 9.11.

The PDMS was prepared by combining the silicone elastomer (SYLGARD 184) with its curing agent at a 10 : 1 mass ratio. This mixture was thoroughly stirred and then degassed in a vacuum chamber for 20 min to prevent air bubbles and leakages. The semi-cured PDMS was then poured over the mold to encapsulate the coil and the associated implant electronics. It was degassed again and then cured in an oven at 70 °C for 2 h. The total thickness of the PDMS-encapsulated coil fabricated by this process was roughly 3 mm.

We settled on 1 MHz as the operating frequency based on the understanding that higher frequencies yield more tissue loss, limiting power transfer capability. The coils for the power transfer link were designed and optimized for operation at 1 MHz as follows: given the size constraint of the human arm, the maximum size of the implant coil is approximately 20 mm in diameter. Thus, we constructed a circular implant coil with a diameter of 20 mm and chose 25 turns to maximize the quality factor at 1 MHz. While the external coil does not have a size limitation, a coil that is larger than the implant coil would have a weak field strength at the center of the implant coil, and a smaller coil would lack sufficient flux

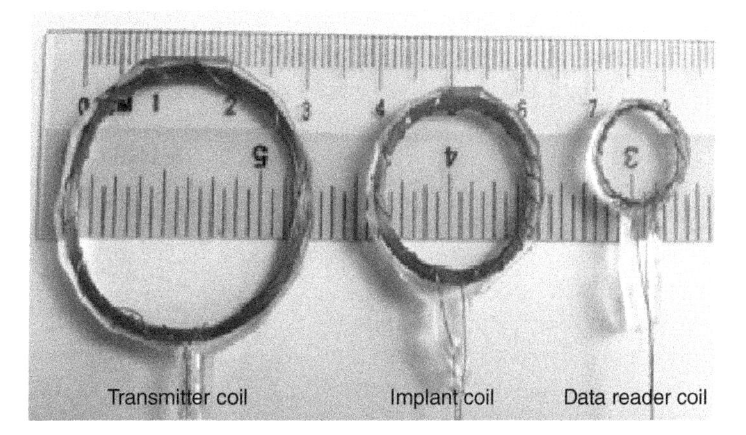

Figure 9.11 The transmitting coil (diameter = 30 mm), implant coil (diameter = 20 mm), and the data reader coil (diameter = 10 mm) encapsulated in PDMS. Source: Jegadeesan et al. [10]/IEEE.

linkage with the implant coil. Therefore, we chose an external coil size of 30 mm to maximize the coupling and maintain sufficiently large field strength. The number of turns was optimized to maximize the quality factor at 1 MHz, which was found to be between 24 and 28 turns. We selected a 25-turn coil, as its inductance value numerically reduced the total number of capacitors required in the load network of the class *E* amplifier. A 10-mm coil with 6 turns was deemed to have sufficient mutual inductance to couple the data modulation from the transmitter coil (L1) and was chosen as the reader coil Ld1 (for the stimulator implant). The coils were characterized using a VNA (R&S ZVL13) and their values are presented in Table 9.3.

The data coils for the recording implant need to have an operating frequency greater than 10.4 MHz to meet a data rate requirement of 1.3 Mb/s, considering the frequency separation between carrier and signal for proper demodulation. Since the data reader coil designed for the stimulator implant has a self-resonant frequency of 75 MHz, it was repurposed as the data link coils for the recording implant. Specific absorption rate (SAR) simulations conducted on a tissue model using the selected coils showed regulatory compliance with a safety margin exceeding two.

9.2.3.3 Stimulator Implant Experiment

The designed power and data links were acutely studied for feasibility in rats. In one experiment, the implant coil was inserted beneath the skin via a small incision made in the rat's stomach area, as illustrated in Figure 9.12(a). The incision was subsequently sutured, leaving two readout terminals protruding. A power of 105 mW was measured across the implant coil when there was a separation of 10 mm between the transmitter and the implant coil. By reducing the separation to 5 mm, i.e., moving the transmitting coil closer to the skin, the power delivered increased to 127 mW. The power link consistently delivered over 100 mW to the implanted coil when the radial misalignment of the powering coil was within 5 mm [see Figure 9.12(b)]. Figure 9.13 displays the power delivered to the implant coil at various transmitter–receiver misalignments for two different implant depths. The PTE of the link for aligned coils (maximum PTE value without rectification) was determined to be 65.8%. Although slightly lower than the PTE at high frequencies, this setup facilitates large power delivery with minimal tissue losses due to heat (Figure 9.14).

The highest amount of heat would be generated at the tissues above and below the implant coil. The temperature at these locations was measured using a K-type thermocouple (sealed with polyimide tape to the implant coil), the terminals of which were drawn out through the skin suture and connected to a thermometer (Fluke 51 II). The temperatures above and below the implant were recorded as T1 and T2, as shown in Figure 9.14. It was observed that there was a temperature increase of less than 0.6 °C while power was continuously delivered for 20 min. This is well within the 2 °C localized temperature rise reported for an equivalent SAR of 4 W/kg [15].

In another experiment, the designed wireless link powered the stimulator implant to provide functional leg movement in the rat. The 4.8 kb/s data link to the implant was tested over the same power link. A scope shot of the data transfer over the same power link is shown in Figure 9.15. The stimulator implant and the class *E* amplifier used for powering are shown in Figure 9.16.

Table 9.3 Wireless implant components and characteristics.

Component	Geometry/specification	Characteristics
L_2 (Stimulator/Recording implant)	25 turns, 20 mm diameter, 32 AWG copper	26.1 μH, 6.85 ohm (at 1 MHz) SRF = 5.3 MHz
L_1 (Stimulator/Recording implant)	25 turns, 30 mm diameter, 32 AWG copper	44.9 μH, 13.9 ohm (at 1 MHz) SRF = 2.7 MHz
L_{D1}, L_{D2} (Stimulator/recording implant)	6 turns, 10 mm diameter 32 AWG copper	1.3 μH, 11.5 ohm (at 43 MHz) SRF = 75 MHz
M1	IRF540	Power FET
Digital logic (stimulator)	MSP430F2132	Microcontroller, UART demodulator
Regulator (stimulator implant) Regulator (recording implant)	LTC3245,	3.3 V regulator, <38 V input
	TPS78833	3.3 V regulator, <12 V input
Rectifier (stimulator/recording implant)	BAS70-04	Dual Schottky diodes
Stimulator	[13, 14], custom stimulator	Four Channel muscle stimulator
Recording amplifier	RHD2216	Up to 16-channel neural signal amplifier
ADC, digital logic	MPS430F2274	8bit ADC, microcontroller
Demodulator	BAS70-04	Asynchronous, single slope demodulator
ADC, digital logic	MSP430F2274	8bit ADC, Microcontroller
Demodulator (stimulator/ recording implant)	BAS70-04, 2-stage filter, LM7239	Single slope synchronous demodulator

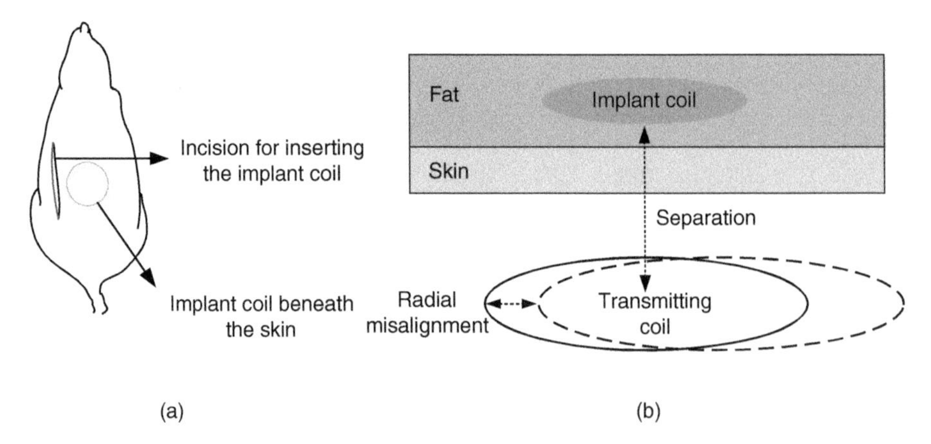

Figure 9.12 (a) Sketch of the rat showing the location of incision and the implant coil, (b) cross-sectional view of the model illustrating the separation and misalignment with respect to implant coil inside rat. Source: Jegadeesan et al. [10]/IEEE.

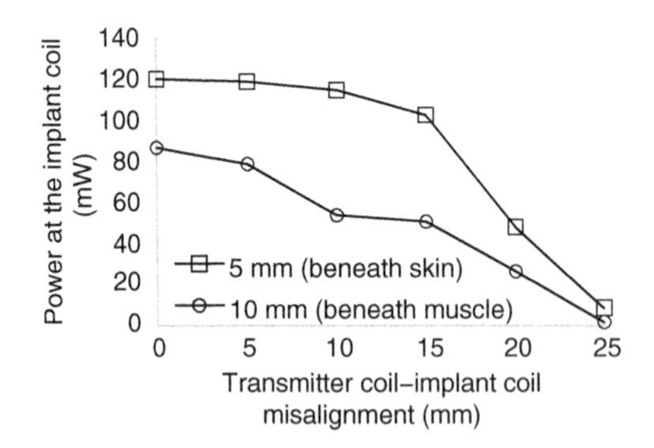

Figure 9.13 Power delivered to the implant coil versus its misalignment with the transmitter for two different implant depths. Source: Jegadeesan et al. [10]/IEEE.

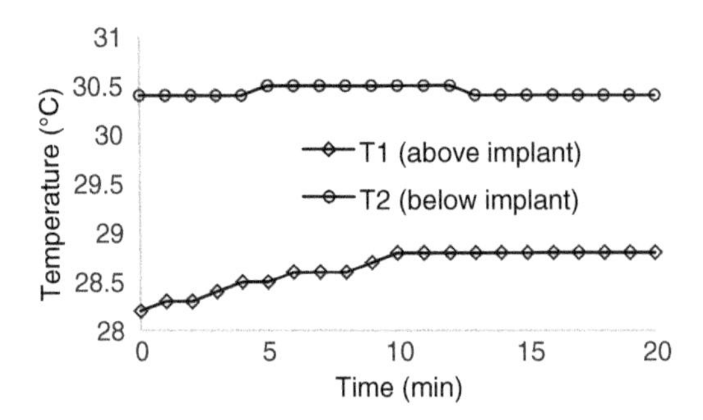

Figure 9.14 Heating of the tissues is characterized by temperature rise of the tissues surrounding implant. Graph shows temperature with time at tissue interface above and below the implant coil. Source: Jegadeesan et al. [10]/IEEE.

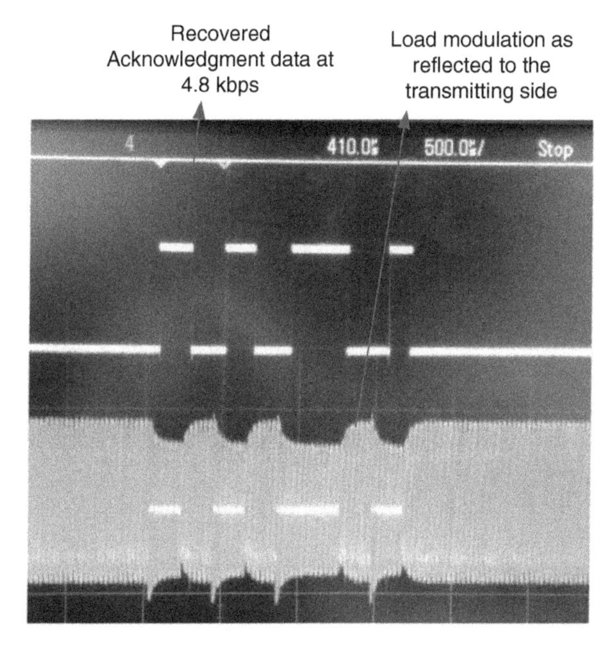

Figure 9.15 Backscattered data waveform captured at the reader coil for the recording implant, showing discernable data at 4.8 kbps. Source: Jegadeesan et al. [10]/IEEE.

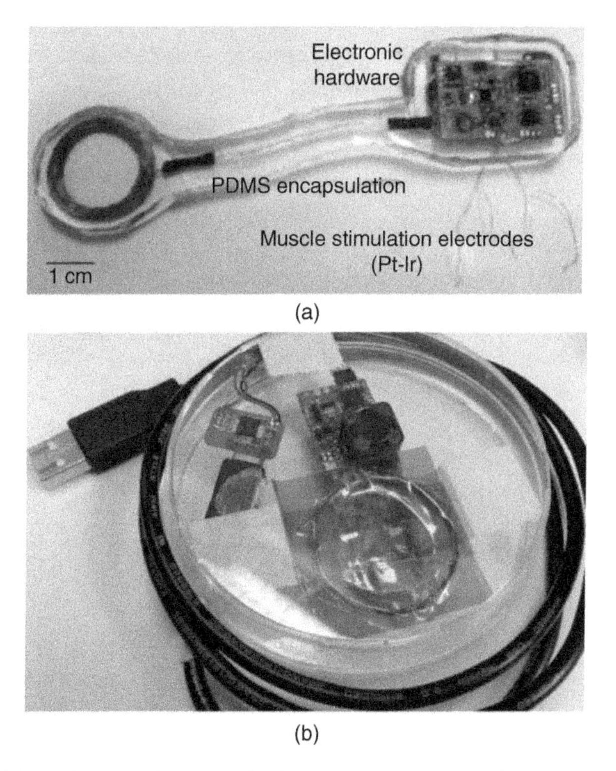

Figure 9.16 (a) Stimulator implant used in the rat experiment with Pt-Ir electrodes, (b) class-E amplifier and external decoder unit powered using USB. Source: Jegadeesan et al. [10]/IEEE.

9.2.4 Safety

9.2.4.1 Biosafety

An often-overlooked factor in the engineering analysis of inductive power transfer schemes is the biosafety of the materials employed in their construction. Since most metals, including copper, are not biocompatible, it is necessary to encapsulate the coils in a material that is safe for biological use. Silicone elastomers, which are generally used as encapsulating materials in FDA-approved implants, were chosen for this purpose. In our case, we used PDMS to encapsulate the coils because it is one of the biocompatible silicones [16]. This PDMS encapsulation not only renders the coils leak-proof but also provides the necessary elasticity for an implantable device. However, it is important to note that this encapsulation does impact the characteristics of the wireless power and data transfer coils.

To precisely analyze the change in the coil properties due to the encapsulation, we characterized PDMS samples using the ring resonator method [17]. The measured values of relative permittivity and loss tangent were found to be 2.8 and 0.012, respectively. These values were then used for optimizing the power link and carrying out SAR simulations to ensure safety.

9.2.4.2 Electrical Safety

Electrical safety is paramount in the design of power transfer systems. Electromagnetic fields can create eddies in dispersive tissues, and if these eddies reach high amplitudes, they can cause electrocution. The impact of these currents varies depending on the current paths, with sensitive areas having lower current thresholds than others. The brain, respiratory muscles, and heart are particularly sensitive to currents, and irreversible damage leading to death can occur if high currents are passed through these areas. Furthermore, tissue heating is another side effect of exposure to electromagnetic waves.

To mitigate these adverse effects, the parameters of the wireless link are iteratively selected to conform to the SAR limits set out in the IEEE standards for safety levels regarding human exposure [15]. The peak localized SAR is restricted to 4 W/kg for the body's extremities (for instance, the pinnae) and 2 W/kg for other regions of the body.

SAR simulations for the stimulator implant were conducted using a high-frequency structural simulator (HFSS) with a transmission power of 180 mW. The simulation setup and the resultant SAR values are presented in Figure 9.17. The tissue models for the simulations were derived from detailed characterization data as outlined in [18].

9.2.5 Near-field Resonant Inductive-coupling Link (NRIC) Versus Near-field Capacitive-coupling Link (NCC)

In Chapter 7, we introduced the NRIC and near-field capacitive-coupling (NCC) wireless charging solutions applied to biomedicine, conducting system design and animal experiments for the subcutaneous implantation of the NCC scheme. In this chapter, we similarly designed an NRIC wireless charging scheme for subcutaneous implantation scenarios and validated it through animal experiments. Based on these experiments and research, we can draw the following conclusions.

NRIC power transfer links utilize tuned resonances to efficiently transfer power. Power transfer can also be made efficient via coil optimization techniques as presented in

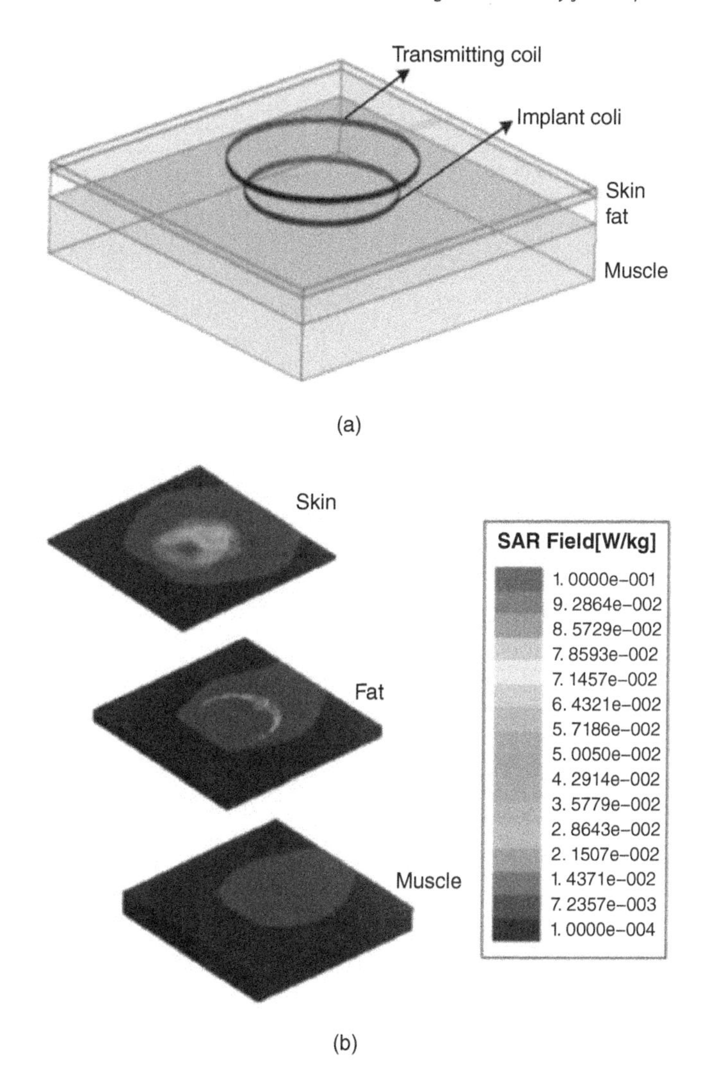

Figure 9.17 (a) Simulation setup for 180-mW transmit power to the implant with skin layer (2.5 mm), fat layer (6 mm), and muscle layer (10 mm), (b) SAR values obtained from the simulation in HFSS V.12. Source: Jegadeesan et al. [10]/IEEE.

Chapter 7. On the other hand, NCC links use capacitive coupling and an optimal band of operating frequencies to transfer power efficiently. Although both schemes serve the same purpose, their performances vary under different circumstances. NCC links are not based on resonance, giving them a broad operating bandwidth that allows for flexible implant implementations. According to our experimental results, in Section 7.5, it is safe to infer that the PTE remains acceptable (>20%) under flexion for the NCC scheme. Conversely, the NRIC link severely detunes due to flexion and operates out of resonance, where efficiencies are below 5% (calculated for a 15 mm × 15 mm implant coil, 5 mm separation, and a bending radius of 10 mm). Under normal operating conditions, both schemes can

Table 9.4 Comparison between NRIC and NCC schemes.

Properties ↓	NRIC link	NCC link
Coupling type	Inductive	Capacitive
Resonance tuned	Yes	No
Operating frequency range for implantable devices	200 kHz to 20 MHz	50 MHz to 200 MHz
Bandwidth	Narrow	Large
Detuning due to flexion	Large	Small
Implant area required for same power delivered (100 mW)	17 mm × 17 mm	Two patches of (20 mm × 10 mm)
Range for operation	Up to 15 mm	Up to 7 mm

deliver similar power levels to the implant. The NRIC link has the advantage of a longer power transfer range compared to the NCC links. However, in many of today's implants [19, 20], the power receivers are positioned just below the skin, which perfectly suits the NCC scheme. Based on our works on NRIC and NCC links, a comparison has been listed above for the two near-field WPT schemes with similar implant dimensions in Table 9.4.

9.3 Co-matching Solution for Neurostimulator Narrow Band Antenna

The neurostimulator, akin to an artificial pacemaker, is a widely utilized implantable medical device in clinical practices for treating neurological disorders, such as Parkinson's disease. Typically encased in a biocompatible metal shell, the neurostimulator is usually placed subcutaneously beneath the clavicle, as illustrated in Figure 9.18. An electrode stemming from the neurostimulator is employed both to record electrical nerve signals and to deliver artificial electrical impulses that stimulate the nerves [21], a critical function

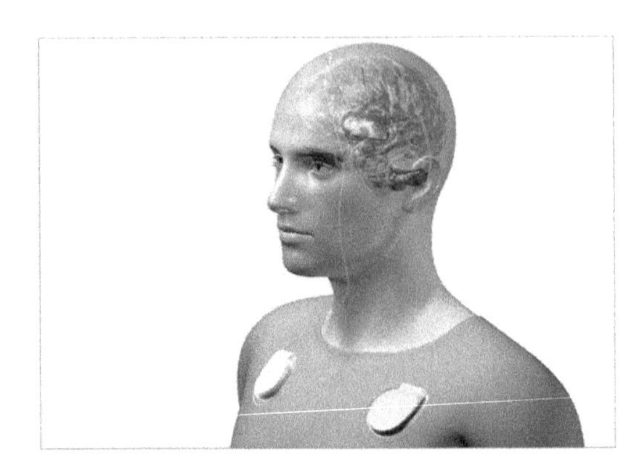

Figure 9.18 Schematic of neurostimulators, placed subcutaneously below the clavicle.

for brain–computer interface (BCI) applications. As a favored method for establishing efficient in-body to off-body links, the radio frequency approach, which offers advantages such as extended communication range and high data rate, is increasingly being adopted in neurostimulators.

Miniaturized antennas for neurostimulators are typically high-Q and narrow bandwidth devices [22–29], and frequency detuning often transpires due to the varying dielectric properties of tissues [30]. In clinical applications, a key challenge lies in the effect of the patient's body shape on antenna tuning. During surgery, physicians position the neurostimulator within the fat and muscle layers for obese and slim patients respectively, as illustrated in Figure 9.19 [32, 33]. Traditional neurostimulator antennas are seldom simultaneously matched in both scenarios, given that the relative permittivity of tissues ranges from 5.58 (fat) to 57.1 (muscle). To tackle this issue, Po et al. [34, 35] have proposed a vector automatic matching system that detects the antenna's complex input impedance and tunes an electronic matching network. Li et al. [36] employed the load–pull analysis method to optimize matching circuits in various tissues. Magill et al. [37, 38] have suggested two potential methods: ensuring a sufficiently wide impedance bandwidth or multiple resonances, to design a robust and tissue-independent implantable antenna capable of operating in the 2.36–2.4 GHz MBAN and 2.4 GHz ISM bands. Lyu et al. [39] reported an effective method to de-Q the implantable antenna to boost the antenna's robustness in tissues. Kiourti and Nikita's research [40] has identified the effects of head anatomy and permittivity on a scalp-implantable antenna. Stability analysis of antennas in tissues has been conducted in the design of capsule antennas [41–47]. A robust, ultra-miniature capsule antenna is resistant to impedance detuning from different surrounding biological environments [48], thanks to the dielectric loading of a high-permittivity capsule shell. Additionally, reconfigurable methods, as reported in [49, 50], are other possible strategies to achieve impedance matching of implantable antennas in diverse tissues.

In this section, we present a co-matching method for designing implantable antennas to address the detuning problem encountered in various tissues, such as fat and muscle. A dual-mode spiral monopole was designed for the neurostimulator. Depending on whether the patient was obese or slim, the device bearing the antenna was implanted into the fat or

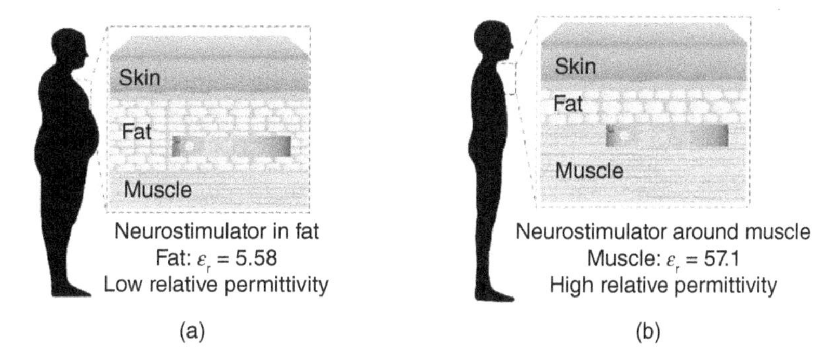

Figure 9.19 (a) Neurostimulator, placed in the fat tissue of obese patients, (b) neurostimulator, placed around the muscle tissue of slim patients. Source: Feng et al. [31]/IEEE.

muscle layer, respectively. Given the stark difference in the relative permittivity of fat and muscle tissues, the antenna experienced different detuning. Leveraging the co-matching technology, we collaboratively tuned the first and third modes of the antenna to the target band in both fat and muscle tissues [31].

9.3.1 Co-matching Antenna Operating Mode

A typical neurostimulator model, comprising a casing, a PCB, a header, and a feedthrough, is depicted in Figure 9.20(a). The neurostimulator measures $65 \times 50 \times 11\ \text{mm}^3$, modeled after the G102RS produced by Beijing PINS Medical Co., Ltd [51, 52]. Inside the biocompatible titanium casing lies a 1 mm-thick PCB, constructed from FR4 ($\varepsilon_r = 4.4$ and $\tan\theta = 0.02$). The implantable antenna is situated within the header, composed of biocompatible silicone rubber ($\varepsilon_r = 3.08$ and $\tan\theta = 0.04$), as measured with the Agilent 85070E dielectric probe kit. In the G102RS model, the antenna compartment measures a mere $25 \times 11\ \text{mm}^2$ with a 5 mm height, as illustrated in Figure 9.20(b) and (c). The antenna

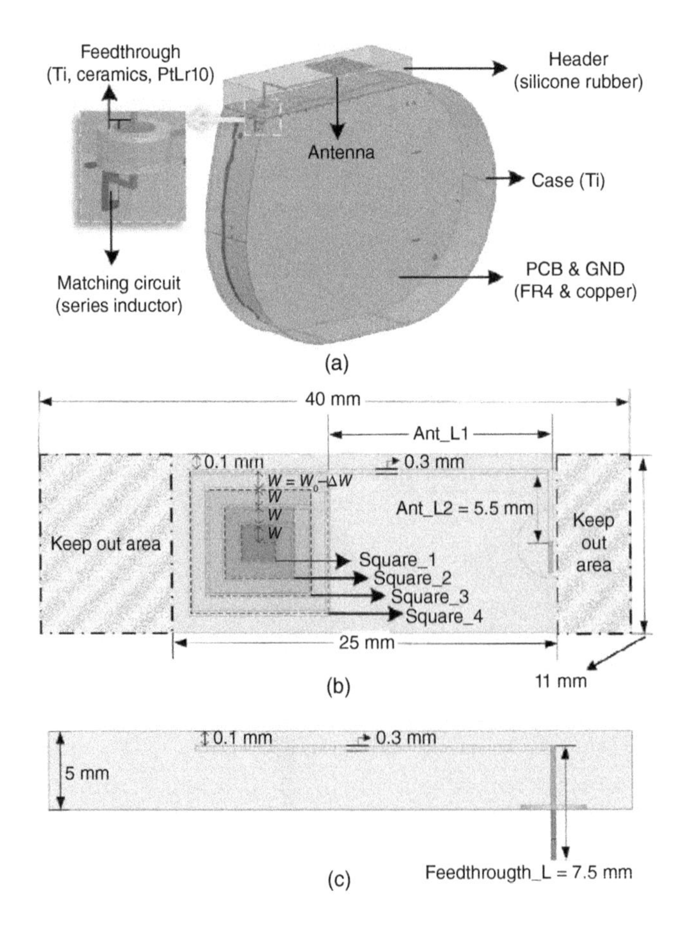

Figure 9.20 (a) Overview of the neurostimulator, (b) top view of the proposed antenna, and (c) front view of the proposed antenna. Source: Feng et al. [31]/IEEE.

design and feed setup principles and solutions were gleaned from mobile phone antenna designs [53, 54]. We designed a monopole antenna with a square planar spiral structure. The yellow conductor in Figure 9.20(b) represents the layout of the planar spiral section. The size of the entire structure was determined by the variable W, with an initial value of $W0 = 1$ mm. For Square_1, Square_2, ..., and Square_n, the lengths are $2W$, $4W$, ..., and $2n \times W$, respectively. A feedthrough wire composed of PtLr10 was utilized to connect the antenna feeding pad to the PCB inside the titanium casing, as shown in Figure 9.20(a) and (c). We set the square planar spiral's turns, n, to 4, at which point the proposed antenna's length approximates the quarter-wavelength of the 403 MHz wave in silicone rubber, as shown in Figure 9.20(b). The commercial software Ansoft High-Frequency Structure Simulator (HFSS) was employed for antenna parameter optimization.

Figure 9.21 presents the parametric studies of the proposed antenna in the air. We exhibit the resonance of the first mode (1/4 wavelength monopole mode) and third mode (3/4 wavelength monopole mode) by adjusting the critical parameters W and Ant_L1. Figure 9.21(a) showcases a parametric study with different spiral pitch W, in the air, while Ant_L1 is kept at 15 mm. Since W significantly affects the electrical length of the antenna, a change in ΔW by 0.25 mm will alter the antenna's electrical length by $(2+4+6+8) \times 4 \times \Delta W = 20$ mm. This value represents the cumulative change in the circumference for the four-turn square planar spiral. Therefore, changes in W dramatically impact the first- and third-mode resonant frequencies of the proposed antenna, which occur concurrently. When we selected a W value from 1 to 1.25 mm, the first-mode

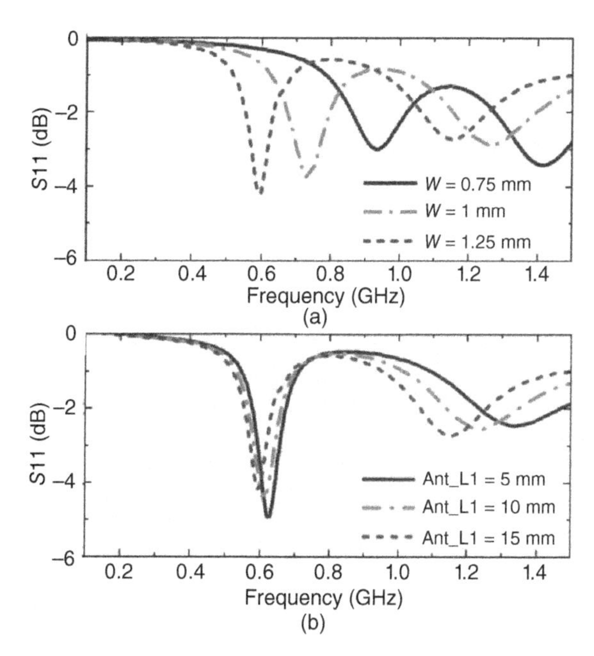

Figure 9.21 (a) Parametric study of the proposed antenna with a different spiral pitch W, in air. Ant_L1 = 15 mm, (b) parametric study of the proposed antenna with a different length Ant_L1, in air. $W = 1.25$ mm. Source: Feng et al. [31]/IEEE.

Table 9.5 Effects of the length Ant_L1 in the resonance frequency.

Ant_LI (mm)	f_1 (GHz)	f_3 (GHz)	f_1/f_3
5	0.62	1.31	0.47
10	0.61	1.20	0.51
15	0.60	1.11	0.54

Note: $W = 1.25$ mm.
Source: Feng et al. [31]/IEEE.

resonant frequency of the proposed antenna offset downward to the Medical Implant Communications Service (MICS) band after the antenna was implanted in fat tissue ($\varepsilon_r = 5.58$, $\sigma = 0.041$ s/m, at 403 MHz [13]).

Figure 9.21(b) showcases a parametric study with varying lengths Ant_L1, in the air, while W remains constant at 1.25 mm. The third-mode resonant frequency of the monopole antenna is more sensitive to changes in electrical length than the first mode. As Figure 9.21(b) illustrates, subtle modifications in the length Ant_L1 significantly affect the third-mode frequency. In our design, we utilized the third mode to match the MICS band when the antenna was placed in muscle tissue ($\varepsilon_r = 57.1$, $\sigma = 0.797$ s/m, at 403 MHz [13]). We chose Ant_L1 = 15 mm due to the proper ratio of the first- and third-mode frequencies f1/f3, which is influenced by the relative permittivity of fat and muscle, as detailed in Table 9.5. An alternative method, enhancing the bandwidth of implantable antennas by merging two close resonances [14, 55, 56], could potentially address the antenna detuning problem in different tissues. However, this method proves ineffective in our case, due to the large difference in relative permittivity.

9.3.2 Antenna Property in Body Phantom

In this section, we explore the effect of the parameter W on antenna matching in human tissues. Figure 9.22 displays a one-layer homogeneous body phantom with dimensions of $140 \times 140 \times 60\ \text{mm}^3$. The body phantom was designed to resemble fat and muscle. The placement of the neurostimulator is presented in Figure 9.22(b) and (c). Figure 9.23 outlines the parameter scan in fat and muscle phantoms. Figure 9.23(a) and (b) display the simulation results of $|S11|$ and the impedance of the proposed antenna in body phantoms at $W = 1$ mm. Similarly, Figure 9.23(c) and (d) correspond to $W = 1.125$ mm, and Figure 9.23(e) and (f) illustrate the results at $W = 1.25$ mm. As the 1/4 wavelength in fat and the 3/4 wavelength in muscle at 403 MHz are approximately equals the electrical length of the antenna, the first- and third-mode resonant frequencies are tuned into the MICS band in fat and muscle tissues, respectively, but are not optimally matched, as shown in Figure 9.23. From Figure 9.23(a), (c), and (e), we can infer that the $\Delta|S11|$ of the proposed antenna at 403 MHz in different tissue phantoms, fat and muscle, hinges on the value of W. If the first- and third-modes of the proposed antenna, in fat and muscle tissues, work cooperatively to achieve similar impedance at the MICS band, then a singular

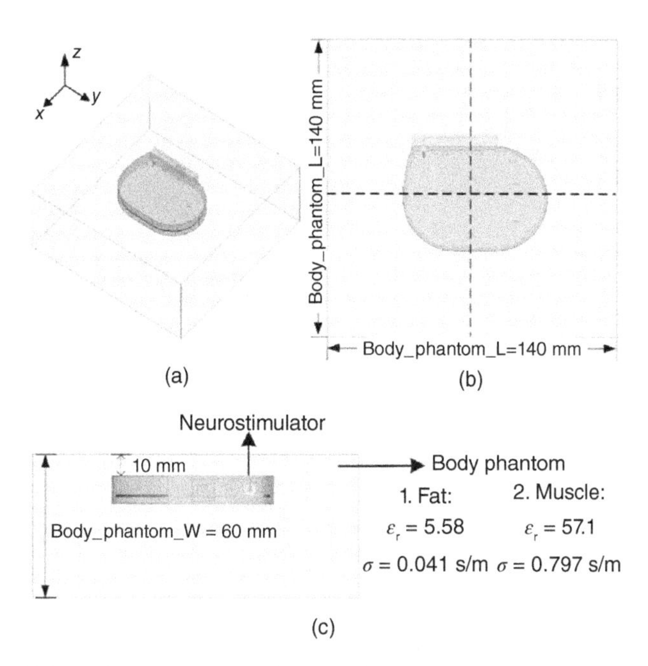

Figure 9.22 (a) Overview of the neurostimulator in one-layer homogeneous body phantom, (b) z-direction view, and (c) x-direction view. Source: Feng et al. [31]/IEEE.

matching circuit can be designed to match both cases synchronously. The co-matching method can be explained as follows:

$$
\begin{aligned}
&\text{s.t.}:\\
&\text{Case 1}: Z_{403\ \text{MHz\_ideal}} = Z_{\text{ant\_1st}}\,(\text{in Fat}) + Z_{\text{circuit\_1st}}\,(\text{in Fat})\\
&\text{Case 2}: Z_{403\ \text{MHz\_ideal}} = Z_{\text{ant\_3rd}}(\text{in Muscle}) + Z_{\text{circuit\_3rd}}\,(\text{in Muscle})\\
&\text{obj}: \qquad (9.1)\\
&\Delta Z = Z_{\text{ant\_1st}}\,(\text{in Fat}) - Z_{\text{ant\_3rd}}\,(\text{in Muscle}) \to (0, 0j)\\
&\text{co-matching with identical matching circuit}:\\
&Z_{\text{circuit\_1st}}\,(\text{in Fat}) \approx Z_{\text{circuit\_3rd}}\,(\text{in Muscle})
\end{aligned}
$$

Figure 9.23(b), (d), and (f), along with Table 9.6, present the analysis of antenna impedance in the body phantom at 403 MHz. Drawing from the analysis method in [36], we examined the constellation of impedance in tissues. The pink area, for instance, in Figure 9.23(d), serves as the constellation of impedance in tissues and is the smallest area on the Smith chart that can encompass the MICS band impedance in both fat and muscle tissues when $W = 1.125$ mm. The pink areas in Figure 9.23(b) and (f) are defined similarly to those in Figure 9.23(d) and can be quantified by the value interval provided in Table 9.6. The smaller the pink area on the Smith chart, the easier it is to match the constellation of impedance with a singular matching circuit. Hence, $W = 1.125$ mm was chosen for cooperative matching due to its offering the closest constellation points, as observed by comparing Figure 9.23(b), (d), and (f).

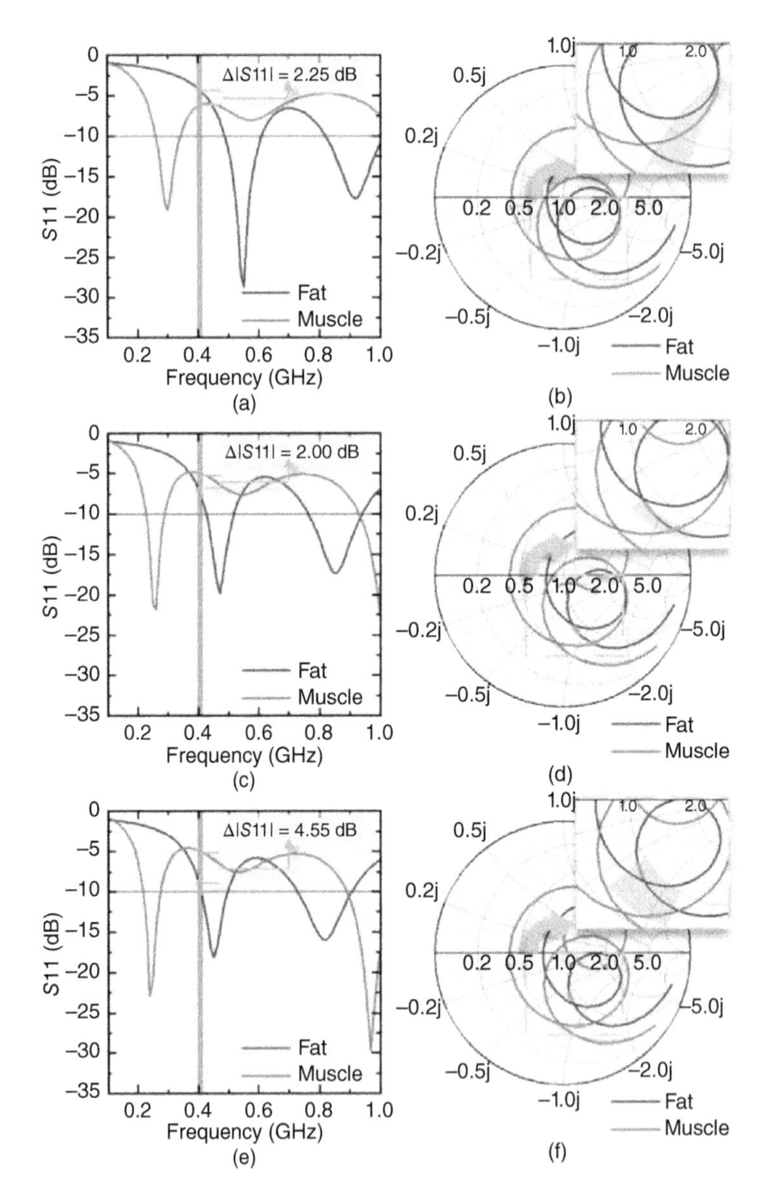

Figure 9.23 Antenna impedance analysis in the body phantom. The change of W exerts a significant influence on the impedance of the proposed antenna at the MICS band: (a) and (b) $W = 1$ mm, (c) and (d) $W = 1.125$ mm, (e) and (f) $W = 1.25$ mm; the smaller the pink area in (b), (d), and (f), the easier the impedance co-matching is Ant_L1 = 15 mm. Source: Feng et al. [31]/IEEE.

9.3.3 Co-matching Circuit Design

Figure 9.24 illustrates the co-matching principle of the impedance constellation. Thanks to the appropriately designed antenna, the real parts of the impedance in the MICS band are positioned within the interval [0.75, 0.875] as per Table 9.2, causing the pink area in Figure 9.24 to traverse along the blue area with a series inductor [53]. Consequently, the

Table 9.6 Antenna impedance analysis in the body phantom.

W (mm)	Normalized impedance (in fat)	Normalized impedance (in muscle)	Δ*Z*	Value interval
1	(0.70, −1.32j)	(2.08, −1.17j)	(−1.38, −0.15j)	([0.625 2.75], [−1.5j −1j])
1.125	(0.82, −0.89j)	(0.83, −1.24j)	(−0.01, 0.35j)	([0.75 0.875], [−1.5j −0.75j])
1.25	(0.89, −0.66j)	(0.64, −1.03j)	(0.25, 0.37j)	([0.625 1.0], [−1.25j −0.625j])

Note: Frequency–403 MHz.
Source: Feng et al. [31]/IEEE.

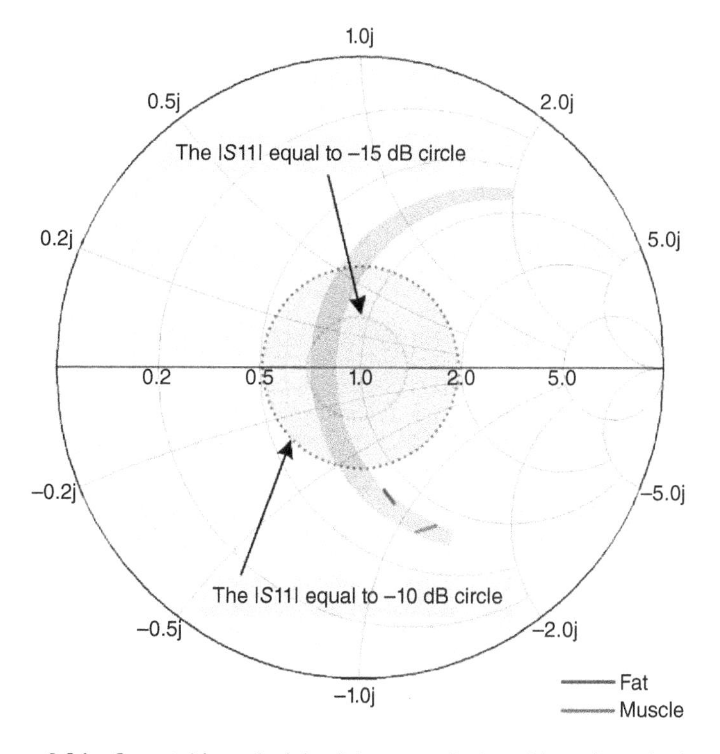

Figure 9.24 Co-matching principle of the constellation of impedance in tissues. Source: Feng et al. [31]/IEEE.

impedance constellation can be readily moved into the ideal matching area, delineated by the interior of the yellow circle, where Δ|*S*11| is less than −15 dB, as depicted in Figure 9.24. The matching results of the proposed antenna are exhibited in Figure 9.25. A series inductor, L_Matching = 21 nH, is employed to co-match the proposed antenna in tissue phantoms. As portrayed in Figure 9.26, we observe 1/4 wavelength mode in fat and 3/4 wavelength mode in muscle, both before and after the matching process. Furthermore,

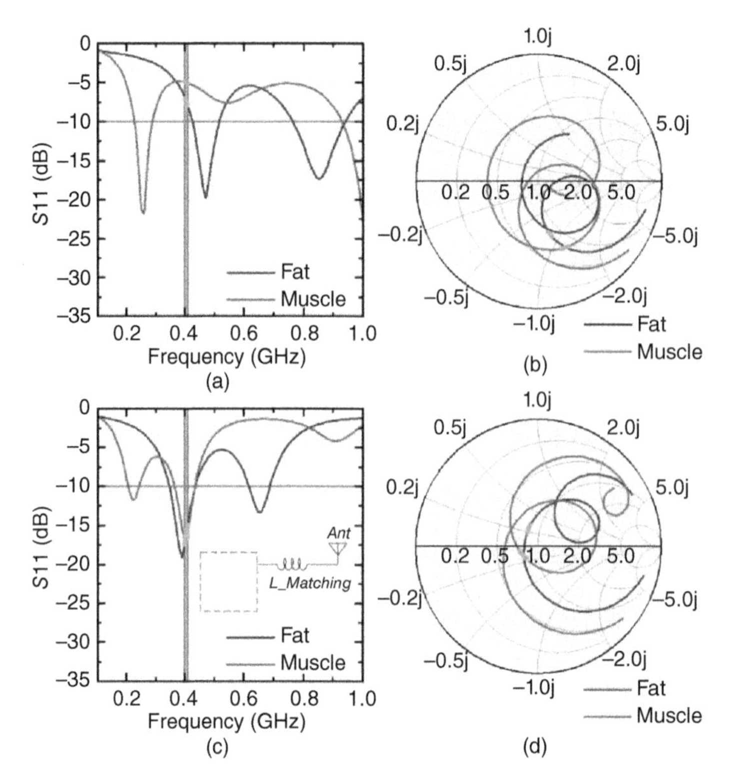

Figure 9.25 Parameters: Ant_L1 = 15 mm and W = 1.125 mm: (a) and (b) Simulated |$S11$| and impedance of the proposed antenna in fat and muscle phantoms, before matching, (c) and (d) using a series inductor, L_Matching = 21 nH, to co-matching the proposed antenna in fat and muscle phantoms. Source: Feng et al. [31]/IEEE.

the current density increases post-matching, signifying a greater amount of energy being delivered from the port.

9.3.4 Fabrication Processing of the Proposed Antenna

Figure 9.27 presents the model of the neurostimulator and its internal components. First, we employed a femtosecond laser (StarCut Tube 2 + 2, manufactured by Rofin) to cut copper sheets and generate the spiral pattern of the antenna, as illustrated in Figure 9.27(a). Next, the antenna was embedded into a grooved housing for secure positioning, as shown in Figure 9.27(b). This grooved housing was created from 3D printed resin material, the relative permittivity of which is similar to that of silicone rubber. A silicone rubber coating was applied to the upper surface of the grooved housing to encapsulate the antenna, maintaining a thickness between 0.1 and 0.3 mm. The antenna was then connected to the feedthrough and the interior of the grooved housing was filled with silicone rubber, as depicted in Figure 9.27(c). In the final step, the header, titanium case, and PCB were assembled together, as illustrated in Figure 9.27(d) and (e). In practical applications, the titanium case would be welded using laser welding. Silicone rubber is used to seal the titanium case, ensuring its water-tightness during the measurement.

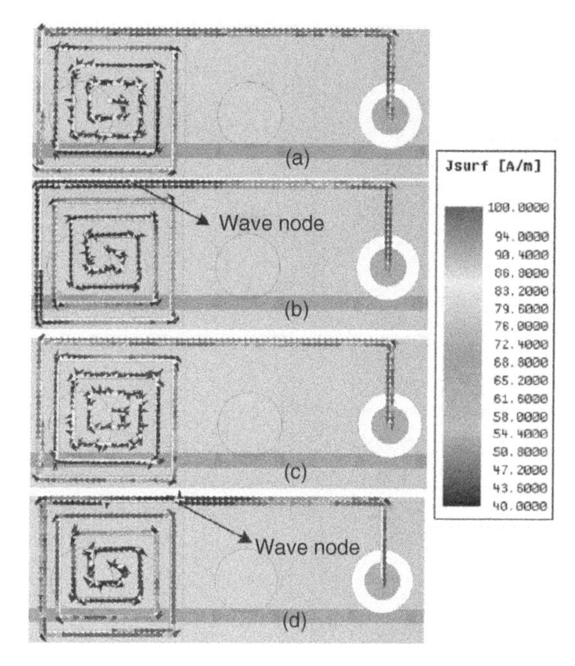

Figure 9.26 Operation mode of the proposed antenna in different tissues at 403 MHz: (a) in fat tissue, before matching, (b) in muscle tissue, before matching, and (c) in fat tissue, after matching, (d) in muscle tissue, after matching. Ant_L1 = 15 mm and $W = 1.125$ mm. Source: Feng et al. [31]/IEEE.

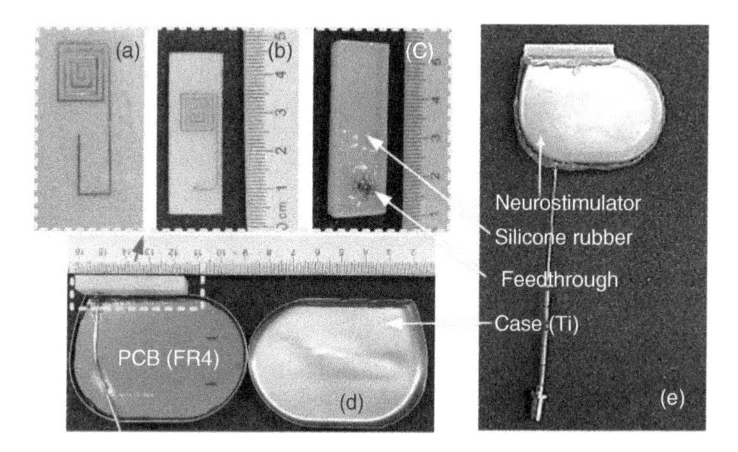

Figure 9.27 (a)–(d) Fabrication processing of the proposed antenna, (e) prototype of the neurostimulator. Source: Feng et al. [31]/IEEE.

9.3.5 Reflection Coefficient and Impedance Measurement

We conducted a reflection coefficient measurement to validate the proposed co-matching method, as presented in Figure 9.28. The neurostimulator was placed in a plastic container filled with liquid phantom and the measurement was taken using an R&S ZVL VNA.

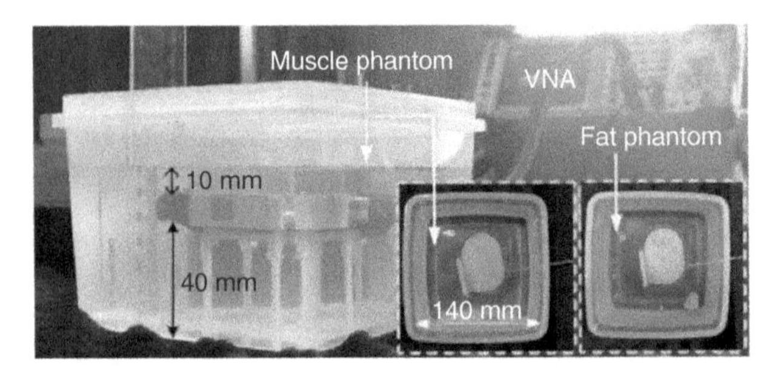

Figure 9.28 |*S*11| and impedance measurement setup. Source: Feng et al. [31]/IEEE.

The liquid phantom, with dimensions of $140 \times 140 \times 60\,\text{mm}^3$, was of the same size as the one-layer phantom in the simulation. Additionally, the liquid phantom was prepared following the recipe provided in [57, 58]. The fat phantom consisted of oil (76.7%) and Triton X-100 (23.3%), while the muscle phantom was composed of deionized water (52.4%), sugar (45.0%), and NaCl (1.4%).

Figure 9.29 presents the measured |*S*11| and impedance prior to matching. Despite a slight leftward shift in resonant frequencies, the impedance in different tissue phantoms at the MICS band still exhibits co-matching characteristics. The value of ΔZ is sufficiently small, particularly the imaginary part of ΔZ, which is virtually 0j, as listed in Table 9.7. In this instance, a matching circuit comprising only a single series inductor was designed. Finally, we employed a 22 nH series inductor for the cooperative matching. The matching results are displayed in Figure 9.30. Despite significant changes in the relative permittivity of the tissue environment, the reflection coefficients correlated well with the simulations. In both fat and muscle phantoms, the reflection coefficients at 403 MHz were found to be −23.05 dB and −23.76 dB, respectively.

9.3.6 Radiation Performance

The indoor path loss is strongly dependent on the environment [59], and indoor measurements cannot yield generalized results. Hence, we measured the communication link |*S*21| between the neurostimulator and the dipole antenna in an open-air environment. As shown in Figure 9.31, an Rx dipole antenna in vertical polarization mode was connected to port-2 of the VNA. The neurostimulator, also in vertical polarization mode and connected to port-1, was placed in a plastic box with a lid, filled with the liquid phantom. Taking into account the noise floor, and without loss of generality, the Rx dipole antenna was set 3 m apart from the neurostimulator, a distance more than four times the wavelength in air at 403 MHz.

In medical data transmission, such as home monitoring, the receiving antenna is always positioned in front of the patients. Hence, in this measurement, we evaluated the |*S*21| in the $+z$ direction, representing the direction facing the patient. The neurostimulator was situated 10 mm from the liquid phantom in the $+z$ direction, which is akin to the distance of the neurostimulator from the body surface. The communication link between two dipole antennas was measured under the same conditions as a frame of reference.

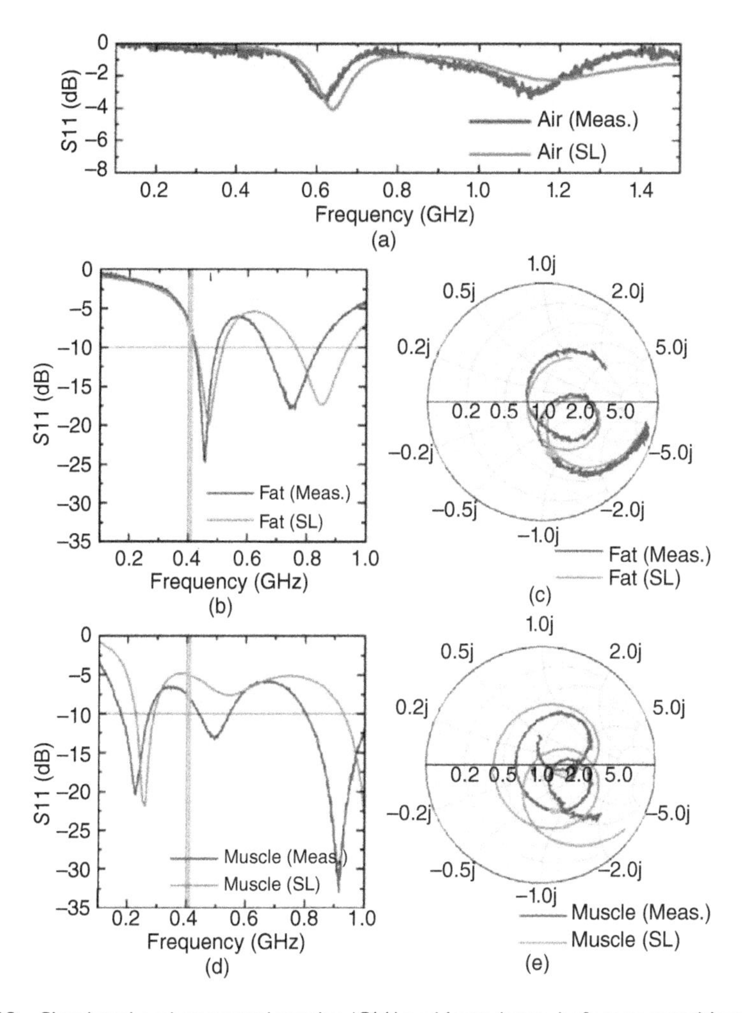

Figure 9.29 Simulated and measured results, |*S*11| and impedance, before co-matching: (a) in air, (b) and (c) in fat phantom, (d) and (e) in muscle phantom. Source: Feng et al. [31]/IEEE.

Table 9.7 Simulated and measured antenna impedance in the body phantom.

	Normalized impedance (in fat)	**Normalized impedance (in muscle)**	**Δ*Z***	**Value interval**
Meas.	(0.72, −0.79j)	(0.91, −0.86j)	(−0.19, 0.07j)	([0.625 1.0], [−0.875j −0.75j])
SL	(0.82, −0.89j)	(0.83, −1.24j)	(−0.01, 0.35j)	([0.75 0.875], [−1.5j −0.75j])

Note: Frequency–403 MHz.

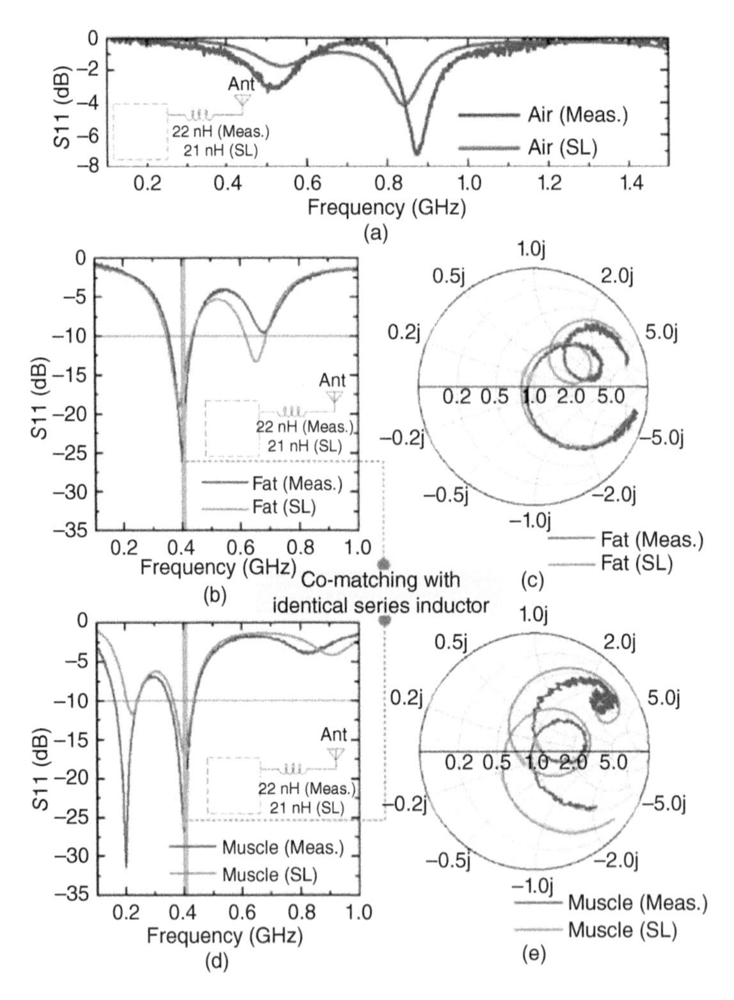

Figure 9.30 Simulated and measured results, |$S11$| and impedance, after co-matching: (a) in air, (b) and (c) in fat phantom, (d) and (e) in muscle phantom. Source: Feng et al. [31]/IEEE.

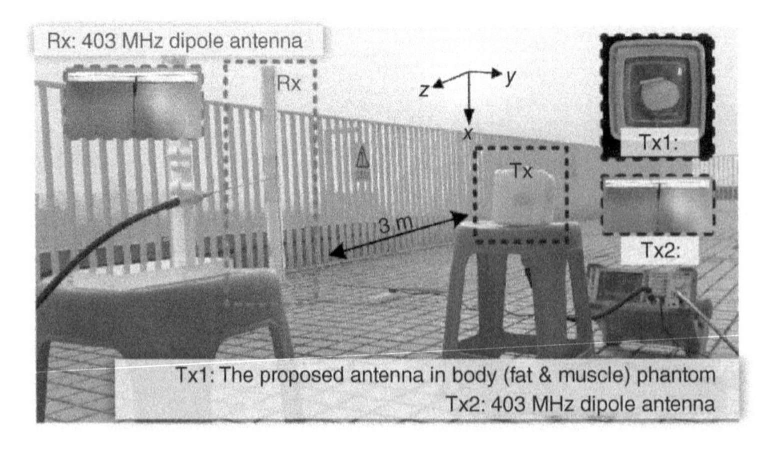

Figure 9.31 |$S21$| measurement setup. Source: Feng et al. [31]/IEEE.

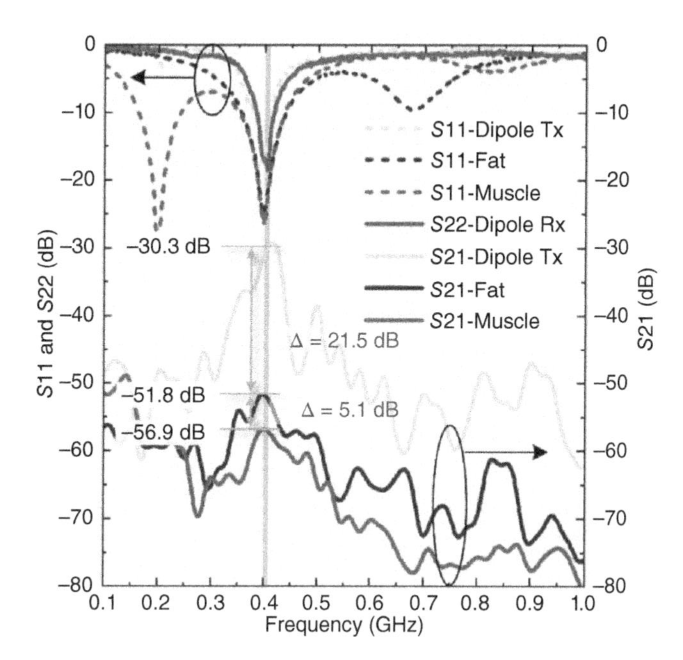

Figure 9.32 Measured |*S*21| in the +*z* direction, zoy plane. Source: Feng et al. [31]/IEEE.

Figure 9.32 presents the measured |*S*21|. The |*S*21| between two dipole antennas is −30.3 dB. Moreover, the 3 m path loss at 403 MHz in free space is 34.12 dB [31]. Thus, the gains of the Tx and Rx dipole antennas are 1.91 dBi. Using the same method, we extracted the approximate gain of the proposed antenna in tissues, considering the gain of the Rx dipole antenna, the path loss, and the measured |*S*21| between the proposed antenna and the Rx dipole antenna. In the +*z* direction of the zoy plane, in the liquid phantom of fat, the Gain = −19.59 dBi, and in the liquid phantom of muscle, the Gain = −24.69 dBi.

9.4 Reconfigurable Antenna for Neurostimulator

In Section 9.3, we applied the dual-mode antenna design method to the design of the neurostimulator antenna and adopted a cooperative impedance matching design scheme to solve the detuning problem of the implantable neurostimulator antenna in two types of human tissues with different dielectric properties. However, in practical applications, due to factors such as individual patient differences, it is difficult to simply place the stimulator in fat or muscle. Moreover, there might be different implantation sites, such as the armpit, that may be influenced by various other factors. The research in this chapter further explores the general solution to the detuning problem of the implantable neurostimulator antenna in human tissues and proposes a design method for the reconfigurable matching antenna of the neurostimulator.

In this section, we will introduce the working principle and design method of the reconfigurable matching antenna of the neurostimulator and implement the system design and test verification on the neurostimulator platform. Through reconfigurable matching design,

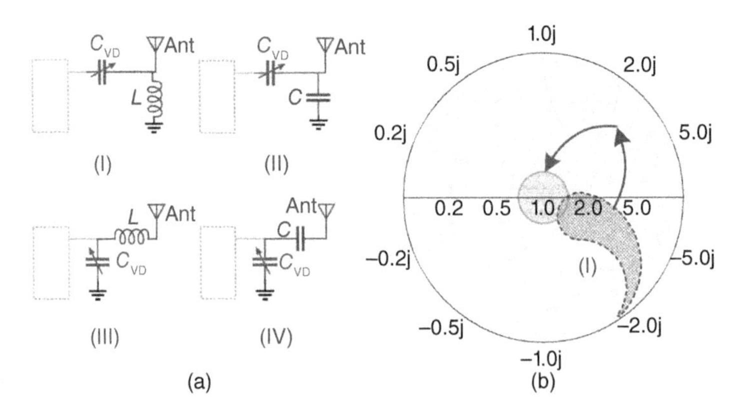

Figure 9.33 (a) Four L-type matching circuits, (b) tuning region of No. (I) L-type matching circuit, plotted in blue. Source: Feng et al. [60]/IEEE.

the implantable antenna of the neurostimulator can maintain stable impedance matching characteristics when the dielectric constant and conductivity of human tissues continuously change within the range of $\varepsilon_r \in [5, 60]$, $\sigma \in [0.04, 1]$ S/m. Therefore, the link efficiency of the implantable antenna is increased by 6–12 dB [60].

9.4.1 Tuning Principle

In mobile antenna designs, components such as varactors [varactor diodes (VDs)] and *p-i-n* diodes are typically employed to alter the electrical properties and hence reconfigure the operating frequencies [50]. Varactors, being voltage-dependent capacitors, are ideal for antenna tuning in a changeable biological environment. Figure 9.33(a) showcases four L-type matching circuits, each composed of a varactor and a lumped inductor or capacitor. Figure 9.33(b), for example, presents the tuning region of the L-type matching circuit No. (I). The green region in the Smith chart denotes the ideal matching area for a load Z_{load} intended for an RF transmitter, as described in (1). The tuning regions for the L-type matching circuit No. (I) are defined in Eqs. (9.2 and (9.3)).

Practically, the impedance trace on the Smith charts allows for a quick definition of the tuning regions. To adjust the antenna in varying biological environments, we need to identify parameters of C_{VD}, L, or C that offer a sufficiently wide tuning region to encompass all the impedances of the antenna in different tissues, as follows:

$$20 \log_{10} | (Z_{\text{load}} - Z_0)/(Z_{\text{load}} + Z_0) | \leq -15\,\text{dB};\; Z_0 = 50 + 0j \tag{9.2}$$

$$Z_{\text{load}} = Z_{\text{Ant}}/(1 - j \cdot Z_{\text{Ant}}/2\pi fL) - j/2\pi fC_{\text{VD}};\; f = 403\,\text{MHz}. \tag{9.3}$$

9.4.2 Antenna Configuration and Design Procedures

In our study, we designed a planar spiral monopole antenna for the neurostimulator without loss of generality. Figure 9.34(a) demonstrates the antenna layout, which occupies a space of $20 \times 10\,\text{mm}^2$ with a clearance height of 5 mm. The total length of the proposed antenna roughly corresponds to the quarter-wavelength of a 403 MHz wave in fat

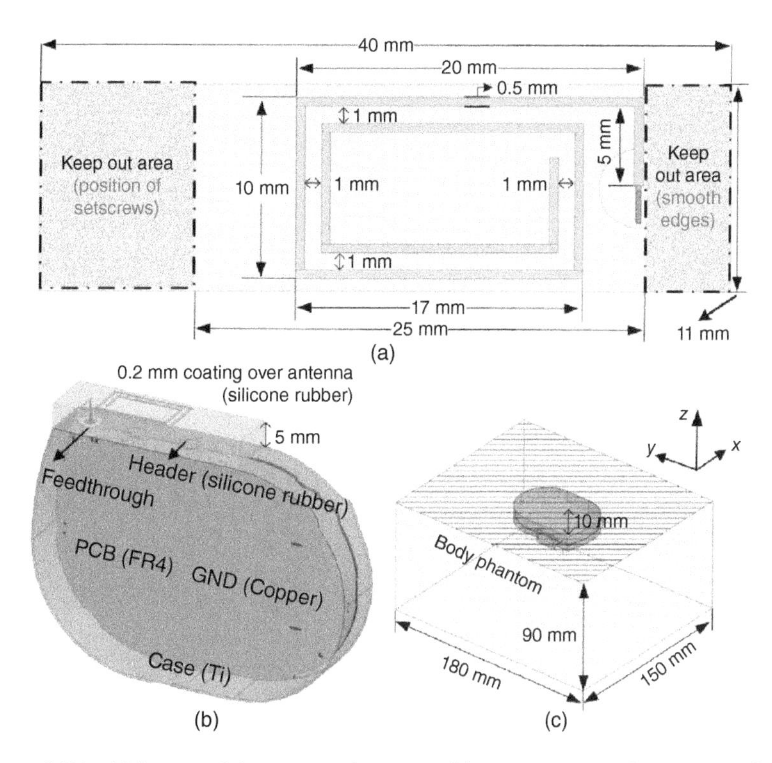

Figure 9.34 (a) Layout of the proposed antenna, (b) components and structures of the neurostimulator, and (c) homogeneous tissue phantom. Source: Feng et al. [60]/IEEE.

tissue (not infiltrated). The simulation setup for the neurostimulator is similar to that in Section 9.3.2, as shown in Figure 9.34(b) and (c).

A myriad of evidence suggests that the spectrum of these dielectric properties should lie within $\varepsilon_r \in [5, 60]$, $\sigma \in [0.04, 1]$ S/m [60]. To discuss the impedance features of the antenna, we utilize five tissue examples with distinct dielectric attributes, including non-infiltrated fat, 40% muscle, 2/3 muscle, skin, and muscle. These tissues' dielectric properties are documented in Table 9.8, symbolizing the most illustrative data sets within the dielectric property range. To optimize the antenna and the reconfigurable matching circuit, we adhered to the subsequent steps:

(1) Step 1: Simulations commenced with a planar spiral monopole antenna for the neurostimulator in non-infiltrated fat tissue.
(2) Step 2: Explored the impedance characteristics of the antenna in various tissues, including non-infiltrated fat, 40% muscle, 2/3 muscle, skin, and muscle tissues.
(3) Step 3: One of the four circuit topologies depicted in Figure 9.33(a) was selected.
(4) Step 4: Circuit parameters were determined with an appropriate tuning region to accommodate all the antenna impedances in different tissues.
(5) Step 5: If the lumped components correlating to the circuit parameters are readily available, the process ends. If not, the circuit parameters are reselected, reverting to Step 4; circuit topologies are reselected, reverting to Step 3; the antenna is redesigned, reverting to Step 1.

Table 9.8 Dielectric properties of tissues at 403 MHz.

	Conductivity (S/m)	Relative permittivity
Fat (not infiltrated)	0.041	5.58
Fat (avg. infiltrated)	0.081	11.62
Skin	0.689	46.72
Muscle	0.797	57.10
Blood	1.351	64.15
2/3 Muscle	0.531	38.07
40% Muscle	0.319	22.84

Source: Adapted Gabriel et al. [18].

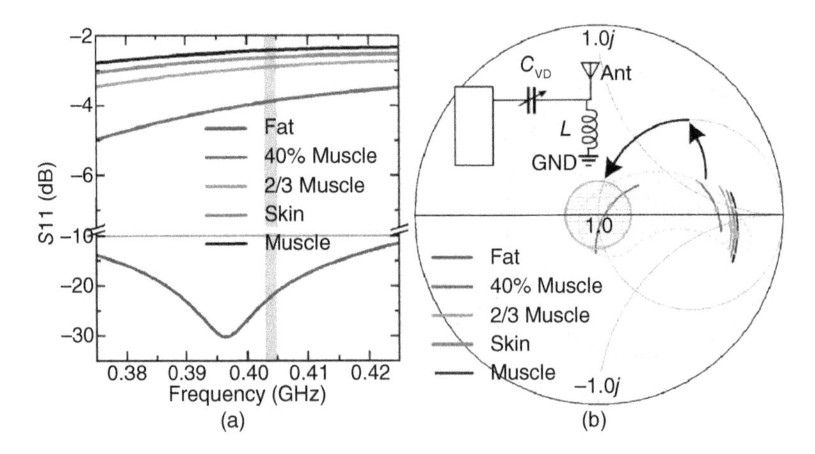

Figure 9.35 (a) and (b) Simulated |$S11$|, and impedance of the proposed antenna in various tissue phantoms. Source: Feng et al. [60]/IEEE.

Following these steps, we discovered a suitably designed layout for the antenna, as demonstrated in Figure 9.34(a). Moreover, as illustrated in Figure 9.35, the unmatched impedance of the antenna in tissues with high permittivity, including 40% muscle, 2/3 muscle, skin, and muscle tissues, led to most of the energy being reflected in the port, culminating in a robust |$S11$|. By comparing the impedance constellation of the antenna in different tissues with the shapes of the tuning regions for the four reconfigurable matching circuits, we opted for the No. (I) L-type matching circuit for reconfigurable matching, and the circuit parameters were set as $C_{VD} \in [3,20]$ pF, $L = 47$ nH. Figure 9.35(b) illustrates the tuning region of the optimized reconfigurable matching circuit, which encompasses the entire impedance constellation of the antenna in different tissues. Figure 9.36 presents the matched antenna impedance in various tissues fine-tuned by the reconfigurable matching circuit. The superior antenna performance in variable biological environments is attributed to the optimized reconfigurable matching circuit.

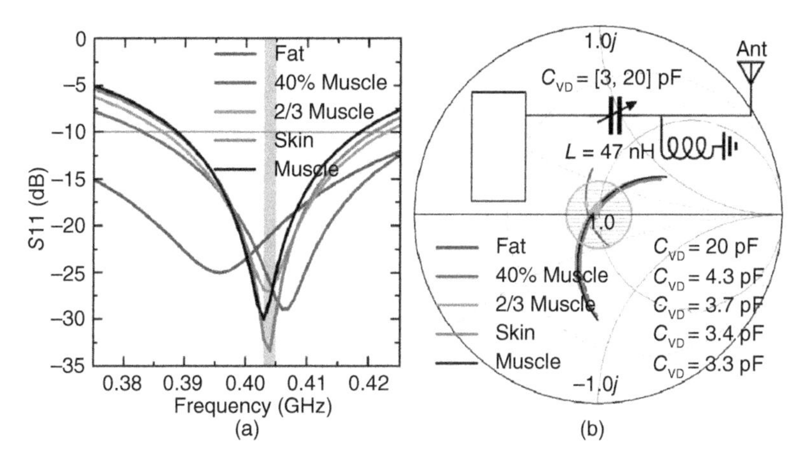

Figure 9.36 (a) and (b) Simulated |*S*11|, and impedance of the proposed antenna after matching by using the reconfigurable matching circuit. Source: Feng et al. [60]/IEEE.

9.4.3 Antenna Manufacturing and Measurement Setup

The proposed antenna was fabricated using a femtosecond laser (StarCut Tube 2 + 2) and assembled within the header, as depicted in Figure 9.37(a) and (b). To secure the antenna, we designed a groove-fitted housing composed of resin material for the header. The interior of this grooved housing was filled with silicone rubber. The resin and the silicone rubber shared identical permittivity. On the upper surface of the housing, we applied a 0.2 mm thick coating of silicone rubber to encapsulate the antenna. Figure 9.37(c) showcases the assembled neurostimulator, which consists of the Ti case, PCB, feedthrough, and header. To guarantee air-tightness during the measurement, we utilized silicone rubber to seal the Ti case, as opposed to laser welding the case, which is the conventional method for actual products.

The antenna impedance was measured in different tissue phantoms. As illustrated in Figure 9.37(d), the antenna, mounted on the neurostimulator, was evaluated in a liquid phantom configured similarly to the simulation. Phantoms mimicking different tissues were created in accordance with the recipes provided in [60]. The detailed compositions are as follows:

(1) Fat (not infiltrated) phantom, by volume: oil (76.7%) and Triton X-100 (23.3%).
(2) 40% muscle phantom, by volume: Triton X-100 (78.5%), deionized water (21.5%), NaCl (4.28 g/L).
(3) 2/3 muscle phantom, by weight: glycerin (71.3%), deionized water (20.1%), and NaCl (4.0%).
(4) Skin phantom, by weight: sugar (56.2%), NaCl (2.3%), and deionized water (41.5%).
(5) Muscle phantom, by weight: sugar (45.0%), NaCl (1.4%), and deionized water (52.4%).

Figure 9.38(b) displays the measured antenna impedance, which aligns well with the simulation results. The entire impedance constellation of the antenna in different tissue

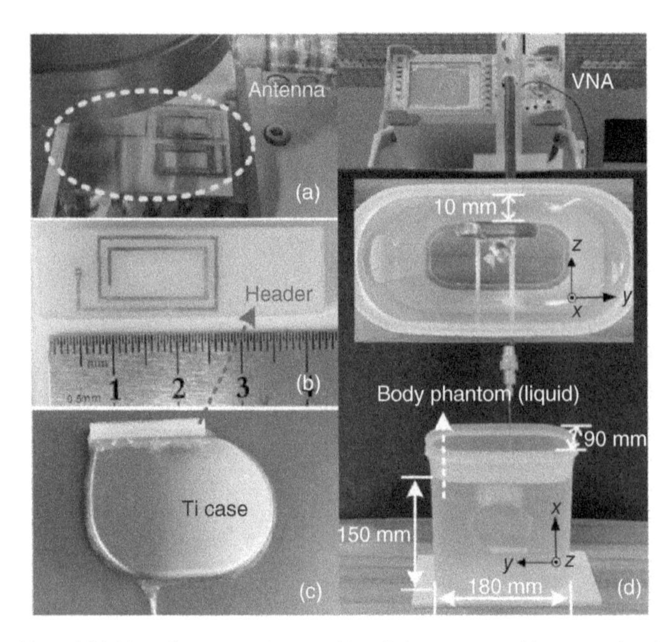

Figure 9.37 (a) and (b) Manufacture and assembly of the antenna, (c) neurostimulator prototype, and (d) |S11| and impedance measurement setup. Source: Feng et al. [60]/IEEE.

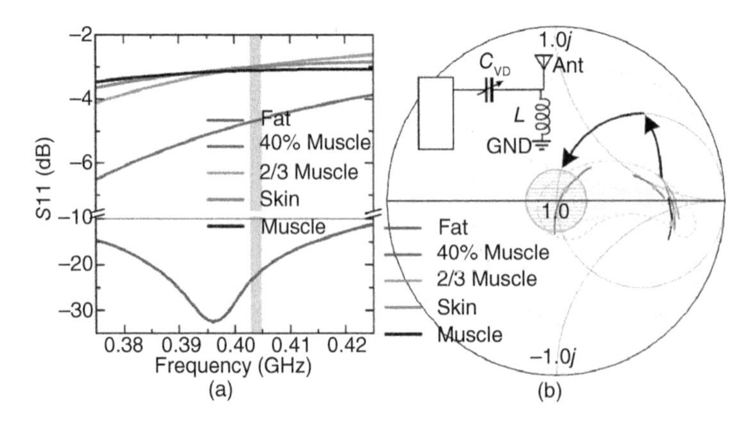

Figure 9.38 (a) and (b) Measured |S11|, and impedance of the proposed antenna in various tissue phantoms. Source: Feng et al. [60]/IEEE.

phantoms was situated within the tuning range of $C_{VD} \in [3,20]$ pF, with a shunt inductor of $L = 47$ nH.

9.4.4 System Design

Based on the design process, we can readily integrate the reconfigurable matching circuits with the neurostimulator platform. A 3.7 V DC battery, commonly used for pacemakers,

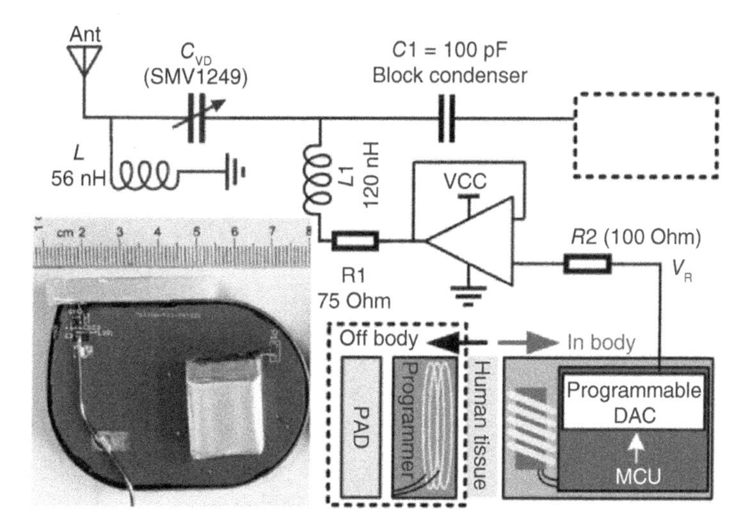

Figure 9.39 Reconfigurable matching circuit schematic and its integrated schematic with the neurostimulator system. Source: Feng et al. [60]/IEEE.

fuels our system. The varactor, SMV1249 (Skyworks), with a variable capacitance range from 37.25 pF (at reverse voltage, $V_R = 0.0$ V) to 2.96 pF (at $V_R = 3.5$ V), satisfies the design requirements. In practice, a 56 nH inductor is used for antenna tuning, instead of the 47 nH inductor used in the simulation. As depicted in Figure 9.39, the varactor is driven by a voltage follower, while inductor L also serves as a DC short inductor. V_R is managed by an on-chip programmable digital-to-analog converter (DAC) within the microcontroller unit (MCU).

9.4.5 Antenna Tuning and Optimized RF Link

We evaluated the matching characteristics $|S11|$ of the proposed antenna. We used the Program System for wireless DAC programming, and the measured $|S11|$ served as the feedback marker to determine the appropriate reverse voltage (V_R) while tuning the antenna. Figure 9.40(a) displays the programmed V_R values when the neurostimulator was immersed in different tissue phantoms. As depicted in Figure 9.40(b) and (c), we also measured the I_{DC} of the DC bias. Thanks to the nanopower operational amplifier LPV521 (Texas Instruments), the maximum IDC remains under 0.5 μA, as shown in Figure 9.40(a). The DC bias loss constitutes less than 0.1% of the total power consumption of the neurostimulator [61]. Figure 9.41 reveals that, despite the variable biological environments, our proposed antenna within the neurostimulator platform maintains a satisfactory match at the MICS band with reflection coefficients below −20 dB.

We measured the $+z$ direction communication link $|S21|$ between the neurostimulator within different tissue phantoms and a dipole antenna in an open rooftop setting, as depicted in Figure 9.42. The neurostimulator and the dipole were vertically polarized and set 3 m apart from each other. For reference, the $|S21|$ between two dipole antennas

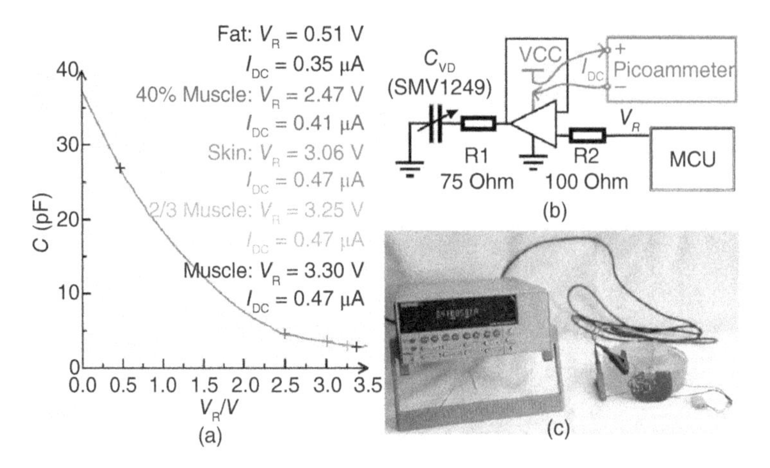

Figure 9.40 (a) Values of V_R and I_{DC} for reconfigurable matching, (b) DC circuit topology of the matching circuit, and (c) IDC measurement setup. Source: Feng et al. [60]/IEEE.

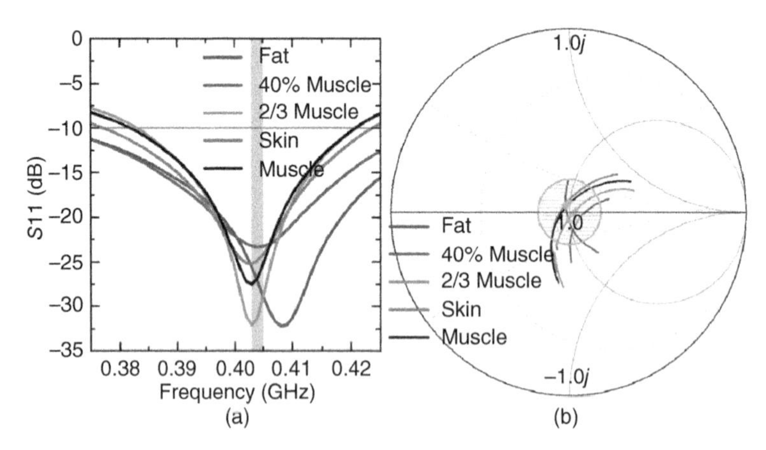

Figure 9.41 (a) and (b) Measured |S11|, and impedance of the proposed antenna after matching by using the RC. Source: Feng et al. [60]/IEEE.

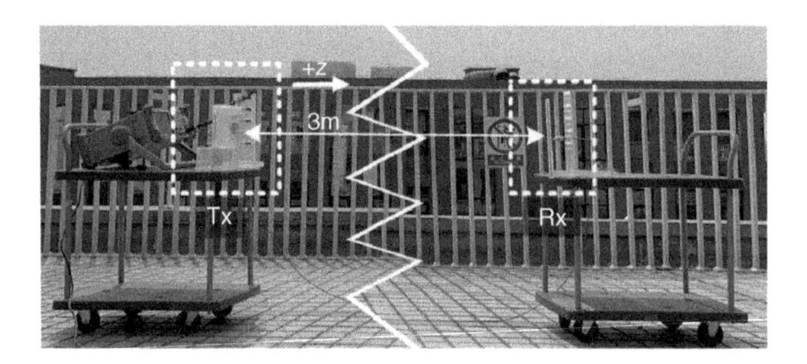

Figure 9.42 |S21| Measurement setup. Source: Feng et al. [60]/IEEE.

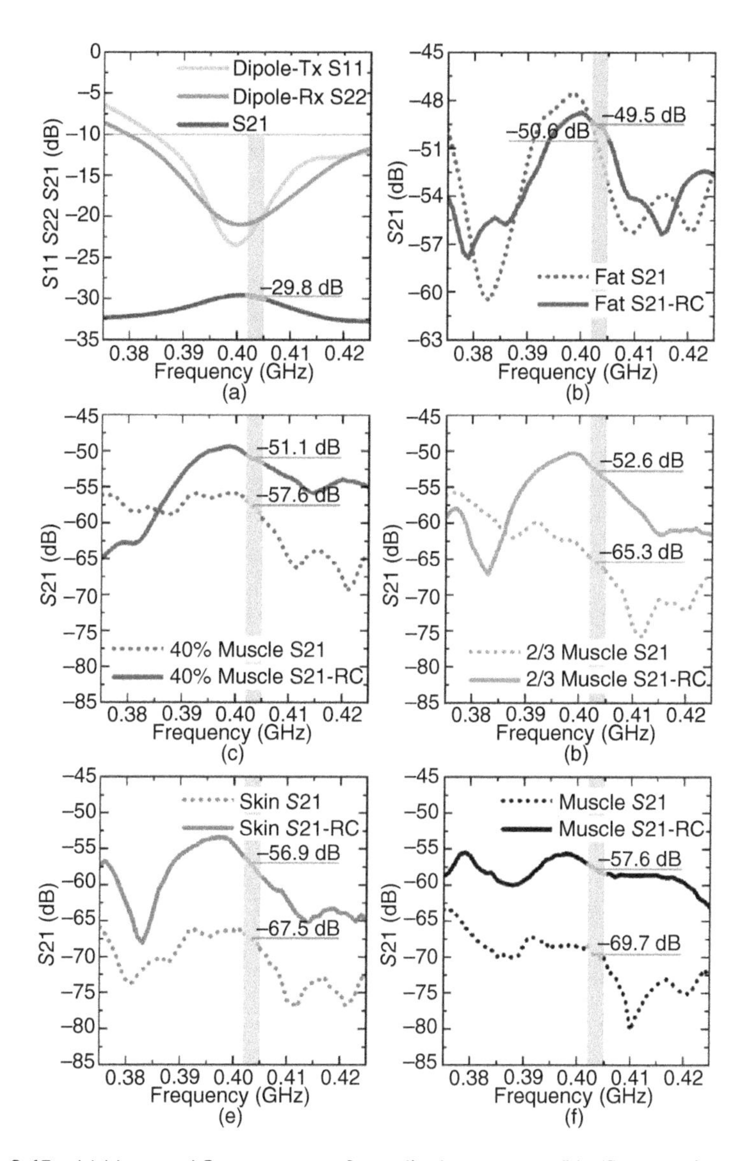

Figure 9.43 (a) Measured S-parameters of two dipole antennas, (b)–(f) comparison of |*S*21|, without or with the RC. The marked values are the |*S*21| at 403 MHz. Source: Feng et al. [60]/IEEE.

was measured in the same environment. To mitigate cable radiation effects, we adopted the approach suggested by El-Saboni et al. [62] and positioned split coaxial ferrite cores on the feed cable. As shown in Figure 9.43, in the fat (not infiltrated) phantom, the impedance of the proposed antenna at the MICS band is consistently matched, leading to a stable link budget. In other tissue phantoms, after being fine-tuned by the reconfigurable circuit (RC), the link budget of the neurostimulator demonstrated significant improvement, with benefits of 6–12 dB achieved.

9.5 Summary

Advancements in wireless power delivery and implantable antenna technologies have significantly elevated the biomedical engineering landscape, paving the way for innovative solutions in neurological treatments and functional restoration. The amalgamation of these wireless capabilities into medical devices not only enhances patient comfort but also mitigates the risk of complications and bolsters treatment outcomes.

In this chapter, our focal point revolved around the application scenarios of Peripheral Nerve Implants and Neurostimulators. We presented a detailed exploration of design solutions for wireless power delivery, wireless telemetry antennas, and system integration. Based on the practical challenges that manufacturers encounter in clinical applications, we have proposed robust and feasible solutions.

When addressing the realm of biomedical and implantable medical device applications, it is clear that methods for wireless power delivery and implantable antenna design are multifarious. With well-defined engineering guidelines at our disposal, the focus of future research should be geared toward practical implementation and confronting the specificities of clinical applications, including considerations of safety, reliability, and seamless interoperability with other medical devices and systems.

To sum up, strides in wireless power delivery and implantable antenna design have laid the groundwork for transformative medical interventions. As we continue to explore and address the practical facets of integrating these technologies, we foresee a further enhancement in the effectiveness and accessibility of implants in clinical practice.

References

1 Kandel, E.R., Markram, H., Matthews, P.M. et al. (2013). Neuroscience thinks big (and collaboratively). *Nat. Rev. Neurosci.* 14 (9): 659–664.

2 Silva, G.A. (2006). Neuroscience nanotechnology: progress, opportunities and challenges. *Nat. Rev. Neurosci.* 7 (1): 65–74.

3 Movassaghi, S., Abolhasan, M., Lipman, J. et al. (2014). Wireless body area networks: a survey. *IEEE Commun. Surv. Tutor.* 16 (3): 1658–1686.

4 Mekhail, N.A., Cheng, J., Narouze, S. et al. (2010). Clinical applications of neurostimulation: forty years later. *Pain Pract.* 10 (2): 103–112.

5 Volkmann, J., Daniels, C., and Witt, K. (2010). Neuropsychiatric effects of subthalamic neurostimulation in Parkinson disease. *Nat. Rev. Neurol.* 6 (9): 487–498.

6 Starnes, K., Miller, K., Wong-Kisiel, L., and Lundstrom, B.N. (2019). A review of neurostimulation for epilepsy in pediatrics. *Brain Sci.* 9 (10): 283.

7 Cruccu, G., Garcia-Larrea, L., Hansson, P. et al. (2016). EAN guidelines on central neurostimulation therapy in chronic pain conditions. *Eur. J. Neurol.* 23 (10): 1489–1499.

8 Barbruni, G.L., Ros, P.M., Demarchi, D. et al. (2020). Miniaturised wireless power transfer systems for neurostimulation: a review. *IEEE Trans. Biomed. Circ. Syst.* 14 (6): 1160–1178.

9 Bihr, U., Liu, T., and Ortmanns, M. (2014). Telemetry for implantable medical devices: Part 3-Data telemetry. *IEEE Solid-State Circ. Magazine* 6 (4): 56–62.

10 Jegadeesan, R., Nag, S., Agarwal, K. et al. (2015). Enabling wireless powering and telemetry for peripheral nerve implants. *IEEE J. Biomed. Health Inform.* 19 (3): 958–970.

11 Keith, M.W., Peckham, P.H., Thrope, G.B. et al. (1989). Implantable functional neuromuscular stimulation in the tetraplegic hand. *J. Hand Surg. Am.* 14: 524–530.

12 Sokal, N.O. and Sokal, A.D. (1975). Class E: a new class of high-efficiency tuned single-ended switching power amplifiers. *IEEE J. Solid-State Circ.* SC-10(3): 168–176.

13 IT'IS Foundation. (2018). *Tissue Frequency Chart.* https://itis.swiss/virtual-population/tissue-properties/database/tissue-frequency-chart/

14 Lee, C.-M., Yo, T.-C., Huang, F.-J., and Luo, C.-H. (2009). Bandwidth enhancement of planar inverted-F antenna for implantable biotelemetry. *Microw. Opt. Technol. Lett.* 51 (3): 749–752.

15 IEEE (2006). *IEEE Standard for Safety Levels With Respect to Human Exposure to Radio Frequency Electromagnetic Fields 3 kHz to 300 GHz*, 1–238.

16 Lee, D. (2012). Biocompatibility of a PDMS-coated micro-device: bladder volume monitoring sensor. *Chin. J. Polym. Sci.* 30 (2): 242–249.

17 Bernard, P.A. and Gautray, J.M. (1991). Measurement of dielectric constant using a microstrip ring resonator. *IEEE Trans. Microw. Theory Tech.* 39 (3): 592–595.

18 Gabriel, S., Lau, R.W., and Gabriel, C. (1996). The dielectric properties of biological tissues—Part III: Parametric models for the dielectric spectrum of tissues. *Phys. Med. Biol.* 4 (11): 2271–2293.

19 Patrick, J.F., Busby, P.A., and Gibson, P.J. (2006). The development of the nucleus freedom cochlear implant system. *Trends Amplif.* 10 (4): 175–200.

20 Zeng, F.G., Rebscher, S., Harrison, W. et al. (2008). Cochlear implants: system design integration and evaluation. *IEEE Rev. Biomed. Eng.* 1: 115–142.

21 Qian, X., Chen, Y., Feng, Y. et al. (2017). A platform for long-term monitoring the deep brain rhythms. *Biomed. Phys. Eng. Express* 3 (1).

22 Huang, W. and Kishk, A.A. (2011). Embedded spiral microstrip implantable antenna. *Hindawi Int. J. Antennas Propag.* 201.

23 Kod, M. et al. (2016). Feasibility study of using the housing cases of implantable devices as antennas. *IEEE Access* 4: 6939–6949.

24 Lee, S., Seo, W., Ito, K., and Choi, J. (2011). Design of an implanted compact antenna for an artificial cardiac pacemaker system. *IEICE Electron. Exp.* 8 (24): 2112–2117.

25 Houzen, T., Takahashi, M., and Ito, K. (2007). Implanted antenna for an artificial cardiac pacemaker system. In: *Proceedings of the Progress in Electromagnetics Research Symposium, Prague, Czech Republic*, 51–54.

26 Choi, J., Kim, U., Lee, S., and Kwon, K. (2012). Design of an implantable antenna for WBAN applications. In: *Proceedings of the iWAT*, 213–216.

27 Huang, H., Chen, P.-Y., Ferrari, M. et al. (2013). Dual band electrically small non-uniform pitch ellipsoidal helix antenna for cardiac pacemakers. In: *Proceedings of the IEEE Radio Wireless Symposium (RWS)*, 325–327.

28 Psenakova, Z., Smondrk, M., Barabas, J. et al. (2016). Simulation and assessment of pacemaker RF exposure (2.4 GHz) by PIFA antenna. In: *Proceedings of the IEEE ELEKTRO*, 569–573.

29 Yang, Z.-J., Zhu, L., and Xiao, S. (2018). An implantable circularly polarized patch antenna design for pacemaker monitoring based on quality factor analysis. *IEEE Trans. Antennas Propag.* 66 (10): 5180–5192.

30 Vidal, N., Curto, S., Lopez-Villegas, J.M. et al. (2012). Detuning effects on implantable antenna at various human positions. In: *Proceedings of the 6th European Conference on Antennas and Propagation (EUCAP)*, 2131–2134.

31 Feng, Y., Li, Y., Li, L. et al. (2019). Tissue-dependent Co-matching method for dual-mode antenna in implantable neurostimulators. *IEEE Trans. Antennas Propag.* 67 (8): 5253–5264.

32 Sherif, C. et al. (2008). Deep brain pulse-generator and lead-extensions: subjective sensations related to measured parameters. *Mov. Disord.* 23 (7): 1036–1041.

33 Bakay, A.E. and Smith, A.P. (2011). Deep brain stimulation: complications and attempts at avoiding them. *Open Neurosurg. J.* 4: 42–52.

34 Po, F.C.W., De Foucauld, E., Delavaud, C. et al. (2008). An vector automatic matching system designed for wireless medical telemetry. In: *Proceedings of the IEEE NEWCAS-TAISA Joint Conference*, 89–92.

35 Po, F.C.W., de Foucauld, E., Morche, D. et al. (2011). A novel method for synthesizing an automatic matching network and its control unit. *IEEE Trans. Circuits Syst. I Reg. Papers* 58 (9): 2225–2236.

36 Li, P., Zhang, L., Liu, F., and Amely-Velez, J. (2015). Optimized matching of an implantable medical device antenna in different tissue medium using load pull measurements. In: *Proceedings of the 86th ARFTG Microwave Measurement Conference*, 1–4.

37 Magill, M.K., Conway, G.A., and Scanlon, W.G. (2015). Robust implantable antenna for in-body communications. In: *Proceedings of the Loughborough Antennas & Propagation Conference (LAPC)*, 1–4.

38 Magill, M.K., Conway, G.A., and Scanlon, W.G. (2017). Tissue-independent implantable antenna for in-body communications at 2.36-2.5 GHz. *IEEE Trans. Antennas Propag.* 65 (9): 4406–4417.

39 Lyu, H., Wang, J., La, J.-H. et al. (2018). An energy-efficient wirelessly powered millimeter-scale neurostimulator implant based on systematic codesign of an inductive loop antenna and a custom rectifier. *IEEE Trans. Biomed. Circuits Syst.* 12 (5): 1131–1143.

40 Kiourti, A. and Nikita, K.S. (2013). Numerical assessment of the performance of a scalp-implantable antenna: effects of head anatomy and dielectric parameters. *Bioelectromagnetics* 34 (3): 167–179.

41 Abadia, J., Merli, F., Zurcher, J.-F. et al. (2009). 3D-spiral small antenna design and realization for biomedical telemetry in the MICS band. *Radioengineering* 18 (4): 359–367.

42 Duan, Z., Guo, Y.-X., Xue, R.-F. et al. (2012). Differentially fed dual-band implantable antenna for biomedical applications. *IEEE Trans. Antennas Propag.* 60 (12): 5587–5595.

43 Nikolayev, D., Zhadobov, M., Karban, P., and Sauleau, R. (2016). Increasing the radiation efficiency and matching stability of in-body capsule antennas. In: *Proceedings of* *the 10th European Conference on Antennas & Propagation (EuCAP), Davos, Switzerland,* 1–5.

44 Bao, Z., Guo, Y.-X., and Mittra, R. (2017). Single-layer dual-/tri-band inverted-f antennas for conformal capsule type of applications. *IEEE Trans. Antennas Propag.* 65 (12): 7257–7265.

45 Psathas, K.A., Keliris, A.P., Kiourti, A., and Nikita, K.S. (2013). Operation of ingestible antennas along the gastrointestinal tract: detuning and performance. In: *Proceedings of the IEEE 13th International Conference in Bioinformatics & Bioengineering (BIBE)*, 1–4.

46 Yun, S., Kim, K., and Nam, S. (2010). Outer-wall loop antenna for ultrawideband capsule endoscope system. *IEEE Antennas Wireless Propag. Lett.* 9: 1135–1138.

47 Bao, Z., Guo, Y.-X., and Mittra, R. (2017). An ultrawideband conformal capsule antenna with stable impedance matching. *IEEE Trans. Antennas Propag.* 65 (10): 5086–5094.

48 Nikolayev, D., Zhadobov, M., Coq, L.L. et al. (2017). Robust ultraminiature capsule antenna for ingestible and implantable applications. *IEEE Trans. Antennas Propag.* 65 (11): 6107–6119.

49 Li, Y., Zhang, Z., Chen, W. et al. (2010). A compact DVB-H antenna with varactor-tuned matching circuit. *Microw. Opt. Technol. Lett.* 52 (8): 1786–1789.

50 Yue, L., Zhang, Z., Chen, W. et al. (2010). A switchable matching circuit for compact wideband antenna designs. *IEEE Trans. Antennas Propag.* 58 (11): 3450–3457.

51 Beijing PINS Medical Co. (2018). *Deep Brain Stimulation System.* http://www.pinsmedical.com/index.php?m=content&c=index&a=lists&catid=5

52 Qian, X., Chen, Y., Ma, B. et al. (2016). Chronically monitoring the deep brain rhythms: From stimulation to recording. *Sci. Bull.* 61 (19): 1522–1524.

53 Zhang, Z. (2011). *Antenna Design for Mobile Devices.* Hoboken, NJ, USA: Wiley pp. 11–13 and 28–33.

54 Sun, L.B., Feng, H., Li, Y., and Zhang, Z. (2018). Compact 5G MIMO mobile phone antennas with tightly arranged orthogonal-mode pairs. *IEEE Trans. Antennas Propag.* 66 (11): 6364–6369.

55 Kiourti, A., Costa, J.R., Fernandes, C.A., and Nikita, K.S. (2014). A broadband implantable and a dual-band on-body repeater antenna: design and transmission performance. *IEEE Trans. Antennas Propag.* 62 (6): 2899–2908.

56 Xu, L.-J., Guo, Y.-X., and Wu, W. (2015). Bandwidth enhancement of an implantable antenna. *IEEE Antennas Wireless Propag. Lett.* 14: 1510–1513.

57 Yilmaz, T., Karacolak, T., and Topsakal, E. (2008). Characterization of muscle and fat mimicking gels at MICS and ISM bands (402–405 MHz and 2.40–2.48 GHz). In: *Presented at the 29th General Association of the International Union of Radio Science.*

58 Hall, P.S. and Hao, Y. (2006). *Antennas and Propagation for Body-Centric Wireless Communications.* Norwood, MA, USA: Artech House ch. 9.

59 Sani, A., Alomainy, A., and Hao, Y. (2009). Numerical characterization and link budget evaluation of wireless implants considering different digital human phantoms. *IEEE Trans. Microw. Theory Techn.* 57 (10): 2605–2613.

60 Feng, Y., Li, Y., Li, L. et al. (2019). Design and system verification of reconfigurable matching circuits for implantable antennas in tissues with broad permittivity range. *IEEE Trans. Antennas Propag.* 68 (6): 4955–4960.

61 Bin-Mahfoodh, M., Hamani, C., Sime, E., and Lozano, A.M. (2003). Longevity of batteries in internal pulse generators used for deep brain stimulation. *Stereotactic Funct. Neurosurg.* 80 (1): 56–60.

62 El-Saboni, Y., Magill, M.K., Conway, G.A. et al. (2017). Measurement of deep tissue implanted antenna efficiency using a reverberation chamber. *IEEE J. Electromagn. RF Microw. Med. Biol.* 1 (2): 90–97.

Index

Note: Page numbers in *italic* and **bold** refers to figures and tables respectively.

Antennas and Wireless Power Transfer Methods for Biomedical Applications, First Edition.
Yongxin Guo, Yuan Feng and Changrong Liu.

h

i

y

z